NONEQUILIBRIUM AND NONLINEAR DYNAMICS IN NUCLEAR AND OTHER FINITE SYSTEMS

Related Titles from AIP Conference Proceedings

NONEQUILIBRIUM AND NONLINEAR DYNAMICS IN NUCLEAR AND OTHER FINITE SYSTEMS

International Conference

Beijing, China   21–25 May 2001

EDITORS

Zhuxia Li
CIAE, Beijing, China

Ke Wu
ITP, Beijing, China

Xizhen Wu
CIAE, Beijing, China

Enguang Zhao
ITP, Beijing, China

Fumihiko Sakata
Ibaraki University, Japan

AMERICAN INSTITUTE OF PHYSICS

Melville, New York, 2001
AIP CONFERENCE PROCEEDINGS ■ VOLUME 597

Editors:

Zhuxia Li and Xizhen Wu
China Institute of Atomic Energy
P. O. Box 275(18)
102413 Beijing
CHINA

E-mail: lizwux@iris.ciae.ac.cn

Ke Wu and Enguang Zhao
Institute of Theoretical Physics
Chinese Academy of Sciences
Haidian Qu, Zhongguan Cun, Nan 4 Jie, 1 Hao
100080 Beijing
CHINA

E-mail: wuke@itp.ac.cn
        Egzhao@itp.ac.cn

Fumihiko Sakata
Department of Mathematical Sciences
Ibaraki University
Mito, Ibaraki 310-8512
JAPAN

E-mail: Sakata@mito.ipc.ibaraki.ac.jp

L.C. Catalog Card No. 2001096513
ISBN 0-7354-0041-5
ISSN 0094-243X
Printed in the United States of America

CONTENTS

PART III.   TRANSPORT THEORY APPROACH TO INTERMEDIATE AND HIGH ENERGY HEAVY ION COLLISIONS

PART IV.   LARGE AMPLITUDE COLLECTIVE MOTION IN FINITE SYSTEMS

Preface

The International Symposium on *Non-equilibrium and Nonlinear Dynamics in Nuclear and Other Finite Systems* was held in Beijing, May 21-25, 2001. A total of about 100 scientists from 11 countries participated in the symposium, which created a truly international atmosphere.

Nowadays, many fields of science seem to share a common goal to explore the dynamics of evolution of matter, which intends to clarify what dynamic relations exist between different levels of nature. Phenomena appearing at high temperature, high density, high-spin, strong isospin asymmetry, and large deformed space in nuclear systems lead us to question non-equilibrium and nonlinear dynamics, which are also considered to be fundamental questions in other fields of physical sciences, such as statistical physics, biological science, science at nano-scale devices, and the study of the early universe.

The purpose of this symposium was to discuss the common interest problems of non-equilibrium and nonlinear dynamics in finite systems. The main topics discussed at the symposium were: (1) phase transition in strong interactions, (2) transport theory approach to intermediate and high energy heavy ion collisions, (3) large amplitude collective motion in nuclear and other finite systems, (4) dynamics of di-nuclear systems and related topics, (5) Microscopic origin of damping phenomena, and (6) single-particle dynamics and complexity. The status of the current research on the above topics was reviewed and extensively discussed. The underlying common physics among a variety of different phenomena was revealed and even novel concepts were clarified. A common view held by many participants was that *non-equilibrium and nonlinear dynamics* will become one of the main themes of physical science in this century.

We thank all the speakers and participants for the valuable talks and the lively discussions during the symposium. We hope that the participants were satisfied by gaining deeper understanding and stimulating new ideas. We believe that the proceedings will certainly remind all the participants of such a nice atmosphere and we hope it will convey important information to those who could not attend the symposium.

We would like to thank the members of the International Advisory Committee, whose opinions and suggestions were invaluable for setting up such a successful program.

We are grateful to the members of the organizing committee for their fruitful collaboration, and particularly the young students and secretaries for their tremendous contributions before and during the meeting.

Finally, we express our deepest appreciation to the National Natural Science Foundation of China and the Chinese Nuclear Physics Society for the funding of the symposium.

Zhuxia Li, Ke Wu,
Xizhen Wu, Enguang Zhao

International Advisory Committee

Y. Abe (Kyoto)   M. Di Toro (Catania)   W. Greiner(Frankfurt)

Bailin Hao (ITP)   K. Kitahara(ICU, Tokyo)   A. Klein(Philadelphia)

J. Randrup (LBL)   F. Sakata (Mito)   Wenqing Shen (INS, Shanghai)

H. Weidenmüller(Heidelberg)   Minghan Ye (CCAST)   Huanqiao Zhang (CIAE)

Zongye Zhang (IHEP, Beijing)   Zhongyuan Zhu (ITP)   Yizhong Zhuo (CIAE)

Local Organizing Committee

Zhuxia Li (CIAE)   Weiping Liu (CIAE)   Zhongcan Ouyang (ITP)

Ke Wu (ITP)   Xizhen Wu (CIAE)   Fengshou Zhang (IMP, Lanzhou)

Enguang Zhao (ITP)   Weiqin Zhao (Chao) (CCAST)   Zhixiang Zhao (CIAE)

Chairman of Symposium
Zhixiang Zhao (CIAE)

Scientific Secretary
Xizhen Wu (CIAE)

Host Institutes

China Institute of Atomic Energy (CIAE)

Faculty of Science, Ibaraki University (Mito, Japan)

China Center of Advanced Science and Technology (CCAST)

Institute of Theoretical Physics, Chinese Academy of Sciences (ITP)

Supported by

National Natural Science Foundation of China

Chinese Nuclear Physics Society

PART I

GENERAL REVIEW OF
NUCLEAR PROPERTIES

Fundamental Issues in the Physics of Elementary Matter:
Cold Valleys and Fusion of Superheavy Nuclei - Hypernuclei – Antinuclei – Correlations in the Vacuum

Walter Greiner

*Institut für Theoretische Physik, Robert-Mayer-Str. 8-10, J.W. Goethe-Universität,
D-60325 Frankfurt, Germany*

Abstract. The extension of the periodic system into various new areas is investigated. Experiments for the synthesis of superheavy elements and the predictions of magic numbers are reviewed. Different ways of nuclear decay are discussed like cluster radioactivity, cold fission and cold multifragmentation, including the recent discovery of the tripple fission of ^{252}Cf. Furtheron, investigations on hypernuclei and the possible production of antimatter–clusters in heavy–ion collisions are reported. Various versions of the meson field theory serve as effective field theories at the basis of modern nuclear structure and suggest structure in the vacuum which might be important for the production of hyper– and antimatter. A perspective for future research is given.

There are fundamental questions in science, like e. g. "how did life emerge" or "how does our brain work" and others. However, the most fundamental of those questions is "how did the world originate?". The material world has to exist before life and thinking can develop. Of particular importance are the substances themselves, i. e. the particles the elements are made of (baryons, mesons, quarks, gluons), i. e. elementary matter. The vacuum and its structure is closely related to that. On this I want to report today. I begin with the discussion of modern issues in nuclear physics.

The elements existing in nature are ordered according to their atomic (chemical) properties in the **periodic system** which was developped by Mendeleev and Lothar Meyer. The heaviest element of natural origin is Uranium. Its nucleus is composed of $Z = 92$ protons and a certain number of neutrons ($N = 128 - 150$). They are called the different Uranium isotopes. The transuranium elements reach from Neptunium ($Z = 93$) via Californium ($Z = 98$) and Fermium ($Z = 100$) up to Lawrencium ($Z = 103$). The heavier the elements are, the larger are their radii and their number of protons. Thus, the Coulomb repulsion in their interior increases, and they undergo fission. In other words: the transuranium elements become more instable as they get bigger.

In the late sixties the dream of the superheavy elements arose. Theoretical nuclear physicists around S.G. Nilsson (Lund)[1] and from the Frankfurt school[2, 3, 4] predicted that so-called closed proton and neutron shells should counteract the repelling Coulomb forces. Atomic nuclei with these special **"magic" proton and neutron numbers** and their neighbours could again be rather stable. These magic proton (Z) and neu-

CP597, *Nonequilibrium and Nonlinear Dynamics in Nuclear and Other Finite Systems,*
edited by Z. Li et al.
© 2001 American Institute of Physics 0-7354-0041-5/01/$18.00

tron (N) numbers were thought to be $Z = 114$ and $N = 184$ or 196. Typical predictions of their life times varied between seconds and many thousand years. Fig.1 summarizes the expectations at the time. One can see the islands of superheavy elements around $Z = 114$, $N = 184$ and 196, respectively, and the one around $Z = 164$, $N = 318$.

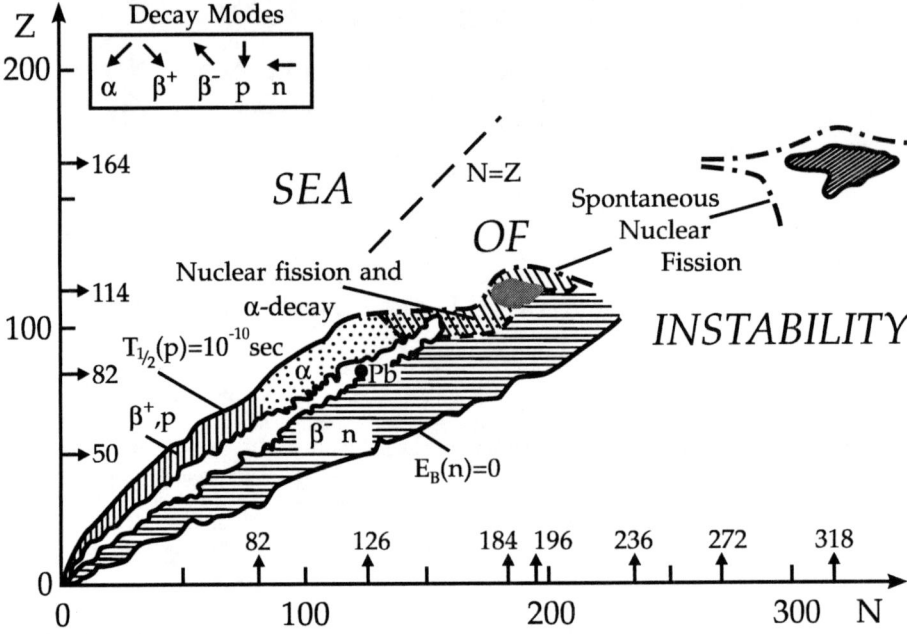

FIGURE 1. The periodic system of elements as conceived by the Frankfurt school in the late sixties. The islands of superheavy elements ($Z = 114$, $N = 184$, 196 and $Z = 164$, $N = 318$) are shown as dark hatched areas.

The important question was how to produce these superheavy nuclei. There were many attempts, but only little progress was made. It was not until the middle of the seventies that the Frankfurt school of theoretical physics together with foreign guests (R.K. Gupta (India), A. Sandulescu (Romania))[6] theoretically understood and substantiated the concept of bombarding of double magic lead nuclei with suitable projectiles, which had been proposed intuitively by the russian nuclear physicist Y. Oganessian[7]. The two-center shell model, which is essential for the description of fission, fusion and nuclear molecules, was developped in 1969-1972 together with my then students U. Mosel and J. Maruhn[8]. It showed that the shell structure of the two final fragments was visible far beyond the barrier into the fusioning nucleus. The collective potential energy surfaces of heavy nuclei, as they were calculated in the framework of the two-center shell model, exhibit pronounced valleys, such that these valleys provide promising doorways to the fusion of superheavy nuclei for certain projectile-target combinations (Fig. 4). If projectile and target approach each other through those **"cold" valleys**, they get only minimally excited and the barrier which has to be overcome (fusion barrier) is lowest (as compared to neighbouring projectile-target combinations). In this way the correct projectile- and target-combinations for fusion were predicted. Indeed, Gottfried

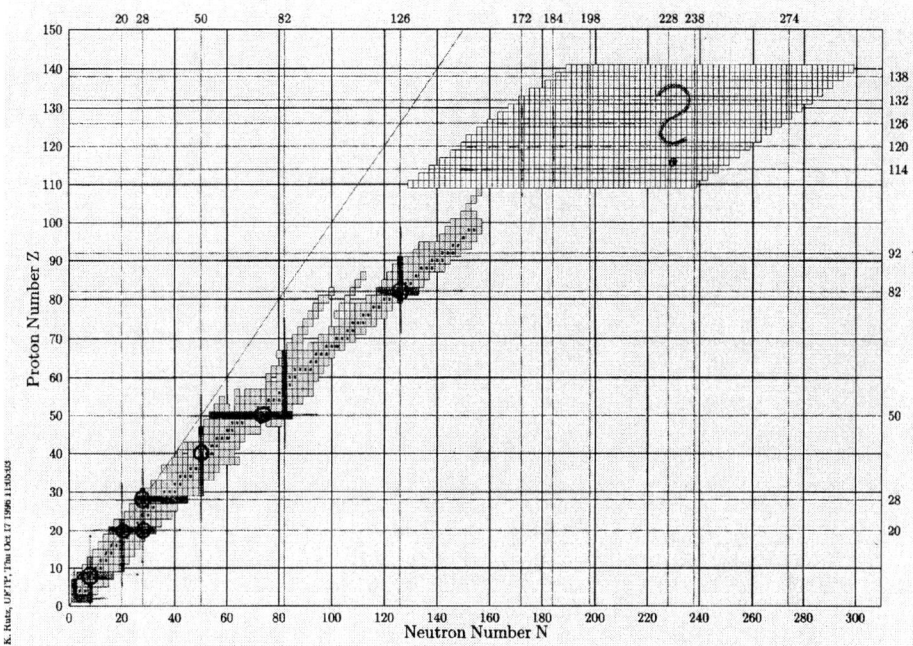

FIGURE 2. The shell structure in the superheavy region around $Z = 114$ is an open question. As will be discussed later, meson field theories suggest that $Z = 120, N = 172, 184$ are the magic numbers in this region.

Münzenberg and Sigurd Hofmann and their group at GSI [9] have followed this approach. With the help of the SHIP mass-separator and the position sensitive detectors, which were especially developped by them, they produced the pre-superheavy elements $Z = 106, 107, \ldots 112$, each of them with the theoretically predicted projectile-target combinations, and only with these. Everything else failed. This is an impressive success, which crowned the laborious construction work of many years. The before last example of this success, the discovery of element 112 and its long α-decay chain, is shown in Fig. 6. Very recently the Dubna–Livermore–group produced two isotopes of $Z = 114$ element by bombarding ^{244}Pu with ^{48}Ca and also $Z = 116$ by ^{48}Ca + ^{248}C m.(Fig. 3). Also these are cold–valley reactions (in this case due to the combination of a spherical and a deformed nucleus), as predicted by Gupta, Sandulescu and Greiner [10] in 1977. There exist also cold valleys for which both fragments are deformed [11], but these have yet not been verified experimentally. The very recently reported $Z = 118$ isotope fused with the cold valley reaction [13] ^{58}Kr + ^{208}Pb by Ninov et al. [14] yields the latest support of the cold valley idea, but this one needs confirmation.

Studies of the shell structure of superheavy elements in the framework of the meson field theory and the Skyrme-Hartree-Fock approach have recently shown that the magic shells in the superheavy region are very isotope dependent [5, 15] (see Fig. 7). **According-ing to these investigations $Z = 120$ being a magic proton number seems to be as**

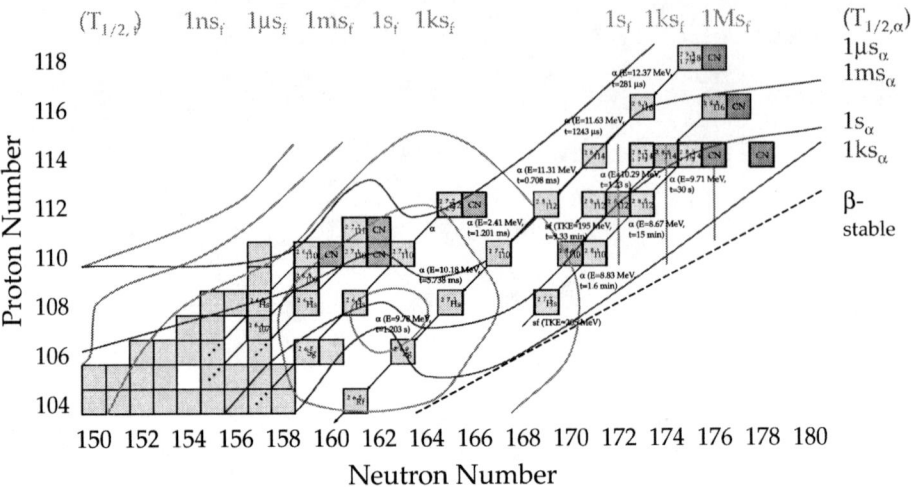

FIGURE 3. The $Z = 106 - 112$ isotopes were fused by the Hofmann–Münzenberg (GSI)–group. The two $Z = 114$ isotopes and the $Z = 116$ isotope were produced by the Dubna–Livermore group. It is claimed that three neutrons are evaporated. Obviously the lifetimes of the various decay products are rather long (because they are closer to the stable valley), in crude agreement with early predictions [3, 4] and in excellent agreement with the recent calculations of the Sobicevsky–group [12]. The recently fused $Z = 118$ isotope by V. Ninov et al. at Berkeley is the heaviest one so far, but needs confirmation.

probable as $Z = 114$. Additionally, recent investigations in a chirally symmetric mean–field theory result also in the prediction of these two magic numbers[42, 44], see also below. The corresponding magic neutron numbers are predicted to be $N = 172$ and - as it seems to a lesser extend - $N = 184$. Deformed calculations within Skyrme-Hartree-Fock (SHF) and Relativistic Mean-Field (RMF) model with the parametrizations SkI4 and PL-40 reveal again different predictions[17]: Though both parametrizations predict $N = 162$ as the deformed neutron shell closure, the deformed proton shell closures are $Z = 108$ (SkI4) and $Z = 104$ (PL-40). Calculations of the potential energy surfaces show single humped barriers[18], their heights and widths strongly depending on the predicited magic number. Additionially, a systematic study of fission barriers[19] ranging from proton number $Z = 108$ to $Z = 120$ reveals systematic differences in barrier heights: Most of the considered Skyrme forces produce axial barriers up to 12 MeV for the nucleus $Z = 120, N = 180$, while relativistic models predict barriers which are smaller by a factor of two. This shows that there are some intrinsic features of the models which get magnified alot for superheavy elements and still have to be understood. The barriers for these superheavy elements are lowered considerably taking into account triaxial degrees of freedom. Thus, this region provides an open field of research. R.A. Gherghescu et al. have calculated the potential energy surface of the $Z = 120$ nucleus. It utilizes interesting isomeric and valley structures (Fig. 8). The charge distribution of the $Z = 120, N = 184$ nucleus indicates a hollow inside. This leads us to suggest that it might be essentially a fullerene consisting of 60 α-particles and one additional binding neutron per alpha. This is illustrated in Fig 5. The protons and neutrons of such a superheavy nucleus are

6

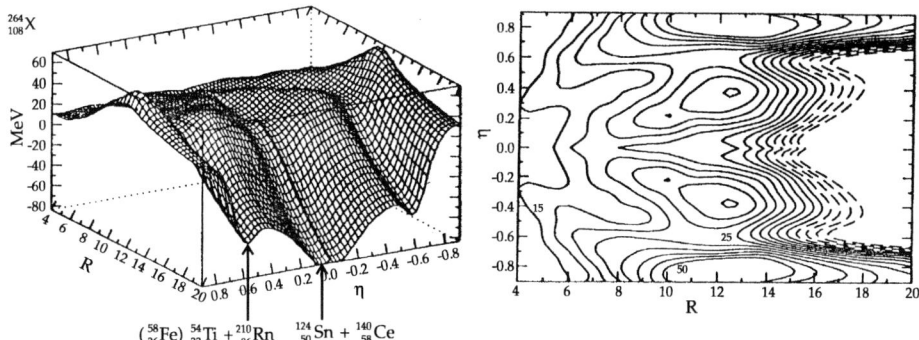

$(^{58}_{26}\text{Fe})$ $^{54}_{22}\text{Ti}$ + $^{210}_{86}\text{Rn}$ $^{124}_{50}\text{Sn}$ + $^{140}_{58}\text{Ce}$

FIGURE 4. The collective potential energy surface of $^{184}114$, calculated within the two center shell model by J. Maruhn et al., shows clearly the cold valleys which reach up to the barrier and beyond. Here R is the distance between the fragments and $\eta = \dfrac{A_1 - A_2}{A_1 + A_2}$ denotes the mass asymmetry: $\eta = 0$ corresponds to a symmetric, $\eta = \pm 1$ to an extremely asymmetric division of the nucleus into projectile and target. If projectile and target approach through a cold valley, they do not "constantly slide off" as it would be the case if they approach along the slopes at the sides of the valley. Constant sliding causes heating, so that the compound nucleus heats up and gets unstable. In the cold valley, on the other hand, the created heat is minimized. The colleagues from Freiburg should be familiar with that: they approach Titisee (in the Black Forest) most elegantly through the Höllental and not by climbing its slopes along the sides.

distributed over 60 α particles and 60 neutrons (forgetting the last 4 neutrons).

The determination of the chemistry of superheavy elements, i. e. the calculation of the atomic structure — which is in the case of element 112 the shell structure of 112 electrons due to the Coulomb interaction of the electrons and in particular the calculation of the orbitals of the outer (valence) electrons — has been carried out as early as 1970 by B. Fricke and W. Greiner[16]. Hartree-Fock-Dirac calculations yield rather precise results.

The potential energy surfaces, which are shown prototypically for $Z = 114$ in Fig 4, contain even more remarkable information that I want to mention cursorily: if a given nucleus, e. g. Uranium, undergoes fission, it moves in its potential mountains from the interior to the outside. Of course, this happens quantum mechanically. The wave function of such a nucleus, which decays by tunneling through the barrier, has maxima where the potential is minimal and minima where it has maxima.

The probability for finding a certain mass asymmetry $\eta = \dfrac{A_1 - A_2}{A_1 + A_2}$ of the fission is proportional to $\psi^*(\eta)\psi(\eta)d\eta$. Generally, this is complemented by a coordinate dependent scale factor for the volume element in this (curved) space, which I omit for the sake of clarity. Now it becomes clear how the so-called **asymmetric** and **superasymmetric** fission processes come into being. They result from the enhancement of the collective wave function in the cold valleys. And that is indeed, what one observes.

For large mass asymmetry ($\eta \approx 0.8$, 0.9) there exist very narrow valleys. They are not as clearly visible in Fig. 4, but they have interesting consequences. Through these narrow valleys nuclei can emit spontaneously not only α-particles (Helium nuclei) but also ^{14}C, ^{20}O, ^{24}Ne, ^{28}Mg, and other nuclei. Thus, we are lead to the **cluster radioactivity**

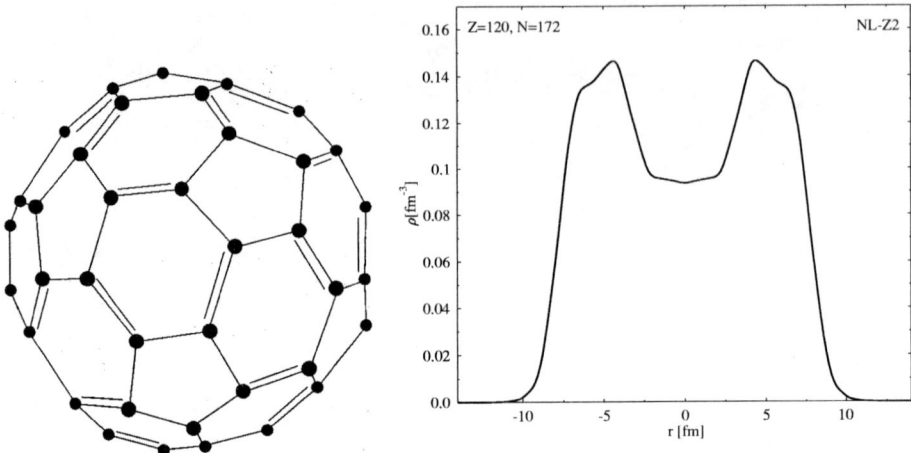

FIGURE 5. Typical structure of the fullerene ^{60}C. The double bindings are illutsrated by double lines. In the nuclear case the Carbon atoms are replaced by α particles and the double bindings by the additional neutrons. Such a structure would immediately explain the semi–hollowness of that superheavy nucleus, which is revealed in the mean–field calculations within meson–field theories. The radial density of the nucleus with 120 protons and 172 neutrons, as emerging from a meson-field calculation with the force NL-Z2 is shown on the right side. Note that the semi-bubble structure is mostly pronounced for this nucleus. When going to higher neutron numbers, this structures becomes less and less.

(Poenaru, Sandulescu, Greiner [20]).

By now this process has been verified experimentally by research groups in Oxford, Moscow, Berkeley, Milan and other places. Accordingly, one has to revise what is learned in school: there are not only 3 types of radioactivity (α-, β-, γ-radioactivity), but many more. Atomic nuclei can also decay through spontaneous cluster emission (that is the "spitting out" of smaller nuclei like carbon, oxygen,...). Fig. 10 depicts some examples of these processes.

The knowledge of the collective potential energy surface and the collective masses $B_{ij}(R, \eta)$, all calculated within the Two-Center-Shell-Modell (TCSM), allowed H. Klein, D. Schnabel and J. A. Maruhn to calculate lifetimes against fission in an "ab initio" way [21]. The discussion of much more very interesting new physics cannot be persued here. We refer to the literature [22, 23, 24, 25, 26, 27, 28].

The "cold valleys" in the collective potential energy surface are basic for understanding this exciting area of nuclear physics! It is a master example for understanding the **structure of elementary matter**, which is so important for other fields, especially astrophysics, but even more so for enriching our "Weltbild", i.e. the status of our understanding of the world around us.

Nuclei that are found in nature consist of nucleons (protons and neutrons) which themselves are made of u (up) and d (down) quarks. However, there also exist s (strange) quarks and even heavier flavors, called charm, bottom, top. The latter has just recently been discovered. Let us stick to the s quarks. They are found in the 'strange' relatives of

$$^{70}\text{Zn} + ^{208}\text{Pb} \rightarrow ^{277}112 + 1n$$

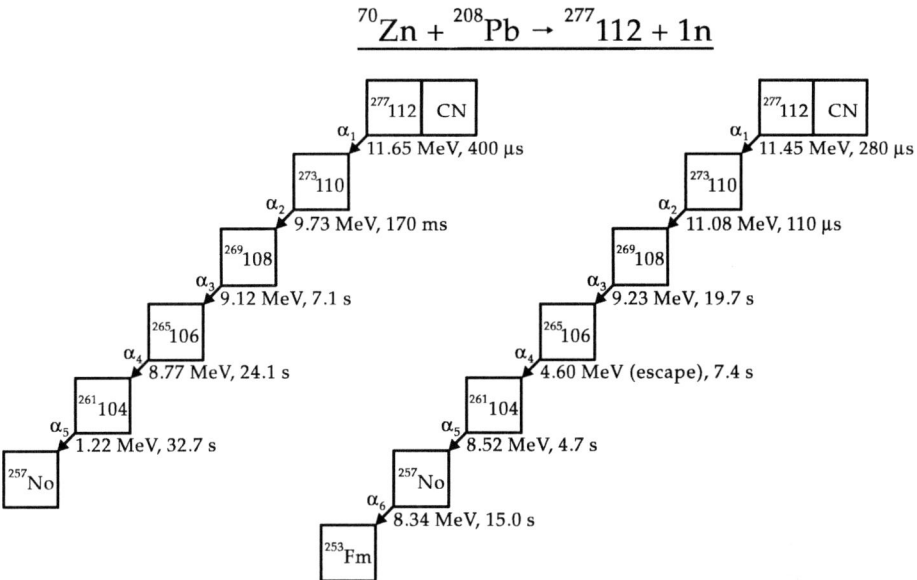

FIGURE 6. The fusion of element 112 with ^{70}Zn as projectile and ^{208}Pb as target nucleus has been accomplished for the first time in 1995/96 by S. Hofmann, G. Münzenberg and their collaborators. The colliding nuclei determine an entrance to a "cold valley" as predicted as early as 1976 by Gupta, Sandulescu and Greiner. The fused nucleus 112 decays successively via α emission until finally the quasi-stable nucleus ^{253}Fm is reached. The α particles as well as the final nucleus have been observed. Combined, this renders the definite proof of the existence of a $Z = 112$ nucleus.

the nucleons, the so-called hyperons (Λ, Σ, Ξ, Ω). The Λ-particle, e. g., consists of one u, d and s quark, the Ξ-particle even of an u and two s quarks, while the Ω (sss) contains strange quarks only.

If such a hyperon is taken up by a nucleus, a **hyper-nucleus** is created. Hyper-nuclei with one hyperon have been known for 20 years now, and were extensively studied by B. Povh (Heidelberg)[31]. Several years ago, Carsten Greiner, Jürgen Schaffner and Horst Stöcker[32] theoretically investigated nuclei with many hyperons, **hypermatter**, and found that the binding energy per baryon of strange matter is in many cases even higher than that of ordinary matter (composed only of u and d quarks). This leads to the idea of extending the periodic system of elements in the direction of strangeness.

One can also ask for the possibility of building atomic nuclei out of **antimatter**, that means searching e. g. for anti-helium, anti-carbon, anti-oxygen. Fig. 11 depicts this idea. Due to the charge conjugation symmetry antinuclei should have the same magic numbers and the same spectra as ordinary nuclei. However, as soon as they get in touch with ordinary matter, they annihilate with it and the system explodes.

Now the important question arises how these strange matter and antimatter clusters can be produced. First, one thinks of collisions of heavy nuclei, e. g. lead on lead, at high energies (energy per nucleon ≥ 200 GeV). Calculations with the URQMD-model

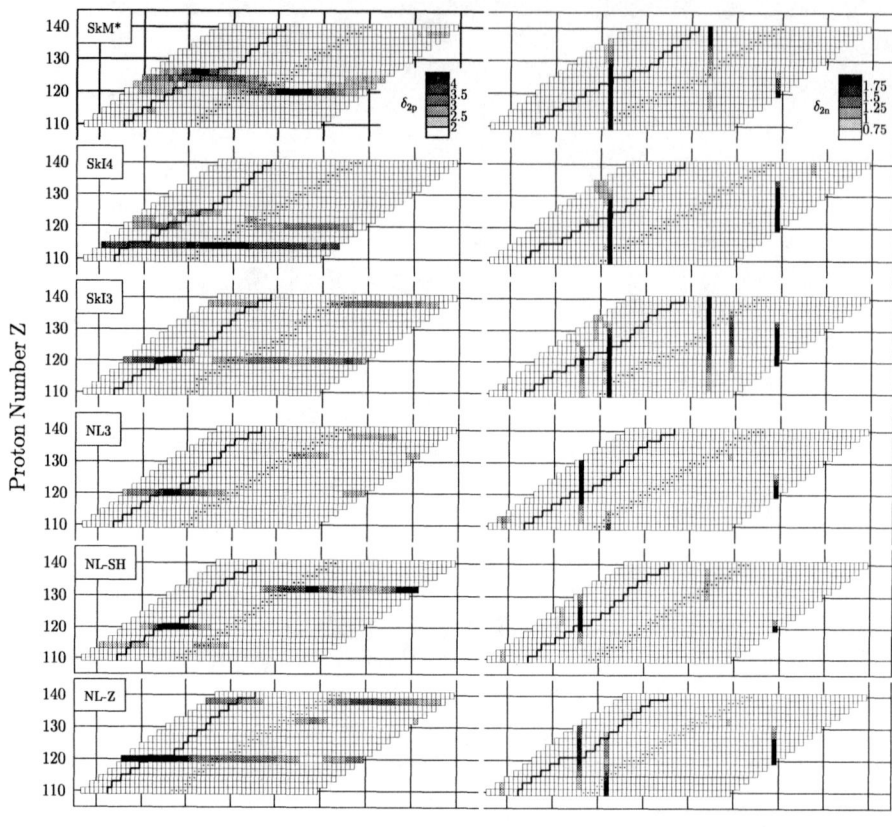

Proton Number Z

Neutron Number N

FIGURE 7. Grey scale plots of proton gaps (left column) and neutron gaps (right column) in the N-Z plane for spherical calculations with the forces as indicated. The assignment of scales differs for protons and neutrons, see the uppermost boxes where the scales are indicated in units of MeV. Nuclei that are stable with respect to β decay and the two-proton dripline are emphasized. The forces with parameter sets SkI4 and NL-Z reproduce the binding energy of $^{264}_{156}108$ (Hassium) best, i.e. $|\delta E/E| < 0.0024$. Thus one might assume that these parameter sets could give the best predictions for the superheavies. Nevertheless, it is noticed that NL-Z predicts only $Z = 120$ as a magic number while SkI4 predicts both $Z = 114$ and $Z = 120$ as magic numbers. The magicity depends — sometimes quite strongly — on the neutron number. These studies are due to Bender, Rutz, Bürvenich, Maruhn, P.G. Reinhard et al. [15].

of the Frankfurt school show that through **nuclear shock waves** [33, 34, 35] nuclear matter gets compressed to 5–10 times of its usual value, $\rho_0 \approx 0.17 \, \text{fm}^3$, and heated up to temperatures of $kT \approx 200$ MeV. As a consequence about 10000 pions, 100 Λ's, 40 Σ's and Ξ's and about as many antiprotons and many other particles are created in a single collision. It seems conceivable that it is possible in such a scenario for some Λ's to get captured by a nuclear cluster. This happens indeed rather frequently for one or two Λ-particles; however, more of them get built into nuclei with rapidly decreasing probability

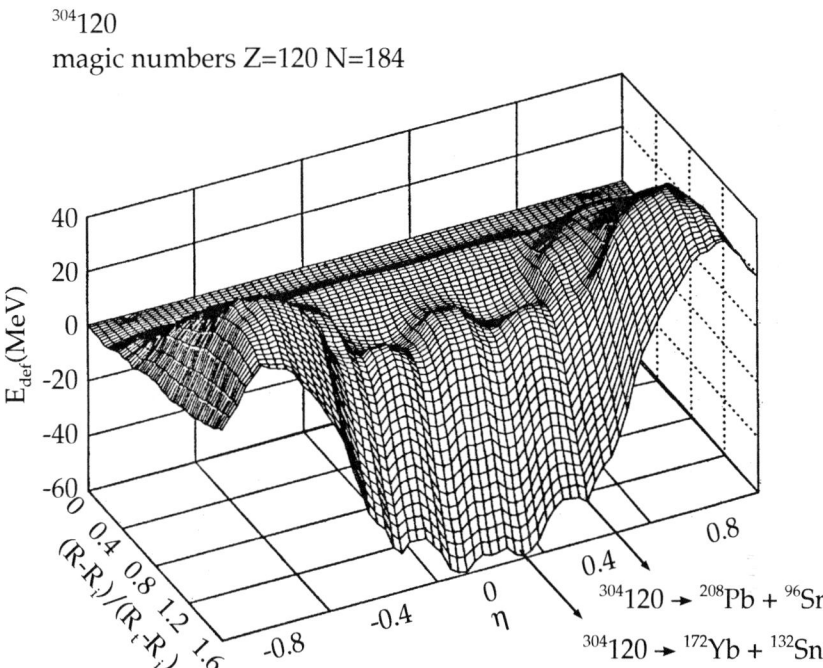

FIGURE 8. Potential energy surface as a function of reduced elongation $(R - R_i)/(R_t - R_i)$ and mass asymmetry η for the double magic nucleus $^{304}120$. $^{304}120_{184}$.

only. This is due to the low probability for finding the right conditions for such a capture in the phase space of the particles: the numerous particles travel with every possible momenta (velocities) in all directions. The chances for hyperons and antibaryons to meet gets rapidly worse with increasing number. In order to produce multi-Λ-nuclei and antimatter nuclei, one has to look for a different source.

In the framework of meson field theory within the mean-field approximation the energy spectrum of baryons in a nucleus has a peculiar structure, depicted in Fig. 12. It consists of an upper and a lower continuum, as it is known from the electrons (see e. g. [30]). The upper well represents the nuclear shell modell potential. It describes the overall structure throughout the nuclear table very well.

Of special interest in the case of the baryon spectrum is the potential well, built of the scalar and the vector potential, which rises from the lower continuum. It is known since P.A.M. Dirac (1930) that the negative energy states of the lower continuum have to be occupied by particles (electrons or, in our case, baryons). Otherwise our world would be unstable, because the "ordinary" particles are found in the upper states which can decay through the emission of photons into lower lying states. However, if the "underworld" is occupied, the Pauli-principle will prevent this decay. Holes in the occupied "underworld" (Dirac sea) are antiparticles. This has been extensively discussed

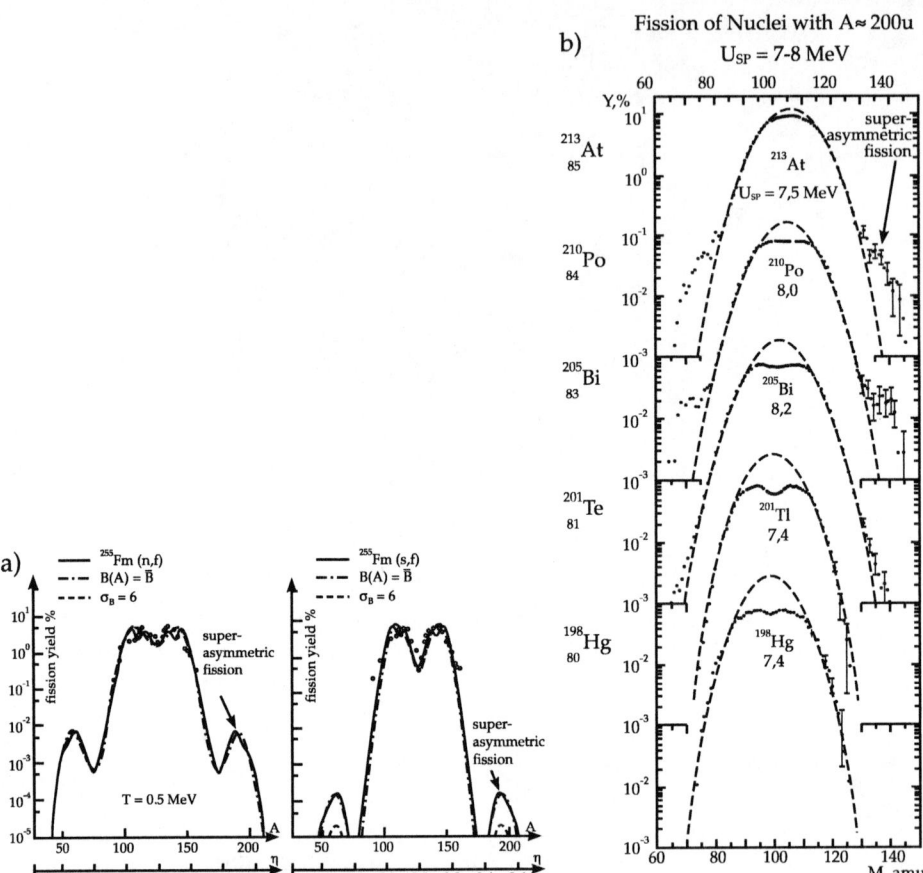

FIGURE 9. Asymmetric (a) and symmetric (b) fission. For domminantly symmetric fissioning nuclei, also superasymmetric fission is recognizable, as it has been observed only a few years ago by the russian physicist Itkis — just as expected theoretically.

in the context of QED of strong fields (overcritical fields, decay of the vacuum from a neutral one into a charged one [30]).

The occupied states of this underworld including up to 40000 occupied bound states of the lower potential well represent the **vacuum**. The peculiarity of this strongly correlated vacuum structure in the region of atomic nuclei is that — depending on the size of the nucleus — more than 20000 up to 40000 (occupied) bound nucleon states contribute to this polarization effect. Obviously, we are dealing here with a **highly correlated vacuum**. A pronounced shell structure can be recognized [36, 37, 38]. Holes in these states have to be interpreted as bound antinucleons (antiprotons, antineutrons). If the primary nuclear density rises due to compression, the lower well increases while the upper decreases and soon is converted into a repulsive barrier (Fig. 13). This compression of nuclear matter can only be carried out in relativistic nucleus-nucleus collision with the

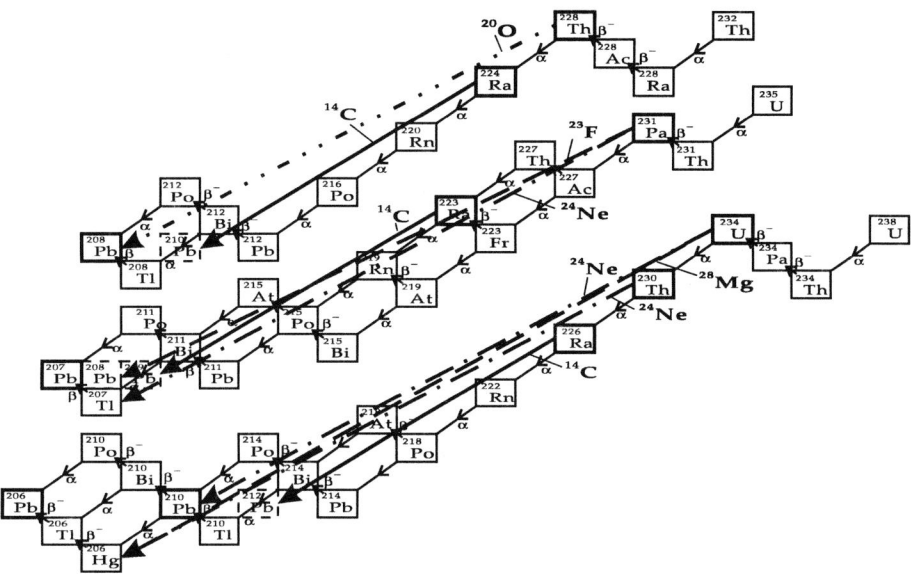

FIGURE 10. Cluster radioactivity of actinide nuclei. By emission of ^{14}C, ^{20}O,..."big leaps" in the periodic system can occur, just contrary to the known α, β, γ radioactivities, which are also partly shown in the figure.

help of shock waves, which have been proposed by the Frankfurt school[33, 34] and which have since then been confirmed extensively (for references see e. g. [39]). These **nuclear shock waves** are accompanied by heating of the compressed nuclear matter. Indeed, density and temperature are correlated in terms of the hydrodynamic Rankine-Hugoniot-equations. Heating as well as the violent dynamics cause the creation of many holes in the very deep (measured from $-M_Bc^2$) vacuum well and an equal number of particles (baryons) in the upper continuum. This is analogous to the dynamical e^+ e^- pair creation in heavy ion collisions. [39]

These numerous bound holes resemble antimatter clusters which are bound in the medium; their wave functions have large overlap with antimatter clusters. When the primary matter density decreases during the expansion stage of the heavy ion collision, the potential wells, in particular the lower one, disappear.

The bound antinucleons are then pulled down into the (lower) continuum. In this way antimatter clusters may be set free. Of course, a large part of the antimatter will annihilate on ordinary matter present in the course of the expansion. However, it is important that this mechanism for the production of antimatter clusters out of the highly correlated vacuum does not proceed via the phase space. The required coalescence of many particles in phase space suppresses the production of clusters, while it is favoured by the direct production out of the highly correlated vacuum. In a certain sense, the highly correlated vacuum is a kind of cluster vacuum (vacuum with cluster structure). The shell structure of the vacuum levels (see Fig. 12) supports this latter suggestion. Fig. 14 illustrates this idea.

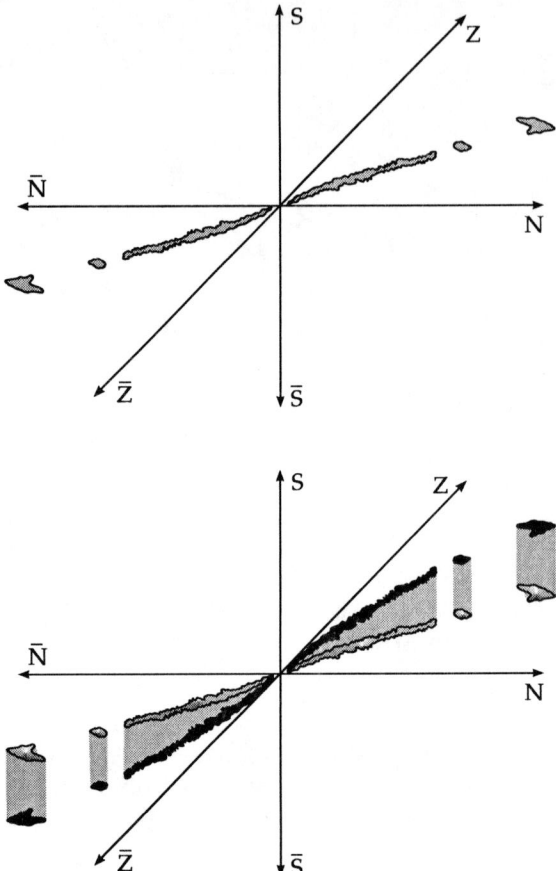

FIGURE 11. The extension of the periodic system into the sectors of strangeness (S, $\bar{S}$) and antimatter ($\bar{Z}$, $\bar{N}$). The stable valley winds out of the known proton (Z) and neutron (N) plane into the S and $\bar{S}$ sector, respectively. The same can be observed for the antimatter sector. In the upper part of the figure only the stable valley in the usual proton (Z) and neutron (N) plane is plotted, however, extended into the sector of antiprotons and antineutrons. In the second part of the figure it has been indicated, how the stable valley winds out of the Z-N-plane into the strangeness sector. This is due to an additional term proportional to $(\frac{A}{A} - \frac{S_0}{A})^2$ in the mass formula.

The mechanism is similar for the production of multi-hyper nuclei (Λ, Σ, Ξ, Ω). Meson field theory predicts also for the Λ energy spectrum at finite primary nucleon density the existence of upper and lower wells. The lower well belongs to the vacuum and is fully occupied by Λ's. Dynamics and temperature then induce transitions (e.g. $\Lambda\bar{\Lambda}$ creation) and deposit many Λ's in the upper well. These numerous bound Λ's (and similarly other hyperons) are sitting close to the primary baryons: in a certain sense a giant multi-Λ hypernucleus has been created. When the system disintegrates (expansion stage) the Λ's distribute over the nucleon clusters (which are most abundant in peripheral

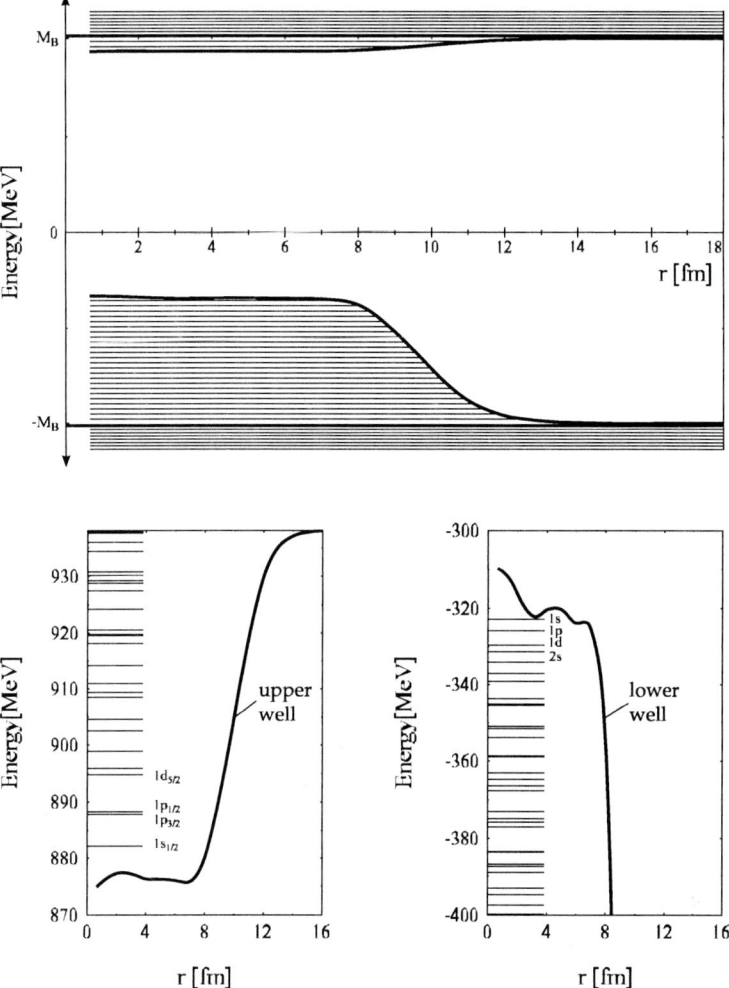

FIGURE 12. Baryon spectrum in a nucleus. Below the positive energy continuum exists the potential well of real nucleons. It has a depth of 50-60 MeV and shows the correct shell structure. The shell model of nuclei is realized here. However, from the negative continuum another potential well arises, in which about 40000 bound particles are found, belonging to the vacuum. A part of the shell structure of the upper well and the lower (vacuum) well is depicted in the lower figures.

collisions). In this way multi-Λ hypernuclei can be formed. Also clusters of hyperons alone($\Lambda, \Sigma, \ldots$) seem possible and quasistable [5, 32] and the Bethe-Weizsäcker mass formula requires at least one additional term proportional to $(f_S - f_{S_0})^2$, where f_S/A is the strangeness content in a hypernucleus.

Of course this vision has to be worked out and probably refined in many respects. This means much more and thorough investigation in the future. It is particularly important

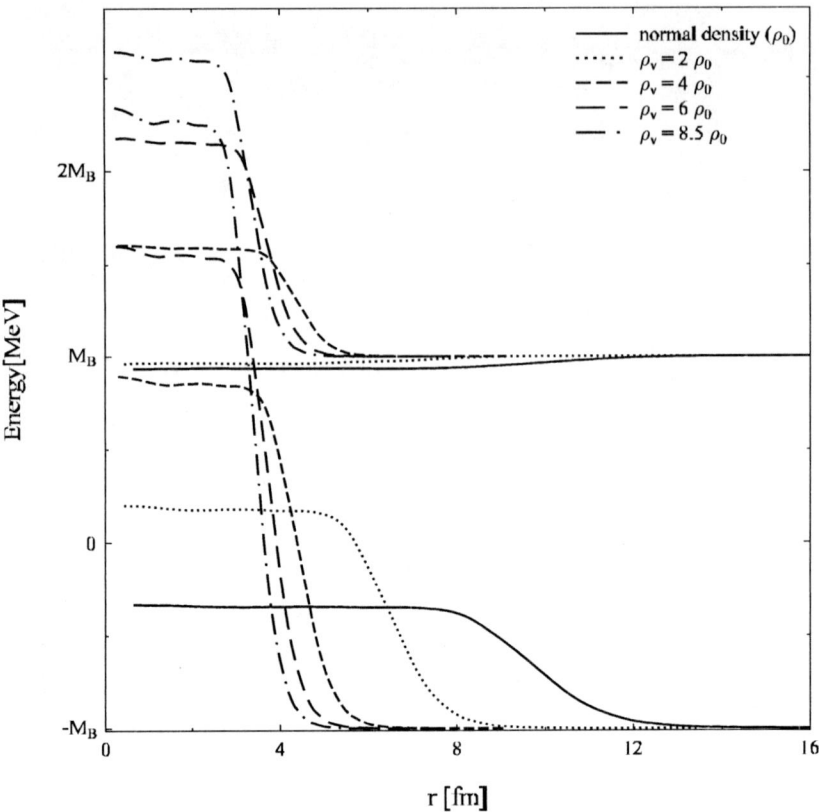

FIGURE 13. The lower well rises strongly with increasing primary nucleon density, and even gets supercritical (spontaneous nucleon emission and creation of bound antinucleons). Supercriticality denotes the situation, when the lower well enters the upper continuum.

to gain more experimental information on the properties of the lower well (how deep is it?, can its shell structure be experimentally investigated?, can its $\rho - T -$ dependence be tested?, ...) by (e, e' p) or (e, e' p p') and also ($\bar{p}_c p_b$, $p_c \bar{p}_b$) reactions at high energy ($\bar{p}_c$ denotes an incident antiproton from the continuum, p_b is a proton in a bound state; for the reaction products the situation is just the opposite)[40]. Also the reaction (p, p' d), (p, p' ^{3}He), (p, p' ^{4}He) and others of similar type need to be investigated in this context. The systematic scattering of antiprotons on nuclei can contribute to clarify these questions: Time-like momentum transfer is required here! The Nambu-Jona-Lasigno (NJL) model seems to give much smaller lower wells, but does not describe the shell model potentials. Studies of I. Mishustin, L. Satarov et al. to improve the NJL model for applications to nuclear und baryon-meson sectors are on the way.

Problems of the meson field theory (e. g. Landau poles) can then be reconsidered. An effective meson field theory has to be constructed. Various effective theories, e. g. of Walecka-type on the one side and theories with chiral invariance on the other side,

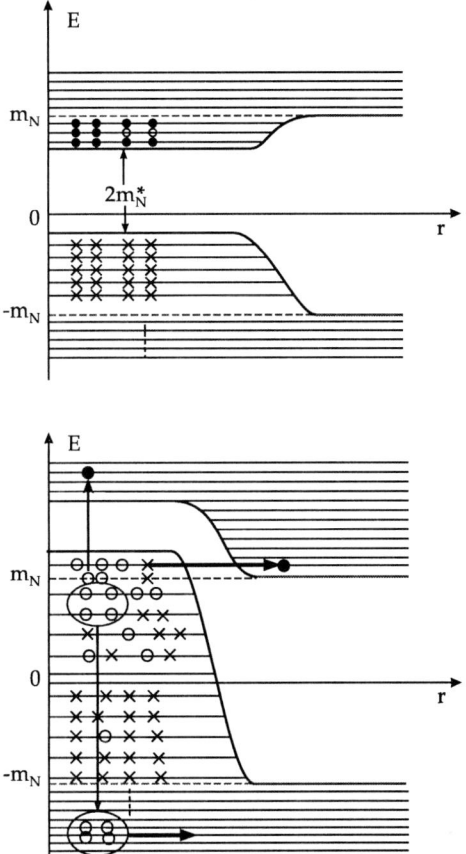

FIGURE 14.
Due to the high temperature and the violent dynamics, many bound holes (antinucleon clusters) are created in the highly correlated vacuum, which can be set free during the expansion stage into the lower continuum. In this way, antimatter clusters can be produced directly from the vacuum. The horizontal arrow in the lower part of the figure denotes the spontaneous creation of baryon-antibaryon pairs, while the antibaryons occupy bound states in the lower potential well. Such a situation, where the lower potential well reaches into the upper continuum, is called supercritical. Four of the bound holes states (bound antinucleons) are encircled to illustrate a "quasi-antihelium" formed. It may be set free (driven into the lower continuum) by the violent nuclear dynamics.

seem to give different strengths of the potential wells and also different dependence on the baryon density [41]. The Lagrangians of the Dürr-Teller-Walecka-type and of the chirally symmetric mean field theories look quantitatively quite differently. We exhibit them — without further discussion — in the following equations:

$$\mathcal{L} = \mathcal{L}_{kin} + \mathcal{L}_{BM} + \mathcal{L}_{vec} + \mathcal{L}_0 + \mathcal{L}_{SB}$$

Non-chiral Lagrangian:

$$\mathcal{L}_{\text{kin}} = \frac{1}{2}\partial_\mu s \partial^\mu s + \frac{1}{2}\partial_\mu z \partial^\mu z - \frac{1}{4}B_{\mu\nu}B^{\mu\nu} - \frac{1}{4}G_{\mu\nu}G^{\mu\nu} - \frac{1}{4}F_{\mu\nu}F^{\mu\nu}$$

$$\mathcal{L}_{\text{BM}} = \sum_B \overline{\Psi}_B[i\gamma_\mu\partial^\mu - g_{\omega B}\gamma_\mu\omega^\mu - g_{\phi B}\gamma_\mu\phi^\mu - g_{\rho B}\gamma_\mu\tau_B\rho^\mu$$

$$-e\gamma_\mu\frac{1}{2}(1+\tau_B)A^\mu - m_B^*]\psi_B$$

$$\mathcal{L}_{\text{vec}} = \frac{1}{2}m_\omega^2\omega_\mu\omega^\mu + \frac{1}{2}m_\rho^2\rho_\mu\rho^\mu + \frac{1}{2}m_\phi^2\phi_\mu\phi^\mu$$

$$\mathcal{L}_0 = -\frac{1}{2}m_s^2 s^2 - \frac{1}{2}m_z^2 z^2 - \frac{1}{3}bs^3 - \frac{1}{4}cs^4$$

Chiral Lagrangian:

$$\mathcal{L}_{\text{kin}} = \frac{1}{2}\partial_\mu\sigma\partial^\mu\sigma + \frac{1}{2}\partial_\mu\zeta\partial^\mu\zeta + \frac{1}{2}\partial_\mu\chi\partial^\mu\chi - \frac{1}{4}B_{\mu\nu}B^{\mu\nu} - \frac{1}{4}G_{\mu\nu}G^{\mu\nu} - \frac{1}{4}F_{\mu\nu}F^{\mu\nu}$$

$$\mathcal{L}_{\text{BM}} = \sum_B \overline{\Psi}_B[i\gamma_\mu\partial^\mu - g_{\omega B}\gamma_\mu\omega^\mu - g_{\phi B}\gamma_\mu\phi^\mu - g_{\rho B}\gamma_\mu\tau_B\rho^\mu$$

$$-e\gamma_\mu\frac{1}{2}(1+\tau_B)A^\mu - m_B^*]\psi_B$$

$$\mathcal{L}_{\text{vec}} = \frac{1}{2}m_\omega^2\frac{\chi^2}{\chi_0^2}\omega_\mu\omega^\mu + \frac{1}{2}m_\rho^2\frac{\chi^2}{\chi_0^2}\rho_\mu\rho^\mu + \frac{1}{2}m_\phi^2\frac{\chi^2}{\chi_0^2}\phi_\mu\phi^\mu + g_4^4(\omega^4 + 6\omega^2\rho^2 + \rho^4)$$

$$\mathcal{L}_0 = -\frac{1}{2}k_0\chi^2(\sigma^2+\zeta^2) + k_1(\sigma^2+\zeta^2)^2 + k_2(\frac{\sigma^4}{2}+\zeta^4) + k_3\chi\sigma^2\zeta - k_4\chi^4$$

$$+\frac{1}{4}\chi^4\ln\frac{\chi^4}{\chi_0^4} + \frac{\delta}{3}\ln\frac{\sigma^2\zeta}{\sigma_0^2\zeta_0^2}$$

$$\mathcal{L}_{\text{SB}} = -\left(\frac{\chi}{\chi_0}\right)^2\left[m_\pi^2 f_\pi\sigma + \left(\sqrt{2}m_K^2 f_K - \frac{1}{\sqrt{2}}m_\pi^2 f_\pi\right)\zeta\right]$$

The non-chiral model contains the scalar-isoscalar field s and its strange counterpart z, the vector-isoscalar fields ω_μ and ϕ_μ, and the the ρ-meson ρ_μ as well as the photon A_μ. For more details see [41]. In contrast to the non-chiral model, the $SU(3)_L \times SU(3)_R$ Lagrangian contains the dilaton field χ introduced to mimic the trace anomaly of QCD in an effective Lagrangian at tree level (for an explanation of the chiral model see [41, 42]).

The connection of the chiral Lagrangian with that of Walecka-type can be established by the substitution $\sigma = \sigma_0 - s$ (and similarly for the strange condensate ζ). Then, neglecting the contribution of the strange scalar field, the difference in the definition of the effective nucleon mass in both models (non-chiral:$m_N^* = m_N - g_s s$, chiral:$m_N^* = g_s\sigma$) can be removed, yielding:

$$m_N^* = g_s\sigma_0 - g_s s \equiv m_N - g_s s \tag{1}$$

for the nucleon mass in the chiral model.

Nevertheless, if the parameters in both cases (e.g. $g_s, g_\omega, g_\rho, m_s, b, c$ in the non-chiral case) are adjusted such that ordinary nuclei (binding energies, radii, shell structure,...) and properties of infinite nuclear matter (equilibrium density, compression constant K,

18

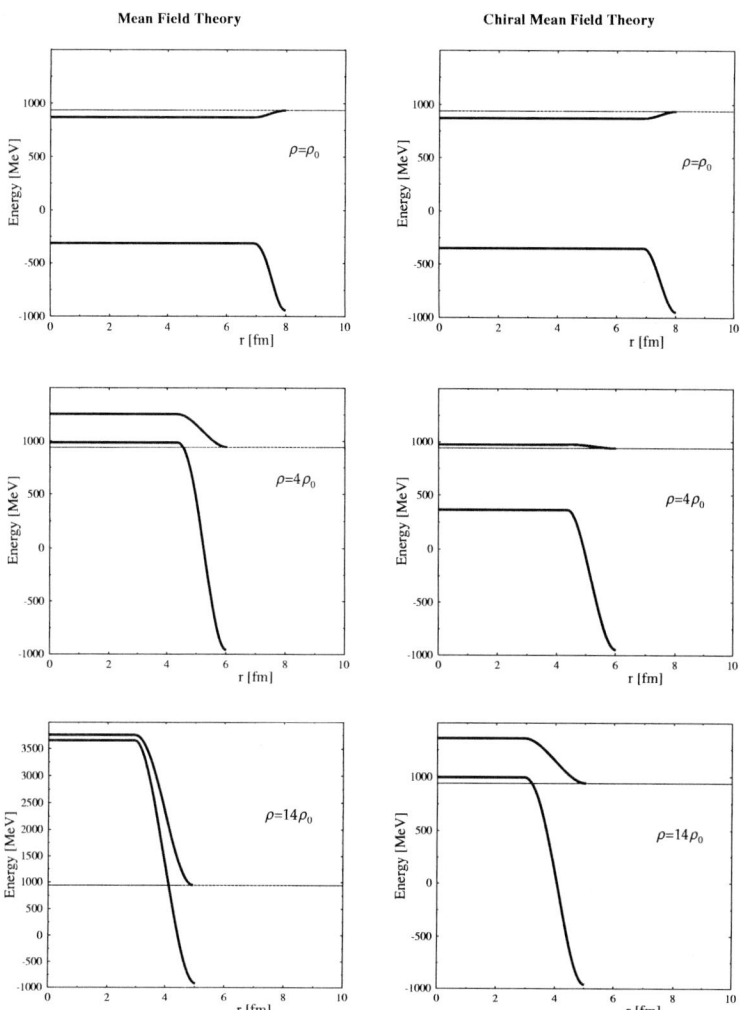

FIGURE 15. The potential structure of the shell model and the vacuum for various primary densities $\rho = \rho_0$, $4\rho_0$, $14\rho_0$. At left the predictions of ordinary Dürr-Teller-Walecka-type theories are shown; at right those for a chirally symmetric meson field theory as develloped by P. Papazoglu, S. Schramm et al. [41, 42]. Note however, that this particular chiral mean–field theory does contain ω^4 terms. If introduced in both effective models, they seem to predict quantitatively similar results.

binding energy) are well reproduced, the prediction of both effective Lagrangians for the dependence of the properties of the correlated vacuum on density and temperature is remarkably different. This is illustrated to some extend in Fig. 15. It seems at first that the chirally symmetric meson field theory predicts much higher primary densities (and temperatures) until the effects of the correlated vacuum are strong enough so that the mechanisms described here become effective. In other words, according to chirally symmet-

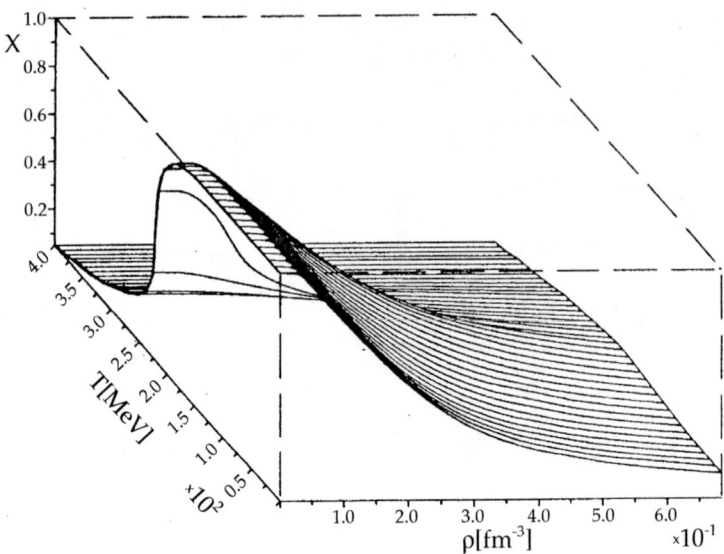

FIGURE 16. The strong phase transition inherent in Dürr-Teller-Walecka-type meson field theories, as predicted by J. Theis et al. [43]. Note that there is a first order transition along the ρ-axis (i.e. with density), but a simple transition along the temperature T-axis. Note also that this is very similar to the phase transition obtained recently from the Nambu-Jona-Lasinio-approximation of QCD [45].

ric meson field theories the antimatter-cluster-production and multi-hypermatter-cluster production out of the highly correlated vacuum takes place at considerably higher heavy ion energies as compared to the predictions of the Dürr-Teller-Walecka-type (D-T-W-MF) meson field theories. However, it should be noted that the introduction of a ω^4-term in the D-T-W-MF model (such a term is present in the chiral model) yields practically the same results in both models. Quite important is the question of the nucleonic substructure (form factors, quarks, gluons) and its influence on the highly correlated vacuum structure. The nucleons are possibly strongly polarized in the correlated vacuum, particularly for high densities and temperatures: the Δ resonance correlations in the vacuum are probably important. Is this highly correlated vacuum state, especially during the compression phase, a preliminary stage of a quark-gluon cluster plasma? To which extent is it similar or perhaps even identical with the standardly described quark-glun plasma? It is well known for more than 10 years that meson field theories predict a phase transition qualitatively and quantitatively similar to that of the quark-gluon plasma [43] — see Fig. 16.

The extension of the periodic system into the sectors hypermatter (strangeness) and antimatter is of general and astrophysical importance. Indeed, microseconds after the big bang the new dimensions of the periodic system we have touched upon, certainly have been populated in the course of the baryo- and nucleo-genesis. Of course, for the creation of the universe, even higher dimensional extensions (charm, bottom, top) come into play, which we did not pursue here. It is an open question, how the depopulation

20

(the decay) of these sectors influences the distribution of elements of our world today. Our conception of the world will certainly gain a lot through the clarification of these questions.

For the Gesellschaft für Schwerionenforschung (GSI), which I helped initiating in the sixties, the questions raised here could point to the way ahead. Working groups have been instructed by the board of directors of GSI, to think about the future of the laboratory. On that occasion, very concrete (almost too concrete) suggestions are discussed — as far as it has been presented to the public. What is necessary, as it seems, is a **vision on a long term basis**. The ideas proposed here, the verification of which will need the **commitment for 2–4 decades of research**, could be such a vision with considerable attraction for the best young physicists. The new dimensions of the periodic system made of hyper- and antimatter cannot be examined in the "stand-by" mode at CERN (Geneva); a dedicated facility is necessary for this field of research, which can in future serve as a home for the universities. The GSI — which has unfortunately become much too self-sufficient — could be such a home for new generations of physicists, who are interested in the **structure of elementary matter**. GSI would then not develop just into a detector laboratory for CERN, and as such become obsolete. I can already see the enthusiasm in the eyes of young scientists, when I unfold these ideas to them — similarly as it was 30 years ago, when the nuclear physicists in the state of Hessen initiated the construction of GSI.

I am grateful to Dipl.-Phys. Thomas Bürvenich for helping me in the technical production of these proceedings.

REFERENCES

1. S.G: Nilsson et al. Phys. Lett. 28 B (1969) 458
    Nucl. Phys. A 131 (1969) 1
    Nucl. Phys. A 115 (1968) 545
2. U. Mosel, B. Fink and W. Greiner, Contribution to "Memorandum Hessischer Kernphysiker" Darmstadt, Frankfurt, Marburg (1966).
3. U. Mosel and W. Greiner, Z. f. Physik 217 (1968) 256, 222 (1968) 261
4. a) J. Grumann, U. Mosel, B. Fink and W. Greiner, Z. f. Physik 228 (1969) 371
    b) J. Grumann, Th. Morovic, W. Greiner, Z. f. Naturforschung 26a (1971) 643
5. W. Greiner, Int. Journal of Modern Physics E, Vol. 5 , No. 1 (1995) 1-90. This review article contains many of the subjects discussed here in an extended version, see also for a more complete list of references.
6. A. Sandulescu, R.K. Gupta, W. Scheid, W. Greiner, Phys. Lett. 60B (1976) 225
    R.K. Gupta, A. Sandulescu, W. Greiner, Z. f. Naturforschung 32a (1977) 704
    R.K. Gupta, A.Sandulescu and W. Greiner, Phys. Lett. 64B (1977) 257
    R.K. Gupta, C. Parrulescu, A. Sandulescu, W. Greiner Z. f. Physik A283 (1977) 217
7. G. M. Ter-Akopian et al., Nucl. Phys. A255 (1975) 509
    Yu.Ts. Oganessian et al., Nucl. Phys. A239 (1975) 353 and 157
8. D. Scharnweber, U. Mosel and W. Greiner, Phys. Rev. Lett 24 (1970) 601
    U. Mosel, J. Maruhn and W. Greiner, Phys. Lett. 34B (1971) 587
9. G. Münzenberg et al. Z. Physik A309 (1992) 89
    S.Hofmann et al. Z. Phys A350 (1995) 277 and 288
10. R. K. Gupta, A. Sandulescu and Walter Greiner, Z. für Naturforschung 32a (1977) 704

11. A. Sandulescu and Walter Greiner, Rep. Prog. Phys 55. 1423 (1992); A. Sandulescu, R. K. Gupta, W. Greiner, F. Carstoin and H. Horoi, Int. J. Mod. Phys. E1, 379 (1992)
12. A. Sobiczewski, Phys. of Part. and Nucl. 25, 295 (1994)
13. R. K. Gupta, G. Münzenberg and W. Greiner, J. Phys. G: Nucl. Part. Phys. 23 (1997) L13
14. V. Ninov, K. E. Gregorich, W. Loveland, A. Ghiorso, D. C. Hoffman, D. M. Lee, H. Nitsche, W. J. Swiatecki, U. W. Kirbach, C. A. Laue, J. L. Adams, J. B. Patin, D. A. Shaughnessy, D. A. Strellis and P. A. Wilk, preprint
15. K. Rutz, M. Bender, T. Bürvenich, T. Schilling, P.-G. Reinhard, J.A. Maruhn, W. Greiner, Phys. Rev. C 56 (1997) 238.
16. B. Fricke and W. Greiner, Physics Lett 30B (1969) 317
    B. Fricke, W. Greiner, J.T. Waber, Theor. Chim. Acta (Berlin) 21 (1971) 235
17. T. Bürvenich, K. Rutz, M. Bender, P.-G. Reinhard, J. A. Maruhn, W. Greiner, Eur. Phys. J. A 3 (1998) 139-147
18. M. Bender, K. Rutz, P.-G. Reinhard, J. A. Maruhn, W. Greiner, Phys. Rev. C 58 (1998) 2126-2132
19. T. Bürvenich, M. Bender, P.-G. Reinhard, J. A. Maruhn, in preparation
20. A. Sandulescu, D.N. Poenaru, W. Greiner, Sov. J. Part. Nucl. 11(6) (1980) 528
21. Harold Klein, thesis, Inst. für Theoret. Physik, J.W. Goethe-Univ. Frankfurt a. M. (1992)
    Dietmar Schnabel, thesis, Inst. für Theoret. Physik, J.W. Goethe-Univ. Frankfurt a.M. (1992)
22. D. Poenaru, J.A. Maruhn, W. Greiner, M. Ivascu, D. Mazilu and R. Gherghescu, Z. Physik A328 (1987) 309, Z. Physik A332 (1989) 291
23. E. K. Hulet, J. F. Wild, R. J. Dougan, R. W.Longheed, J. H. Landrum, A. D. Dougan, M. Schädel, R. L. Hahn, P. A. Baisden, C. M. Henderson, R. J. Dupzyk, K. Sümmerer, G. R. Bethune, Phys. Rev. Lett. 56 (1986) 313
24. K. Depta, W. Greiner, J. Maruhn, H.J. Wang, A. Sandulescu and R. Hermann, Intern. Journal of Modern Phys. A5, No. 20, (1990) 3901
    K. Depta, R. Hermann, J.A. Maruhn and W. Greiner, in "Dynamics of Collective Phenomena", ed. P. David, World Scientific, Singapore (1987) 29
    S. Cwiok, P. Rozmej, A. Sobiczewski, Z. Patyk, Nucl. Phys. A491 (1989) 281
25. A. Sandulescu and W. Greiner in discussions at Frankfurt with J. Hamilton (1992/1993)
26. J.H. Hamilton, A.V. Ramaya et al. Journ. Phys. G 20 (1994) L85 - L89
27. B. Burggraf, K. Farzin, J. Grabis, Th. Last, E. Manthey, H. P. Trautvetter, C. Rolfs, Energy Shift of first excited state in 10Be ?, accepted for publication in Journ. of. Phys. G
28. P. Hess et al., Butterfly and Belly Dancer Modes in 96Sr + 10Be + 146Ba, in preparation
29. E.K. Hulet et al. Phys Rev C 40 (1989) 770.
30. W. Greiner, B. Müller, J. Rafelski, QED of Strong Fields, Springer Verlag, Heidelberg (1985). For a more recent review see W. Greiner, J. Reinhardt, Supercritical Fields in Heavy–Ion Physics, Proceedings of the 15th Advanced ICFA Beam Dynamics Workshop on Quantum Aspects of Beam Physics, World Scientific (1998)
31. B. Povh, Rep. Progr. Phys. 39 (1976) 823; Ann. Rev. Nucl. Part. Sci. 28 (1978) 1; Nucl. Phys. A335 (1980) 233; Progr. Part. Nucl. Phys. 5 (1981) 245; Phys. Blätter 40 (1984) 315
32. J. Schaffner, Carsten Greiner and H. Stöcker Phys. Rev. C45 (1992) 322; Nucl. Phys. B24B (1991) 246; J. Schaffner, C.B. Dover, A. Gal, D.J. Millener, C. Greiner, H. Stöcker: Annals of Physics235 (1994) 35; C. Greiner and J. Schaffner, Physics of Strange Matter for Relativistic Heavy-Ion Collision Int. J. Mod. Phys. E5, 239-300 (1996)
33. W. Scheid and W. Greiner, Ann. Phys. 48 (1968) 493; Z. Phys. 226 (1969) 364
34. W. Scheid, H. Müller and W. Greiner Phys. Rev. Lett. 13 (1974) 741
35. H. Stöcker, W. Greiner and W. Scheid Z. Phys. A 286 (1978) 121
36. I. Mishustin, L.M. Satarov, J. Schaffner, H. Stöcker and W.Greiner Journal of Physics G (Nuclear and Particle Physics) 19 (1993) 1303
37. P.K. Panda, S.K. Patra, J. Reinhardt, J. Maruhn, H. Stöcker, W. Greiner, Int. J. Mod. Phys. E 6 (1997) 307
38. N. Auerbach, A. S. Goldhaber, M. B. Johnson, L. D. Miller and A. Picklesimer, Phys. Lett. B182 (1986) 221
39. H. Stöcker and W. Greiner, Phys. Rep. 137 (1986) 279.
40. J. Reinhardt and W. Greiner, to be published.
41. P. Papazoglou, D. Zschiesche, S. Schramm, H. Stöcker, W. Greiner, J. Phys. G 23 (1997) 2081;

P. Papazoglou, S. Schramm, J. Schaffner-Bielich, H. Stöcker, W. Greiner, Phys. Rev. C 57 (1998) 2576.

42. P. Papazoglou, D. Zschiesche, S. Schramm, J. Schaffner–Bielich, H. Stöcker, W. Greiner, nucl–th/9806087, accepted for publication in Phys. Rev. C.

43. J. Theis, G. Graebner, G. Buchwald, J. Maruhn, W. Greiner, H. Stöcker and J. Polonyi, Phys. Rev. D 28 (1983) 2286

44. P. Papazoglou, PhD thesis, University of Frankfurt, 1998; C. Beckmann et al., in preparation

45. S. Klimt, M. Lutz, W. Weise, Phys. Lett. B249 (1990) 386.

Nuclear Equation of State

Paweł Danielewicz

*National Superconducting Cyclotron Laboratory and Department of Physics and Astronomy,
Michigan State University, East Lansing, MI 48824, USA,
and Gesellschaft für Schwerionenforschung mbH, D-64291 Darmstadt, Germany*

Abstract. Nuclear equation of state plays an important role in the evolution of the Universe, in supernova explosions and, thus, in the production of heavy elements, and in stability of neutron stars. The equation constrains the two- and three-nucleon interactions and the quantum chromodynamics in nonperturbative regime. Despite the importance of the equation, though, its features had remained fairly obscure. The talk reviews new results on the equation of state from measurements of giant nuclear oscillations and from studies of particle emission in central collisions of heavy nuclei.

INTRODUCTION

An equation of state (EOS) is a nontrivial relation between thermodynamic variables characterizing a medium. While the term is used in its singular form in nuclear physics, actually different relations are of interest, such as between pressure p and baryon density ρ and temperature T, $p(\rho, T)$, or chemical potential μ and T, $p(\mu, T)$, between energy density e and ρ and T, $e(\rho, T)$, etc. Some of the relations are fundamental under certain conditions, i.e. all other relations may be derived from them (such as from $e(\rho)$ at $T = 0$).

The nuclear EOS is of interest because it affects the fate of the Universe at times $t \gtrsim 1\,\mu$s from the Big Bang and because its features are behind the supernova explosions. Moreover, its features ensure the stability of neutron stars. Through its effects on the evolution of the Universe, on supernovae explosions, and on neutron-star collisions, the EOS affects nucleosynthesis. Moreover, the EOS impacts central reactions of heavy nuclei. Finally, the form of the EOS constraints hadronic interactions and the nonperturbative quantum chromodynamics (QCD).

IMPORTANCE OF EOS

Different regimes for the strongly interacting are conveniently assessed in the $\mu - T$ plane, see Fig 1. Along the $T = 0$ axis, at $\mu \approx 930$ MeV, we have the matter in heavy nuclei. The matter in the interior of neutron stars corresponds to higher chemical potentials, in combinations with low temperatures. The matter in the early Universe evolved along the temperature axis, at low baryon number content, and thus at low μ. Different regions of the plane are explored at different accelerators. In the early Universe and likely at the higher-energy accelerators, the matter crosses the transition between the hadronic matter and quark-gluon plasma. The transition is observed in numerical lattice QCD calculations as a rapid change in energy density in the temperature region

CP597, *Nonequilibrium and Nonlinear Dynamics in Nuclear and Other Finite Systems,*
edited by Z. Li et al.
© 2001 American Institute of Physics 0-7354-0041-5/01/$18.00

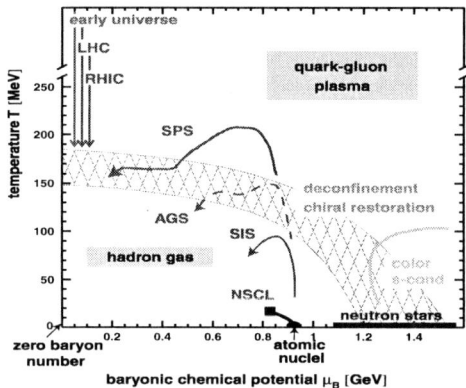

FIGURE 1. Strongly interacting matter in the $\mu - T$ plane, after [1]*.

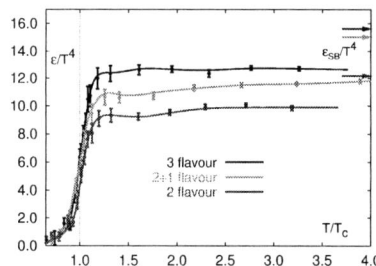

FIGURE 2. Energy in baryonless matter vs T, from calculations of Ref. [2].

of $T_c \sim 170$ MeV, cf. Fig. 2. The numerical calculations are carried out on a lattice of a finite size and it can be difficult to establish whether one deals just with a transitional behavior or with a phase transition and, if so, of what order. Whether or not there is a first-order phase transition is of importance for the early Universe.

Early Universe

Associated with a first-order phase transition is the surface tension σ and a possibility of supercooling. For sufficiently high σ, the early Universe might supercool down to temperatures as low as half of the critical temperature T_c, cf. Fig. 3. The large surface tension would lead to a wide separation, by as much as $\ell \sim 1$m, of the forming hadronic bubbles and, eventually, as the hadronic bubbles grow and begin to fill all space, of the remnant quark-gluon bubbles, cf. Fig. 4. The separation would produce large nonuniformities, characterized by masses $M \sim 10^{18}$ kg (i.e. of a medium size asteroid), in the distribution of the baryon number following the hadronization, with the baryon number concentrated in the regions that hadronize last. The excess baryon number would get trapped in the quark-gluon bubbles, because the baryon number costs

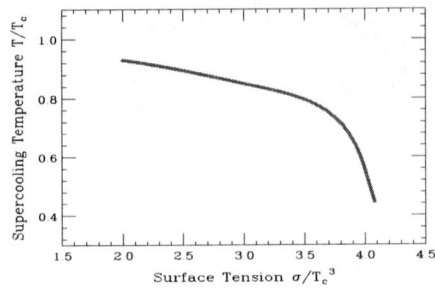

FIGURE 3. Supercooling tension for the confinement phase transition, after [3].

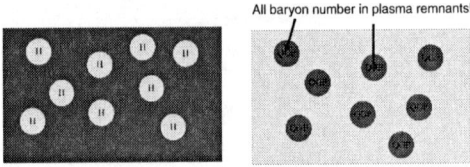

FIGURE 4. Phase bubbles in the region of the confinement phase transition, after [3].

little in the quark-gluon phase, with quarks being massless, and a lot in the hadronic phase, with massive baryons. An analogous situation takes place when seawater freezes. Then the salt appears in the areas that are last to freeze, Fig. 5. There are some cautioning theoretical and experimental indications, though, regarding the scenario, that the surface tension might not be very large between in the quark-gluon and hadron phases.

Supernova Explosions

Type II supernova explosions are the source of at least half of the nuclei heavier than iron around us. Only very massive stars, of masses $M \gtrsim 8 M_\odot$, explode. Generally, the more massive a star, the shorter it lives, burning faster due to higher density and temperature in its interior. A star starts out burning hydrogen, then helium and successively heavier nuclei; at each stage the products are accumulated. After a given fuel runs out, the gravitation compresses the star core raising temperature and the next fuel ignites with its burning preventing further compression. When the core consists of iron only, the burning stops. It is then up to the electron pressure (such as resisting the compression of solids) to prevent the gravitational collapse of the core. However, the electron

FIGURE 5. Seawater analogy.

26

pressure fails when the core exceeds the threshold Chandrasekhar mass. This is seen by examining the contributions to the energy from gravity and from an ultrarelativistic electron gas:

$$E = -\frac{3}{5}G\frac{(N m_N)^2}{R} + N_e\left(\frac{3}{4}p_F c + \frac{3m_e^2 c^4}{2p_F c} + \ldots\right) + \ldots$$

$$= \frac{1}{R}\left(\frac{3}{5}G(Nm_N)^2 + \frac{3}{4}\hbar c\left(\frac{9\pi}{4}\right)^{1/3}N_e^{4/3}\right) + O(R). \tag{1}$$

The electron Fermi momentum is proportional to the cube root of electron density and, thus, is inversely proportional to core radius, $p_F \propto \rho_e^{1/3} \propto 1/R$. Both the gravitational and electron energies are then inversely proportional to the radius, but the electron energy grows only as the number of electrons to the 4/3 power while the gravitational energy as the square power of the nucleon number. For the electron number equal to half the nucleon number, $N_e = N_N/2$, the gravity wins over electrons for core mass

$$M > M_{th} = \left(\frac{5}{6}\frac{\hbar c}{G}\right)^{3/2}\frac{3\pi^{1/2}}{m_N^2} \sim 1.5 M_\odot. \tag{2}$$

When the iron core exceeds the threshold mass, a gravitational collapse of the core starts and progresses till the nuclear densities are reached. The nuclear matter is more incompressible than the electron gas – what starts as an implosion gets reversed at the nuclear densities into an explosion. From the center of star a shock wave moves out, see the schematic view in Fig. 6, while at the center a so-called protoneutron star forms at a density of the order of that in nuclei. Inside, as the electron Fermi energy exceeds the proton-neutron mass difference, the process of neutronization takes place, $e^- + p \to \nu_e + n$. Additionally, thermal neutrinos are copiously produced. In the meantime, the shock moving through the infalling material stalls outside of the protostar and gets, most likely, revived by the neutrinos coming out from the center. Aside from propelling the shock, the neutrinos drive the neutron wind from the center within which copper, nickel, zinc and other elements form. Eventually, the shock reaches the star surface producing a magnificant display in the sky and throwing 7 $M_\odot$ of material space. The properties of nuclear matter, where the collapse reverses and that is the site of neutrino production, are, however, generally not well known.

Neutron Stars

The protoneutron star eventually turns into a black hole or into a neutron star. Which is the case depends on the properties of nuclear matter, Fig. 7. Dependent on those properties are also the characteristics of the forming neutron star and, in particular, the density profile and radius, see Fig. 8. In astrophysical modelling of neutron stars or of supernova explosions, a host of nuclear EOS is employed, such as those in Fig. 9, in terms of the dependence on pressure on energy density. Some EOS are excluded by causality (those with high p) and some by known masses of existing neutron stars (those

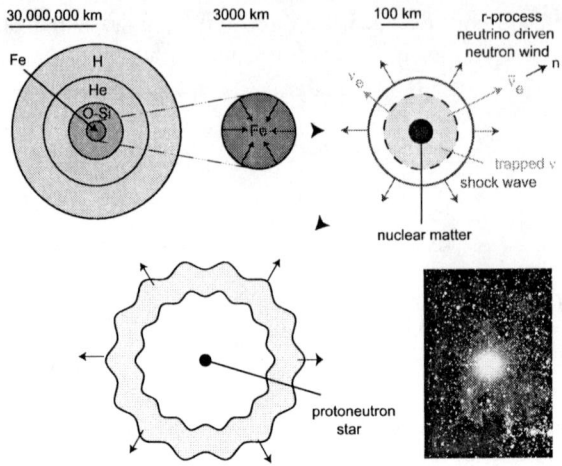

FIGURE 6.   Supernova explosion.

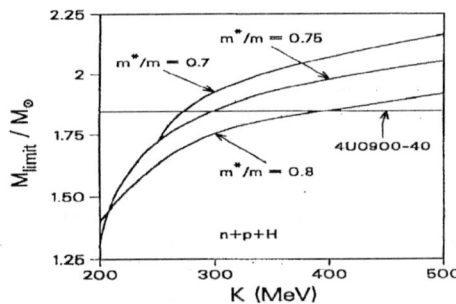

FIGURE 7.   Limiting neutron star mass as a function of the compression modulus of the corresponding symmetric matter [5]*.

with low p). This still leaves a wide range of possibilities; there are EOS taken from nonrelativistic and relativistic calculations and some of the EOS incorporate different types of phase transitions.

A possible site for the synthesis of heavy elements, other than supernova explosions, are mergers of neutron stars. These mergers shed much more matter into space if the nuclear EOS is relatively soft than when it is stiff, Fig. 10.

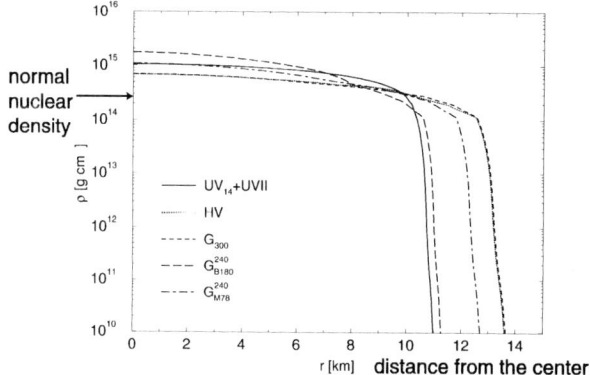

FIGURE 8. Density profile of a neutron star of mass $M = 1.4 M_\odot$, after [4].

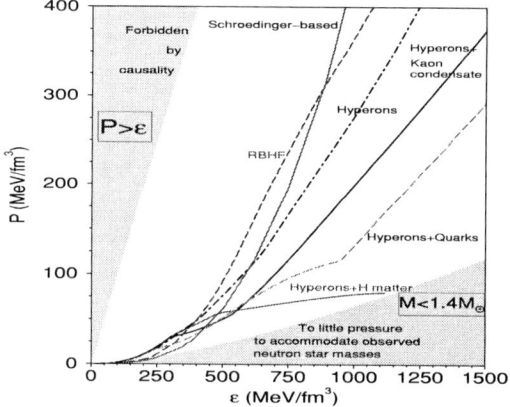

FIGURE 9. Pressure-energy relations [6] for nuclear matter, employed in astrophysical calculations.

ELEMENTARY FEATURES OF THE NUCLEAR EOS

Energy Minimum

The advances in the determination of the nuclear EOS have been, generally, difficult. The elementary information comes from the Weizsäcker binding-energy formula and from the systematics of nuclear density profiles. The Weizsäcker formula separates out the contributions to the energy associated with nuclear interactions and the interior and surface of nuclei, the contributions associated with isospin asymmetry and with Coulomb interactions, and the shell correction,

$$
\begin{aligned}
-B(A,Z) &= -16\,\text{MeV}\,A + a_s A^{2/3} \\
&\quad + a_a \frac{(A-2Z)^2}{A} + a_c \frac{Z(Z-1)}{A^{1/3}} - B_{p,s}.
\end{aligned}
\tag{3}
$$

29

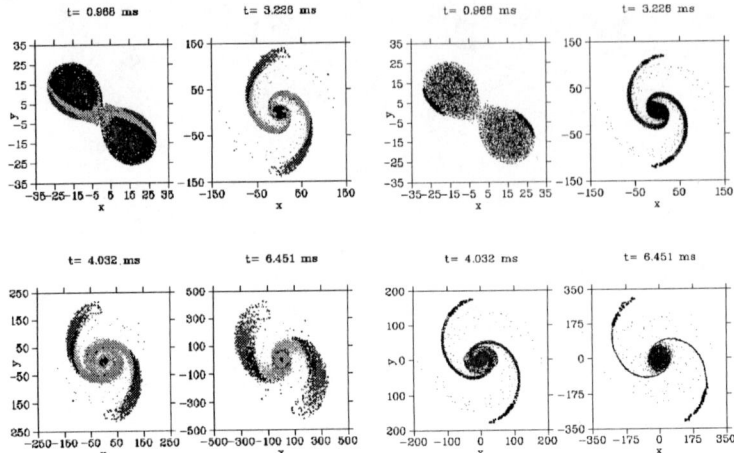

FIGURE 10. Neutron star mergers for soft (left panels) and stiff (right panels) nuclear EOS, after [7].

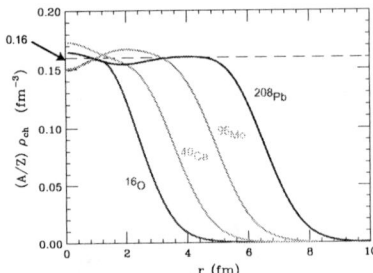

FIGURE 11. Nuclear densities deduced from electron scattering.

Nuclear densities, obtained from charge densities multiplied by mass to charge number ratio, are seen to reach the same value, $\rho_0 = 0.16$ fm$^{-3} \simeq 1/(6$ fm$^3)$, for a wide range of nuclear masses, see Fig. 11. We conclude that the energy per nucleon in a uniform symmetric nuclear matter at $T = 0$, in the absence of Coulomb interactions, has a minimum at the normal density ρ_0 with the energy value, relative to nucleon mass, of -16 MeV, from the volume term in the binding formula, see Fig. 12. As, obviously, the binding energy approaches zero for separated nucleons at $\rho \to 0$, we actually know two points in the $(T = 0)$ dependence of the energy per nucleon, $E/A \equiv e/\rho$, on density.

The next nontrivial feature of the energy per nucleon is its curvature in the dependence on ρ, around ρ_0. This curvature is commonly quantified in terms of the so-called nuclear incompressibility, with an unusal numerical factor:

$$K = 9\rho_0^2 \frac{d^2}{d\rho^2}\left(\frac{E}{A}\right) = R^2 \frac{d^2}{dR^2}\left(\frac{E}{A}\right). \tag{4}$$

The factor stems from the fact that the nuclei were first considered as sharp-edged

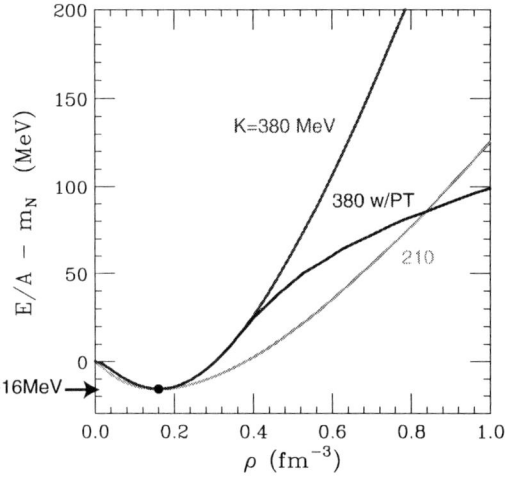

FIGURE 12. Energy per nucleon vs density in nuclear matter.

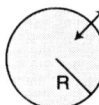

FIGURE 13. Oscillating nuclei were first considered as sharp-edged spheres with the energy changing as a function of the radius.

spheres with the energy changing as a function of the radius (Fig. 13). To get an idea of what might be expected for the incompressibility, one might just run a parabola through the two known points on the curve of $\frac{E}{A}(\rho)$. The then resulting incompressibility has a value of $K \sim 290$ MeV. If the actual incompressibility turns out to be below this benchmark value, we may consider the nuclear EOS to be soft, and stiff if the opposite is the case.

Microscopic Calculations

To get the features of the nuclear EOS outside of the minimum, one might turn to microscopic calculations, such as within Brueckner and variational frameworks. These calculations utilize elementary nucleon-nucleon interactions constrained by nucleon-nucleon interactions and by deuteron properties. However, the nonrelativistic calculations with only nucleon-nucleon interaction miss the known position of the minimum in the nuclear EOS; the minimae line up along the so-called Coester line (Fig. 14) in the energy vs density or Fermi momentum, with the change of the version of the interaction. The relativistic calculations line up along another Coester line that passes closer to the true minimum; aside from relativity, though, those calculations are generally more

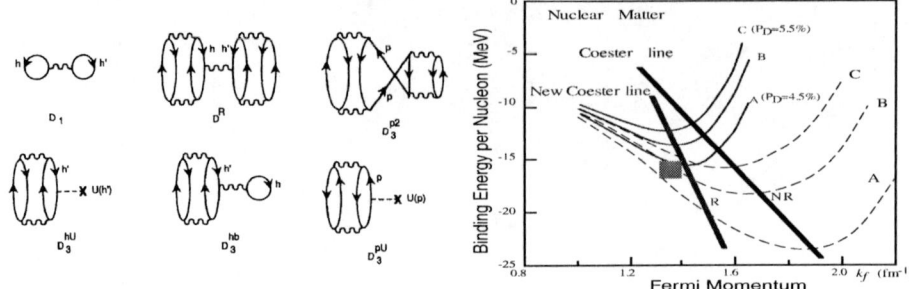

FIGURE 14. Left: Diagrams for different terms in the energy per nucleon in many-body calculations [8]. Right: Binding energy vs Fermi momentum in many-body calculations, after [9].

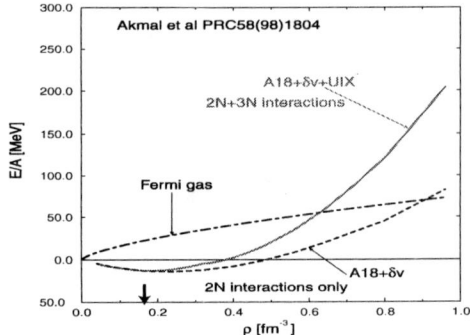

FIGURE 15. Energy per nucleon in nuclear matter as a function of density, from a variational calculation of Ref. [10] with two- and three-nucleon interactions.

primitive than the nonrelativistic ones.

To get the right position of the minimum in the EOS, Fig. 15, it is necessary to incorporate three-nucleon interactions in the microscopic calculations. These interactions are not well constrained by scattering, hampering the predictive power of the theory. In this situation, one may want to turn to experiment to get the information on the EOS away from the normal density.

INCOMPRESSIBILITY - GETTING OUT OF THE MINIMUM

The simplest way to determine the incompressibility experimentally may seem to induce volume oscillations in a nucleus. This could be done by scattering α particles off a nucleus, Fig. 16. For the lowest excitation, the excitation energy E^*, deduced from the final α energy, would be related to the classical frequency through $E^* = \hbar\Omega$, and the latter would be related to K. Let us examine the classical energy of an oscillating nucleus:

$$E_{tot} = \int d\mathbf{r}\, \rho \frac{m_N v^2}{2} + \frac{1}{2} A K (R - R_0)^2$$

$$= \frac{A m_N \langle r^2 \rangle_A \dot{R}^2}{2} + \frac{1}{2} A K (R - R_0)^2, \tag{5}$$

where we use the fact that, for a nucleus uniformly changing its density, the velocity is proportional to the radius, $v = \dot{R}(r/R)$. We then obtain the energy of a simple harmonic oscillator; the frequency is a square root of the spring constant divided by mass constant, yielding:

$$E^* = \hbar \sqrt{\frac{K}{m_N \langle r^2 \rangle_A}}. \tag{6}$$

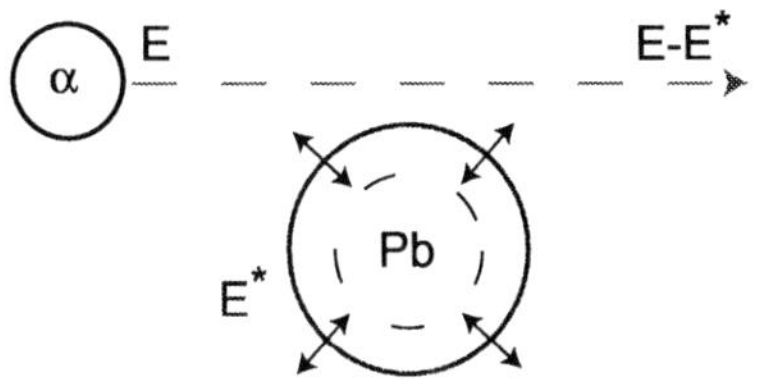

FIGURE 16. Volume oscillations induced by alpha scattering.

There are complications regarding this reasoning. Thus, the nucleus is not a sharp-edged sphere and the Coulomb interactions play a role in the oscillations as well as nuclear interactions. These effects may be taken care of by using an incompressibility constant characteristic for a nucleus, $K \rightarrow K_A$, and isolating different contributions in an analogy to those for the binding energy:

$$K_A = K + K_s A^{-1/3} + K_a \left(\frac{N - Z}{A} \right)^2 + K_c \frac{Z(Z-1)}{A^{4/3}} + \dots. \tag{7}$$

With the corrections, it turns out that the incompressibilities for medium to heavy nuclei are about 2/3 of the incompressibility for infinite nuclear matter, e.g. $K^{Pb} \sim 0.64 K$; $K^{Sm} \simeq 0.67 K$.

However, there are more problems. Thus, the density oscillations lie high up in the excitation energy and get broadened up. This may be remedied by employing a sum rule (notably, sum rules are often robust tools in helping to link simple classical considerations with the characteristics of quantum states):

$$\hbar \sqrt{\frac{K_A}{m_N \langle r^2 \rangle_A}} = \sqrt{\frac{\langle E^{*3} \rangle_{0^+ spectrum}}{\langle E^* \rangle_{0^+ spectrum}}}, \tag{8}$$

i.e. the incompressibility may be obtained from dividing the third by the first moment of the spectrum. An alternative is to use a microscopic theory, with an effective interaction, to describe both the excitation spectrum and the incompressibility for infinite matter.

The final complication is that other types of oscillations, than that changing the density, are excited in scattering, such as the oscillation of protons vs neutrons and

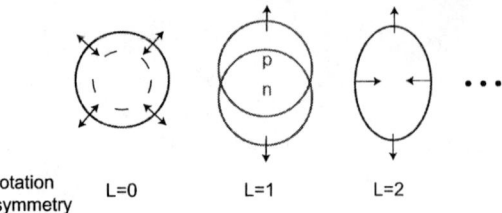

rotation symmetry   L=0   L=1   L=2

FIGURE 17. Different collective oscillations transform differently under rotations.

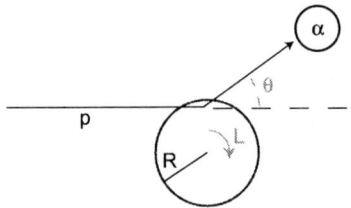

FIGURE 18. Delivering angular momentum to a target.

the quadrupole shape oscillation, cf. Fig. 17. However, those oscillations transform differently under rotations and, correspondingly, the elementary excitations for those oscillations are characterized by different angular momenta, with the uniform density changes characterized by $L = 0$. It is possible to isolate the $L = 0$ excitations by analyzing scattering at the very forward angles, Fig. 18. When the alpha particle scatters off a nucleus it transfers linear and angular momenta to the nucleus. The angular momentum is limited by the product of the linear momentum transfer and the distance over which the transfer occurs, i.e. roughly the sum of projectile and target radii. At high beam energies and small angles we get

$$L < |p - p'|R \approx p\theta R. \qquad (9)$$

Excitations characterized by $L \geq 1\hbar$ may suppressed by looking at scattering into the angles $\theta < \frac{\hbar}{pR}$, i.e. within the first diffraction peak.

Scattering of alpha particles from different targets has been carefully studied in recent years simultaneously as a function of excitation energy and scattering angle, allowing to isolate the contributions of $L = 0$ excitations [11, 12], see Fig. 19 for data from a samarium target. For the shown excitation energy of 16.5 MeV, a pronounced $L = 0$ peak is evident at low scattering angles. The $L = 0$ excitation strength is next shown for the samarium target in Fig. 20. A peak is evident at the excitation energy of 15.5 MeV, yielding an incompressibility of samarium $K^{Sm} = \frac{E^{*2}m_N\langle r^2\rangle_A}{\hbar^2} = 138\,\text{MeV}$, and of nuclear matter $K = K^{Sm}/0.67 \sim 210\,\text{MeV}$. However, explorations with microscopic models produce different results for K_A/K. In particular, relativistic models can yield results in the range $K \sim (250 - 270)$ MeV [13]. Generally, the results are, though, on the soft side of the incompressibility.

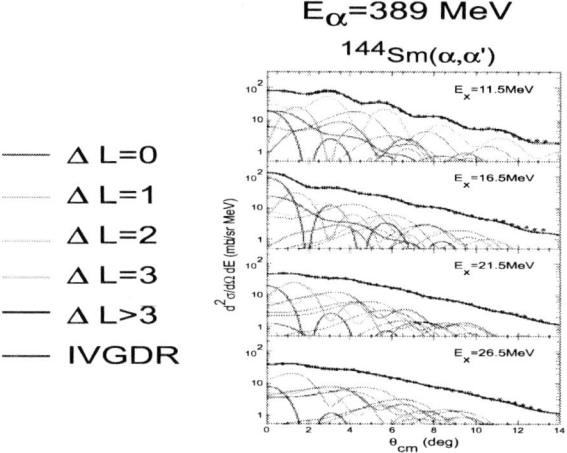

FIGURE 19. Alpha scattering cross section from ^{144}Sm, determined in Ref. [12].

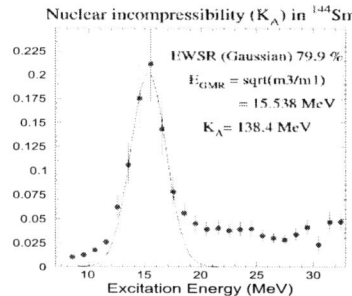

FIGURE 20. 0^+-strength function in ^{144}Sm, determined in Ref. [12].

EOS AT SUPRANORMAL DENSITIES FROM FLOW

Features of EOS at supranormal densities can be inferred from flow produced in collisions of heavy nuclei at high energies. At low impact parameters, in those collisions, macroscopic regions of high density are formed. The collective flow, that can be quantitatively assessed in collisions, is the particle motion characterized by space-momentum correlations of dynamic origin. The flow can provide information on the pressure generated in the collision.

To see how the flow relates to pressure, we may look at the hydrodynamic Euler equation for the nuclear fluid, an analog of the Newton equation, in a local frame where the collective velocity vanishes, $v = 0$:

$$(e+p)\frac{\partial}{\partial t}\vec{v} = -\vec{\nabla}p. \tag{10}$$

The collective velocity becomes an observable at the end of the reaction. In comparing

35

to the Newton equation, we see that the pressure $p = \rho^2 \frac{\partial(e/\rho)}{\partial \rho}|_{s/\rho}$ plays the role of a potential for the hydrodynamic motion, while the density of enthalpy $w = e + p$ plays the role of a mass. In fact, at moderate energies, the enthalpy density is practically the mass density, $w \approx \rho\, m_N$. We see from the Euler equation that the collective flow can tell us about the pressure in comparison to enthalpy. In establishing the relation, we need to know the spatial size where the pressure gradients develop and this will determined by the nuclear size. However, we also need the time during the hydrodynamic motion develops and this can represent a problem.

The equilibrium required for hydrodynamics is not quite achieved in reactions and, thus, transport theory is actually required to establish links between the EOS and observables; the hydrodynamics just yields important insights. The reacting system in the transport theory relying on Boltzmann equation is described in terms of the phase-space distribution functions f for different particles. In particular, the system energy is a functional of the distributions, $E\{f\}$, and can be parametrized to yield different EOS in equilibrium. The distributions follow a set of the Boltzmann equations with single-particle energies that are functional derivatives of the energy, $\varepsilon = \delta E / \delta f$:

$$\frac{\partial f}{\partial t} + \frac{\partial \varepsilon}{\partial \mathbf{p}} \frac{\partial f}{\partial \mathbf{r}} - \frac{\partial \varepsilon}{\partial \mathbf{r}} \frac{\partial f}{\partial \mathbf{p}} = I, \tag{11}$$

where I is the collision integral.

The first observable that one may want to consider to extract the information on EOS is the net radial or transverse collective energy. That energy may reach as much as half of the total kinetic energy in a reaction. Despite its magnitude, the energy is not useful for extracting the information on EOS because of the lack of information on how long the energy develops. Large pressures acting over a short time can produce the same net collective energy as low pressures acting over a long time. This makes appearent the need for a timer in reactions.

The role of the timer in reactions may be taken on by the so-called spectators. The spectator nucleons are those in the periphery of an energetic reaction, weakly affected by the reaction process, proceeding virtually at undisturbed original velocity, see Fig. 21. Participant nucleons, on the other hand, are those closer to the center of the reaction, participating in violent processes, subject to matter compression and expansion in the reaction. As the participant zone expands, the spectators, moving at a prescribed pace, shadow the expansion. If the pressures in the central region are high and the expansion is rapid, the anisotropies generated by the presence of spectators are going to be strong. On the other hand, if the pressures are low and, correspondingly, the expansion of the matter is slow, the shadows left by spectators will not be very pronounced.

There are different types of anisotropies in the emission that the spectators can produce. Thus, throughout the early stages of a collisions, the particles move primarily along the beam axis in the center of mass. However, during the compression stage, the participants get locked within a channel, titled at an angle, between the spectator pieces, cf. Fig. 21. As a consequence, the forward and backward emitted particles acquire an average deflection away from the beam axis, towards the channel direction. Another anisotropy may be observed for particles emitted in the transverse directions with zero longitudinal velocity. The region with compressed matter is open to the vacuum in

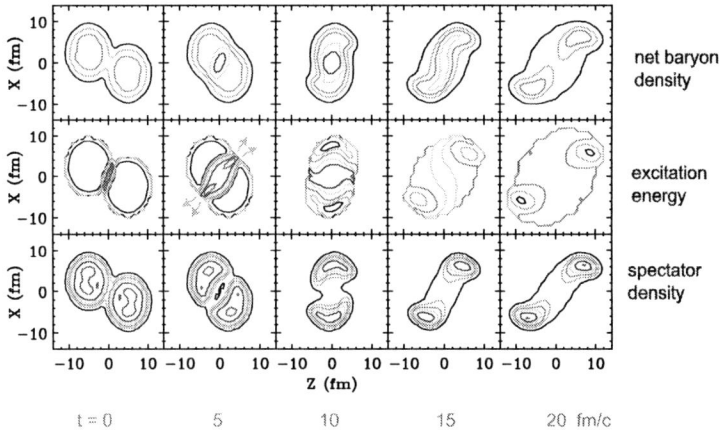

FIGURE 21. Reaction-plane contour plots for different quantities in a ^{124}Sn + ^{124}Sn reaction at 800 MeV/nucleon and $b = 6$ fm, from transport simulations by Shi [14].

the direction perpendicular to the reaction plane. However, in the direction within the reaction plane the region is shadowed by the participants. Thus, more particles are expected to be transversally emitted from the participant region perpendicular than within the direction plane. The anisotropy should be stronger the faster the expansion of the compressed matter.

The different anisotropies have been quantified experimentally over a wide range of bombarding energies. Figure 22 shows the measure of the sideward forward-backward deflection in Au + Au collisions as a function of the beam energy, with symbols representing data. Lines represent simulations assuming different EOS. On top of the figure, typical maximal densities are indicated which are reached at a given bombarding energy. Without interaction contributions to pressure, the simulations labelled cascade produce far too weak anisotropies to be compatible with data. The simulations with EOS characterized by the incompressibility $K = 167$ MeV yield adequate anisotropy at lower beam energies, but too low at higher energies. On the other hand, with the EOS characterized by $K = 380$ MeV, the anisotropy appears too high at virtually all energies. It should be mentioned that the incompressibilities should be considered here as merely labels for the different utilized EOS. The pressures resulting in the expansion are produced at densities significantly higher than normal and, in fact, changing in the course of the reaction.

Figure 23 shows next the anisotropy of emission at midrapidity or zero longitudinal velocity in the c.m., cf. Fig. 24, with symbols representing data and lines representing simulations. Again, we see that without interaction contributions to pressure, simulations cannot reproduce the measurements. The simulations with $K = 167$ MeV give too little pressure at high energies, and those with $K = 380$ MeV generally too much. A level of discrepancy is seen between data from different experiments.

We see that no single EOS allows for a simultaneous description of both types of anisotropies at all energies. In particular, the $K = 210$ MeV EOS is best for the sideward anisotropy, and the $K = 300$ MeV EOS is the best for the other, so-called

37

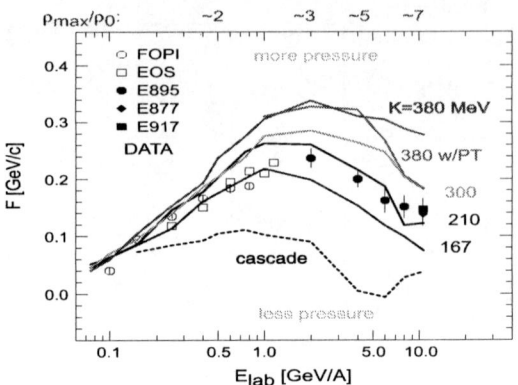

FIGURE 22. Sideward flow excitation function for Au + Au. Data and transport calculations are respresented, respectively, by symbols and lines [15].

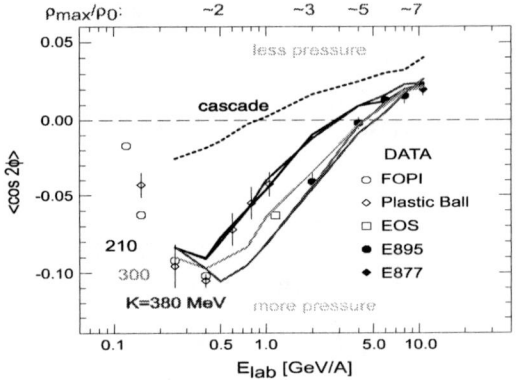

FIGURE 23. Elliptic flow excitation function for Au + Au. Data and transport calculations are respresented, respectively, by symbols and lines [15].

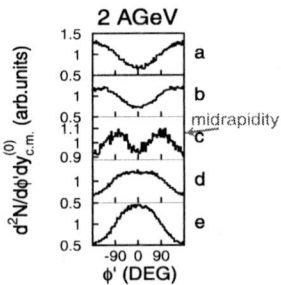

FIGURE 24. Azimuthal distribution of protons from Au + Au collisions at 2 GeV/nucleon in different rapidity intervals [16].

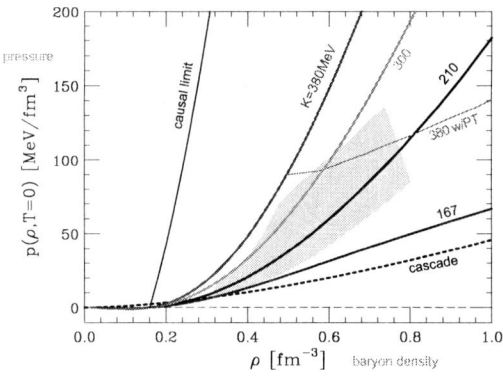

FIGURE 25. Constraints from flow on the $T = 0$ pressure-density relation, indicated by the shaded region [15].

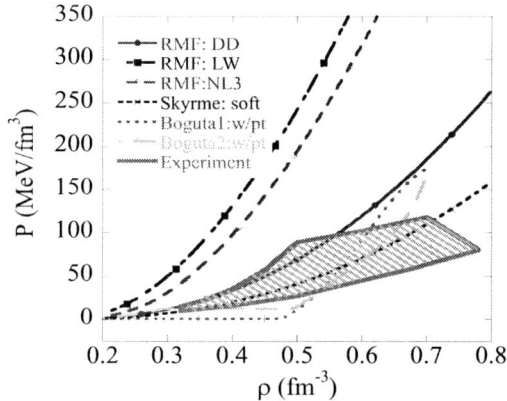

FIGURE 26. Impact of the constraints on models for EOS [15].

elliptic, anisotropy. We can use the discrepancy between the conclusions drawn from the two types of anisotropies as a measure of inaccuaracy of the theory and draw broad boundaries on pressure as a function of density from what is common in conclusions based on the two anisotropies. To ensure that the effects of compression dominate in the reaction over other effects, we limit ourselves to densities higher than twice the normal. The boundaries on the pressure are shown in Fig. 25 and they eliminate some of the more extreme models for EOS utilized in nuclear physics, such as the relativistic NL3 model and models assuming a phase transition at relatively low densities, cf. Fig. 26.

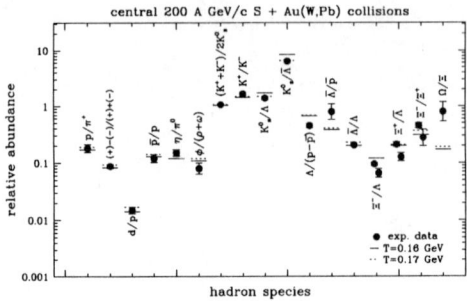

FIGURE 27. Relative particle abundancies in measurements (symbols) and calculated in the thermal freeze-out model (lines) in Ref. [17].

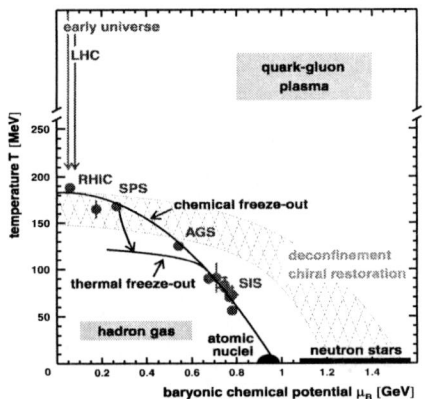

FIGURE 28. Freeze-out temperature and baryon chemical potential*.

HIGH-*T* LOW-ρ LIMITS OF THE HADRONIC WORLD

In central reactions of medium to heavy nuclei, over a broad range of bombarding energies, it is found that hadronic yields are consistent with thermal equilibrium at definite T and μ when interactions appear to stop [17, 18]. This is illustrated in Fig. 27 showing measured particle yields and those calculated assuming thermal equilibrium. The results indicate that, at the deduced temperatures and chemical potentials, the spectrum of hadrons is close to that in free space and, thus, the phase transition to quark-gluon phase has not been crossed. The boundaries of the hadronic world, staked out in this fashion, are shown in Fig. 28.

CONCLUSIONS AND OUTLOOK

Nuclear EOS ties together different areas of physics. Progress on the EOS has been made in different directions. Data on giant monopole resonances (and also on giant vector resonances) have been collected with significant background reductions and high resolution both in the energy and angle direction, allowing for improved determinations of the nuclear incompressibility. Anisotropies of flow from central reactions allow to constrain the EOS at supranormal densities. The parameters of freeze-out in reactions allow to stake out the limits of the hadronic world. Additional sources of information on EOS that I had no chance to talk about include measurements of neutron-star properties, studies of nuclear systematics and lattice QCD calculations. Unconquered EOS frontiers include the dependence of EOS on the isosopin degree of freedom and the detection of the quark-gluon plasma. The first frontier is, in particular, to be tackled at the NSCL coupled-cyclotrons and at the proposed RIA accelerator. In the baryonless regime, the second frontier is pursued at RHIC. However, the baryon-rich regime awaits stepped-up dedicated studies with good resolution in bombarding energy in the range of (2-40) GeV/nucleon.

ACKNOWLEDGMENTS

Information provided by M. Itoh on giant resonances is gratefully acknowledged. This work was partially supported by the National Science Foundation under Grant PHY-0070818.

*Original figures from Refs. [1] and [4], respectively, have been utilized, with permission from Elsevier Science.

REFERENCES

1. J. Stachel, Nucl. Phys. A654, 119c (1999).
2. F. Karsch, hep-lat/0106019.
3. B. Kämpfer et al., nucl-th/0011088.
4. Ch. Schaab et al., Nucl. Phys. A605, 531 (1996).
5. N. K. Glendenning, Phys. Rev. C 37, 2733 (1988).
6. F. Weber, Pulsars as Astrophysical Laboratories for Nuclear and Particle Physics, IoP Publishing, Bristol (1999).
7. S. Rosswog et al., Astronomy and Astrophysics 341, 499 (1999).
8. M. Baldo et al., Phys. Rev. C 41, 1748 (1990).
9. R. Brockmann and R. Machleit, Phys. Rev. C 42, 1965 (1990).
10. A. Akmal, V. R. Pandharipande, and D. G. Ravenhall, Phys. Rev. C 58, 1804 (1998).
11. D. H. Youngblood, Nucl. Phys. A687 1c (2001).
12. M. Itoh et al., Nucl. Phys. A687, 52c (2001).
13. D. Vretenar et al., Nucl. Phys. A621, 853 (1997).
14. L. Shi, P. Danielewicz and R. Lacey, Phys. Rev. C 64, 034601 (2001).
15. P. Danielewicz, W. Lynch and R. Lacey (2001), to be published.
16. C. Pinkenburg et al., Phys. Rev. Lett. 83, 1295 (1999).
17. P. Braun-Munzinger et al., Phys. Lett. B 365, 1 (1996).
18. J. Cleymans and K. Redlich, Nucl. Phys. A661, 379c (1999).

PART II

PHASE TRANSITIONS IN STRONG INTERACTIONS

Hadron Ratios: Chiral Symmetry Restauration vs. Nonequilibrium Quark Dynamics

K. Paech*, M. Bleicher*, S. Scherer*, D. Zschiesche*, H. Stöcker* and W. Greiner*

*Institut für Theoretische Physik, J.W. Goethe-Universität, Robert-Mayer-Str. 8-10, D-60054 Frankfurt am Main, Germany

Abstract.
Hadron production in recent nucleus-nucleus collisions experiments at the SPS and at RHIC have been analyzed in various thermal model fits. Here it is shown that the fit cannot be unique: the chemical freeze-out close to the phase transition to a QGP should be accompanied by shifting baryon masses in medium, which can yield considerably smaller break-up temperatures. Also a simple microscopic quark molecular dynamics model yields hadron ratios in nice agreement with those data. Hence, additional information, e.g. from HBT, is needed to determine the freeze-out parameters from data.

INTRODUCTION

The relativistic heavy ion collider (RHIC) at Brookhaven National Laboratory and the Pb+Pb program at the CERN-SPS aim to explore the phase diagram of hot and dense matter near the quark gluon plasma (QGP) phase transition [1]. The QGP is a hypothetical state in which the individual hadrons dissolve into a gas of free (or almost free) quarks and gluons in strongly compressed and heated matter.

Ideal gas model calculations have been used for a long time to calculate particle production in relativistic heavy ion collisions, e.g. [2, 3, 4, 5, 6, 7, 8]. Fitting the particle ratios as obtained from those ideal gas calculations to the experimental measured ratios at SIS, AGS and SPS for different energies and different colliding systems yields a curve of chemical freeze-out in the $T - \mu$ plane. The obvious question arises, to what extent do the deduced temperature and chemical potentials depend on the model employed.

PARTICLE PRODUCTION

The influence of changing hadron masses and effective deserves special attention [9, 10, 11, 12]. This is of special importance for the formation of a deconfined phase, i.e. the quark-gluon plasma. As deduced from lattice data [13], the critical temperature for the onset of a deconfined phase coincides with that of a chirally restored phase. Chiral effective models of QCD therefore can be utilized to give insights on the chiral signals from a quark-gluon plasma formed in heavy-ion collisions.

CP597, *Nonequilibrium and Nonlinear Dynamics in Nuclear and Other Finite Systems,*
edited by Z. Li et al.
© 2001 American Institute of Physics 0-7354-0041-5/01/$18.00

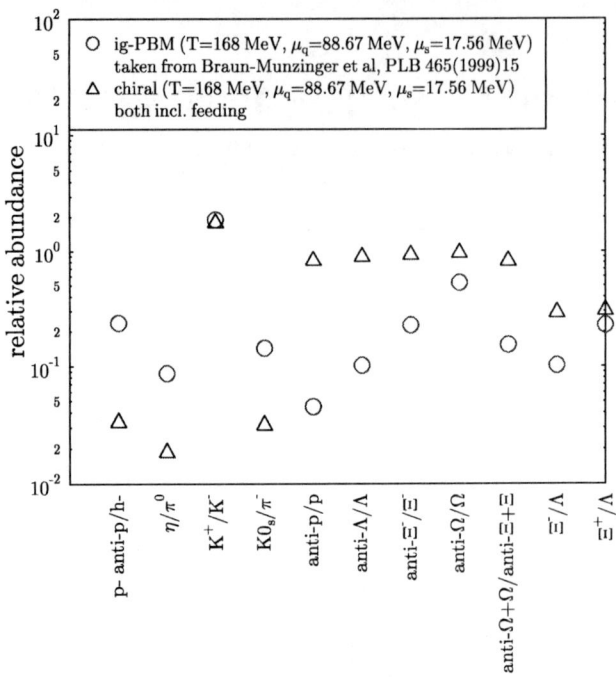

FIGURE 1. Comparison of a chiral calculation with [5] for $T = 168$MeV, $\mu_q = 88.67$MeV. The huge differences result from the dropping baryonic masses, depending on the strangeness content of the particles, and from the change of the effective potentials in the chirally restored phase.

Therefore we compare experimental data for Pb+Pb collisions at SPS with the ideal gas calculations and results obtained from a chiral SU(3) model [12, 14]. This effective hadronic model predicts a chiral phase transition at $T \approx 150$MeV. Furthermore the model predicts changing hadron masses and effective chemical potentials, due to strong scalar and vector fields in hot and dense hadronic matter, which are constrained by chiral symmetry from the QCD Lagrangean.

In [5] the ideal gas model was fitted to particle ratios measured in Pb+Pb collisions at SPS. The lowest χ^2 is obtained for $T = 168$MeV and $\mu_q = 88.67$MeV. Using these values as input for the chiral model leads to dramatic changes of the particle ratios. As can be seen in Figure 1.

There are two main reasons for the strong deviations from the ordinary hadron gas model:

- First, since the chosen temperature lies above the chiral phase transition temperature of the SU(3) chiral model, the effective masses of the baryons are lowered dramatically (see Figure 2).
- The second reason are the strongly changed effective potentials. While the chemical potential of the proton with the chosen parameters is $\mu_p = 266$MeV in the ideal gas model, the effective chemical potential of the proton in the chiral model is

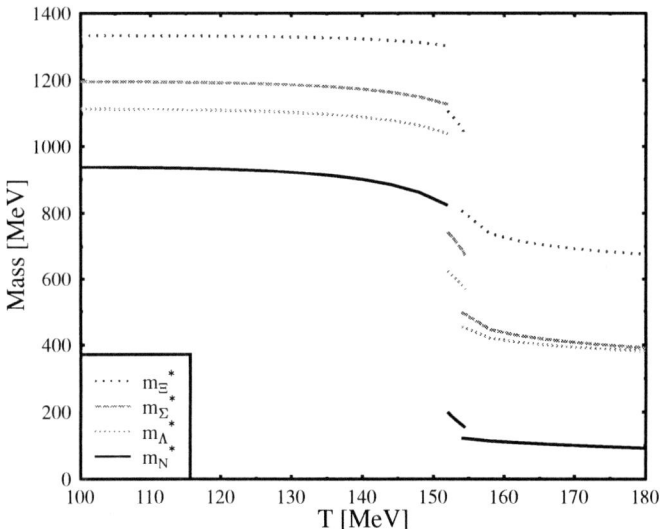

FIGURE 2. Baryon masses as function of temperature for vanishing chemical potential. One sees that a phase-transition occurs at $T_c \approx 150 MeV$. There the masses drop significantly and then nearly saturate. The baryon mass above T_c scales with the strangeness content of the corresponding baryon.

only $\mu_p \approx 8 MeV$, due to the reduction by the vector-field. Such a small effective chemical potential leads to nearly equal particle-antiparticle numbers and therefore to particle-antiparticle ratios ≈ 1.

Obviously, the freeze-out temperature and chemical potential have to be readjusted to account for the in-medium effects of the hadrons in the chiral model.

We call the best fit the parameter set that gives a minimum in the value of χ^2, with

$$\chi^2 = \sum_i \frac{\left(r_i^{exp} - r_i^{model}\right)^2}{\sigma_i^2}. \tag{1}$$

Here r_i^{exp} is the experimental ratio, r_i^{model} is the ratio calculated in the model and σ_i represents the error in the experimental data points as quoted in [5]. We included the following ratios in the fit procedure: $\frac{\bar{p}-p}{h^-}$, $\frac{\eta}{\pi^0}$, $\frac{K^+}{K^-}, \frac{K_s^0}{\pi^-}$, $\frac{\bar{p}}{p}, \frac{\bar{\Lambda}}{\Lambda}, \frac{\bar{\Xi}}{\Xi}, \frac{\bar{\Omega}}{\Omega}, \frac{\bar{\Omega}+\Omega}{\bar{\Xi}^-+\Xi^-}, \frac{\Xi^-}{\Lambda}, \frac{\Xi^+}{\Lambda}$. The resulting values of χ^2 for different $T - \mu$ pairs are shown in Figure 3. In all calculations μ_s was chosen such that the overall net strangeness f_s is zero. The best values for the parameters are $T = 144$ MeV and $\mu_q \approx 95$ MeV.

While the value of the chemical potential does not change much compared to the ideal gas calculation, the value of the temperature is lowered by more than 20 MeV. Furthermore Figure 3 shows, that the dropping effective masses and the reduction of the effective chemical potential make the reproduction of experimentally measured particle ratios as seen at CERN-SPS within this model impossible for $T > T_c$.

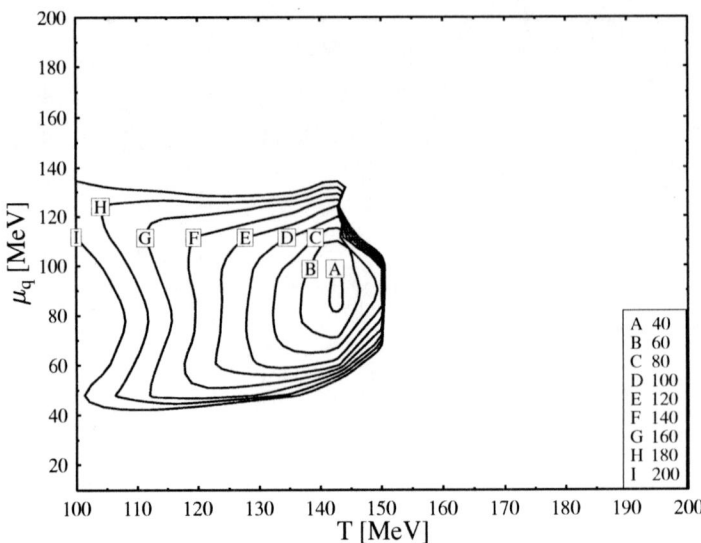

FIGURE 3. χ^2 for the SU(3) chiral model. Data are taken from [5]. The best fit parameters are $T = 144 MeV$ and $\mu_q \approx 95 MeV$. μ_q is chosen such that $f_s = 0$. For $T > T_c$ no agreement with data from CERN-SPS lead on lead collisions can be obtained.

Only for temperatures below T_c reasonable fits are obtained. Using the best fit parameters the particle ratios, as calculated in the chiral model, are compared with the ideal gas calculation of [5] and data as compiled in [5] (Figure 4). Satisfactory agreement over several magnitudes are obtained. The χ^2 values of the chiral SU(3) model is $\chi^2_{chiral} = 26.5$. This is larger than the value in the ideal gas model of [5] ($\chi^2_{ig} = 13$). Note that in [5] weak decays are accounted for. If we use our ideal gas calculation without feeding from weak decays we obtain only slightly changed best values ($T = 168, \mu_q = 82$ MeV) compared to [5] and $\chi^2_{ig-FFM} = 21.6$. This shows that the chiral and ideal gas analysis using the same feeding procedures yield comparable agreement with data concerning the value of χ^2. Figure 4 shows that there is satisfactorily agreement between data and experiment.

We want to emphasize that, in spite of the strong assumption of thermal and chemical equilibrium, the obtained values for T and μ differ significantly depending on the underlying model, i.e. whether and how effective masses and effective chemical potentials are accounted for. In the chiral SU(3) model, we conclude that the observed particle ratios as measured at CERN-SPS do not signal the freeze-out from a chirally restored quark-gluon phase. Note that we assume implicitly, that the particle ratios are determined by the medium effects and freeze out during the late stage expansion - no flavor changing collisions occur anymore, but the hadrons can take the necessary energy to get onto their mass shell by drawing energy from the fields. Rescattering effects will alter our conclusion but are presumably small when the chemical potentials are frozen.

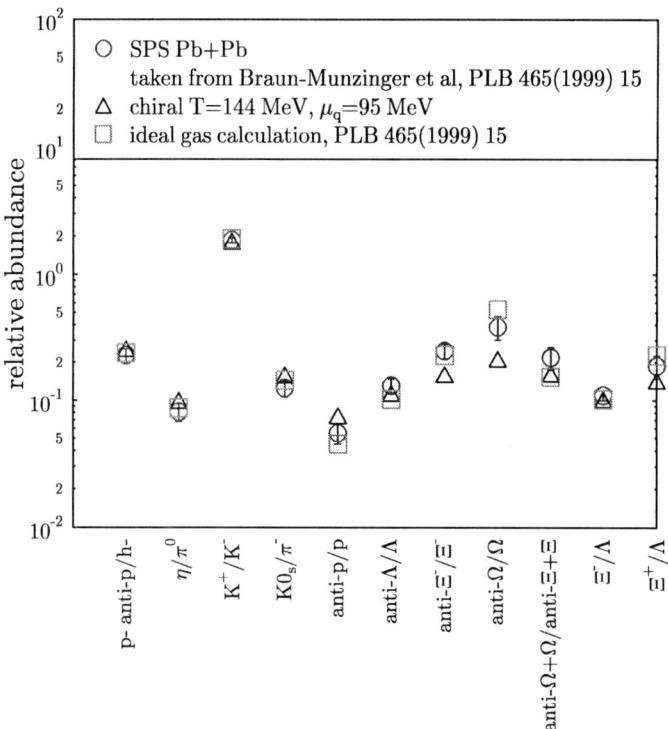

FIGURE 4. Comparison of the chiral model for $T = 144$ MeV with data and ideal gas results from [5].

INSIGHTS FROM QUARK MOLECULAR DYNAMICS

Further insights about the possible formation of deconfined matter can be obtained from the Quark Molecular Dynamics Model (qMD) [15] which explicitly includes quark degrees of freedom. The qMD can provide us with detailed information about the dynamics of the quark system and the parton-hadron conversion. Particle Ratios as well as correlations between the quarks clustering to build new hadrons can be studied [16].

One can apply the qMD model to study the expansion and hadronization of the fireball created in heavy ion collisions between S and Au nuclei at the SPS.

The initial state of the quark dynamics is generated by a transport simulation with hadronic and string degrees of freedom (in this paper the UrQMD model [17] is used—other models can also be applied). The initial nuclei are propagated within the transport simulation until a critical energy density is reached. In the investigated collisions, we assume the dissolution of the hadronic degrees of freedom at an energy density of 1 GeV/fm^{-3} in line with recent lattice calculations [19]. This energy density is reached at full overlap, i.e. 1.5 fm/c after the begin of the interaction. At this time all hadrons (mesons and baryons) are dissolved and all newly produced qq pairs from the decaying colour fields are used as input of the quark Molecular Dynamics model.

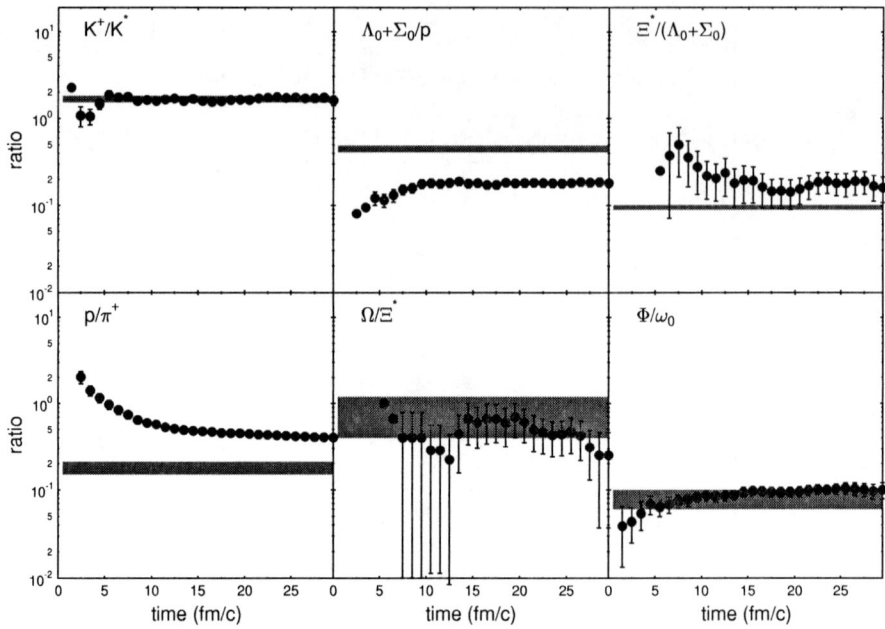

FIGURE 5. Time evolution of the particle ratios in central S+Au collisions at SPS (200 GeV/N). The shaded areas indicate data taken from [21].

First we want to investigate the hadro-chemical evolution and the behaviour of hadron ratios in the qMD model. Figure 5 displays the time development of several hadron ratios in the course of the hadronization (shaded areas denote the measured ratios). The particle ratios stay nearly constant during the parton-hadron conversion. This may explain the success of statistical model analysis [20]. However, baryons seem to be produced in the later stage of the hadronization. The discrepancies between the ratios of p/π^+ and $(\Lambda_0 + \Sigma_0)/p$ obtained from our approach and the measured data can be understood from the missing hadronic rescattering in our approach. This can be seen by comparing the measured hadron ratios in Figure 6 not only with the qMD calculation (diamonds) but also to ratios obtained with full UrQMD simulations (open circles).

While the proton to antiproton ratio remains unaltered as compared to the UrQMD calculation alone (and nearly one order of magnitude lower than the experimental value), ratios involving protons or antiprotons alone or the net proton number are by factors 2–5 higher in qMD than UrQMD, yielding a better fit to data for the $\bar{p}/\pi^-$ ratio, but an increasing overestimation for the p/π^+ and the $(p - \bar{p})/h^-$ ratio. Simultaneously, the ratio of $\Lambda/(p - \bar{p})$ drops against the value from UrQMD. Both trends can be understood as consequences of hadronic rescattering in the UrQMD model. Hadronic rescattering lowers the number of antiprotons in the final state, and, by inelastic baryon-baryon and antikaon-baryon collisions, yields a systematic population of hyperons at the expense of

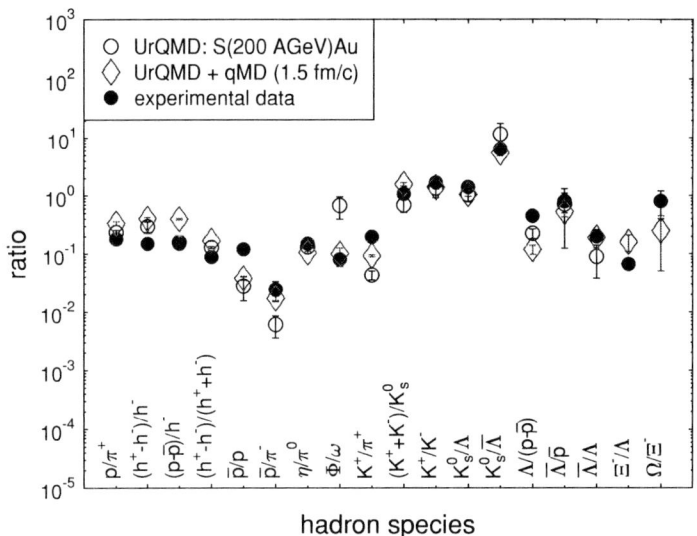

FIGURE 6. Particle ratios in central S+Au collisions at SPS (200 GeV/N). Experimental data are taken from [21].

protons. These channels are not implemented in the present version of our qMD model. In qMD, delta resonances for example (which are abundant around the first 10 fm/c of the collision) cannot create hyperons and kaons by inelastic collisions, but eventually yield always nucleons and pions. As this simplification of the combined UrQMD+qMD approach with respect to a full UrQMD treatment shifts the calculated ratios away from the known experimental values, this can be seen as a clear sign for the importance of hadronic rescattering.

THE PARTON-HADRON CONVERSION

Secondly, the qMD approach can also provide us with detailed information about the dynamics of the quark system and the parton-hadron conversion. Whenever a new hadron is formed, the correlation between the quarks clustering to build this hadron can be studied. The mean value of the path length these quarks have travelled within the quark phase from their points of origin until the clustering point and the distance in space between the origins of the involved quarks can be studied.

Figure 7 shows (for S+Au collisions at SPS energies of 200 GeV/N) the number distribution for the mean path travelled by quarks forming a hadron (a) from the same initial hadron (solid line) and (b) from different initial hadrons (dotted line).

One sees that the reclustering in (a) is quite quick: in this case a hadron is decomposed into two or three quarks, these quarks propagate a short distance of about 2.2 fm ("dif-

51

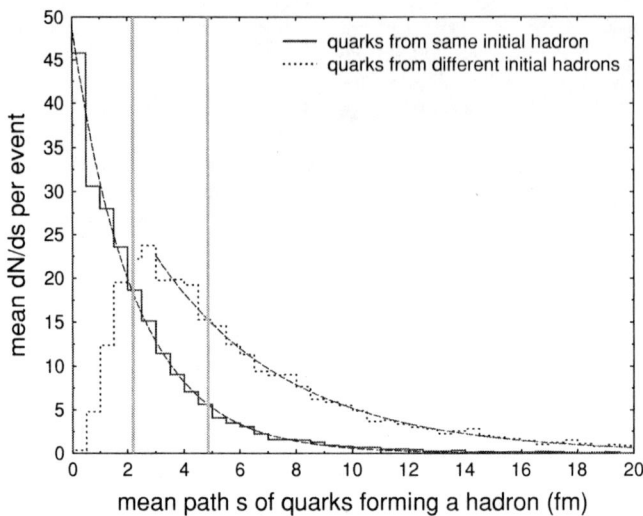

FIGURE 7. Hadronization in S+Au collisions at SPS (200 GeV/*N*): Number density distribution of mean diffusion path of quarks forming a hadron from the same initial hadron (solid line) and from different initial hadrons (dashed line) within qMD. Fitting the decay profiles yields diffusion lengths of 2.2 fm and 4.8 fm, respectively.

fusion length" – marked by the grey box) and rehadronize again. The hadron formation follows an exponential decay of the quark cluster to hadrons with a decay length equal to the diffusion length of 2.2 fm.

On the other hand, the rearrangement of quarks to form new, different hadrons happens on a length scale of about 3 fm. Following this rearrangement, the clusters decay exponentially with a decay length of 4.8 fm.

A measure of the relative mixing within the quark system – and also for thermalization which means homogenisation of phase space population and loss of correlations – are the relative number of hadrons formed by quarks stemming from the same hadron (rehadronization – these hadrons are dubbed "direct hadrons") versus hadrons formed by quarks stemming from different hadrons ("mixed hadrons"). The ratio r of mixed hadrons to the total number of hadrons formed from the quark system is $r = 0.574 \pm 0.008$ for the S+Au collision at 200 GeV/N. The error is statistics only. A value of $r = 1$ would indicate complete rearrangement of quarks and thus complete loss of correlations in the quark system. Considering the presumed transition to the quark-gluon plasma in Pb+Pb collisions at 160 GeV/N, one would expect a larger value of $r \approx 1$.

We see that while global observables do not show large differences between hadronic calculations from UrQMD alone and our combined description using quark molecular dynamics, the easy rearrangement of quarks does indeed play a role and may hint to state of deconfined matter at the CERN-SPS. Note that this rearrangement may be observed with the help of balance functions [22]. Such an analysis would, however, require better statistics. Fluctuations might also reveal additional information on i the quark-hadron phase transition [23].

CONCLUSION

The latest data of CERN-SPS will be interesting to follow closely. Simple energy densities estimated from rapidity distributions and temperatures extracted from particle spectra indicate that initial conditions could be near or just above the domain of deconfinement and chiral symmetry restoration. Still the quest for an *unambiguous* signature remains open.

We have shown that the thermal model fits cannot be unique: the chemical freezeout close to the phase transition to a QGP should be accompanied by shifting baryon masses in medium, which can yield considerably smaller break-up temperatures. Further, the nonequilibrium quark Molecular Dynamics model yields hadron ratios in nice agreement with the data.

REFERENCES

1. S. Bass, M. Gyulassy, H. Stöcker, W. Greiner, J. Phys. G25 (1999) R1
2. D. Hahn, H. Stöcker, Nucl. Phys. A452 (1986) 723
3. D. Hahn, H. Stöcker, Nucl. Phys. A476 (1988) 718
4. C. Spieles, H. Stöcker, C. Greiner, Eur. Phys. J. C2 (1998) 351
5. P. Braun-Munzinger, J. Heppe, J. Stachel, Phys. Lett. B465 (1999) 15
6. J. Rafelski, J. Letessier, nucl-th/9903018 (1999)
7. F. Becattini, J. Cleymans, A. Keranen, E. Suhonen, K. Redlich, Phys. Rev. C64 (2001) 024901
8. G.D. Yen, M.I. Gorenstein, Phys. Rev. C59 (1999) 2788
9. H. Stöcker, W. Greiner, Z. Phys. A286 (1978) 121
10. J. Theis, H. Stöcker, J. Polonyi, Phys. Rev. D28 (1983) 2286
11. J. Schaffner, I.N. Mishustin, L.M. Satarov, H. Stöcker, W. Greiner, Z. Phys. A341 (1991) 47
12. D. Zschiesche, P. Papazoglou, S. Schramm, C. Beckmann, J. Schaffner-Bielich, H. Stöcker, W. Greiner, Springer Tracts in Modern Physics 163 (2000) 129
13. F. Karsch, hep-lat/9903031 (1998)
14. P. Papazoglou, D. Zschiesche, S. Schramm, J. Schaffner-Bielich, H. Stöcker, W. Greiner, Phys. Rev. C59 (1999) 411
15. M. Hofmann, S. Scherer, M. Bleicher, L. Neise, H. Stöcker, W. Greiner, Phys. Lett. B478 (2000) 161
16. S. Scherer, M. Hofmann, M. Bleicher, L. Neise, H. Stöcker, W. Greiner, New J. Phys. 3 (2001) 8
17. S. Bass, M. Belkacem, M. Bleicher, M. Brandstetter, L. Bravina, C. Ernst, L. Gerland, M. Hofmann, S. Hofmann, J. Konopka, G. Mao, L. Neise, S. Soff, C. Spieles, H. Weber, L. Winckelmann, H. Stöcker, W. Greiner, C. Hartnack, J. Aichelin, N. Amelin, Prog. Part. Nucl. Phys. 41 (1998) 225
18. M. Bleicher, E. Zabrodin, C. Spieles, S.A. Bass, C. Ernst, S. Soff, L. Bravina, M. Belkacem, H. Weber, H. Stocker, W. Greiner, J. Phys. G25 (1999) 1859
19. M. Okamoto (for the CP-PACS Collaboration), Phys. Rev. D60 (1999) 094510
20. J. Rafelski, J. Letessier, Proceedigns of the 15th Winter Workshop on Nuclear Dynamics, Park City, January 1999, W Bauer and G Westfall, Eds. *Preprint* hep-ph/9902365
21. P. Braun-Munzinger, J. Stachel, J.P. Wessels, N. Xu, Phys. Lett. B365 (1996) 1
22. S.A. Bass, P. Danielewicz, S. Pratt, Phys. Rev. Lett. 85 (2000) 2689
23. M. Bleicher, M. Belkacem, C. Ernst, H. Weber, L. Gerland, C. Spieles, S.A. Bass, H. Stöcker, W. Greiner, Phys. Lett. B435 (1998) 9
24. M. Bleicher, S. Jeon, V. Koch, Phys. Rev. C62 (2000) 061902

Probing Hadronization with Strangeness

Steffen A. Bass*, Adrian Dumitru†, Pawel Danielewicz** and Scott Pratt**

*Department of Physics, Duke University & RIKEN-BNL Research Center, Brookhaven National
Laboratory
†Department of Physics, Columbia University
**Department of Physics and Astronomy and National Superconducting Cyclotron Laboratory,
Michigan State University

Abstract. A novel state of matter has been hypothesized to exist during the early stage of relativistic heavy ion collisions, with normal hadrons not appearing until several fm/c after the start of the reaction. Radial flow of multi-strange baryons is shown to be a sensitive probe to the expansion of the deconfined phase prior to hadronization. Using a hybrid macroscopic/microscopic transport model we show that if at hadronization the system has been significantly out of chemical equilibrium, hadronic rescattering cannot drive the system towards full chemical equilibration. Furthermore, we suggest a novel model-independent observable, balance functions,to evaluate correlations between charges and their associated anti-charges It is shown that balance functions are extremely sensitive to the time-scale of hadronization: late-stage hadronization is characterized by tightly correlated charge/anti-charge pairs when measured as a function of relative rapidity.

INTRODUCTION

Relativistic heavy ion collisions produce mesoscopic regions of enormous energy density, perhaps surpassing 3 GeV/fm^3 in Pb collisions at the CERN SPS [1, 2] with even higher energy densities expected at RHIC. At such energies hadronic degrees of freedom should be replaced by quark-gluon degrees of freedom. Several experimental measurements have been proposed as signals to the quark-gluon plasma[3]. Among these signals is an expected enhancement in strange-quark production which should take place 5-10 fm/c into the collision when the local temperature has dropped to near 160 MeV, but the system is still far from freeze-out. Strangeness enhancement has indeed been observed in heavy ion collisions [4], but alternative hadronic explanations have also been put forward assuming early-stage hadronization with medium modifications, referred to as color ropes [5, 6] or baryon junctions [7]. In this paper the use of balance functions is proposed as a means to determine whether quark production occurred at early times, $\tau < 1$ fm/c, or according to a late-stage hadronization scenario, see e.g. [8, 9].

MULTISTRANGE BARYONS AS PROBES OF QGP EXPANSION

In this and the following section we use boost-invariant hydrodynamics to model a first order phase transition from a QGP to a hadronic fluid [10], and combine it with a non-equilibrium microscopic transport calculation [11] for the later, purely hadronic, stages

CP597, *Nonequilibrium and Nonlinear Dynamics in Nuclear and Other Finite Systems*,
edited by Z. Li et al.
© 2001 American Institute of Physics 0-7354-0041-5/01/$18.00

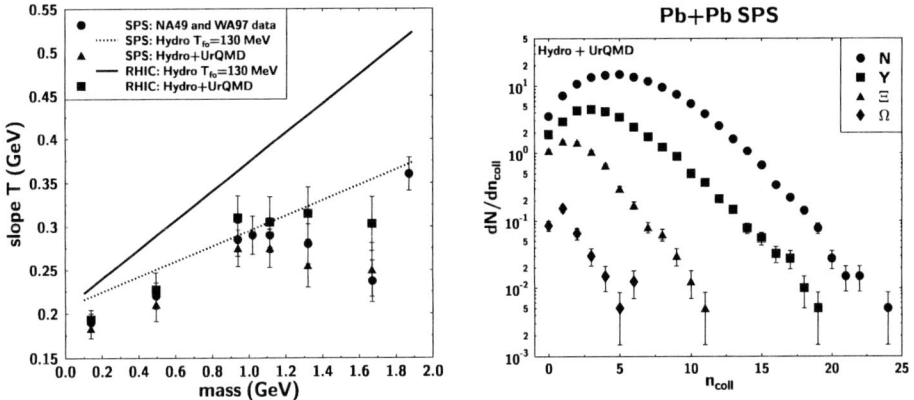

FIGURE 1. Left: Inverse slopes of the m_T-spectra of π, K, p, $\Lambda + \Sigma^0$, $\Xi^0 + \Xi^-$, and Ω^- at $y_{c.m.} = 0$. Right: collision number distributions for N, Y, Ξ and Ω baryons.

of the reaction. With this ansatz we are able to self-consistently calculate the freeze-out of the system: no decoupling hypersurface is imposed by hand, but the space-time points are rather determined by an interplay between the (local) expansion scalar ∂u [12] (where u is the collective flow four-velocity), the relevant elementary cross sections, and the equation of state (EoS), which actually changes dynamically as more and more hadron species decouple. A detailed description of this novel hybrid approach towards the dynamics of ultra-relativistic heavy-ion collisions can be found in ref. [13].

The left frame of fig. 1 displays the inverse slope parameters T obtained by an exponential fit to $dN_i/d^2 m_T dy$ in the range $m_T - m_i < 1$ GeV for SPS and RHIC and compares them to SPS data [4]. The trend of the data, namely the "softer" spectra of Ξ's and Ω's as compared to a linear $T(m)$ relation is reproduced reasonably well. This is in contrast to "pure" hydrodynamics with kinetic freeze-out on a common hypersurface (e.g. the $T = 130$ MeV isotherm), where the stiffness of the spectra increases linearly with mass as denoted by the lines in fig. 1. When going from SPS to RHIC energy, the model discussed here generally yields only a slight increase of the inverse slopes, although the specific entropy is larger by a factor of 4-5 ! The reason for this behavior is the first-order phase transition that softens the transverse expansion considerably.

The reason for the softening of the spectra is that the hadron gas emerging from the hadronization of the QGP is almost "transparent" for the multiple strange baryons. This can be seen in the right frame of fig. 1 which shows the collision number distributions for N, Y, Ξ and Ω in Pb+Pb at the SPS: Whereas Ω's suffer on average only one hadronic interaction, N's and Λ's suffer far more collisions with other hadrons before they freeze-out.

Thus, we may conclude that the spectra of Ξ's and especially Ω's are practically unaffected by the hadronic reaction stage and closely resemble those on the phase boundary. They therefore act as probes of the QGP expansion prior to hadronization and can be used to measure the expansion rate of the deconfined phase.

THE MYSTERY OF CHEMICAL EQUILIBRATION

The success of statistical models in describing the (strange) hadron abundances and ratios at the CERN/SPS [14] and the extracted γ_s values close to 1 have led to the common belief that chemical freeze-out in heavy-ion reactions at the SPS occurred very close to – or even at hadrochemical equilibrium – and that this state most likely has been created by a hadronizing QGP [15, 16, 17].

However, the estimated chemical equilibration times may not be sufficiently rapid to cause a saturation in the production of strange quarks before hadronization: calculations based on boost-invariant hydrodynamics, augmented with rate-equations for strange quark production, [18, 19, 20], pQCD rate-equations [21] or the Parton Cascade Model [22] all indicate that strangeness saturation cannot be achieved during realistic life-times of the deconfined phase. It has been suggested (e.g. in [18, 19]) that the system would be driven towards and come close to chemical equilibrium in the subsequent hadronic phase – a scenario which would help to bridge the gap between the calculations indicating insufficient equilibration time in the plasma phase and the SPS data apparently close to chemical equilibrium at chemical freeze-out.

Here, we attempt to quantify this scenario by studying the hadronic reaction dynamics and chemical evolution of a hadronizing QGP in and out of chemical equilibrium. The question is how much of the hadronic final state is governed by the conditions at hadronization (e.g. local chemical equilibrium) and how much the hadronic interactions may be able to balance variations in the initial conditions (i.e. a possible chemical non-equilibrium) by pushing the hadronic system towards chemical equilibrium prior to the break-up and chemical freeze-out stage of the reaction.

In the default, "chemical equilibrium", mode of our calculation, the mixed phase of the system is described *locally* within the hydrodynamical framework, and it is assumed by construction that the hadronic and quark-gluon components are in thermal and chemical equilibrium with each other. However, on the outer side of the hadronization hypersurface the microscopic model is applied and the hadron "fluid" is allowed to depart from local equilibrium as the stationary final state is approached.

For the "chemical non-equilibrium" mode we show results obtained by assuming $\gamma_s = 50\%$ undersaturation of strangeness on the hadronization hypersurface. In these calculations, the hadron ensemble has been generated as

$$E_i \frac{dN_i}{d^3 p \, d\sigma \cdot p} = \mathcal{N}_i \gamma_s^{l_i} f(p \cdot u), \tag{1}$$

where l_i is the sum of the number of s and $\bar{s}$ quarks in hadrons of species i. It is clear that this ad-hoc prescription does *not* represent a solution of hydrodynamics. Once the EoS together with the initial conditions on some spacelike hypersurface (in our case $\tau = \tau_i$) have been specified, the laws of local energy-momentum and current conservation determine the solution in the forward light-cone uniquely. This is in contrast to "fireball" fits, where one *is allowed* to chose the temperature and the chemical potentials (and also possible velocity fields) independently since there is no energy-momentum tensor or current that needs to be conserved.

For the above-mentioned prescription baryon number is violated by $\sim 5\%$, as it is caused by the rescaling of the strange baryons, which are not too many. Strangeness

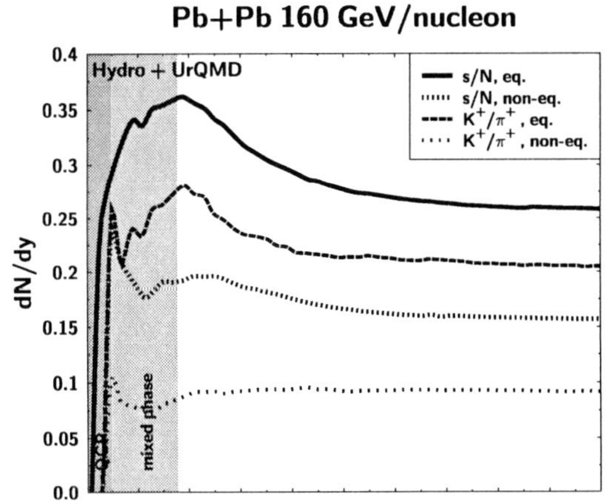

Pb+Pb 160 GeV/nucleon

FIGURE 2. Time-evolution of the K^+/π^+ and strangeness/particle ratios for the hadronic phase of a QGP hadronizing in or out of chemical equilibrium.

conservation is violated to a lesser degree, as the fraction of hadrons with $l_i \geq 2$ is very small. Furthermore, we have conserved the transverse energy per unit of rapidity,

$$\frac{dE_\perp}{dy} = \sum_i \int d^2 p_\perp \sqrt{m_i^2 + p_\perp^2} E_i \frac{dN_i}{d^3 p} \quad , \tag{2}$$

by chosing the normalization factor $\mathcal{N}_\pi$ in eq. (1) such that $dE_\perp/dy$ is the same as for $\gamma_s = 1$ ($\mathcal{N}_i = 1$ for all other hadron species). That is, we have compensated the loss of transverse energy by scaling the pion multiplicity upwards. Therefore, a calculation with the prescription (1) provides useful insight into the rate of chemical reactions in the hadronic stage relative to the expansion rate, albeit a small degree of violation of current conservation and energy-momentum conservation is inevitable.

Figure 2 shows the time-evolution of the total strangeness per particle and the K^+/π^+ ratio for both initial conditions. Coming out of a QGP in local chemical equilibrium, both the strangeness per particle ratio as well as the K^+/π^+ ratio build up rapidly during the mixed phase, reach their maximum at the end of the mixed phase and then exhibit a monotonic decrease until they freeze-out at approximately 35–40 fm/c. The monotonic decrease observed is due mostly to resonance decays – strangeness annihilation only plays a minor role.

For the non-equilibrium initial conditions, however, the ratios behave very differently in that their absolute maximum is obtained at the beginning of the mixed phase. The strangeness per particle then decreases rapidly, but reverts and shows a second maximum at the end of the mixed mixed phase before slightly decreasing towards its freeze-out value. The K^+/π^+ ratio, on the other hand, after going through its minimum in the middle of the mixed phase slowly but monotonously increases until freeze-out. Compared to the behavior of the K^+/π^+ ratio with chem. equilibrium initial conditions this clearly demonstrates the attempt of the hadronic phase to push the system closer towards chemical equilibrium. However, it is obvious that the hadronic phase cannot account for chemical equilibration if that state of equilibrium has not already been obtained (or at least approached) in the plasma phase.

Thus, while the data clearly point towards chemical equilibration during the evolution of the heavy-ion reaction, it is by no means clear through which reaction mechanisms this state may have been obtained – even under the assumption of deconfinement.

CLOCKING HADRONIZATION WITH BALANCE FUNCTIONS

Balance functions, which are being introduced in this section, offer a unique model-independent formalism to probe the time-scales of a deconfined phase and subsequent hadronization. Late-stage production of quarks could be attributed to three mechanisms: formation of hadrons from gluons, conversion of the non-perturbative vacuum energy into particles, or hadronization of a quark gas at constant temperature. Hadronization of a quark gas should approximately conserve the net number of particles due to the constraint of entropy conservation. Since hadrons are formed of two or more quarks, creation of quark-antiquark pairs should accompany hadronization. All three mechanisms for late-stage quark production involve a change in the degrees of freedom. Therefore, any signal that pinpoints the time where quarks first appear in a collision would provide valuable insight into understanding whether a novel state of matter has been formed and persisted for a substantial time. The fact that the hadronic phase has a higher concentration of charges than the QGP phase at the same entropy has been discussed in the context of charge fluctuations in [23].

The link between balance functions and the time at which quarks are created has a simple physical explanation. Charge-anticharge pairs are created at the same location in space-time, and are correlated in rapidity due to the strong collective expansion inherent to a relativistic heavy ion collision. Pairs created earlier can separate further in rapidity due to the higher initial temperature and due to the diffusive interactions with other particles. The balance function, which describes the momentum of the accompanying antiparticle, quantifies this correlation.

The balance functions employed here are similar to observables used to investigate hadronization in jets produced in $p\bar{p}$ or e^+e^- collisions [24, 25]. The balance function describes the conditional probability that a particle in the bin p_1 will be accompanied by a particle of opposite charge in the bin p_2. We define the balance function,

$$B(p_2|p_1) \equiv \frac{1}{2}\{\rho(b,p_2|a,p_1) - \rho(b,p_2|b,p_1) \\ + \rho(a,p_2|b,p_1) - \rho(a,p_2|a,p_1)\}, \tag{3}$$

where $\rho(b,p_2|a,p_1)$ is the conditional probability of observing a particle of type b in bin p_2 given the existence of a particle of type a in bin p_1. The label a might refer to all negative kaons with b referring to all positive kaons, or a might refer to all hadrons with a strange quark while b refers to all hadrons with an antistrange quark. The conditional probability $\rho(b,p_2|a,p_1)$ is generated by first counting the number $N(b,p_2|a,p_1)$ of pairs that satisfy both criteria and dividing by the number $N(a, p_1)$ of particles of type a that satisfy the first criteria.

$$\rho(b, p_2, a, p_1) = \frac{N(b, p_2 | a, p_1)}{N(a, p_1)}. \tag{4}$$

Both sums run over all events, though pairs only involve particles from the same event.

An example of binning might be that p_1 refers to a measurement anywhere in the detector, while p_2 refers to the relative rapidity $|y_b - y_a|$. Then the balance function would be a function of Δy only, and would represent the probability that the balancing charges were separated by Δy (in our formalism we include a division by Δy to express $B(\Delta y)$ as a density).

The balance function is normalized to unity if a/b refer to all particles with a positive/negative globally conserved charge.

$$\sum_{p_2} B(p_2 | p_1) = \frac{1}{2} \{ M_b - (M_b - 1) + M_a - (M_a - 1) \} = 1, \tag{5}$$

where M_a and M_b are the average multiplicities of the a and b particles. The normalization derives from the fact that for every extra positive charge there exists one extra negative charge. If the acceptance measures only a fraction of the charge, e.g. only kaons are measured and the strangeness in hyperons is excluded, the balance function would sum to that fraction. Balance functions can exploit any conserved charge: electric charge, strangeness, baryon number or charm. The first two terms in Eq. (3) constitute the balance functions defined in several analyses of $e^+ e^- \rightarrow$ jets. By adding the last two terms the normalization properties are retained even for the case where there is a non-zero net charge, $M_a - M_b \neq 0$.

If many charges are present in the event, the balance function represents the subtraction of two large numbers. However, large multiplicities also imply a large number of pairs from which to calculate the balance function. Since the number of uncorrelated pairs rises as the square of the multiplicity M, the statistical error in calculating the numerators of the conditional probabilities, which rises as the square root of the number of pairs, increases linearly with M. Since the denominator also rises linearly with M, the statistical error in the balance function is independent of multiplicity and is principally determined by the number of events:

$$\sigma_B \propto \frac{1}{\sqrt{N_{\text{ev}}}}. \tag{6}$$

Thus, the baryon-antibaryon balance function which might involve a few dozen antibaryons would require the same number of events as the electric-charge balance function which might be constructed from a thousand particles. Typically, 10^5 events are required to determine a balance function with statistical fluctuations at the level of 10^{-2}.

Balance functions probe the dynamics of charge-anticharge pairs by quantifying the degree to which the charges are correlated in momentum space given the constraint of being created at the same space-time point in a system exhibiting strong position-momentum correlations such as a relativistic collision where source velocities might

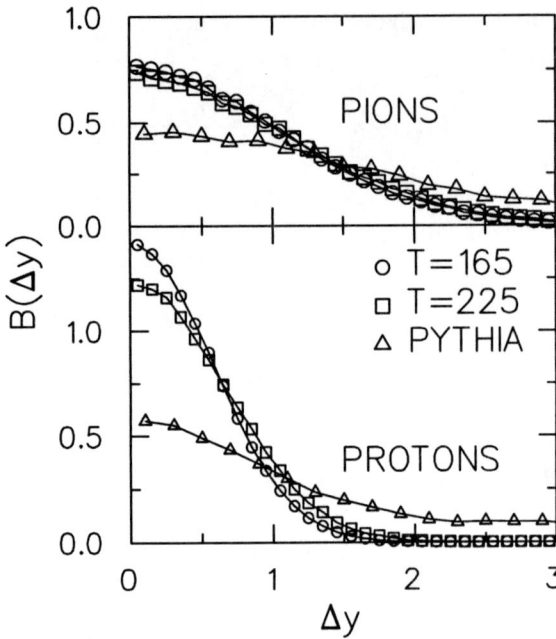

FIGURE 3. Balance functions as predicted in a simple Bjorken thermal model are shown for two temperatures, 225 MeV and 165 MeV. Since heavier particles from cooler systems have smaller thermal velocities, they are more strongly correlated in rapidity and result in narrower balance functions. Also shown are balance functions as predicted by PYTHIA where the shape of the balance function is largely determined by string phenomenology.

span several units of rapidity. In a globally equilibrated system with no collective flow, there would exist no correlation between the balancing charges, and the numerator in Eq. (4) would factorize. The width of the balance function would then correspond to the extent of single-particle emission in momentum space.

To illustrate the way in which balance functions quantify the charge-anticharge correlations, we consider a Bjorken boost-invariant parameterization [26] of a source expanding along the z axis with a collective velocity proportional to the position, $v_{coll} = z/t$. All intrinsic variables, such as density or temperature, depend only on the proper time $\tau = (t^2 - z^2)^{1/2}$. We first consider only direct production of hadrons, as the possibility of hadrons coalescing from quarks is discussed later in the paper. Particles and antiparticles of mass m are generated in pairs at the same point in space-time following a local thermal distribution, and the relative rapidities are used to generate balance functions. The characteristic width of the balance function is determined by the ratio of the temperature to the mass. Non-relativistically, $\sigma_y = (2T/m)^{1/2}$, and heavier particles are characterized by narrower balance functions. For particles with masses much less than the temperature, the balance functions become independent of the temperature.

Figure 3 displays balance functions assuming a Bjorken parameterization of an expanding pion gas and an expanding proton gas, for two temperatures, 225 MeV and

165 MeV. Clearly, the balance functions of the more massive particles are sensitive to the temperature. This suggests that the strangeness and baryon balance functions should provide more insight than the electric-charge balance function which would be largely dominated by pions.

Balance functions in heavy ion collisions should be compared to those from pp collisions at the same $\sqrt{s}$ where hadronization is nearly instantaneous. Charged-pion balances measured in e^+e^- collisions as a function of the rapidity defined along the jet axis have been reasonably explained by the string hadronization dynamics of the Lund model [27], e.g. as implemented in PYTHIA [28]. Thermally generated balance functions are compared to predictions of PYTHIA for pp collisions at $\sqrt{s} = 200$ GeV in Fig. 3. The PYTHIA balance functions tend to be broader than those that are thermally generated, especially for the more massive protons and kaons. Assuming that experimental balance functions in pp collisions would be well described by similar string dynamics, Fig. 3 suggests that narrower balance functions might indeed point to thermal production at a lower temperature and thus at later times in the evolution of the heavy ion reaction.

The overall width of the balance function in relative rapidity is a combination of the thermal rapidity spread σ_{therm} and the effect of diffusion in η of both particles:

$$\sigma_y^2 = \sigma_{\text{therm}}^2 + 4\beta \ln(\tau/\tau_0), \tag{7}$$

with β related to the diffusion constant of the system. Due to cooling, the width σ_{therm} falls with time which provides a competition between diffusion which stretches the balance function, and cooling which narrows it. If the production occurs at early times, then $\ln(\tau/\tau_0)$ is large and the effect of collisions is to significantly broaden the balance function.

Some hadrons will contain coalesced quarks that were created at early times. The thermal contribution to σ_y described in Eq. (7) should be unaffected by the past history of the constituent quarks. However, the diffusive contribution might significantly depend on the fact that the charge moved as a free quark rather than as a hadron during it's early history. Balance functions constructed from hadrons can thus provide meaningful information regarding the creation and mobility of the constituent quarks.

Rescattering and annihilation should also affect balance functions. To quantitatively illustrate this effect, we model a pair of particles produced at an initial proper time τ_0 that collide N_{coll} times before disassociating at a final time τ_f. Each collision is assumed to completely reorient the particle with the local collective velocity. The collision times are chosen randomly such that the number of collisions as a function of $\ln(\tau)$ is uniform. The temperature is chosen to vary linearly with the proper time, cooling from 225 MeV at $\tau = 1$ fm/c to 120 MeV at $\tau = 15$ fm/c. Figure 4 shows the K_+K_- balance function with $N_{\text{coll}} = 0$ and $N_{\text{coll}} = 10$ assuming kaons are created at $\tau = 1$ fm/c and cease to collide at $\tau_f = 15$fm/c. In this case collisions clearly broaden the balance function.

Annihilations should also broaden the balance function. Annihilation forms new correlated pairs with the surviving partners of the annihilated particles, which tend to be less correlated than the original pairs. Annihilation combined with an equal amount of creation does not affect the balance function since the relative rapidities of formed and annihilated pairs should be identical. Figure 4 illustrates the effects of annihilation by considering the same case described above, but with the additional assumption that half

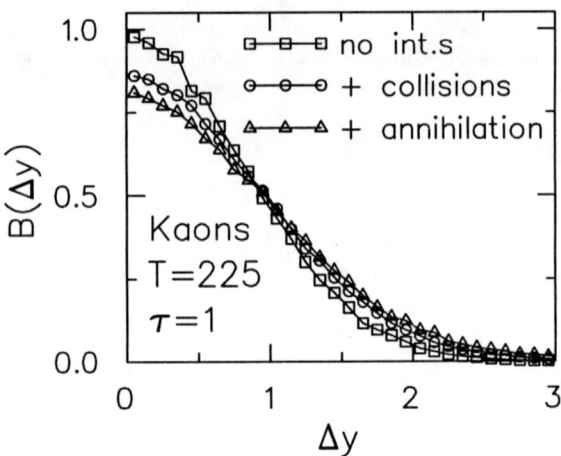

FIGURE 4. Kaon balance functions are shown assuming an initial local temperature of 225 MeV and a production time of 1 fm/c. The balance function is broadened by the inclusion of randomizing collisions and annihilation.

the particles disappear due to annihilation. In hadronic models of heavy ion collisions, the number of both antibaryons and strange particles tend to decrease with time due to cooling, which should result in broadened balance functions.

The simple calculations presented here sidestep two issues: correlations from decays such as $\phi \rightarrow K^+K^-$, and experimental acceptance problems. Both problems can be addressed by modeling constrained by the multitude of other observables measured in a heavy ion collision. Although some open questions remain, it seems clear that the canonical picture of a heavy-ion reaction, quark-gluon plasma formation followed by late-stage hadronization, should have a clear signature in the balance functions. Compared to pp collisions, one expects the peak in the balance function in nucleus-nucleus collisions to be narrower near $\Delta y = 0$ due to the contribution of late-stage production of quark pairs, while the tails of balance function should become broader reflecting the extra diffusion of charge in the early stages of the collision. Finally, we remark that we have barely explored the possibilities of balance functions. The rich nature of the binnings $(p_2|p_1)$ should provide a powerful means for resolving many of the issue regarding creation and diffusion of quarks and hadrons in relativistic heavy ion collisions.

Recently, the balance function formalism has been applied to to first RHIC data by the STAR collaboration [29]. The (preliminary) analysis for K^+K^- balance functions in rapidity space indicates rather narrow balance functions - their width being almost a factor of two smaller than a HIJING calculation subject to the same experimental cuts and resolution. These data therefore point towards a strong degree of collectivity and a late hadronization time.

ACKNOWLEDGMENTS

We are grateful to T. Sjöstrand for providing valuable references. S.A.B. was supported under DOE grant Contract No. DE-FG02-96ER40945 P.D. and S.P were supported by the National Science Foundation, grant PHY-00-70818. A.D. received support from DOE grant Contract No. De-FG-02-93ER-40764.

REFERENCES

1. T. Alber et al. Phys. Rev. Lett. **74** (1995) 1303.
2. M .Aggarwal et al. Nucl. Phys. **A610** (1996) 200c.
3. J. Harris and B. Müller, Ann. Rev. Nucl. Part. Sci. **46** (1996) 71.
   S.A. Bass, M. Gyulassy, H. Stöcker and W. Greiner, J. Phys. **G25** (1999) R1.
4. E. Andersen et al. (WA97 collaboration), Phys. Lett **B433** (1998) 209.
   R. Lietava et al. (WA97 collaboration), Journal of Physics **G25** (1999) 181.
   S. Margetis et al. (NA49 collaboration), Journal of Physics **G25** (1999) 189.
   D. Evans et al. (WA85 and WA94 collaborations), Journal of Physics **G25** (1999) 209.
5. T. S. Biro, H. B. Nielsen, J. Knoll, Nucl. Phys. **B245** (1984) 449.
   J. Knoll, Z. Phys. **C38** (1988) 187.
6. H. Sorge, M. Berenguer, H. Stöcker, W. Greiner, Phys. Lett. **B289** (1992) 6.
7. S. E. Vance and M. Gyulassy, Phys. Rev. Lett. **83** (1999) 1735.
8. J. Rafelski, B. Müller, Phys. Rev. Lett. **48** (1982) 1066; Erratum-ibid.**56** (1986) 2334.
9. P. Koch, B. Müller, J. Rafelski, Phys. Rep. **142** (1986) 167.
10. A. Dumitru and D.H. Rischke, Phys. Rev. **C59** (1999) 354.
11. S. A. Bass *et al.*, Progr. Part. Nucl. Physics Vol. **41** (1998) 225;
    M. Bleicher *et al.*, J. Phys. **G25** (1999) 1859
12. I. Mishustin and L. Satarov, Sov. J. Nucl. Phys. **37** (1983) 532;
    C.M. Hung and E. Shuryak, Phys. Rev. **C57** (1998) 1891.
13. S.A. Bass and A. Dumitru, Phys. Rev. **C61** (2000) 064909.
14. J. Letessier, A. Tounsi and J. Rafelski. Phys. Lett. **B292** (1992) 417;
    J. Sollfrank, M. Gadzicki, U. Heinz and J. Rafelski. Z. Phys. **C61** (1994) 659;
    P. Braun-Munzinger, J. Stachel, J. P. Wessels and N. Xu. Phys. Lett. **B365** (1996) 1
    F. Becattini, J. Cleymans, A. Keranen, E. Suhonen and K. Redlich, Phys. Rev. **C64** (2001) 024901.
15. see http://cern.web.cern.ch/CERN/Announcements/2000/NewStateMatter/
16. U. Heinz and M. Jacob, nucl-th/0002042.
17. P. Braun-Munzinger, Nucl. Phys. **A663-664** (2000) 183.
18. J. Kapusta and A. Mekjian. Phys. Rev. **D33** (1986) 1304.
19. T. Matsui, B. Svetisky and L. D. McLerran. Phys. Rev. **D34** (1986) 2047.
20. D.M. Elliott and D.H. Rischke. Nucl. Phys. **A671** (2000) 583.
21. T. Biro, E. van Doorn, B. Müller, M. H. Thoma and X. N. Wang. Phys. Rev. **C48** (1993) 1275.
22. K. Geiger and J. I. Kapusta. Phys. Rev. **D47** (1993) 4905.
23. S. Jeon and V. Koch, Phys. Rev. Lett. **85** (2000) 2076;
    M. Asakawa, U. Heinz and B. Müller, Phys. Rev. Lett. **85** (2000) 2072.
24. D. Drijard et al., Nucl. Phys. **B155** (1979) 269.
    D. Drijard et al., Nucl. Phys. **B166** (1980) 233.
    I.V. Ajinenko et al., Z. Phys. **C43** (1989) 37.
25. R. Brandelik et al., Phys. Lett. **B100** (1981) 357; H. Aihara et al., Phys. Rev. Lett. **53** (1984) 2199;
    P.D. Acton et al., Phys. Lett. **B305** (1993) 415.
26. J.D. Bjorken, Phys. Rev. **D27** (1983) 140.
27. B. Anderson et al. Nucl. Phys. **B281** (1987) 289.
28. H.-U. Bengtsson and T. Sjöstrand, Comp. Phys. Com. **46** (1987) 43.
29. M. Belt-Tonjes for the STAR Collaboration, 17th Winter Workshop on Nuclear Dynamics, March 2001, Park City, Utah, USA.

A new dibaryon candidate $(\Omega\Omega)_{0^+}$

Z.Y. Zhang, Y.W. Yu* and C.R. Ching, T.H. Ho†

*Institute of High Energy Physics, 100039 Beijing, P.R. China
†Institute of Theoretical Physics, 100081 Beijing, P.R. China

Abstract. The structure of a new dibaryon $(\Omega\Omega)_{0^+}$ is predicted in the framework of the chiral SU(3) quark model by solving a resonating group method (RGM) equation. The binding energy of this dibaryon is around 100 MeV, the mean squared root of the distance between two Ωs is 0.84 fm and the preliminary estimated mean life time is about 10^{-10} sec. All these interesting properties, and also the two negative charge units it carries could make it to be easily identified experimentally in the heavy ion collision process.

INTRODUCTION

As is well known, in the ordinary strong interation world there are only baryons consisting of three valence quarks and mesons of quark-antiquark pair. However, for more than twenty years people have discusseed and explored the possible existence of some exotic multi-quark states and gluonic states. Not only because it provides a good place to examine the Quantum Chromodynamics (QCD) theory and to display the quark-gluon behavior in short distance, but also the very existence of such systems would open a new area for studying many new physical phenomena we have not known before. The theoretical investigations have concentrated mainly on H dihyperon [1,2,3,4] and d' particle [5], and up to now there is no convincing experimental evidence for the existence of these particles. This indicates, in seeking for multi-quark system, or so-called dibaryon, one probably should go beyond these few candidates to the multi-strangeness systems.

Recently we have developed a chiral SU(3) quark model [6,7] in which the coupling between the chiral fields and quarks is considered to describe the medium range nonperturbative QCD effect. Since this model reproduced the energies of the baryon ground states, the nucleon-nucleon $(N-N)$ scattering phase shifts and the hyperon-nucleon $(Y-N)$ cross sections correctly, making an extrapolation in the same framework without introducing any new parameters to predict dibaryons' structure is feasible and reliable to some extent. Using this model, [8], we found that some six quark states with high strange number have more attraction from the chiral quark coupling. Of particular interest is the $(\Omega\Omega)_{0^+}$. It is a deeply bound state and its mean life time is quite long because it can only undergo through weak decay. Since this dibaryon has charge -2 and enough long mean life time, and therefore it could be a new interesting candidate of dibaryon.

CP597, Nonequilibrium and Nonlinear Dynamics in Nuclear and Other Finite Systems,
edited by Z. Li et al.
© 2001 American Institute of Physics 0-7354-0041-5/01/$18.00

A BRIEF INTRODUCTION OF CHIRAL $SU(3)$ QUARK MODEL

In the chiral SU(3) quark model, the coupling between chiral fields and quarks is introduced to describe the non-perturbative QCD effect, and in order to study the systems with strangeness, the idea of the $SU(2)$ σ model is generalized to the flavor $SU(3)$ case. With this generalization, the interacting Hamiltonian between quarks and chiral fields is now written as:

$$H_I^{ch} = g_{ch} F(q^2) \bar{\psi} (\sum_{a=0}^{8} \sigma_a \lambda_a + i \sum_{a=0}^{8} \pi_a \lambda_a \gamma_5) \psi, \tag{1}$$

where $\sigma_0 ... \sigma_8$ are the scalar nonet fields and $\pi_0 ... \pi_8$ the pseudo-scalar nonet fields. The chiral field coupling constant g_{ch} can be fixed by the relation with $g_{NN\pi}$ and $\frac{g_{NN\pi}^2}{4\pi}$ is taken to be the experimental value.

Then the total Hamiltonian of the chiral $SU(3)$ quark model is obtained by adding V_{ij}^{ch} as:

$$H = \sum_i T_i - T_G + \sum_{i<j} V_{ij}, \tag{2}$$

and

$$V_{ij} = V_{ij}^{conf} + V_{ij}^{OGE} + V_{ij}^{ch}. \tag{3}$$

Where $\sum_i T_i - T_G$ is the kinetic energy of the system, and V_{ij} is the interaction between two quarks. V_{ij}^{conf} is the confinement potential, taken as quadratic form, V_{ij}^{OGE} the OGE interaction and V_{ij}^{ch} includes scalar meson exchange and pseudo-scalar meson exchange potentials. Their expressions can be found in Ref.[6].

We fixed the parameters by fitting the baryons' masses and their stability conditions. Once the parameters are determined, all the Octet and Decuplet baryons' masses can be reproduced in our model [8]. Equipped with this model with all the parameters thus determined, we studied the two baryon systems on quark level dynamically by solving the resonating group method (RGM) equation of the Hamiltonian Eq.(2). In the RGM calculation, the trial wave function is taken to be

$$\Psi_{ST} = \sum_i c_i \Psi_{ST}^{(i)}(\vec{s}_i), \tag{4}$$

with

$$\Psi_{ST}^{(i)}(\vec{s}_i) = \mathcal{A}(\phi_A(\vec{\xi}_1, \vec{\xi}_2)_{S_A T_A} \phi_B(\vec{\xi}_4, \vec{\xi}_5)_{S_B T_B} \chi(\vec{R}_{AB} - \vec{s}_i) \mathcal{R}_{CM}(\vec{R}_{CM}))_{ST}, \tag{5}$$

where A and B describe two clusters, and ϕ, χ and $\mathcal{R}$ represent internal, relative and center of mass motion wave functions respectively. $\vec{s}_i$ is the generator coordinate and $\mathcal{A}$ is the anti-symmetrization operator,

$$\mathcal{A} = 1 - \sum_{i \in A, j \in B} P_{ij}, \tag{6}$$

where P_{ij} is the permutation operator of ith and jth quarks.

65

The results of this dynamical calculation show that all the calculated $N-N$ scattering phase shifts, the $Y-N$ scattering and reaction cross-sections are in agreement with the experimental data. Then we extend our study to the dibaryon systems by solving a bound state equation. We found that among all possible candidates, of particular interest is the two Ω dibaryon $(\Omega\Omega)_{0+}$ [8].

MAIN PROPERTIES OF $(\Omega\Omega)_{0+}$ DIBARYON

We calculated the eigenenergy of the two Ω system by taking the same set of parameters which we used in the $N-N$ and $Y-N$ scattering calculations [6]. The results showed, that $(\Omega\Omega)_{0+}$ is a deeply bound state with small quark distribution distance.

$$B_{(\Omega\Omega)_{0+}} = -(E_{(\Omega\Omega)_{0+}} - 2M_\Omega) = 116 MeV, \tag{7}$$

and

$$rms = \sqrt{\frac{1}{6} < (\vec{r}_i - \vec{R}_{CM})^2 >} = 0.627 fm, \tag{8}$$

where $B_{(\Omega\Omega)_{0+}}$ denotes the binding energy and rms the root mean squared radius of quark. Moreover, the relative motion wave function between two Ωs, $\chi(R)$ is separated and drawn in Fig.1, from which the root of mean squared distance between two Ωs can be obtained.

$$RMS = \sqrt{< R_{\Omega\Omega}^2 >} = 0.84 fm. \tag{9}$$

It is much smaller than the size of deuteron.

The mean life time of $(\Omega\Omega)_{0+}$ can be estimated using the standard weak interaction theory, since there is no strong decay for such system with such a big strangeness quantum number. There are two different kinds of decay modes: three-particles decay mode and two-particle decay mode, namely,

$$(\Omega\Omega)_{0+} \longrightarrow \Omega^- + \Xi^- + \pi^0,$$

$$(\Omega\Omega)_{0+} \longrightarrow \Omega^- + \Xi^0 + \pi^-,$$

and

$$(\Omega\Omega)_{0+} \longrightarrow \Omega^- + \Xi^-.$$

A rough estimation shows that the mean life time of $(\Omega\Omega)_{0+}$ is in the order of 10^{-10} sec.

These results have shown that $(\Omega\Omega)_{0+}$ is a very interesting dibaryon, not only because of its huge binding energy , but also of the quite short distance between $\Omega-\Omega$. This means that its binding feature is very different with deuteron. It looks like a six quark state. This property, combined with the long life-time, makes it most favorable candidate for looking for dibaryon.

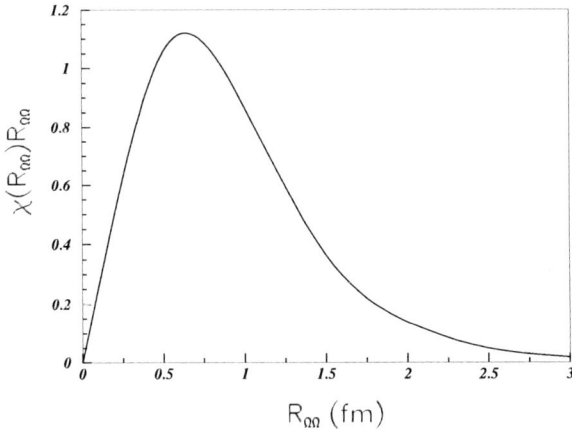

FIGURE 1. The relative motion wave function of $\Omega - \Omega$

AN ANALYSIS TO DISCUSS WHY $(\Omega\Omega)_{0^+}$ IS DEEPLY BOUND

To understand the very physics underlying the curious features, and to physical states of two baryons and the symmetry bases (group chain examine to what extent the results are dependent on the model with all its classification bases) of six quark system parameters, we have carried out an analysis to see the reason. [9], one sees that among all 280 physical bases only

First, let us discuss the symmetry property of this system. As is well known, the compositeness of the baryon and the color magnetic interaction in the OGE potential govern the short range region. In the $(\Omega\Omega)_{0^+}$ case, although the OGE interaction offers repulsion, but its symmetry property is very special. According to the transformation relations between the physical states of two baryons and the symmetry bases of six quark system [9], one sees that among all 280 physical bases only six of them, namely, $(\Delta\Delta)_{ST=30}$, $(\Delta\Delta)_{ST=03}$, $(\Delta\Sigma^*)_{ST=3\frac{1}{2}}$, $(\Delta\Sigma^*)_{ST=0\frac{5}{2}}$, $(\Xi^*\Omega)_{ST=0\frac{1}{2}}$, and $(\Omega\Omega)_{ST=00}$, have the largest component of $[6]_r$ symmetry in the orbital space of the symmetry bases. And what is remarkable, that $(\Omega\Omega)_{0^+}$ is the only state among them, which is stable under the strong interaction.

On the other hand, the eigenvalue of the normalization kernel in the spin-flavor-color space $< \mathcal{A}^{\sigma f c} >=< 1 - \sum_{i\in A, j\in B} P_{ij}^{\sigma f c} >$ can serve as a direct measure on how the Pauli principle works in the relevant state [10]. When $< \mathcal{A}^{\sigma f c} >= 0$, it is a forbidden state for forming $[6]_r$ symmetry in the orbital space as a consequence of the Pauli exclusion principle. $< \mathcal{A}^{\sigma f c} >= 1$ means the quark exchange effect between two baryons is unimportant, the two baryons can be regarded as two clusters without any quark exchange. When $< \mathcal{A}^{\sigma f c} >$ larger than 1, the quark exchange effect is cardinal, and tends to drag two clusters closer. Our calculation showed that for $(\Omega\Omega)_{0^+}$, $< \mathcal{A}^{\sigma f c} >= 2$, so that $(\Omega\Omega)_{0^+}$ has relatively high antisymmetry in the spin-flavor-color space and symmetry in the orbital space, therefore is favored for forming the $[6]_r$ symmetry basis.

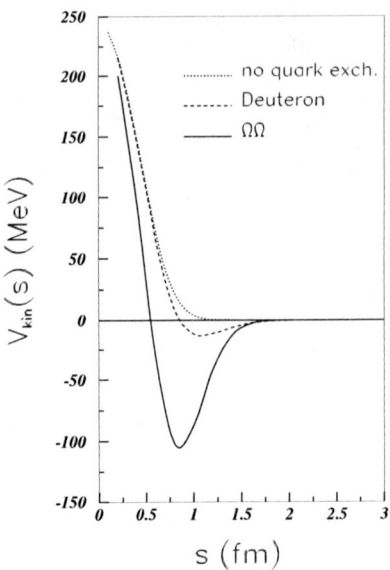

FIGURE 2. The GCM matrix elements of the kinetic energy

TABLE 1. Binding Energies and RMS of $(\Omega\Omega)_{0^+}$ for different chiral-quark models

	Chiral SU(3) Quark Model	$\pi, K, \eta, \eta' + \sigma_0$ Model	Chiral SU(2) Quark Model
	(I)	(II)	(III)
$B_{(\Omega\Omega)_{0^+}}(MeV)$	116	74	54
$RMS(fm)$	0.84	0.92	0.98

This effect can be seen more clearly in the generator coordinate method (GCM) matrix elements of the kinetic energy $V_{kin}(s)$. In Fig.2, the solid curve represents the case of $(\Omega\Omega)_{0^+}$, and the dashed one is for the case of deuteron (only central part). For comparison, the corresponding potential without quark exchange is also shown as dotted curve. From Fig.2, one sees that, in contrary to the deuteron case, the $(\Omega\Omega)_{0^+}$ state has very strong attraction in the kinetic energy. This is because $< \mathcal{A}^{\sigma fc} >_{\Omega\Omega}= 2$, but $< \mathcal{A}^{\sigma fc} >_{deu}= 10/9$, the attraction arising from the quark exchange in the kinetic energy part for $(\Omega\Omega)_{0^+}$ is quite large, but for deuteron it is very small. Since the kinetic energy itself is spin-flavor-color independent, the big difference between $V_{kin}^{deuteron}(s)$ and $V_{kin}^{\Omega\Omega}(s)$ is solely arising from the symmetry property.

We have also performed a simplified model calculation, in which only the kinetic energy $\sum_i t_i - T_G$ and color confinement potential V_{ij}^{conf} between quarks are taken into account. The result showed, that $(\Omega\Omega)_{0^+}$ is bound with a binding energy as large as $B_{(\Omega\Omega)_{0^+}} = 17MeV$. Since the color confinement potential almost does not contribute to

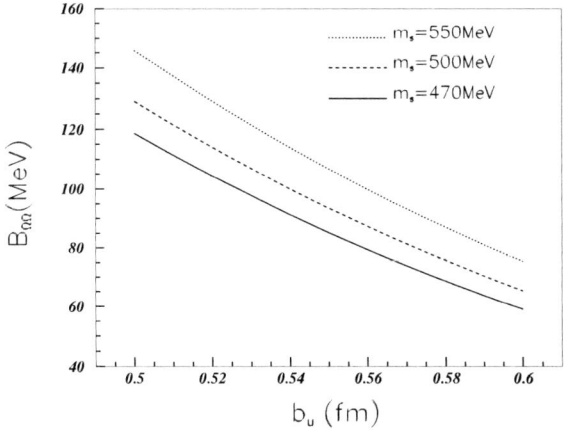

FIGURE 3. Binding energy of $(\Omega\Omega)_{0^+}$ vs b_u and m_s

the interaction between two color singlet clusters, this binding is purely resulted from the quark exchange effect of the kinetic energy part, i.e. from the symmetry property of the system.

For the chiral field coupling effect, first we have made a chiral $SU(2)$ quark model calculation to see the contribution from the σ_0 exchange quantitatively. The binding energy of $(\Omega\Omega)_{0^+}$ in this case is larger than $50MeV$ (see case (III) in Table 1), while for deuteron the binding energy is only of $2\ MeV$ as the experimental data have dictated. That means, even in the framework of SU(2), the theoretical calculation predicts that the binding energy of $(\Omega\Omega)_{0^+}$ is almost twenty times bigger than that of deuteron's. Then we include the exchange of all nonet pseudo-scalar mesons plus σ_0. The result shows, that when π, K, η, η' and σ_0 chiral fields are added, $B_{(\Omega\Omega)_{0^+}} = 74MeV$ (see Table 1 case (II)). Finally, with the inclusion of the exchange of all nonet pseudo-scalar and scalar mesons in our calculation, i.e. the chiral SU(3) quark model, the binding becomes even more stronger. One sees that besides σ_0, the effect from the other mesons also enhance the binding in the $(\Omega\Omega)_{0^+}$ system. The total binding energy reaches $116\ MeV$ when all scalar and pseudo-scalar mesons are included (see Table 1, case (I)). All these analyses point to the fact, that both the specific symmetry property of the system $(\Omega\Omega)_{0^+}$ in the short range, and the nonperturbative QCD effect described by chiral-quark coupling in the medium range, are responsible for forming a deeply bound state of $(\Omega\Omega)_{0^+}$.

We also studied the influence of the input parameters. The results are shown in Fig. 3, in which the solid curve represents the case of $m_s = 470MeV$, the dashed curve $m_s = 500MeV$, and dotted one $m_s = 550MeV$. One sees that with the relevant parameters varying in a maximum allowable region, the binding energy $B_{(\Omega\Omega)_{0^+}}$ are between 60 and $140MeV$. In general, when b_u is taken to be $0.50 - 0.55fm$, and m_s to be $470 - 520MeV$, the data of $N - N$ scattering and the $Y - N$ cross section can be roughly fitted, and given the parameters in this range, the variations of $B_{(\Omega\Omega)_{0^+}}$ are limited, and the binding energy is $100 \pm 15MeV$. This means that our result is not very sensitive to the used parameters.

CONCLUSIONS

In the preceding sections we have shown, that the $(\Omega\Omega)_{0^+}$ is a deeply bound dibaryon with the binding energy, which is about forty times bigger than that of the deuteron's. Its root mean square radius of quark is only about half of that of the deuteron. Such remarkable features predominantly are stemmed from its special symmetry property. This can be seen by recalling that the kinetic energy alone contributes $17MeV$ to the binding energy. The symmetry property further coupled with meson exchange potentials, and finally resulted in about $100MeV$ binding energy for $(\Omega\Omega)_{0^+}$. In this meaning, $(\Omega\Omega)_{0^+}$ is not a normal two-baryon's bound state, but rather a six-quark state. At the same time, the deuteron is another kind of bound state : its symmetric structure prevents the exchange of quarks between the two baryons, and the individuality of the baryons preserves. Consequently, deuteron is a loosely bound state. The symmetry property analysis is quite general, and is irrelevant to any specific model, but the quark model of baryon.

However, as far as the concrete value of the binding energy is concerned, then it is obvious, that it depends on the input parameters, or in other words, it is model-dependent. But in using the chiral $SU(3)$ quark model, our approach is mostly phenomenological one. By fitting with a large body of the experimental data concerning the baryon and hyperon structures and scattering, the set of parameters we have chosen is optimized and the range of variations is also known. It seems, that in a reasonably wide range of variations of the the parameters, our numerical results do not alter very much. And the most preferable value we obtained for the binding energy is $100MeV \pm 15MeV$.

REFERENCES

1. R.J. Jaffe, *Phys. Rev. Lett.*, **38**, 1977, 195.
2. K. Shimizu and M. Koyama, *Nucl. Phys.*,**A646**, 1999, 211.
3. P.N. Shen, Z.Y. Zhang, Y.W. Yu, X.Q. Yuan and S. Yang, "H Dihyperon in Quark Cluster Model" *J. Phys.*, **G-Nu 25**, 1999.
4. K. Imai, *Nucl. Phys.*, **A553**, 1993, 667.
5. A.J. Buchmann et al,, *Prog. Part. Nucl. Phys.*, **36** 1996, 383.
6. Z.Y. Zhang, Y.W. Yu, P.N. Shen, L.R. Dai, *Nucl. Phys.*, **A625**, 1997, 59.
7. P.N. Shen, Y.B. Dong, Y.W. Yu, Z.Y. Zhang and T.S.H. Lee, *Phys. Rev.*, **C55**, 1997, 2024.
8. Y.W.Yu, Z.Y.Zhang and X.Q.Yuan, *Commun.Theor.Phys.*, **31** 1999, 1.
   Z.Y. Zhang, Y.W. Yu and X.Q.Yuan, *Nucl. Phys.*, **A670**, 2000, 178c.
   Z.Y. Zhang, Y.W. Yu, C.R. Ching, T.H. Ho and Z.D. Lu *Phys. Rev.*, **C61**, 2000, 065204.
   Q.B. Li, P.N. Shen, Z.Y. Zhang and Y.W. Yu *Nucl. Phys.*, **A683**, 2001, 487.
9. M. Harvey, *Nucl. Phys.*, **A352**, 1981, 301.
   F. Wang, J.L. Ping and T. Goldman, *Phys. Rev.*, **C51**, 1995, 1648.
10. K.Shimizu, *Prog. Phys.*, **52**, 1989, 1.

Strangeness equilibration in heavy ion collisions

Che Ming Ko, Zi-wei Lin, and Subrata Pal

Cyclotron Institute and Physics Department, Texas A&M University, College Station, Texas 77843, USA

Abstract. In terms of both kinetic and transport models, we discuss the chemical equilibration of strange particles in heavy ion collisions at both SIS, where their abundance is rare, and at SPS and RHIC, where the abundance is small for some strange particles but large for others.

INTRODUCTION

Recent analyses have shown that most hadrons measured in heavy ion collisions can be described by statistical models based on the grand canonical ensemble for abundant particles [1] and the canonical ensemble for rare particles [2]. On the other hand, the chemical equilibration time for strange particles such as kaons in a hot dense matter has been shown to be an order of magnitude longer than the heavy ion collision time [3]. One suggestion [4] for resolving this puzzle is that the kaon equilibration time can be significantly shortened if the kaon mass is reduced in dense matter as a result of the large attractive scalar interaction and the diminishing repulsive vector interaction due to the assumption of vector decoupling. However, there are other explanations that are provided by dynamical models based on either the kinetic theory [5] or the transport equation [6].

In the kinetic theory, it has been shown that the equilibrium time for rare particles carrying $U(1)$ charge and described by the canonical ensemble due to $U(1)$ charge conservation, is much shorter than what is expected from the grand canonical ensemble, and the equilibrium multiplicity in the canonical ensemble is also much lower than that given by the grand canonical ensemble.

For heavy ion collisions at SIS, which are below the threshold for strange particle production in nucleon-nucleon collisions, both kaons and antikaons are rare. In the transport model, it is found that kaons are far from chemical equilibrium at the initial high density stage, but approach equilibrium during the expansion stage of the collisions when the production rate is small and becomes comparable to the annihilation rate. In contrast, antikaons approach chemical equilibrium much earlier but eventually fall out of equilibrium as a result of their large annihilation cross sections in nuclear matter.

For strange particle production from heavy ion collisions at ultrarelativistic collisions at SPS and RHIC, kaons and antikaons are relatively abundant but multistrange baryons such as cascades and omegas are rare. Studies based on the transport model show that most strange particles such as the hyperons and especially the kaons and antikaons approach equilibrium before the system freezes out.

CP597, Nonequilibrium and Nonlinear Dynamics in Nuclear and Other Finite Systems,
edited by Z. Li et al.

In this talk, we shall discuss the results from both the kinetic model and the transport model.

KINETIC MODEL

The rate equation for a binary process $a_1 a_2 \leftrightarrow b_1 b_2$ with $a \neq b$ in the kinetic model is usually given by

$$\frac{d\langle N_{b_1} \rangle}{d\tau} = \frac{G}{V} \langle N_{a_1} \rangle \langle N_{a_2} \rangle - \frac{L}{V} \langle N_{b_1} \rangle \langle N_{b_2} \rangle, \tag{1}$$

where $G \equiv \langle \sigma_G v \rangle$ and $L \equiv \langle \sigma_L v \rangle$ give the momentum-averaged cross sections for the gain process $a_1 a_2 \to b_1 b_2$ and the loss process $b_1 b_2 \to a_1 a_2$, respectively. N_k represents the total number of particles k, and V is the proper volume.

For processes in which b_1 and b_2 are constrained by $U(1)$ charge conservation, such as kaon production/annihilation via $\pi^+ \pi^- \leftrightarrow K^+ K^-$, the correlation between particles b_1 and b_2 requires that the general rate equation for the average number of $b_1 b_2$ pairs should be written as

$$\frac{d\langle N \rangle}{d\tau} = \frac{G}{V} \langle N_{a_1} \rangle \langle N_{a_2} \rangle - \frac{L}{V} \langle N^2 \rangle. \tag{2}$$

For abundant production of $b_1 b_2$ pairs, where $\langle N \rangle \gg 1$, one has $\langle N^2 \rangle \approx \langle N \rangle^2$ and (2) reduces to the standard form. However, for rare production of $b_1 b_2$ pairs, where $\langle N \rangle \ll 1$, one has instead $\langle N^2 \rangle \approx \langle N \rangle$, which reduces (2) to the following form:

$$\frac{d\langle N \rangle}{d\tau} \approx \frac{G}{V} \langle N_{a_1} \rangle \langle N_{a_2} \rangle - \frac{L}{V} \langle N \rangle. \tag{3}$$

Thus, in the limit $\langle N \rangle \ll 1$, the absorption term depends on the pair number only linearly, instead of quadratically for the limit $\langle N \rangle \gg 1$.

In the limit when $\langle N \rangle \gg 1$, the standard rate equation (1) is valid and has the following well-known solution:

$$\langle N \rangle^{GC}(\tau) = N_{eq}^{GC} \tanh \left(\tau / \tau_0^{GC} \right), \tag{4}$$

where the equilibrium value for the number of $b_1 b_2$ pairs N_{eq}^{GC} and the relaxation time constant τ_0^{GC} are given by

$$N_{eq}^{GC} = \sqrt{\varepsilon}, \quad \tau_0^{GC} = \frac{V}{L\sqrt{\varepsilon}}, \tag{5}$$

respectively, with $\varepsilon \equiv G\langle N_{a_1} \rangle \langle N_{a_2} \rangle / L$.

In the special case where particle momentum distributions are thermal, the gain (G) and loss (L) terms just represent the thermal averages of the production and absorption cross sections, and their ratio is

$$\frac{G}{L} = \frac{d_{b_1} \alpha_{b_1}^2 K_2(\alpha_{b_1}) d_{b_2} \alpha_{b_2}^2 K_2(\alpha_{b_2})}{d_{a_1} \alpha_{a_1}^2 K_2(\alpha_{a_1}) d_{a_2} \alpha_{a_2}^2 K_2(\alpha_{a_2})}, \tag{6}$$

where d_k's denote the degeneracy factors, $\alpha_k \equiv m_k/T$, and K_2 represents the modified Bessel function. The equilibrium value for the number of $b_1 b_2$ pairs in (5) now reads as

$$N_{eq}^{GC} = \frac{d_{b_1}}{2\pi^2} VT^3 \alpha_{b_1}^2 K_2(\alpha_{b_1}). \tag{7}$$

Thus it is described by the Grand Canonical (GC) result with vanishing chemical potential due to our requirement of the $U(1)$ charge neutrality of the system.

In the opposite limit where $\langle N \rangle \ll 1$, the time evolution is described by (3), which has the following solution:

$$\langle N \rangle^C(\tau) = N_{eq}^C \left(1 - e^{-\tau/\tau_0^C} \right), \tag{8}$$

with the equilibrium value and relaxation time given by

$$N_{eq}^C = \varepsilon, \quad \tau_0^C = \frac{V}{L}. \tag{9}$$

With a thermal momentum distribution, the equilibrium value of $b_1 b_2$ pair multiplicity has the following form:

$$N_{eq}^C = \left[\frac{d_{b_1}}{2\pi^2} VT^3 \alpha_{b_1}^2 K_2(\alpha_{b_1}) \right] \left[\frac{d_{b_2}}{2\pi^2} VT^3 \alpha_{b_2}^2 K_2(\alpha_{b_2}) \right]. \tag{10}$$

This equation thus demonstrates the locality of the $U(1)$ charge conservation. With each particle b_1, a particle b_2 with the opposite charge is produced in the same event in order to conserve charge locally. This is the result expected from the Canonical (C) formulation of conservation laws.

(7) and (10) show that the equilibrium multiplicity in the canonical formulation for rare particles is much lower than what is expected from the grand canonical result,

$$N_{eq}^C = (N_{eq}^{GC})^2 \ll N_{eq}^{GC}. \tag{11}$$

One also notes that the volume dependence in the two cases is different. The particle density in the GC (abundant) limit is independent of V, whereas in the opposite canonical (rare) limit the density scales linearly with V.

Secondly, the relaxation time for a canonical system is far shorter than what is expected from the grand canonical result,

$$\tau_0^C = \tau_0^{GC} N_{eq}^{GC} \ll \tau_0^{GC}, \tag{12}$$

due to small number of particles ($N_{eq}^{GC} \ll 1$). For example, the total number of produced kaons in Au+Au collisions at 1 GeV/A is of the order of 0.02. Thus the canonical relaxation time is a factor of 7 shorter than what is expected from the grand canonical formulation.

TRANSPORT MODEL

The above idea based on the kinetic model can be quantitatively studied using transport models, which take into account consistently the effects due to finite size and time in heavy ion collisions and have been very successful in studying heavy ion collisions at various energies [7].

Kaon and antikaon equilibration at SIS

A useful transport model for describing heavy ion collisions at SIS energies is the relativistic transport model RVUU [8]. In this model, the nuclear potential is taken from the nonlinear Walecka model, so it has both an attractive scalar and a repulsive vector part. The attractive scalar potential allows one to treat consistently the change of nucleon mass in nuclear matter. In dense matter, the nucleon mass is reduced and the energy is in the scalar field. As the system expands, the nucleon regain its mass from the scalar field energy. Kaons are produced from both baryon-baryon [9] and meson-baryon [10] interactions. The produced kaons together with their partners, mainly hyperons, not only undergo elastic scatterings with baryons [11] but are also affected by mean-field potentials. The kaon potential is taken from the chiral Lagrangian including both scalar and vector interactions [12] with their strengths determined from experimental observables such as the kaon yield and collective flow in heavy ion collisions [11, 13]. The repulsive vector potential needed for understanding the experimental results thus does not support the suggestion of [4] that the vector potential vanishes at high densities.

To study the chemical equilibration of kaons, kaon annihilation by hyperon needs to be included. Because of strangeness conservation, a kaon is produced together with a hyperon. As the kaon production probability is much less than one in heavy ion collisions at subthreshold energies, there is only one hyperon in an event in which a kaon is produced. Kaon annihilation can thus occur only when there is a collision between the same pair of kaon and hyperon that is produced in the baryon-baryon or meson-baryon interaction. As shown in the kinetic model discussed above, the annihilation between such a pair of particles that are produced simultaneously as a result of the $U(1)$ charge conservation would lead to an equilibration described by the canonical ensemble.

To illustrate how kaons approach kaon chemical equilibration in heavy ion collisions at SIS, we consider Ni+Ni collisions at $1A$ GeV and impact parameter $b = 0$ fm, which is below the threshold for both kaon and antikaon production in nucleon-nucleon interactions. In the top left panel of Fig. 1, the time evolution of the kaon abundance for the scenarios with (solid curve) and without (dotted curve) kaon annihilation are given. It is seen that including kaon annihilation by hyperon reduces the final kaon yield by only about 10%. As shown in the lower left panel of Fig. 1, the kaon production rate (solid curve) is appreciable only when the nuclear density (thick solid curve) is high. The effect due to kaon annihilation is better illustrated by the kaon annihilation rate (dashed curve) shown in the lower left panel of Fig. 1. One sees that the annihilation rate is negligible during the high density stage when most kaons are produced. This result thus justifies the neglect of kaon annihilation in previous studies [14, 15, 16, 17]. Although the kaon

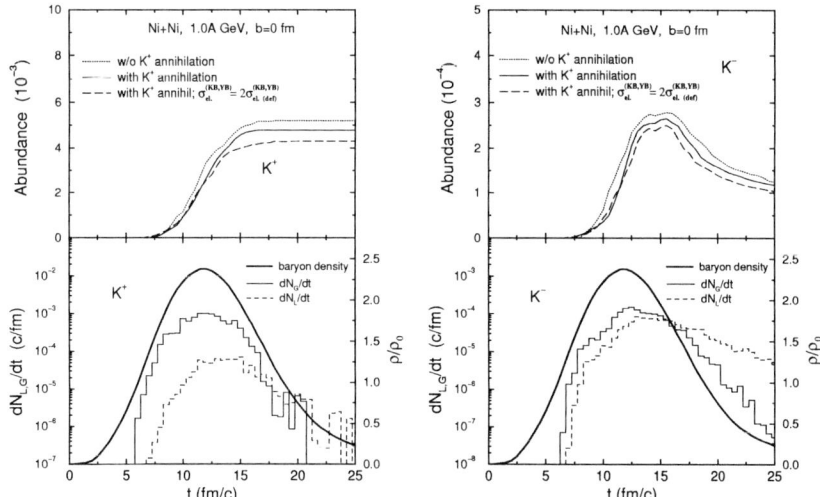

FIGURE 1. Top panel: Time evolution of kaon and antikaon abundances in Ni+Ni collisions at $1A$ GeV and impact parameter $b = 0$. Dotted curve is the result without kaon and antikaon absorption while solid and dashed curves are results with kaon and antikaon absorption from using default KB and YB elastic scattering cross sections and twice the cross sections, respectively. Bottom panel: Time evolution of kaon and antikaon production rates (solid curve) and absorption rates (dashed curve) as well as the central baryon density (thick solid curve).

annihilation rate is small, it becomes comparable to the kaon production rate at about 17 fm/c, indicating that the kaon yield eventually approaches chemical equilibrium. The baryon density at which kaons reach chemical equilibrium is about 1.2 ρ_0, where $\rho_0 \approx 0.16$ fm^{-3} is the normal nuclear matter density.

According to the kinetic model, the chemical equilibration time in the canonical formalism is given by $\tau_0^C = V/L = N_K/(dN_L/dt)$, where V is the volume of the region where kaon annihilation occurs, L is the momentum averaged cross section for kaon annihilation, and N_K and dN_L/dt are the kaon number and its absorption rate, respectively. At time $t = 12$ fm/c when the kaon absorption rate is largest, one has from Fig. 1, $N_K = 4.2 \times 10^{-3}$ and $dN_L/dt = 7.0 \times 10^{-5}$, which give a kaon chemical equilibration time of 60 fm/c if the system is prevented from expanding from $t = 12$ fm/c. On the other hand, the thermal average of the kaon annihilation cross section $\sigma_{KY \to \pi N}$ is about 0.25 fm^2 at temperature $T = 75$ MeV, and it changes by less than 20% for $50 < T < 100$ MeV due to the exothermic nature of the annihilation process. The above value for the kaon chemical equilibration time implies that the effective volume in which kaon annihilation occurs is about 15 fm^3. Since the chemical equilibration time for kaons is much longer than the heavy ion collision time, one would normally expect them not to reach chemical equilibrium during the collisions. However, since the kaon production rate decreases strongly as the temperature decreases due to the large threshold of the production process, its value can thus become comparable to the annihilation rate when the system expands and cools. When this happens for a later stage of heavy ion collisions, kaons can then reach chemical equilibrium.

The effect due to kaon annihilation depends on the magnitude of kaon and hyperon elastic scattering cross sections with other particles. If there are no such scatterings, e.g, if these cross sections are set to zero, then the produced kaon and hyperon would simply move away from each other without further interactions, leading to a result similar to that without kaon annihilation. On the other hand, larger kaon and hyperon scattering cross sections with other particles would force them into a smaller region, thus increasing the kaon annihilation rate. This is demonstrated in the top left panel of Fig. 1 by the dashed curve, which is obtained by taking the kaon and hyperon scattering cross sections with other baryons to be twice the default values. It is seen that these larger cross sections indeed further reduce the kaon yield.

Antikaon production has also been studied within the relativistic transport model, where antikaons are produced not only from baryon-baryon and meson-baryon interactions [18] but also from meson-hyperon interactions [19]. As first pointed out in [20], antikaon production in heavy ion collisions at subthreshold energies is mainly due to the meson-hyperon interactions. For antikaon annihilation, it is dominated by the reaction $\bar{K}N \to \pi Y$. To account for the observed enhancement of antikaon production in heavy ion collisions at subthreshold energies requires, however, a dropping of antikaon mass due to medium effects [11, 21, 22].

The results for antikaon production in Ni+Ni collisions at $1A$ GeV and impact parameter $b = 0$ fm are also shown in Fig. 1. In the top right panel, the time evolution of antikaon abundance is shown for the cases with (solid curve for default kaon and hyperon elastic scattering cross sections and dashed curve for twice the default cross sections) and without (dotted curve) kaon annihilation. The reduction of antikaon yield when kaon annihilation is allowed is due to the reduction in the production probability of hyperons, which contribute most to antikaon production. The time evolution of the antikaon production and annihilation rates are shown in the bottom right panel of Fig. 1 by the solid and dashed curves, respectively. Similar to kaon production, one sees that antikaons are mostly produced in the high density stage of heavy ion collisions. Because of the large baryon density, their annihilation rate through the reaction $\bar{K}N \to \pi Y$ is larger than the kaon annihilation rate. As a result, antikaons approach chemical equilibrium even in the earlier stage of heavy ion collisions. However, they eventually fall out of equilibrium as shown in the bottom right panel of Fig. 1.

Multistrange baryon equilibration at SPS and RHIC

In heavy ion collisions at ultrarelativistic heavy ion collisions at SPS and RHIC, kaons and antikaons are relatively abundant while multistrange baryons such as cascades and omegas are rare. To study the production of these particles, a multiphase transport model (AMPT) has been developed. In the AMPT model, the initial conditions are obtained from the HIJING model [23] by including the nuclear shadowing effect on minijet partons via the gluon recombination mechanism of Mueller-Qiu [24]. After the colliding nuclei pass through each other, the Gyulassy-Wang model [25] is then used to generate the initial space-time information of partons. Subsequent time evolution of the parton phase-space distribution is modeled by Zhang's Parton Cascade (ZPC) [26], which at present includes only the gluon elastic scattering. After minijet partons stop interacting, they combine with their parent strings and are then converted to hadrons using the Lund string fragmentation model [27] after an average proper formation time of 0.7 fm/c. Dynamics of the resulting hadronic matter is described by a relativistic transport model (ART) [28].

The parameters in the AMPT model are fixed using the experimental data from central Pb+Pb collisions at center of mass energy of 17A GeV at SPS [29]. Specifically, one includes in the Lund string fragmentation model the popcorn mechanism for baryon-antibaryon production in order to describe the measured net baryon rapidity distribution. Also, to account for the pion and enhanced kaon yields, values of the parameters in the string fragmentation function have been modified. The same parameters are then used to study heavy ion collisions at RHIC energies. The model is able to describe the PHOBOS data [30] for the pseudorapidity distribution of charged particles [31]. Also, the predicted $\bar{p}/p$ and K^-/π^- ratios are consistent with the data from the STAR [32] and BRAHMS [33] Collaborations. Furthermore, it gives an elliptic flow [34] which is comparable to that measured in the STAR experiment [35].

Multistrange baryon production is included through the strangeness-exchange reactions,

$$\bar{K}\Lambda \leftrightarrow \Xi\pi, \quad \bar{K}\Sigma \leftrightarrow \Xi\pi, \quad \text{and} \quad \bar{K}\Xi \leftrightarrow \Omega\pi. \tag{13}$$

Since there is no experimental information on their cross sections, it has been assumed that they are the same as the cross section for $\bar{K}N \to \Sigma\pi$, which are known empirically and have been parameterized in [19]. Specifically, all cross sections are taken to have the same value at center-of-mass energies that are the same amount above the corresponding threshold energies. Preliminary results from an effective hadronic model that is based on SU(3) flavor symmetry indeed show that the cross sections for these reactions have comparable values [36].

In Fig. 2, we show the time evolution of the abundance of midrapidity kaons (including K^*), antikaons (including $\bar{K}^*$), Λ, Σ, Ξ^-, and Ω^- in Pb+Pb collisions at SPS energy of $\sqrt{s} = 17A$ (left panel) and RHIC energy of 130A GeV (right panel) at an impact parameter $b = 0 - 3$ fm. As is evident from the figure, most multistrange baryons are produced within 10 fm/c after the initial contact of the colliding nuclei when the energy density is high. Compared to the initial yield obtained from HIJING, the dynamical evolution of the system leads to about 30% and 70% increase in the Λ and Ξ^- production, while most of the Ω^-s are produced from hadronic rescattering. Therefore the observed enhancement of multistrange baryons at SPS [37] can be largely accounted by the strangeness-exchange reactions in the hadronic matter. The Ω^- production in the AMPT model is, however, a factor of two smaller compared to the data. Using a rate equation approach, it was demonstrated [38] that multimesonic reactions $\bar{Y} + N \leftrightarrow n\pi + n_Y K$ may lead to an enhanced antistrange hyperon $\bar{Y}$ production in a purely hadronic scenario. Similarly, including multiparticle reactions, such as $3\pi + 2\bar{K} \leftrightarrow \Xi + \bar{N}$ and $2\pi + 3\bar{K} \leftrightarrow \Omega + \bar{N}$, could lead to a better reproduction of the experimental data for multistrange baryons. Also, the process $\bar{K}(\bar{K}^*) + \bar{K}(\bar{K}^*) \leftrightarrow \Xi + \bar{N}$ also contributes to Ξ production.

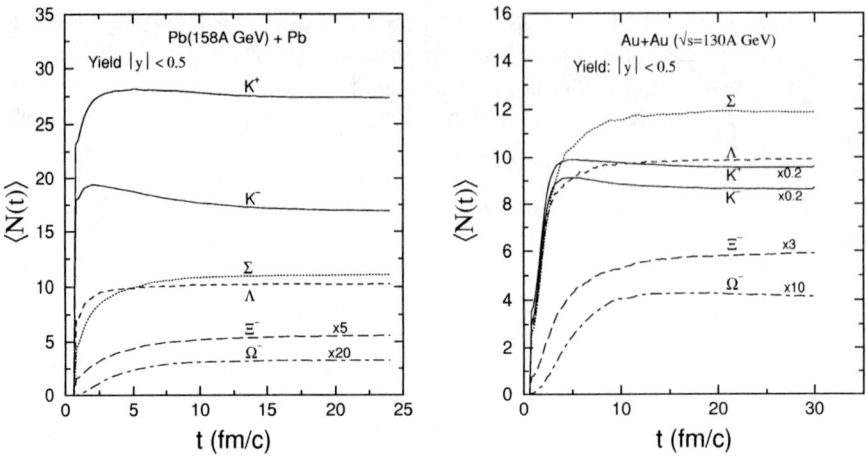

FIGURE 2. Time evolution of midrapidity hadrons for Pb+Pb collisions at SPS energy of $\sqrt{s} = 17A$ (left panel) and Au+Au collisions at RHIC at $130A$ GeV (right panel) at an impact parameter of $b \leq 3$ fm in the AMPT model.

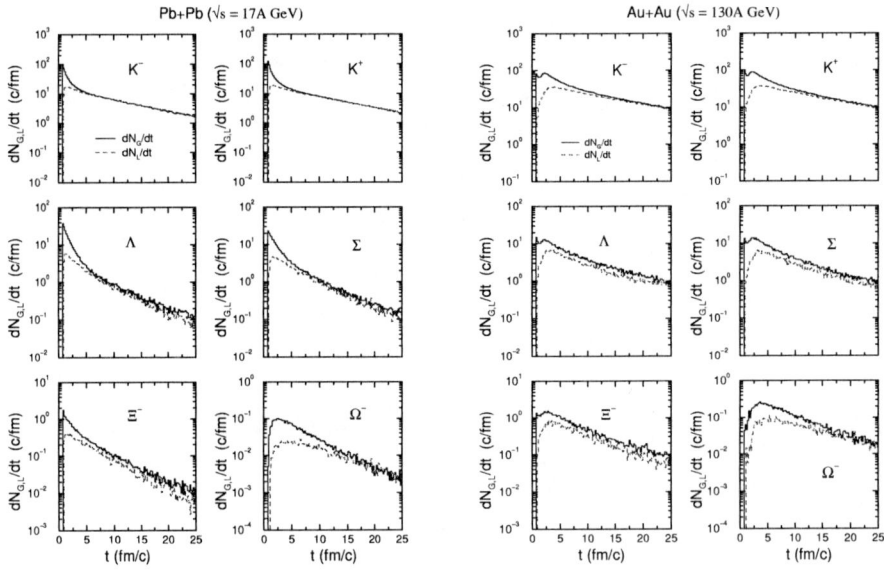

FIGURE 3. Production and absorption rates of strange particles in heavy ion collisions at SPS (left panel) and RHIC (right panel) as functions of time.

To see if strange hadrons reach chemical equilibrium in the collisions, we show in Fig. 3 the time evolution of their production (solid lines) and absorption rates (dashed lines). It is seen that most strange particles such as the hyperons and especially the kaons and antikaons approach equilibrium as their production and absorption rates become comparable before the system freezes out.

The enhanced yield of multistrange baryons obtained through the strangeness-exchange reactions between antikaon and baryons relies on the initial antikaon and hyperon abundance in the hadronic matter. The multiphase transport model reproduces their enhanced yield by modifying the parameters for string fragmentation. Since the energy density associated with the initial string matter is high, particularly at RHIC, the modification introduced in the AMPT model may simply be a way to mimic the effect of quark-gluon plasma formation in the initial stage of the collisions. As suggested in [39], strangeness production is significantly enhanced in the quark-gluon plasma than in the hadronic matter.

SUMMARY

To summarize, the kinetic model leads naturally to the grandcanonical equilibrium for abundant particles and the canonical equilibrium for rare particles. It further shows that the equilibrium time for rare particles carrying $U(1)$ charge is much shorter than what is expected from the grand canonical ensemble, and their equilibrium multiplicity in the canonical ensemble is much lower than that given by the grand canonical ensemble.

With finite size and time effects consistently included, transport models provide a realistic description of the chemical equilibration of strange particles in heavy ion collisions. For heavy ion collisions at SIS energies below the threshold for strange particle production in nucleon-nucleon interactions, one finds that both kaons and antikaons are largely produced during the high density stage of the collisions when the system has not reached thermal equilibrium. Because of their large annihilation cross sections in dense nuclear matter, antikaons are near chemical equilibrium much earlier. For kaons, their abundance at the high density stage is far from equilibrium, and it only becomes close to the equilibrium value during the expansion stage of heavy ion collisions when the production rate is small and comparable to the annihilation rate.

For heavy ion collisions at ultrarelativistic heavy ion collisions, a multiphase transport model, that includes both an initial partonic matter and a final hadronic matter, has been used to study the production of multistrange baryons from the strangeness-exchange reactions in the hadronic matter. The cross sections for these reactions are assumed to be the same as the empirically known cross sections for antikaon-nucleon to hyperon-pion reactions. For heavy ion collisions at SPS energies, one finds that these reactions lead to an enhanced production of multistrange particles, comparable to that observed in experiments. For heavy ion collisions at RHIC, a similar enhancement is obtained from this model. In both cases, most strange particles such as the hyperons and especially the kaons and antikaons approach equilibrium before the system freezes out.

ACKNOWLEDGMENTS

This work was supported by the National Science Foundation under Grant Nos. PHY-9870038 and PHY-0098805, the Welch Foundation under Grant No. A-1358, and the

Texas Advanced Research Program under Grant No. FY99-010366-0081.

REFERENCES

1. P. Braun-Munzinger, I. Heppe and J. Stachel, Phys. Lett. B 465, 15 (1999).
2. J. S. Hamieh, K. Redlich and A. Tounsi, Phys. Lett. B 486, 61 (2000).
3. J. E.L. Bratkovskaya *et al.*, Nucl. Phys. A 675, 661 (2000).
4. G.E. Brown, M. Rho, and C. Song, Nucl. Phys. A 690, 184 (2001).
5. C. M. Ko, V. Koch, Z. W. Lin, K. Redlich, M. Stephanov, and X. N. Wang, Phys. Rev. Lett. 86, 5438 (2001).
6. S. Pal, C. M. Ko, and Z. W. Lin, Phys. Rev. C, in press; nucl-th/0105036; S. Pal, C. M. Ko, and Z. W. Lin, nucl-th/0106073.
7. C.M. Ko and G.Q. Li, J. Phys. G 22, 1673 (1996); W. Cassing and E.L. Bratkovskaya, Phys. Rep. 308, 65 (1999); S. Bass *et al.*, Prog. Part. Nucl. Phys. 42, 313 (1999).
8. C.M. Ko, Q. Li, and R. Wang, Phys. Rev. Lett. 59, 1084 (1987); C.M. Ko, and Q. Li, Phys. Rev. C 37, 2270 (1988); Q. Li, J.Q. Wu, and C.M. Ko, Phys. Rev. C 39, 849 (1989); C.M. Ko, Nucl. Phys. A 495, 321c (1989).
9. G.Q. Li and C.M. Ko, Nucl. Phys. A 594, 439 (1995).
10. K. Tsushima, S.W. Huang, and A. Faessler, Phys. Lett. B 337, 245 (1994); J. Phys. G 21, 33 (1995).
11. G.Q. Li, C.H. Lee, and G.E. Brown, Nucl. Phys. A 625, 372 (1997).
12. D. B. Kaplan and A. E. Nelson, Phys. Lett. B 175 57 (1986); *ibid.* 192, 193 (1987).
13. S. Pal, C. M. Ko, Z. W. Lin, and B. Zhang, Phys. Rev. C 62, 061903 (2000).
14. J. Randrup and C.M. Ko, Nucl. Phys. A 343, 519 (1980).
15. X.S. Fang, C.M. Ko, and Y.M. Zheng, Nucl. Phys. A 556, 499 (1993).
16. J. Aichelin and C.M. Ko, Phys. Rev. Lett. 55, 2661 (1985); X.S. Fang, C.M. Ko, G.Q. Li, and Y.M. Zheng, Phys. Rev. C 49, 1139 (1994); Nucl. Phys. A 575, 766 (1994); G.Q. Li and C.M. Ko, Phys. Lett. B 349, 405 (1995).
17. S.W. Huang *et al.*, Phys. Lett. B 298, 41 (1993); T. Maruyama *et al.*, Nucl. Phys. A 573, 653 (1994); C. Hartnack *et al.*, *ibid.* 580, 643 (1994).
18. A. Sibirtsev, W. Cassing, and C.M. Ko, Z. Phys. A 358, 101 (1997).
19. J. Cugnon, P. Deneye, and J. Vandermeulen, Phys. Rev. C 41, 1701 (1990).
20. C.M. Ko, Phys. Lett. B 120, 294 (1983); *ibid.* 138, 361 (1984).
21. G.Q. Li, C.M. Ko, and X.S. Fang, Phys. Lett. B 329, 149 (1994).
22. W. Cassing *et al.*, Nucl. Phys. A 614, 415 (1997).
23. M. Gyulassy and X.N. Wang, Comp. Phys. Comm. 83, 307 (1994).
24. A.H. Mueller and J. Qiu, Nucl. Phys. B 268, 427 (1986); J. Qiu, Nucl. Phys. B 291, 746 (1987).
25. M. Gyulassy and X.N. Wang, Nucl. Phys. B 420, 583 (1994).
26. B. Zhang, Comp. Phys. Comm. 109, 193 (1998).
27. B. Andersson, G. Gustafson, G. Ingelman and T. Sjöstrand, Phys. Rep. 97, 31 (1983); T. Sjöstrand, Comp. Phys. Comm. 82, 74 (1994).
28. B.A. Li and C.M. Ko, Phys. Rev. C 52, 2037 (1995).
29. H. Appelshäuser *et al.* (NA49 Collaboration), Phys. Rev. Lett. 82, 2471 (1999).
30. B.B. Back *et al.* (PHOBOS Collaboration), Phys. Rev. Lett. 85, 3100 (2000); B.B. Back *et al.* (PHOBOS Collaboration), nucl-ex/0106006.
31. Z. Lin, S. Pal, C.M. Ko, B.A. Li, and B. Zhang, Phys. Rev. C 64, 011902 (2001).
32. C. Adler *et al.* (STAR Collaboration), Phys. Rev. Lett. 86, 4778 (2001).
33. I.G. Bearden *et al.* (BRAHMS Collaboration), nucl-ex/0106011.
34. Z. Lin and C. M. Ko, nucl-th/0108039.
35. K.H. Ackermann *et al.* (STAR Collaboration), Phys. Rev. Lett. 86, 402 (2001).
36. C. H. Li, C. M. Ko, and Z. W. Lin, in preparation.
37. E. Anderson *et al.* (WA97 Collaboration), Phys. Lett. B 433, 209 (1998); *ibid.* 449, 401 (1999).
38. C. Greiner and S. Leupold, nucl-th/0009036.
39. J. Rafelski and B. Müller, Phys. Lett. B 101, 111 (1982).

Pion Enhancement and Chiral Symmetry

P.Zhuang, Z.Yang and M.Huang

Physics Department, Tsinghua University, Beijing 100084, China

Abstract. The pion production by sigma decay and its relation with chiral symmetry restoration in a hot matter are investigated. It is pointed out that the nonthermal decay into pions of sigma mesons which are popularly produced in chiral symmetric phase leads to a peak in the pion spectra in relativistic heavy-ion collisions.

INTRODUCTION

As is well known, the ultimate goal of relativistic heavy ion collisions is to study the deconfinement process in moving from a hadron gas to a quark-gluon plasma and the chiral transition from the chiral symmetry breaking phase to the phase in which it is restored. In recent years some signatures of the chiral transition have been proposed, such as the excess of pions due to a rapid thermal and chemical equilibration in the symmetric phase[1], the excess of low energy photon pairs by pion annihilation in hot and dense medium[2], the dilepton enhancement from leptonic decay of sigma[3], and the enhancement of the continuum threshold in the scalar channel[4]. Most of the signatures are associated with the changes of sigma properties at finite temperatures and densities[5, 6]. With increasing temperature and density sigma changes its character from a resonance with large mass to a bound state with small mass. Around the critical point of the chiral transition sigma is one of the most numerous species, since it is nearly massless. If the transition happens at the freeze-out these light sigmas will lead to large event-by-event fluctuations[7] of final state pions due to the σ-exchange in the pion thermodynamic potential[8].

A direct signature of these light sigmas in hot and dense matter is a nonthermal excess[8] of pions due to the decay process $\sigma \to 2\pi$. With the expansion of the hadronic system produced in relativistic heavy ion collisions, the in-medium sigma mass rises towards its vacuum value and eventually exceeds the $\pi\pi$ threshold. As the $\sigma\pi\pi$ coupling is large, the decay proceeds rapidly. When this process occurs before freeze-out, the produced pions will be thermalized in the heat bath. However, when the decay happens after freeze-out, the generated pions do not have a chance to thermalize. Thus, it is expected[8, 9] that the resulting pion spectrum will have a nonthermal enhancement at low momentum.

CP597, *Nonequilibrium and Nonlinear Dynamics in Nuclear and Other Finite Systems*, edited by Z. Li et al.

PION ENHANCEMENT RATES

The dependence on temperature T and baryon density n of non-strange hadron masses is well parameterized as[10]

$$\frac{m(T,n)}{m(0,0)} = \left(1 - \left(\frac{T}{T_d}\right)^{1/3}\right)\left(1 - 0.2\frac{n}{n_0}\right) , \tag{1}$$

where T_d is the critical temperature of chiral phase transition, and n_0 the baryon density of normal nuclear matter. From this parameterization, the temperature and density effects are quite different: The temperature dependence is very weak at low temperatures ($T/T_d < 1/2$) while the hadron masses decrease linearly in density in the whole density region. This difference between temperature and density effects is reflected in the pion spectra in high energy nuclear collisions at SPS and at RHIC. At RHIC where the central region is almost baryon free and the temperature effect is dominant, the sigma and pion masses are nearly constants when the temperature of the system is much below the critical one. Therefore, the pions produced from nonthermal sigma decay after freeze-out will have almost the same momentum $p_0 = \sqrt{\frac{m_\sigma^2(0,0)}{4} - m_\pi^2(0,0)}$. Aside from an overall enhancement of pions due to the thermal sigma decay, there will be also a peak in the pion spectra due to the nonthermal sigma decay. This peak will disappear partly when the density effect can not be neglected, since in this case the sigma mass varies significantly even after the freeze-out.

We now illustrate the above statement by using a simple statistic model. The measured final-state pions include the direct pions emitted from the hot system and the pions from sigma decay,

$$N = N_{dir} + N_{dec} . \tag{2}$$

N_{dir} is determined by thermodynamics at freeze-out, and N_{dec} is twice the number of sigmas at chiral phase transition where sigma has minimum mass and is then most numerous,

$$N_{dir} = V_f n_\pi(T_f) ,$$
$$N_{dec} = 2N_\sigma(T_d) = 2V_d n_\sigma(T_d) , \tag{3}$$

where T_f and V_f are the temperature and volume of the system at freeze-out, T_d and V_d the temperature and volume at chiral phase transition, $n_\pi(T_f)$ and $n_\sigma(T_d)$ the pion and sigma number densities at T_f and T_d,

$$n_\pi(T_f) = \int \frac{d^3\mathbf{p}}{(2\pi)^3} \frac{3}{e^{\varepsilon_\pi/T_f} - 1} ,$$
$$n_\sigma(T_d) = \int \frac{d^3\mathbf{p}}{(2\pi)^3} \frac{1}{e^{\varepsilon_\sigma/T_d} - 1} , \tag{4}$$

with the pion and sigma energies $\varepsilon_\pi(T_f) = \sqrt{m_\pi^2(T_f) + p^2}$ and $\varepsilon_\sigma(T_d) = \sqrt{m_\sigma^2(T_d) + p^2}$.

In central region of relativistic heavy-ion collisions where there is almost no net baryon and one considers the temperature effect only, the Bjorken's scaling hydrodynamics[11] is often used to describe the space-time evolution of the produced hot system. In this scenario the temperature T_f, volume V_f and proper time τ_f at freeze-out and the corresponding quantities T_d, V_d and τ_d at chiral phase transition have simple relations

$$T_f = T_d \left(\frac{\tau_d}{\tau_f} \right)^{\frac{1}{3}} \quad , V_f = V_d \frac{\tau_f}{\tau_d} . \tag{5}$$

For central $A - A$ collisions with impact parameter $b = 0$, the number of participant nucleons and the volume at freeze-out can be simply taken as $\langle N_{part} \rangle = 2A$ and $V_f = A v_f$. To compare with the experimental data at RHIC, we calculate the pseudorapidity density of charged pions per participant pair in the mid-rapidity range $|\eta| < 1$,

$$\frac{dN}{d\eta} \bigg|_{\Delta\eta} / 0.5 \langle N_{part} \rangle = \frac{1}{2} \frac{2}{3} v_f \left(1 + 2 \left(\frac{T_f}{T_d} \right)^3 \frac{n_\sigma(T_d)}{n_\pi(T_f)} \right) n_\pi(T_f) \bigg|_{\Delta\theta} , \tag{6}$$

where $n_\pi(T_f)\big|_{\Delta\theta}$ is the pion number density in the angle interval $2 \arctan e^{-1} = \theta_1 < \theta < \theta_2 = 2 \arctan e$,

$$n_\pi(T_f) \bigg|_{\Delta\theta} = \int_{\Delta\theta} \frac{d^3 \mathbf{p}}{(2\pi)^3} \frac{3}{e^{\varepsilon_\pi/T_f} - 1} , \tag{7}$$

and the factor of $2/3$ is the charged to total pion ratio.

The pseudorapidity density (6) depends on the volume parameter v_f and the two temperature scales T_f and T_d which characterize the collective effect of the system at freeze-out and at chiral transition. From the lattice simulations of QCD which make measurements of the Polyakov loop as well as the scalar quark density, the deconfinement and chiral transitions coincide at a temperature of about $T_d = 170 \, MeV$[12]. In the QCD phase diagram[13] the thermal freeze-out temperature at RHIC is about $T_f = 120 \, MeV$. As is well known, the pion mass is approximately temperature independent in the whole chiral breaking phase, we can take $m_\pi(T_f) = m_\pi(0) = 140 \, MeV$ at freeze-out and $m_\sigma(T_d) = 2m_\pi(T_d) = 2m_\pi(0)$ at chiral phase transition. With the given T_f and T_d, the ratio of the pions produced by sigma decay to the direct pions is

$$r = 2 \left(\frac{T_f}{T_d} \right)^3 \frac{n_\sigma(T_d)}{n_\pi(T_f)} = 52\% , \tag{8}$$

this shows a significant influence of the chiral phase transition on the final state pions in relativistic heavy-ion collisions. The volume parameter v_f is certainly colliding energy dependent, it can be fixed by experimental data. From the comparison of the pseudorapidity density (6) with the data[14] at RHIC, we have $v_f = 116$ and $154 \, fm^3$ corresponding to $\sqrt{s} = 56$ and $130 \, A \, GeV$, respectively.

In above study of pion multiplicity we did not distinguish the nonthermal sigma decay from thermal sigma decay. However, as we discussed qualitatively, the thermal and nonthermal pions have very different momentum spectra. The pions produced by sigma decay in the temperature region $T_f < T < T_d$ have time to thermalize before freeze-out, they leads to an enhancement of thermal pions, while in the interval $T < T_f$

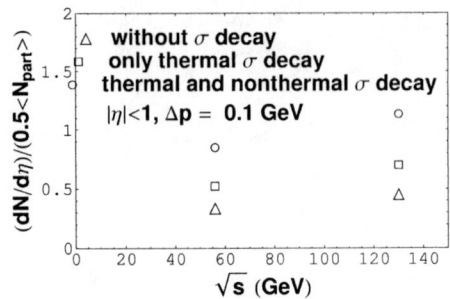

FIGURE 1. The pseudorapidity density (9) normalized per participant pair in the mid-rapidity region $|\eta| < 1$ and in the symmetric momentum window around p_0 with width $\Delta p = 100 MeV$ for three cases: without sigma decay, with only thermal sigma decay, and with both thermal and nonthermal sigma decay for thermal fraction $\beta = 0.5$.

the produced pions do not get a chance to thermalize, they have almost the same momentum $p_0 = \sqrt{\frac{m_\sigma^2(0)}{4} - m_\pi^2(0)}$ in the rest frame of sigma due to the weak temperature dependence of non-strangle hadron masses at low temperatures. Therefore, if we take a narrow momentum window around p_0, only a small part of the thermal pions including the direct pions and thermal pions from sigma decay distribute in the window, but all the nonthermal pions are deposited in the window. The contribution from sigma decay to final state pions is therefore greatly amplified in the window. This strong pion enhancement can be taken as a signature of chiral phase transition in relativistic heavy-ion collisions.

Considering a symmetric window around p_0 and taking the thermal fraction of the pions produced by sigma decay to be β, the pseudorapidity density in the mid-rapidity region and in the momentum window is

$$\frac{dN}{d\eta}\Big|_{\Delta\eta,\Delta p}/0.5\langle N_{part}\rangle = \frac{1}{2}\frac{2}{3}v_f \times$$

$$\left(\left(1 + 2\beta \left(\frac{T_f}{T_d}\right)^3 \frac{n_\sigma(T_d)}{n_\pi(T_f)}\right) n_\pi(T_f)\Big|_{\Delta\theta,\Delta p} + 2(1-\beta)\left(\frac{T_f}{T_d}\right)^3 \frac{n_\sigma(T_d)}{n_\pi(T_f)} n_\pi(T_f)\Big|_{\Delta\theta}\right) \quad (9)$$

with the pion number density

$$n_\pi(T_f)\Big|_{\Delta\theta,\Delta p} = \int_{\Delta\theta,\Delta p} \frac{d^3\mathbf{p}}{(2\pi)^3} \frac{3}{e^{\varepsilon_\pi/T_f} - 1} \quad (10)$$

in the window.

In our numerical calculation we take the sigma mass in the vacuum $m_\sigma(0) = 600\,MeV$, the corresponding momentum of the nonthermal pions is then $p_0 = 265\,MeV$. For a symmetric momentum window around p_0 with width $\Delta p = 100\,MeV$, the pseudorapidity density is shown in Fig.1 for three cases, without sigma decay, only thermal sigma decay, and both thermal and nonthermal sigma decay with thermal fraction $\beta = 0.5$. For the second case, namely all the sigmas decay before the freeze-out, the pion density

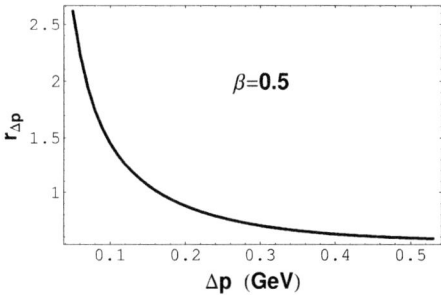

FIGURE 2. The window width dependence of the enhancement factor (11) for $\beta = 0.5$.

increases 52%, as discussed above. However, when the nonthermal decay is taken into account, the pion density increases 145% in the window. 82% of the enhancement comes from the nonthermal decay.

The enhancement depends on the thermal fraction β. Even for $\beta = 0.9$, namely only 10% of the sigmas decay after the freeze-out of the system, the enhancement is 71% which is still significant.

The enhancement is certainly window dependent. Fig.2 shows this dependence of the enhancement factor

$$r_{\Delta p} = 2\beta \left(\frac{T_f}{T_d}\right)^3 \frac{n_\sigma(T_d)}{n_\pi(T_f)} + 2(1-\beta)\left(\frac{T_f}{T_d}\right)^3 \frac{n_\sigma(T_d)}{n_\pi(T_f)} \frac{n_\pi(T_f)\big|_{\Delta\theta}}{n_\pi(T_f)\big|_{\Delta\theta,\Delta p}} \tag{11}$$

for $\beta = 0.5$. It decreases very fast with increasing Δp for small windows, then varies smoothly, and reaches the limit value $r = 52\%$ finally. Even for a window with $\Delta p = 500\ MeV$, the sigma decay brings a pion enhancement factor 59%.

PION SPECTRA IN NJL MODEL

The two-flavor NJL model is defined through the lagrangian density,

$$L_{NJL} = \bar{\psi}(i\gamma^\mu \partial_\mu - m_0)\psi + G[(\bar{\psi}\psi)^2 + (\bar{\psi}i\gamma_5\tau\psi)^2], \tag{12}$$

where only the scalar and pseudoscalar interactions corresponding to σ and π mesons, respectively, are considered, ψ and $\bar{\psi}$ are the quark fields, the operate τ is the $SU(2)$ isospin generator, G is the coupling constant with dimension GeV^{-2}, and m_0 is the current quark mass.

In an expansion in the inverse number of colors, $1/N_c$, the zeroth order (Hartree) approximation for quarks together with the first-order (RPA) approximation for mesons gives a self-consistent treatment of the quark-meson plasma in the NJL model. In the chiral limit the tricritical point P is located at $T_P = 79\ MeV, \mu_P = 280\ MeV$ in the $T - \mu$ plane. In the case of explicit chiral symmetry breaking the endpoint of the first-order transition line is shifted with respect to the tricritical point P towards large μ, its position is $T_E = 23\ MeV, \mu_E = 325\ MeV$.

With the decay rate $\Gamma_{\sigma \to 2\pi}(T, \mu)$ calculated at the one-loop level[9] and the relation (5) between temperature and time, the number of σ's present at time t is determined by

$$\frac{dN_\sigma(t)}{dt} = -\Gamma(t)N_\sigma(t) , \tag{13}$$

and is therefore

$$N_\sigma(t) = N_\sigma(t_d)e^{-\int_{t_d}^{t} \Gamma(t')dt'} , \tag{14}$$

where $N_\sigma(t_d)$ is the maximum number at the beginning time t_d, $N_\sigma(t_d) = V(t_d)n_\sigma(t_d)$ with the σ energy $\varepsilon_\sigma = \sqrt{m_\sigma^2 + \mathbf{p}^2}$ and the volume $V(t_d)$ of the system at time t_d.

Since each sigma yields two pions, the number of pions generated by σ decay at time t is related to the number of sigma by $N_\pi(t) = 2(N_\sigma(t_d) - N_\sigma(t))$. As the time $t \to \infty$, $N_\sigma(t \to \infty) = 0$, sigma disappears, and the pion number reaches its maximum $N_\pi(t \to \infty) = 2N_\sigma(t_d)$.

The conventional definition for freeze-out is that the time between two successful scattering events is larger than the collective expansion time scale[13], $\tau_{scatt} \geq \tau_{exp}$. τ_{scatt} can be calculated[15] from the thermally averaged $\pi\pi$ cross section[16] in the NJL model, and $\tau_{exp} = 1/\partial_\mu u^\mu$ is simply reduced to $\tau_{exp} = t$ in the scaling hydrodynamics[11]. Thus the freeze-out time t_f and the freeze-out temperature T_f are determined by $\tau_{scatt} = t_f$ and the expansion mechanism. With the two temperature scales T_d and T_f we can divide the total pions generated by σ decay into thermal and nonthermal pions. For $T_d > T_f$, the produced pions in the interval $T_f < T < T_d$ have time to thermalize before the freeze-out, they lead to an enhancement of thermal pions, while in the interval $T < T_f$ the produced pions do not get a chance to thermalize, they result in a nonthermal enhancement of pions with low momentum. For $T_d < T_f$, σ decay yields no thermal pions, all the resulted pions are nonthermal pions.

With the in-medium pion momentum $p = \sqrt{\frac{m_\sigma^2}{4} - m_\pi^2}$, the expansion mechanism, and the pion distribution, we calculate the normalized momentum spectra of the total pions including the direct pions, the thermal pions from sigma decay, and the nonthermal pions from sigma decay,

$$\frac{1}{N_{total}} \frac{dN_{total}}{dp} = \frac{1}{n_\pi^{dir}(t_f) + 2_{t_f}^{t_d} n_\sigma(t_d)} \times \tag{15}$$

$$\left(\left(1 + \frac{t_d}{t_f} \frac{n_\sigma(t_d)}{n_\pi^{dir}(t_f)} \frac{N_\pi(t_f)}{N_\sigma(t_d)}\right) \frac{dn_\pi^{dir}(t_f, p)}{dp} + 2\frac{t_d}{t_f} n_\sigma(t_d) \frac{N_\sigma(t)}{N_\sigma(t_d)} \Gamma(t) \frac{dt}{dp} \theta(t - t_f) \right) ,$$

where $n_\pi^{dir}(t_f)$ is the pion density (4) at time t_f. The distribution (15) is shown in Fig.3. The thermal peak is always smooth, but the nonthermal peak is always sharp, as we pointed out above.

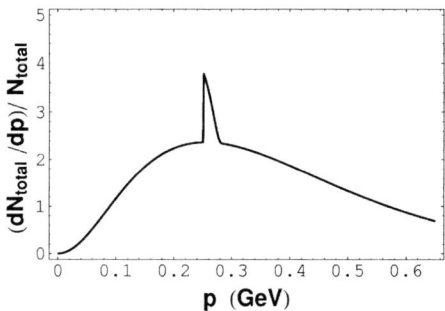

FIGURE 3. The normalized momentum spectra of the total pions at $\mu = 0$.

CONCLUSIONS

we have investigated the effect of chiral phase transition on final state pion spectra. The sigma enhancement at chiral phase transition leads to a significant enhancement of pions due to the sigma decay in chiral symmetry breaking phase. Especially, the sigma decay after the freeze-out of the system results in a crucial enhancement in a small momentum window around $p_0 = \sqrt{\frac{m_\sigma^2}{4} - m_\pi^2} \sim 260\,MeV$. This anomalous enhancement comes from the temperature effect which does not affect the sigma mass at low temperatures. For density effect, the sigma mass drops down rapidly even at low densities, the pions from sigma decay after the freeze-out distribute also in a wide momentum region, then the pion enhancement becomes weak in the window. Therefore, in the central region of heavy-ion collisions at RHIC where there is almost no density effect, the pion enhancement in a small window around the nonthermal momentum p_0 may be considered as a signature of the chiral phase transition. Although the sigma width and the sigma momentum which are not taken into account in our estimation will smooth the sharp peak around p_0, we expect that the dynamics of chiral phase transition (here via temperature dependent masses) plays still an essential role in the pion spectra.

acknowledgments: This work was supported in part by the NSFC under grant number 19925519, and by the Major State Basic Research Development Program under contract number G2000077407.

REFERENCES

1. Chungsik Song, Volker Koch, Phys. Lett. **B404**(1997)1.
2. M.K.Volkov et. al, Phys. Lett. **B424**(1998)235.
3. H.Arthur Weldon, Phys. Lett. **B274**(1992)133.
4. S.Chiku and T.Hatsuda, Phys. Rev. **D57**(1998)R6.
5. M.Rho, Proceedings of the Hirschegg conference, Austria, January, 16-21,1995.
6. P.Zhuang, J.Hüfner, and S.P.Klevansky, Nucl. Phys. **A576**(1994)525.
7. R.Rapp et. al, Phys. Rev. Lett.**81**(1998)53
8. M.Stephanov, K.Rajagopal and E.Shuryak, Phys. Rev. **D60**(1999)114028.
9. P.Zhuang and Z.Yang, Phys. Rev. **D63**(2001)016004.

10. C.M.Ko et. al, Phys. Rev. Lett.**66**(1991)2577
11. J.D.Bjorken, Phys. Rev. **D27**(1983)140.
12. F.Karsch, Nucl. Phys. **B**(Proc. Suppl.)83-84(2000)14.
13. U.Heinz, hep-ph/0009170.
14. B.B.Back et. al. (PHOBOS Collaboration), hep-ex/0007036.
15. P.Zhuang et. al, Phys. Rev. **D51**(1995)3728.
16. E.Quack et. al, Phys. Lett. **B348**(1995)1; M.Huang, P.Zhuanf and W.Chao, Phys. Lett. **B465**(1999)55.

The effect of the effective nucleon-nucleon-ρ-meson coupling density dependence on liquid-gas phase transition in hot asymmetric nuclear matter

Ru-Keng Su, Wei Liang Qian

Department of Physics, Fudan University, Shanghai 200433, P.R.China

Abstract. By using the Furnstahl, Serot and Tang's model, the effect of density dependence of the effective nucleon-nucleon-ρ-meson (NNρ) coupling on the liquid-gas phase transition in hot asymmetric nuclear matter is investigated. A limit pressure p_{lim} has been found. We found that the liquid-gas phase transition cannot take place if $p > p_{lim}$. The binodal surface for density dependent NNρ coupling situation is addressed.

It is generally recognized that the liquid-gas (LG) phase transition of one component system is of first order. The chemical potential continues at the phase transition point but its first order derivatives, namely, entropy and volume, are discontinuous. But for a multi-components or multi-conserved charges system, as was pointed out by Müller and Serot [1], because of the greater dimensionality of the binodal surface, the LG phase transition can be of second order, i.e., the entropy continues but the second order derivatives of chemical potential (for example, capacity) are discontinuous. An asymmetric nuclear matter has two components of proton and neutron, and two conserved charges of baryon number and the third component of isospin, will undergo a continuous second order phase transition.

Obviously, because of charge independence, the basic difference between proton and neutron be isospin. The isospin dependent interactions of nucleon-nucleon-isovector mesons play the key role to address the LG phase transition. As was pointed out by our previous papers [2, 3, 4], if one employed the isospin independent model, for example, Welacka model [5] or Zimanyi-Moszkowski model [6], to investigate asymmetric nuclear matter, a lot of difficulties, e.g. Coulomb instability and negative asymmetric parameter in the vapor phase, will emerge. To overcome these difficulties, the isospin vector ρ-meson must be introduced. It can be shown that the chemical potentials of proton and neutron may depend on the third component of isospin when NNρ interaction exists. A model without isospin vector ρ-meson, or even if it has ρ-meson, but the chemical potentials of the proton or neutron are still independent of the NNρ interaction because the third component I_3 of isospin be zero such as in a symmetric nuclear matter, the LG phase transition is still of first order.

In fact, the chemical potentials of proton and neutron not only depend on I_3 but also on the effective NNρ coupling g_ρ. Then the effective coupling g_ρ is also essential for studying the LG phase transition because the chemical potentials determine the binodal

CP597, *Nonequilibrium and Nonlinear Dynamics in Nuclear and Other Finite Systems*,
edited by Z. Li et al.

surface directly. It can been shown [7, 8, 9] that the effective couplings depend on the density and temperature in a hot and dense nuclear matter. By using the Thermo Field Dynamics (TFD)[5] to calculate the three-lines vertices Feynman diagrams of NNρ interactions, we found that the effective coupling of g_ρ decreases as the nucleon density increases. In an asymmetric nuclear matter, one can easily prove that the chemical potentials of nucleons have a term which is proportional to g_ρ^2 and I_3. This term has opposite signs for proton and neutron due to their different third component of isospin. Obviously, if g_ρ depends on density, this term will change and then the chemical potentials of proton and neutron, as well as the binodal surface of LG phase transition will also be changed. The objective of this paper is to investigate the effect of the density dependence of g_ρ on LG phase transition. We employ a model suggested by Furnstahl, Serot and Tang (FST) [11, 12, 13, 14] recently to study this problem. This model is an extension of quantum hadrodynamics and has been proven to be successful to explain many experimental properties of both nuclear matter and the finite nuclei in mean field approximation. The Lagrangian density of FST model under mean field approximation is

$$
L_{MFT} = \overline{\Psi}\left[i\gamma^\mu\partial_\mu - (M - g_s\phi_0) - g_v\gamma^0 V_0 - \frac{1}{2}g_\rho\tau_3\gamma^0 b_0\right]\Psi \tag{1}
$$
$$
+\frac{1}{2}m_v^2 V_0^2\left(1+\eta\frac{\phi_0}{S_0}\right) + \frac{1}{4!}\zeta(g_v V_0)^4 + \frac{1}{2}m_\rho^2 b_0^2
$$
$$
-H_q\left(1-\frac{\phi_0}{S_0}\right)^{4/d}\left[\frac{1}{d}ln\left(1-\frac{\phi_0}{S_0}\right)-\frac{1}{4}\right]
$$

where g_s, g_v g_ρ are, respectively, the couplings of light scalar meson σ, vector meson ω and isovector meson ρ fields to the nucleon, ϕ_0, V_0, b_0 are the expectation values $\phi_0 \equiv <\phi>$, $<V_\mu> \equiv \delta_{\mu 0}V_0$, $<b_{\mu 3}> \equiv \delta_{\mu 0}b_0$. The scalar fluctuation field ϕ is related to S by $S(x) = S_0 - \phi(x)$ and H_q is given by $m_s^2 = 4H_q/(d^2 S_0^2)$, d the scalar dimension. By using the standard technique of statistical mechanics, we get the thermodynamic potential Ω as[15]

$$
\Omega = V\left\{H_q\left[\left(1-\frac{\phi_0}{S_0}\right)^{4/d}\left(\frac{1}{d}ln\left(1-\frac{\phi_0}{S_0}\right)-\frac{1}{4}\right)+\frac{1}{4}\right]\right. \tag{2}
$$
$$
-\frac{1}{2}m_\rho^2 b_0^2 - \frac{1}{2}\left(1+\eta\frac{\phi_0}{S_0}\right)m_v^2 V_0^2 - \frac{1}{4!}\zeta(g_v V_0)^4\right\}
$$
$$
-2k_B T\left[\sum_{k,\tau}ln\left(1+e^{-\beta(E^*(k)-v_\tau)}\right)+\sum_{k,\tau}ln\left(1+e^{-\beta(E^*(k)+v_\tau)}\right)\right]
$$

where $\beta = 1/k_B T$ and the quantity v_i $(i = n, p)$ is related to the usual chemical potential μ_i by the equations

$$
v_n = \mu_n - g_v V_0 + \frac{g_\rho^2 \rho_3}{4m_\rho^2} \tag{3}
$$

$$\nu_p = \mu_p - g_v V_0 - \frac{g_\rho^2 \rho_3}{4m_\rho^2} \tag{4}$$

where $\rho_3 = \rho_p - \rho_n$ and the third component of isospin $I_3 = (N_p - N_n)/2 = V\rho_3/2$. The third term of the right hand side of Eq(3) and Eq(4) depends on ρ_3 and g_ρ^2. They have opposite signs and play the essential role to determine the LG phase transition.

Having obtained the thermodynamic potential, all other thermodynamical quantities, for example, pressure $p = -\Omega/V$, can be calculated. The two-phase coexistence equations are

$$\mu_i^L \left(T, \rho_i^L \right) = \mu_i^V \left(T, \rho_i^V \right) \tag{5}$$

$$p^L \left(T, \rho_i^L \right) = p^V \left(T, \rho_i^V \right) \tag{6}$$

where subscripts of one phase L and V stand for liquid and vapor, respectively. The stability conditions are given by [1]

$$\rho \left(\frac{\partial p}{\partial \rho} \right)_{T,\alpha} = \rho^2 \left(\frac{\partial^2 \mathcal{F}}{\partial \rho^2} \right)_{T,\alpha} > 0 \tag{7}$$

$$\left(\frac{\partial \mu_p}{\partial \alpha} \right)_{T,p} < 0 \text{ or } \left(\frac{\partial \mu_n}{\partial \alpha} \right)_{T,p} > 0 \tag{8}$$

where $\mathcal{F}$ is the density of free energy, $\alpha = (\rho_n - \rho_p)/\rho$ the asymmetric parameter, and $\rho = \rho_n + \rho_p$.

The numerical calculatioins have been done by adopting the parameters set T 1 of FST model [11, 12, 13] The parameters of set T1 are

$$\begin{aligned} g_s^2 &= 99.3, \quad g_v^2 = 154.5, \quad g_\rho^2 = 70.2 \\ m_s &= 509 MeV, \quad S_0 = 90.6 MeV \\ \zeta &= 0.0402, \quad \eta = -0.496, \quad d = 2.70 \end{aligned} \tag{9}$$

Our results for $g_\rho = (70.2)^{1/2}$=constant are shown in Fig.1 and Fig.2 by solid curves. The Gibbs conditions (5) and (6) for phase equilibrium demand equal pressures and chemical potentials for two phase with different concentrations. The collection of all such pairs $\alpha_1 (T, p)$ and $\alpha_2 (T, p)$ form the binodal suface. In Fig.1, the chemical isobar vs. α curves at fixed temperature T=10MeV and p=0.100MeV$(fm)^{-3}$ are labeled by A and A' for neutron and proton respectively. The two desired solutions form the edges of a rectangle and can be found by means of the geometrical construction shown in Fig.1 [1]. The critical curves with T=10MeV and $p_{crit} = 0.165$MeV$(fm)^{-3}$ are shown in Fig.1 by B and B' where the chemical potential curve arrive at a inflection point and the rectangle is degenerate to a line vertical to the α axis. The behaviour of the nuclear matter under isothermal compression, or in other words, the section of binodal surface at finite temperature T=10MeV are shown in Fig.2. The physical behaviour of this processes has been discussed by ref.[1]. Assume that the system is initially prepared with $\alpha = 0.6$ (gas), during the compression, the two-phase region is encountered at point A, and the liquid phase emerges at point B. The gas phase evolves from A to D, while

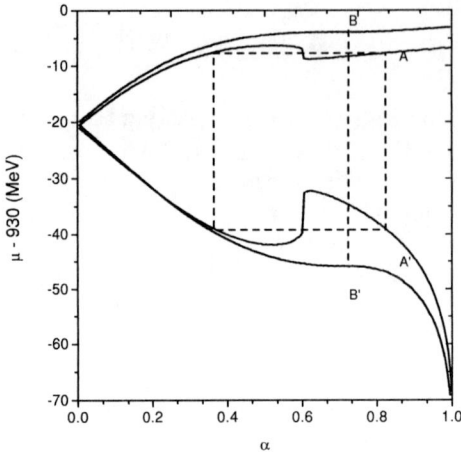

FIGURE 1. The chemical isobar as a function of α at fixed temperature T=10MeV and the geometrical construction used to obtain the asymmetry parameters in the two coexisting phases. B, B' correspond to the critical curves.

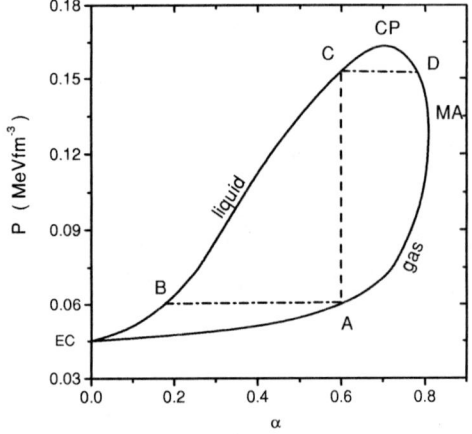

FIGURE 2. The section of binodal surface at T=10MeV, where CP, EC, and MA stand for critical point, equal concentratioin and maximal asymmetry respectively.

the liquid phase evolves from B to C. The system leaves the region of instability at point C, while the original gas phase is about to disappear. The critical point (CP), the point of equal concentration (EC) and the maximal asymmetry (MA) are indicated in Fig.2.

Now we are in a position to extend our discussion to the case of g_ρ density dependence. In fact, the effective masses of nucleons, effective masses or screening masses of mesons, and the effective coupings of NN-mesons are all dependent on density and temperature. By using the TFD to calculate the three-lines vertex diagrams for effective

NNρ couplings, we get [7, 8, 9, 15, 16, 17]

$$\frac{\left(-\frac{1}{2}\right)^3 g_\rho^3}{(2\pi)^4} \int d^4k \gamma^\mu \Delta\left(p'-k\right) \gamma^\sigma \Delta\left(p-k\right) \gamma_\mu D\left(k\right) \left(\tau^j \tau_i \tau^j\right) \tag{10}$$

where

$$\Delta(k) = (\gamma \cdot k + m_N) \left[\frac{1}{k^2 - m_N^2 + i\varepsilon} + \frac{i\pi}{E_N(k)} \delta(k_0 - E_N(k)) n_k \right. \tag{11}$$

$$\left. + \frac{i\pi}{E_N(k)} \delta(k_0 + E_N(k)) \bar{n}_k \right]$$

$$D(k) = \frac{1}{k^2 - m_\rho^2 + i\varepsilon} - 2\pi i n_\rho(k) \delta\left(k^2 - m_\rho^2\right) \tag{12}$$

with

$$E_N(k) = \sqrt{m_N^2 + k^2} \tag{13}$$

$$\omega(k) = \sqrt{m_\rho^2 + k^2}$$

$$n_k = \frac{1}{\exp\left[\beta\left(E_N(k) + \mu\right)\right] + 1}$$

$$\bar{n}_k = \frac{1}{\exp\left[\beta\left(E_N(k) - \mu\right)\right] + 1}$$

$$n_\rho = \frac{1}{\exp\left[\beta\left(\omega(k)\right)\right] - 1}$$

from Fig.3. The dependence of effective coupling $g_\rho(\beta, \rho)$ on temperature and density reads

$$g_\rho(\beta, \rho) = g_\rho^0 \left(1 + \Lambda_\rho\right) \tag{14}$$

where

$$\Lambda_\rho = \Lambda_\rho^1 + \Lambda_\rho^2 \tag{15}$$

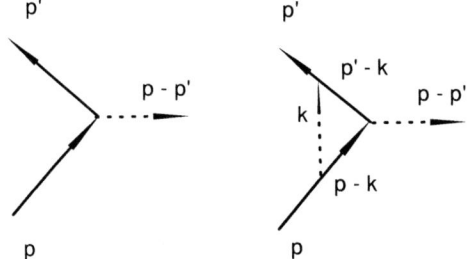

FIGURE 3.   The three line vertice for NNρ coupling.

$$\Lambda_\rho^1 = \frac{g_\rho^2}{8(2\pi^2)} \int d|\vec{k}| |\vec{k}|^2 \left[\frac{4m_N^2 + 2m_\rho^2 - 4E_\rho^2 - 4m_N E_\rho}{\left(m_\rho^2 - 2m_N E_\rho\right)^2} \right. \tag{16}$$

$$\left. + \frac{4m_N^2 + 2m_\rho^2 - 4E_\rho^2 + 4m_N E_\rho}{\left(m_\rho^2 + 2m_N E_\rho\right)^2} \right] \frac{n_B(E_\rho)}{E_\rho}$$

$$\Lambda_\rho^2 = \frac{g_\rho^2}{8(2\pi^2)} \int d|\vec{k}| \left[\frac{8m_N E_N - 4E_N^2}{2\left(2m_N^2 - m_\rho^2 - 2m_N E_N\right)} n_F(E_N) \right. \tag{17}$$

$$\left. + \frac{-8m_N E_N - 4E_N^2}{2\left(2m_N^2 - m_\rho^2 + 2m_N E_N\right)} \overline{n_F}(E_N) \right] \frac{1}{E_N}$$

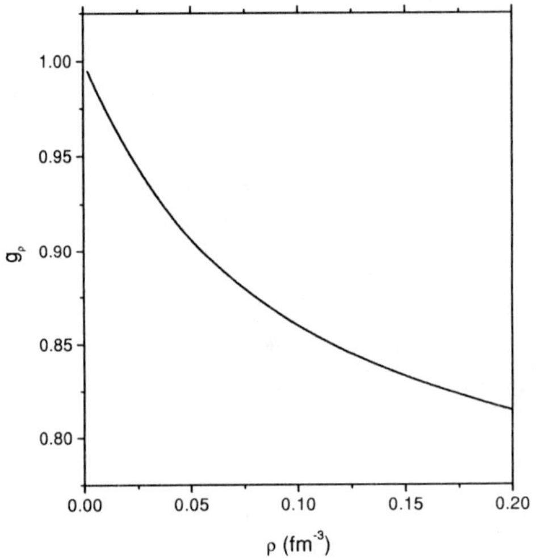

FIGURE 4. The polimonial fit for the coupling g_ρ as a function of ρ.

Our result for $g_\rho(\rho)$ is shown in Fig.4 where we fix the temperature T=10 MeV. The effect of density dependence of g_ρ on the LG phase transition are shown in Fig.5, Fig.6 and Fig.7. We see from Fig.5 and Fig.6 that the chemical potential of neutron μ_n increases rapidly with density. It passes through an inflection point and becomes monotoneous when pressure increases. But the shape of μ_p vs. α curves change slowly. Then we get a limit pressure p_{lim}, when $p > p_{lim}$, the rectangle cannot be found and the coexistenced equations have no solution. The last rectangle in the chemical isobar vs. α curves for T=10MeV, and p_{lim}=0.115MeV$(fm)^{-3}$ is shown in Fig.5 by dashed lines, where α_1=0.568 and α_3 =0.704 correpond to the maximum and the minimum of μ_n respectively. The pair α_1=0.568 and α_2=0.846 form the end of the binodal surface, as shown in Fig.7. The curve for T=10 MeV, but p=0.1402MeV$(fm)^{-3}$ $(p > p_{lim})$ is

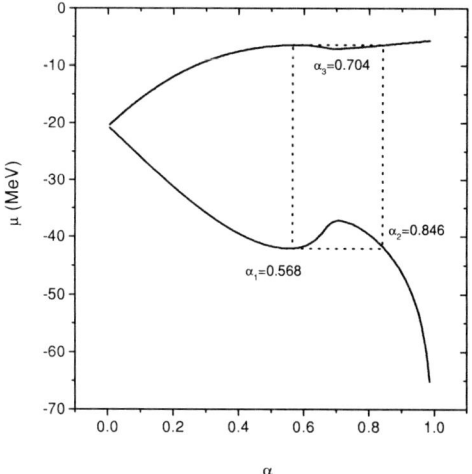

FIGURE 5. The chemical isobar as a function of α for T=10 MeV and p=0.115 MeV$(fm)^{-3}$.

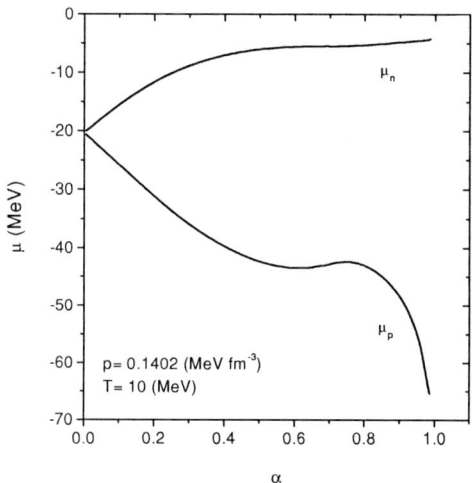

FIGURE 6. The chemical isobar as a function of α for T=10 MeV and p=0.1304 MeV$(fm)^{-3}$.

shown in Fig.6. We see that μ_n becomes monotoneous at this pressure, and no rectangle can be found.

The section of binodal surface for T=10MeV is shown in Fig.7. We see from Fig.7 that the curve will cut off at limit temperature p_{lim} clearly. The total asymmetric parameter α is divided into four regions, namely, $[0, \alpha_1]$, $[\alpha_1, \alpha_3]$, $[\alpha_3, \alpha_2]$ and $[\alpha_2, \alpha_{max}]$. The physical behaviour of isothermal compression in different regions are different. The results are

(1) If the system is initially prepared with $0 < \alpha < \alpha_1$, the precess of isothermal

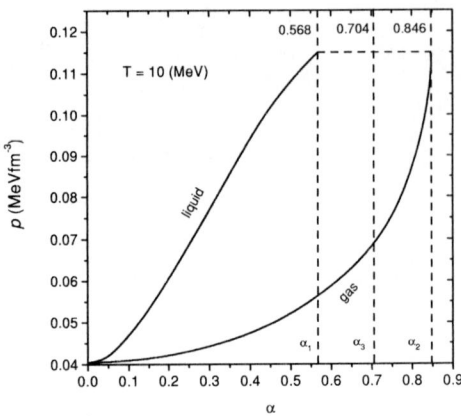

FIGURE 7. The section of binodal surface for T=10MeV. α_1, α_2 correspond to the maximum and minimum of μ_n respectively.

compression is similar to that of the case with constant g_ρ. It begin at gas phase, suffers a second order LG phase transition and ends at liquid phase.

(2) If the initial α is located at the region $\alpha_2 < \alpha < \alpha_{max}$, the system enters and leaves the two-phase region on the same branch, so the system remain in the same gase phase. As was pointed out by Müller and Serot [1], this retrograde condensation is unique to the binary system and dose not occur in one-component systems.

(3) Suppose that the initial α is located at the region $\alpha_1 < \alpha < \alpha_3$. Since α_1 and α_3 correspond to the maximum and minimum of μ_n, we find $(\partial\mu_n/\partial\alpha)_{T,p} < 0$ in this region and the stability condition Eq.(8) will be destroied. The system begins at gas phase, enters a two-phase region and becomes instable at the limit pressure.

(4) If the initial α is located at the region $\alpha_3 < \alpha < \alpha_2$, the behaviou of the system is similar to that of the case (3), except it will be ended to a stable phase at the limit pressure because the stability condition $(\partial\mu_n/\partial\alpha)_{T,p} > 0$ is satisfied.

In summary, we have shown that the density dependence of effective NNρ coupling is important for the LG phase transition. A limit pressure p_{lim} has been found for a fixed temperature and the LG phase transition cannot take place in asymmetric nuclear matter if $p > p_{lim}$. Of course, for a fixed pressure, we can also get a limit temperature. This conclusion is similar to that of the Coulomb instability [2, 3, 4] of nuclei. The basic difference is that instead of finite nuclei, our conclusion has be found to be suitable for asymmetric nuclear matter. Finally, we would like to emphasize that the isospin effect is very important for the LG phase transition of asymmetric nuclear matter.

This work was supported in part by NNSF of China and the Foundation of Education Ministry of China.

REFERENCES

1. H.Müller and B.D. Serot, Phys. Rev. C52 2072 (1995)

2. H.Q. Song, Z.X. Qian and R.K. Su, Phys. Rev. C47 2001(1993)
3. H.Q. Song, Z.X. Qian and R.K. Su, Phys. Rev. C49 2924(1994)
4. P. Wang, Z.W. Chong, R.K. Su and P.K.N. Yu Phys. Rev. C59 928(1999)
5. B.D. Serot and J.D. Walecka, Advances in Nuclear Physics, edited by J.W.Negels and E.Vogt Plenam, New York, 1986), Vol. 16. P.1
6. J. Zimanyi and J.A. Moszkowski, Phys. Rev. C42 1416 (1990)
7. S. Gao, Y.J. Zhang and R.K. Su, Nucl. Phys. A593 362 (1995)
8. Z.X. Qian, C.G. Su and R.K. Su, Phys. Rev. C47 877 (1993)
9. R.K. Su, G.T. Zheng and G.G. Siu, J.Phys. G19 79 (1993)
10. H. Umezawa, H. Matsumoto and M. Tachiki, Thermo Field Dynamics and Condensed States (North-Holland, Amsterdam, 1982)
11. R.J. Furnstahl, H.B. Tang and B.D. Serot, Phys. Rev. C52 1368 (1995)
12. R.J. Furnstahl, B.D. Serot and H.B. Tang, Nucl. Phys. A598 539 (1996)
13. R.J. Furnstahl, B.D. Serot and H.B. Tang, Nucl. Phys. A615 441 (1997)
14. B.D. Serot and J.D. Walecka, Int. Jour. of Mod. Phys. E6 515 (1997)
15. L.L. Zhang, H.Q. Song, P. Wang and R.K. Su, Phys. Rev. C59 3292 (1999)
16. Y.J. Zhang, S. Gao and R.K. Su, Phys. Rev. C56 3336 (1997)
17. S. Gao, Y. J. Zhang and R.K. Su Phys. Rev. C53 1098 (1996)

Strange hadronic matter in the FST model

H.Q. Song[1,2],L.L.Zhang[2],P.Wang[3],R.K.Su[1,3]

1.CCAST(World Laboratory), P.O.Box 8730, Beijing 100080, China
2. Shanghai Institute of Nuclear Research, Chinese Academy of Sciences
P.O.Box 800204, Shanghai 201800, China
3. Department of Physics, Fudan University, Shanghai 200433, China

Abstract. The strange hadronic matter with nucleon, Λ-hyperon and Ξ-hyperon is studied by using the FST model in a mean-field approximation. The saturation properties and stabilities of the strange hadronic matter are discussed.

INTRODUCTION

Strangeness opens a new dimension to nuclear structure. In recent years, exploring nuclear system with strangeness has received increasing interest. Such system has many astrophysical and cosmological implications and is indeed interesting by itself. For instance, the core of neutron stars may contain a high fraction of hyperons. There are two kinds of strange matter: strange quark matter and strange hadronic matter. On one hand, it has been speculated that states of quark matter ("strangelets") with large strangeness per baryon might be more stable than the normal nuclei. On the other hand, strange hadronic matter or hypernuclei have also been investigated. In this case, the strange quarks are localized within individual hyperons, which are assumed to retain their identity in the bound system. In this work, we will focus our attention on the properties of infinite strange hadronic matter. For a review about strange matter in general, one can see Ref.[1].

From a theoretical point of view, the discussions rely on the models at our disposal for many-body system with strong interactions. A discussion by Ikeda, Bando, and Motoba[2] was based on the Brueckner theory with different Nijmegen potentials. The results arising from these different potentials vary too much to give a definite conclusion. On the other hand, relativistic mean-field theories appear to yield reasonable answers within a restricted parameter space. Barranco et al.[3]adopted a derivative coupling model (known as the ZM model) to discuss the saturation properties of multi-lambda matter. The results suggest that multi-lambda systems are stable against particle emission. In a recent paper, Schulze et al.[4] studied the properties of multi-lambda nuclear matter by using an extended BHF formalism with the Paris nucleon-nucleon potential and the latest version of the Nijmegen soft-core hyperon-nucleon potential. In particular, they evaluated the maximum strangeness content reserving binding. Vidana et al.[5] developed a formalism for microscopic Brueckner-type calculations of dense nuclear matter with strangeness, allowing for

CP597, Nonequilibrium and Nonlinear Dynamics in Nuclear and Other Finite Systems,
edited by Z. Li et al.

any concentration of the different baryon species. The properties of the strange hadronic matter have also been studied by Zhang et al.[6]using the quark-meson coupling (QMC) model and by Wang et al.[7] using the modified QMC model.

It is believed that the descriptions of nuclear matter and finite nuclei are ultimately governed by the physics of low-energy quantum chromodynamics(QCD). In the absence of direct derivations from QCD, the effective hadronic models should be constrained by the underlying symmetries of QCD. Furnstahl, Serot and Tang worked out a relativistic hadronic model[8](refereed as the FST model), which incorporates nonlinear chiral symmetry, broken scale invariance and the phenomenology of vector dominance. This model has successfully reproduced the properties of nuclear matter, finite nuclei and Coulomb instability of hot nuclei[9]. It is therefore interest to use this model to study the strange hadronic matter. As proposed by Schaffner et al.[10], we will also include σ^* and ϕ mesons in the model to describe the strong interaction between hyperons. In order to keep the chiral symmetry exactly, one has to extend the chiral transformation of the hadron fields from SU(2) to SU(3). The correct chiral treatment has been explored by Parazoglou and collaborators in a series of papers[11-13]. They proposed a chiral Lagrangian based on a linear realization of chiral SU(3) symmetry[12]. However, it has been shown that the central potentials of the hyperons come out too large. More recently, they constructed a chiral SU(3) model[13] based on a nonlinear realization of chiral symmetry, which overcomes the difficulty mentioned above. They then used this model to study the properties of finite nuclei. Since there is much to do in order to apply this model to study the strange hadronic matter including hyperons, we will do so in a next work.

Considering reactions, $\Lambda + \Lambda \rightarrow \Xi^- + p, \Lambda + \Lambda \rightarrow \Xi^0 + n$ and their reverses, one has to consider also the mixture of the cascades Ξ^- and Ξ^0 in the strange matter, besides lambdas. For simplicity, we assume that Ξ^- and Ξ^0 will appear in the strange matter with equal amount. We will, therefore, use a single symbol Ξ for these particles. In order to compare present results with our previous work[7], we will discuss multi-lambda nuclear matter first. Then, we take the mixture of the Ξ hyperon into account under the condition of chemical equilibrium for reactions mentioned above. In this work, we will not consider the mixture of the Σ hyperons. The reason is twofold. First, the Σ potential in the nuclear matter is rather uncertainly predicted, ranging from completely unbound[14]to $U_\Sigma = -25 \pm 5$MeV[15]. As pointed out by Balberg et al.[16], systems involving Σ's together with nucleons or Λ's generally will be unstable with respect to the strong decays $\Sigma N \rightarrow \Lambda N$ or $\Sigma \Lambda \rightarrow \Xi N$. Secondly, the Q values for the strong transitions $\Sigma N \rightarrow \Lambda N$, $\Sigma \Sigma \rightarrow \Lambda \Lambda$, $\Sigma \Lambda \rightarrow \Xi N$ and $\Sigma \Xi \rightarrow \Lambda \Xi$ are about 78, 156, 50 and 80MeV, respectively[17]. To Pauli block these processes, we need a rather high density of Λ. On the other hand, the Q value of $\Xi N \rightarrow \Lambda \Lambda$ is only about 28 MeV. This process can be Pauli blocked by relatively low Λ density.

THE FST MODEL

The original FST model[8] is extended by including Λ and Ξ hyperons in the system and introducing a new hyperon-hyperon interaction mediated by two additional strange mesons σ^* and ϕ which couple only to hyperons. In the FST model, a new scaler field $S(x)=S_0-\sigma(x)$ is introduced instead of the usual sigma meson, where $\sigma(x)$ is the fluctuation of $S(x)$ related to a equiriblium value S_0. There is no contribution from the π meson to a unpolarized system. In a mean field approximation, the Lagarangian density reads:

$$L_{MFT} = \sum_j \overline{\psi}_j (i\gamma^\mu \partial_\mu - g_{\omega j}\gamma^0 V_0 - g_{\phi j}\gamma^0\phi_0 - M_j + g_{Sj}\sigma_0 + g_{\sigma^* j}\sigma_0^*)\psi_j$$
$$+ \frac{1}{2}\left(1+\eta\frac{\sigma_0}{S_0}\right)m_\omega^2 V_0^2 + \frac{1}{4!}\varsigma(g_{\omega N}^2 V_0)^2 + \frac{1}{2}m_\phi^2\phi_0^2 + \frac{1}{2}m_\sigma^2.\sigma_0^{*2} - H_q\left(1-\frac{\sigma_0}{S_0}\right)^{4/d}\left[\frac{1}{d}\ln\left(1-\frac{\sigma_0}{S_0}\right)-\frac{1}{4}\right] \quad (1)$$

where g_{ij} are the coupling constants of the baryons to the meson fields. H_q is linked to the mass of the light scalar S by the relation $m_S^2 = 4H_q/(d^2 S_0^2)$.

By using the standard technique, we obtained the average energy density ε as

$$\varepsilon = \sum_i \frac{\gamma_i}{(2\pi)^3}\int_0^{k_{Fi}} d^3 k E_i^*(k) + g_{\omega N}V_0\rho_{BN} + (g_{\omega\Lambda}V_0 + g_{\phi\Lambda}\phi_0)\rho_{B\Lambda} + (g_{\omega\Xi}V_0 + g_{\phi\Xi}\phi_0)\rho_{B\Xi}$$
$$+ H_q\left\{\left(1-\frac{\sigma_0}{S_0}\right)^{\frac{4}{d}}\left[\frac{1}{d}\ln\left(1-\frac{\sigma_0}{S_0}\right)-\frac{1}{4}\right]+\frac{1}{4}\right\} - \frac{1}{2}\left(1+\eta\frac{\sigma_0}{S_0}\right)m_\omega^2 V_0^2 - \frac{1}{4!}\varsigma g_{\omega N}^4 V_0^4$$
$$- \frac{1}{2}m_\phi^2\phi_0^2 + \frac{1}{2}m_\sigma^2.\sigma_0^{*2} \quad (2)$$

where the Fermi momentum k_{Fi} is related to the baryon density by $\rho_{Bi} = \gamma_i k_{Fi}^3/6\pi$. The spin-isospin degeneracy $\gamma_i=4$ for nucleons and Ξ hyperons and $\gamma_i=2$ for Λ hyperons. And $E_i^*(k) = \sqrt{M_i^{*2}+k^2}$ with $M_i^* = M_i - g_{si}\sigma_0 - g_{\sigma^*i}\sigma_0^*$ (i= Λ, Ξ), $M_i^* = M_i - g_{si}\sigma_0$ (i=N) being the effective baryon mass. Considering the reactions, $\Lambda + \Lambda \rightarrow \Xi^- + p$, $\Lambda + \Lambda \rightarrow \Xi^0 + n$ and their reverses, one should use a chemical equilibriun condition: $2\nu_\Lambda - \nu_N - \nu_\Xi = 0$, where $\nu_i = \sqrt{k_{Fi}^2 + M_i^{*2}}$. We also define a strangeness fraction : $f_S \equiv (\rho_{B\Lambda} + 2\rho_{B\Xi})/\rho_B$, where $\rho_B = \rho_{BN} + \rho_{B\Lambda} + \rho_{B\Xi}$. Given ρ_B and f_s, we determine ρ_{BN}, $\rho_{B\Lambda}$ and $\rho_{B\Xi}$ by the above three equations.

THE PARAMETERS

Now we discuss the parameters. The couplings between nucleons and the ordinary mesons are taken from Ref.[8], where the parameters are determined by fitting to finite nuclei. In this work, the parameter sets T1 and T3 in Ref.[8] are used and quoted in Table 1. Since the σ^* and ϕ mesons do not couple to the nucleon, the coupling constants of nucleon are completely fixed. But we still have eight coupling constants to be determined, namely $g_{\omega\Lambda}$, $g_{\omega\Xi}$, $g_{S\Lambda}$, $g_{S\Xi}$, $g_{\sigma^*\Lambda}$, $g_{\sigma^*\Xi}$, $g_{\phi\Lambda}$ and $g_{\phi\Xi}$. As for the determination of the coupling constants $g_{S\Lambda}$ and $g_{\omega\Lambda}$, there is some uncertainty. Rufa et al.[18] pointed out that the spectroscopic data on lambda nuclei can not determine the coupling constants unambiguously. As done in Ref.[3], in order to fix these two parameters $g_{S\Lambda}$ and $g_{\omega\Lambda}$, we have used only one physical constraint, namely, the energy E_Λ of one single lambda in symmetric nuclear matter at saturation. According to the analysis of Bouyssy[19] and of Hausmann and Weise[20], we have taken E_Λ= -28 MeV. So, there is one free parameter left. By using the OZI rule[21], we set $g_{\omega\Lambda}/g_{\omega N} = 2/3$ and then the parameters $g_{S\Lambda}$ is fixed. Similarly, we have used the energy of one single Ξ in symmetric nuclear matter, E_Ξ=-18MeV[22] and set $g_{\omega\Xi}/g_{\omega N} = 1/3$ to determine the coupling constant $g_{S\Xi}$. For the ϕ meson, we use the quark model relationships[10] $g_{\phi\Xi} = 2g_{\phi\Lambda} = -2\sqrt{2}\,g_{\omega N}/3$. We treat the couplings of σ^* phenomenologically so as to satisfy potential depths $U_\Lambda^{(\Xi)} = U_\Xi^\Xi = 40\,\text{MeV}$ of a Λ or Ξ in a Ξ "bath" with $\rho_\Xi \approx \rho_0$. The resultant parameters are also listed in Table 1. The bare masses of baryons and mesons are M_N=939MeV, M_Λ=1116MeV, M_Ξ=1318.1MeV, m_ω=783MeV, m_{σ^*}=975MeV and m_ϕ=1020MeV.

Table 1. Parameter sets, with the values for S_0 and scalar mass m_S, in MeV

Set	g_{SN}^2	m_S MeV	$g_{\omega N}^2$	S_0 MeV	ζ	η	d	$g_{S\Lambda}^2$	$g_{S\Xi}^2$	$g_{\sigma^*\Lambda}^2$	$g_{\sigma^*\Xi}^2$
T1	99.3	509	154.5	90.6	0.0402	–0.496	2.70	37.32	9.99	48.31	154.62
T3	109.5	508	178.6	89.8	0.0346	–0.160	3.50	41.51	11.12	53.87	175.24

RESULTS AND DISCUSSIONS

Now we discuss the properties of the strange hadronic matter. Following the usual way, we subtract the fermion masses in the total energy of the strange matter given by Eq.(2) and study the binding energy per baryon expressed as:

$$E/B = \varepsilon/\rho_B - M_N(1 - Y_\Lambda - Y_\Xi) - M_\Lambda Y_\Lambda - M_\Xi Y_\Xi \qquad (3)$$

where $Y_\Lambda = \rho_{B\Lambda}/\rho_B$ and $Y_\Xi = \rho_{B\Xi}/\rho_B$ are the hyperon fractions in the matter. In order to compare the present results with those in other models, we study the system of symmetric nuclear matter plus lambda with a ratio $f_s = Y_\Lambda$ first. In Fig.1, we have plotted for different f_s values the energy per baryon versus the baryon density with parameter set T1. The most notable feature is that the saturation curves become deeper first and then shallower. The lowest minimum occurs aroud $Y_\Lambda = 0.1$. This result is quite similar to those in the ZM [3], the usual QMC[6] and the modified QMC[7] models. Compared to the ordinary symmetric nuclear matter, the matter with $Y_\Lambda = 0.1$ has additional binding energy of about 0.6MeV. It means that this system is more stable than the usual nuclear matter. This gain in binding energy is also comparable with those in the ZM model, the usual QMC model, the modified QMC model or the

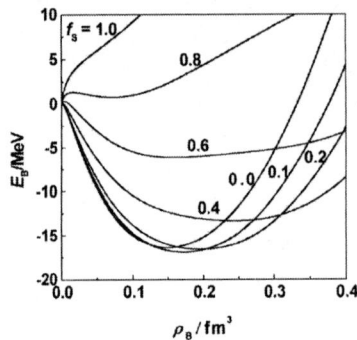

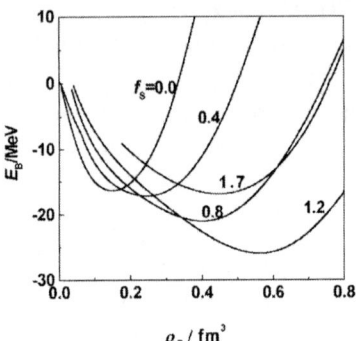

FIGURE 1. Energy per baryon versus the baryon density in the nucleon-lambda mixture with various values of Y_Λ, calculated with the coupling set T1.

FIGURE 2. The same as in Fig.1, but in the nucleon-lambda-cascade mixture.

BHF calculation given by Schulze et al.[4] or by Vidana et al.[5], but much less than the 4 MeV obtained by Ikeda et al.[2] using D version of the Nijmegen potential. Another feature of significance is that with the increase of Y_Λ,the saturation density increases from 0.148 fm^{-3} for ordinary nuclear matter to a maximum value of about 0.24fm^{-3}, which is reached for $Y_\Lambda \approx 0.4$. From this value on, the saturation point moves downward in density, disappearing at $Y_\Lambda \approx 0.8$. The E/B curve for a fixed value of Y_Λ has a negative minimum at least up to $Y_\Lambda =0.75$. It means that the system containing up to 75% of lambdas is still stable against particle emission.

Now we study the saturation properties of a strange hadronic matter system containing nucleons, Λ and Ξ hyperons. We show in Fig.2 the binding energy per baryon versus baryon density ρ_B at various strange fraction f_S, calculated with coupling set T1. The gross trend of the curves does not change in comparison with those in the system without the Ξ hyperons. But in this case the system gets its maximum binding energy around $f_S \approx 1.2$ and has an additional binding energy of about 9.0 MeV. The inclusion of Ξ largely increases the binding energy of the system and thus the system is more stable. This is caused by the strong attraction between the cascades. Furthermore, the density corresponding to the minimum point shifts from $\rho_B = 0.17$ fm^{-3} in the N-Λ matter to $\rho_B =0.56$ fm^{-3}. Up to $f_S = 1.8$, there still exits a negative minimum on the E/B-ρ_B curve for each f_S value. This means that with Ξ hyperons included in, the strange hadronic matter can possess a larger strangeness.

To see the stability of the system against f_S, we minimize the E/B with respect to ρ_B at each strangeness fraction f_S. As a function of the strangeness fraction f_S, we present in Fig.3(a) the minimized E/B, in Fig.3(b) the corresponding baryon density ρ_B and in Fig.3(c) the corresponding fractions of the lambda Y_Λ and of the cascade Y_Ξ, calculated with parameter sets T1 and T3.   As a comparison, we also show in the same figures the results for the nucleon-lambda matter. We examine the results with set T1 first. One can easily learn that the minimum energy per baryon shifts from

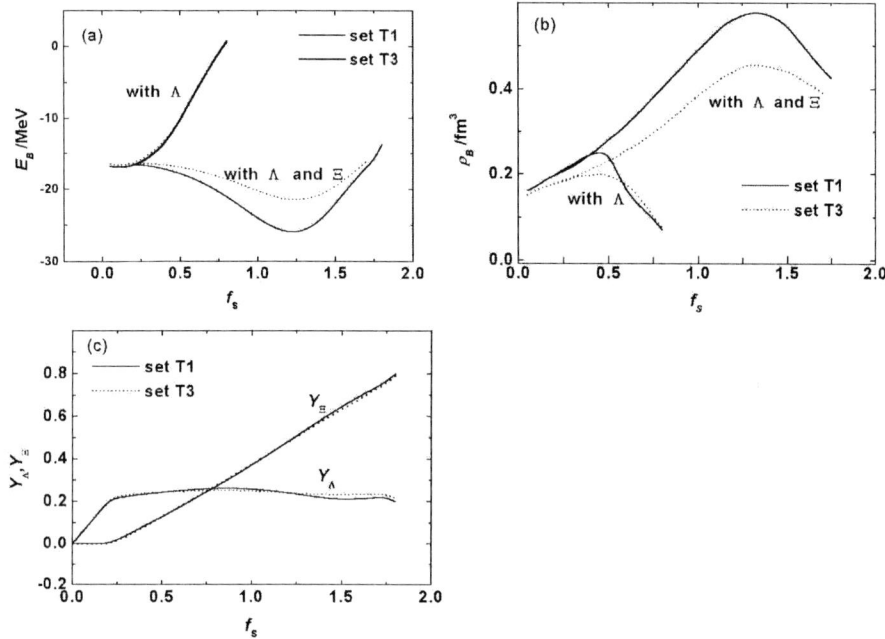

FIGURE 3. (a) The minimized energy per baryon in the N- Λ - Ξ mixture and in the N- Λ mixture as a function of strangeness fraction f_S, calculated with the coupling set T1 and T3. (b) The corresponding baryon densities. (c) The corresponding lambda fraction Y_Λ and cascade fraction Y_Ξ in the N- Λ - Ξ mixture.

$(E/B, f_s) \approx (-16.9 MeV, 0.1)$ for the nucleon-lambda matter to $(E/B, f_s) \approx (-26.0 MeV, 1.23)$ for the nucleon-lambda-cascade matter. Compared to the nucleon-lambda mixture, system gets an additional binding energy per baryon of about 9MeV. The baryon density corresponding to the minimum energy is about 0.56fm^{-3}(see Fig.3(b)). From Fig.3(c), one can find that in the low f_S region (say f_S<0.5), Y_Λ is larger than Y_Ξ. This means that in this region the system is in favour of containing more lambdas than cascades. Above $f_S \approx 0.20, Y_\Xi$ increases linearly, crossing the Y_Λ curve around f_S=0.75 and dominating in the large f_S region. We can learn from Fig.3 that there may be a stable state with a quite large strangeness of $f_S \approx 1.23$ and a quite high baryon density of $\rho_B \approx 0.56$fm^{-3}, where the component of the cascade dominates and even exceeds the component of the nucleon. This result is quite similar to that given by Schaffner et al.[10] for heavy strange baryonic system and that given in the MQMC model[7]. In Fig.3(c),we found that Y_Ξ vanishes for $f_S \leq 0.2$, where there is no positive solution for Y_Ξ under the condition of chemical equilibrium. We thus conclude that Λ and Ξ can not coexist in this situation and the system have nucleons and Λ hyperons only. The results with set T3 are qualitatively similar but quantitatively quite different to those with set T1. The lowest minimum shifts to $(E/B, f_s, \rho_B) \approx (-21.5MeV, 1.25, 0.45 fm^{-3})$.

Finally, we will discuss the role of the strange mesons in the saturation properties of the strange hadronic matter. We show the binding energy per baryon against baryon density in Fig.4, calculated with coupling set T1 but without the strange mesons. One can find that the saturation curve becomes much shallower than the corresponding one in Fig.2, where the strange mesons are included in the calculations. The lowest minimum appears around $f_S=0.1$ instead of $f_S=1.2$ in Fig.2. This means that the strange mesons play an important role in mediating the interaction between the hyperons.

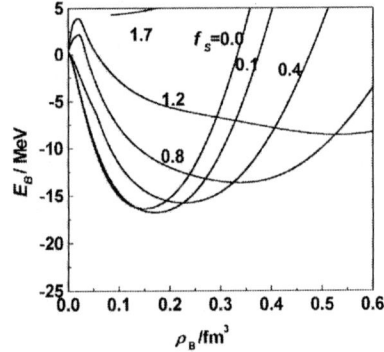

FIGURE 4. The same as Fig.2 but without the σ^* and ϕ mesons.

In this work, the excited baryons are not included. Since these excited baryons have larger masses, they contribute only in the very high density region at zero temperature. The second reason is that their coupling constants are not well known[23]. For simplicity, only the symmetric case for nucleons and for the Ξ hyperons are studied here. The charge asymmetry will reduce the binding energy of the system[5]. For the reason mentioned in the introduction, the Σ hyperons are not included in this work. The inclusion of the Σ hyperons will increase the binding energy[17]. In a complete calculation they have to be included. The work on this problem is in progress.

In summary, we have used the FST model to discuss the properties of strange hadronic matter. The result indicates a quite large strangeness fraction region where the matter is stable against particle emission. Although it is not the absolute ground state, it may be relevant for the discussion of high energy hyperon production experiments well above threshold. In the large f_S region, the Ξ component dominates, resulting in a deep minimum in the E/B-f_S curve, say for parameter set T1, with $(E/B, f_S) \approx (-26.0MeV, 1.23)$. This main feature remains qualitatively, although there is a quite large difference quantitatively in different coupling sets. The strange mesons play an important role in mediating the strong interaction between the hyperons.

ACKNOWLEDGMENTS

This work was supported in part by National Natural Science Fundation of China under No.10075071 and 19975010, Foundation of Education Department, Chinese Academy of Sciences under contract No.KJ951-A1-410 and the Major State Basic Research Development Programe in China under contract No.G200077400.

REFERENCES

1. Greiner, C., J. Phys. **G25** (1999)389.
2. Ikeda, K., Bando, H. and Motoba, T., Prog. Theor. Phys. Supplement **81**(1985)147.
3. Barranco, M., Lombard, R.J., Marcos, S. and Moszkowski, S.A., Phys. Phys. **C44** (1991)178.
4. Schulze, H.-J., Baldo, M., Lombardo, U., Cugnon, J. and Lejeune, A., Phys. Rev. **C57**(1998)704.
5. Vidana, I., Polls, A., Ramos, A., Hjorth-Jensen, M. And Stoks, V.G.J., Phys. Rev. **C61** (2000) 025802-1.
6. Zhang, L.L., Song, H.Q. and Su, R.K., J. Phys. **G23**(1997)557.
7. Wang, P., Su, R.K., Song, H.Q. and Zhang, L.L., Nucl. Phys. **A653**(1999)166.
8. Furnstahl, R.J., Tang, H.B. and Serot, Phys. Rev. **C52** (1995)1368.
   Furnstahl, R.J., Serot, B.D. and Tang, H.B., Nucl. Phys. **A598**(1996) 539.
   Furnstahl, R.J., Serot, B.D. and Tang, H.B., Nucl. Phys. **A615**(1997) 441.
9. Zhang, L.L., Song, H.Q., Su, R.K. and Wang, P., Phys. Rev. **C59**(1999)3292.
10. Schaffner, J., Dover, C.B., Gal, A., Greiner, C., Millener, D.J. and Stoecher, H., Ann. Phys. **235** (1994) 35.
11. Parazoglou, P., Zschiesche, D., Schramm, S., Stoecker, H. and Greiner, W., J.Phys. **G23**(1997) 2081.
12. Parazoglou, P., Schramm, S., Scharffner-Bielich, J., Stoecker, H. and Greiner, W., Phys. Rev. **C57** (1997)2756.
13. Parazoglou, P., Zschiesche, D., Schramm, S., Scharffner-Bielich, J., Stoecker, H. and W. Greiner, Phys. Rev. **C59** (1999)411.
14. Mares, J., Triedman, E., Gal, A. and Jennings, B.K., Nucl. Phys. **A594**(1995)311.
15. Dover, C.B., Millener, D.J. and Gal, A., Phys. Rep. **184**(1989)1.
16. Balberg, S., Gal, A. and Scharffner, J., Progr. Theor. Phys. Suppl. **117**(1994)325.
17. Stoks, V.G. and Lee, T.-S.H., Phys. Rev. **C60** (1999)024006-1.
18. Rufa, M., Schaffner, J., Maruhn, J., Stoecker, H., Greiner, W. and Reinhard, P.G., Phys. Rev. **C42** (1990)2469.
19. Bouyssy, A., Nucl. Phys. **A290**(1977)429.
20. Hausmann, R. and Weise, W., Nucl. Phys. **A491**(1989)601.
21. Dover, C.B. and Gal, A., Pro. Part. Nucl. Phys., ed. D.Wilkinson, Vol.12 (Pergamon, Oxford, 1984); Jinnings, B.K., Phys. Lett. **B246**(1990)325.
22. Fukuda, T., Higashi, A., Matsuyama, Y., Nagoshi, C., Nakano, J., Sekimoto, M., Tlusty, P., Ahn, J.K., En'yo, H., Funahashi, H. et al., Phys.Rev. **C58**(1998)1306.
23. Li, Zhuxia, Mao, Guangjun, Zhuo, Yizhong and Greiner, Walter, Phys.Rev. **C56** (1997)1570.

Energy and centrality dependences of charged multiplicity density in relativistic nuclear collisions

Ben-Hao Sa*,§,¶, A. Bonasera†, An Tai** and Dai-Mei Zhou‡

*China Institute of Atomic Energy, P.O.Box 275(18), Beijing 102413, China
†Laboratorio Nazionale del Sud, Instituto Nazionale Di Fisica Nucleare, v. S. Sofia 44 95132 Catania, Italy
**Department of Physics and Astronomy, University of California, at Los Angeles, Los Angeles, CA 90095 USA
‡Institute of Particle Physics, Huazhong Normal University, Wuhan, 430079 China
§CCAST (World Lab.), P. O. Box 8730 Beijing, 100080 China
¶Institute of Theoretical Physics, Academia Sinica, Beijing 100080 China

Abstract. Using a hadron and string cascade model, JPCIAE, the experimental data of charged particle pseudorapidity density at mid-rapidity in both the relativistic $p + \bar{p}$ collisions and the $Au + Au$ collisinos at $\sqrt{s_{nn}}$=130 GeV were reproduced fairly well without retuning the model parameters. The predictions for full RHIC energy $Au + Au$ collisions and for $Pb + Pb$ collisions at the LHC energy were given. Participant nucleon distributions were calculated based on different methods. It was found that the number of participant nucleons, $< N_{part} >$, is not a well defined variable both experimentally and theoretically. Thus, it is inappropriate to use charged particle pseudorapidity density per participant pair as a function of $< N_{part} >$ for distinguishing various theoretical models.

The main focus of the Relativistic Heavy-Ion Collider (RHIC) at Brookhaven National Laboratory (BNL) is to explore the phase transition related to the quark deconfinement and the chiral symmetry restoration. The first available experimental data were the energy dependence of charged particle pseudorapidity density in $Au + Au$ collisions at $\sqrt{s_{nn}}$=56 and 130 GeV from the PHOBOS collaboration [1]. After that, the PHENIX collaboration published their data of centrality dependence of the charged particle pseudorapidity density in $Au + Au$ collisions at $\sqrt{s_{nn}}$=130 GeV [2].

It has been predicted that the rare high charged multiplicity could indicate the onset of the Quark-Gluon-Plasma (QGP) phase, since the extra entropy in the QGP phase could manifest itself as a huge number of produced particles in the final state [3, 4, 5]. On the other hand, in [6] the centrality dependence of charged multiplicity has been proposed to provide information on the relative importance of soft versus hard processes in particle production and therefore provide a means of distinguishing various theoretical models for particle production.

The EKRT model (pQCD calculation with assumption of gluon saturation) [7], the HIJING model [6], the KN model (the conventional eikonal approach and the high density QCD) [8], and the Dual Parton Model [9] were used to investigate the centrality dependence. It was found that the experimental observation, the charged particle pseudo-

CP597, Nonequilibrium and Nonlinear Dynamics in Nuclear and Other Finite Systems,
edited by Z. Li et al.
© 2001 American Institute of Physics 0-7354-0041-5/01/$18.00

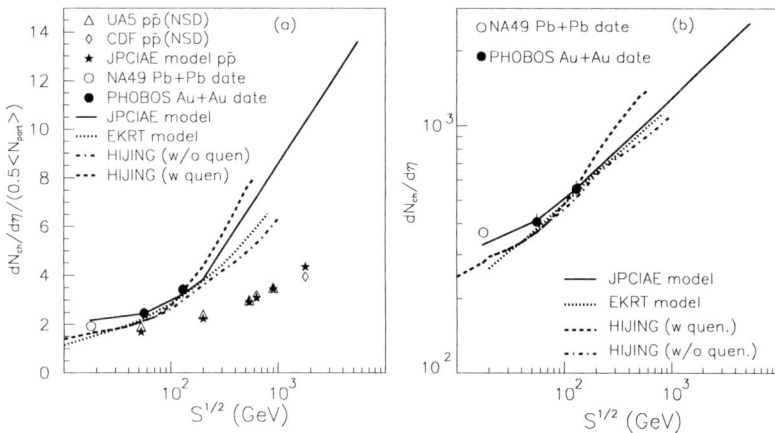

FIGURE 1. The energy dependence of the charged particle pseudorapidity density at mid-pseudorapidity in relativistic $p + \bar{p}$ and $A + A$ collisions.

rapidity density per participant pair slightly increasing with $< N_{part} >$, was reproduced by [6, 8, 9], but contradicted the results of [7].

In this paper a hadron and string cascade model, JPCIAE [10], was employed to study this issue further. Within the framework of this model the experimentally measured energy dependence of the charged particle mid-pseudorapidity density per participant pair both in relativistic $p + \bar{p}$ and $Au + Au$ collisions at RHIC was reproduced fairly well without retuning the model parameters. The predictions for the full RHIC energy $Au + Au$ collisions and for $Pb + Pb$ collisions at the LHC energy were also given. In studying centrality dependence the focus was put on the calculations of $< N_{part} >$, its definition and uncertainty. Both the PHENIX [2] and the PHOBOS [11] observations that the charged particle mid-pseudorapidity density per participant pair slightly increases with $< N_{part} >$ could be reproduced fairly well by JPCIAE. However, we indicated that it is not suitable to use the charged particle mid-pseudorapidity density per participant pair as a function of $< N_{part} >$ to constrain theoretical models for particle production, because $< N_{part} >$ is not a well defined variable both experimentally and theoretically.

The JPCIAE model was developed based on PYTHIA [12]. In the JPCIAE model the nucleons in a colliding nucleus are distributed randomly in the sphere of the nucleus with a radius of $1.12 A^{1/3}$ fm. Each nucleon is given a beam momentum in z direction and zero initial momentum in x and y directions. What followed is the conventional two-body scattering cascade processes untile the collision list is empty. It should be expressed here that for each collision pair of hadrons, if its CMS energy is larger than a given cut, we assume that strings are formed after the collision and PYTHIA is used to deal with particle production. Otherwise, the collision is treated as a two-body collision [13, 14]. The cut (=4 GeV in the program) was chosen by observing that JPCIAE correctly reproduces charged multiplicity distributions in AA collisions. JPCIAE model is not a simple superposition of nucleon-nucleon collisions since the rescatterings among participant, spectator nucleons and produced particles are taken into account. We refer

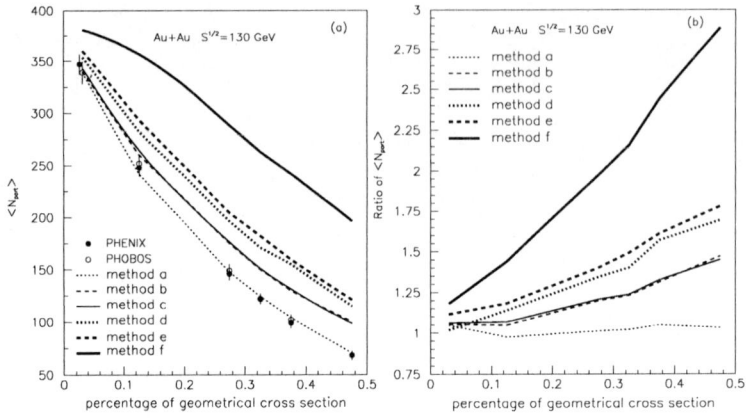

FIGURE 2. The number of participant nucleons $< N_{part} >$, (a), and the ratios of the different curves in (a) to the PHENIX data, (b), as a function of the percentage of geometrical cross section.

to [10] for more details of the JPCIAE model.

In Fig.1(a) the experimental data of charged particle pseudorapidity density per participant pair at mid-pseudorapidity in relativistic $p + \bar{p}$ (open triangles and rhombuses with error bar) and $A + A$ collisions (open circles with error bar for $Pb + Pb$ at SPS and full circles with error bar for $Au + Au$ at RHIC) [1] are compared with JPCIAE model (full stars for $p + \bar{p}$ and the solid curve for $A + A$ collisions), HIJING model (dotted-dash curve without jet quenching and dashed curve with jet quenching) [6] and EKRT model (dotted curve) [7]. The data of both $p + \bar{p}$ and $A + A$ collisions at relativistic energies were reproduced fairly well by JPCIAE model without retuning model parameters. Fig. 1 (b) is the same as (a) but the vertical coordinate here is the charged particle pseudorapidity density itself. The JPCIAE model predictions for full RHIC energy $Au + Au$ collisions and for $Pb + Pb$ collisions at the LHC energy in both panels supply a benchmark for QGP formation since the QGP phase is not included in the JPCIAE model.

Studying centrality dependence the focus was put on $< N_{part} >$. In the fixed target experiments the participant nucleons from the projectile nucleus with atomic number A, for instance, is estimated from

$$N_{part}^{p} = A * (1 - \frac{E_{ZDC}}{E_{beam}^{kin}})$$ (1)

where E_{ZDC} is the energy deposited in the Zero Degree Calorimeter and E_{beam}^{kin} is the kinetic energy of beam [15]. However, in the collider experiments, in order to obtain $< N_{part} >$ one has to relate the measurables to Monte Carlo simulations. In PHENIX, for instance, simulations for the response of the Beam-Beam Counter and the ZDC were used to calculate $< N_{part} >$ via a Glauber model [2]. In PHOBOS $< N_{part} >$ is derived relating HIJING simulations to the signals in the paddle counter [11].

On the theory side, first, $< N_{part} >$ could be calculated geometrically (referred to as method a later) [16], number of participant nucleons from the projectile nucleus, for

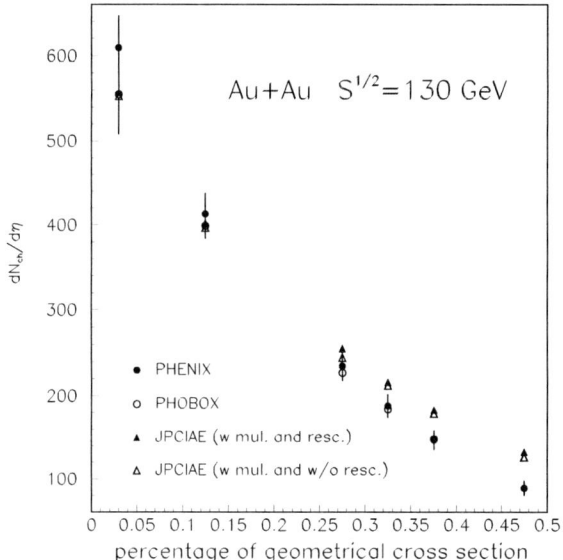

FIGURE 3. The charged particle pseudorapidity density at mid-pseudorapidity in $Au + Au$ collisions at $\sqrt{s}_{nn}$=130 GeV as a function of the percentage of geometrical cross section.

instance, reads

$$N_{part}^{p}(b) = \rho_0 \int dV \, \theta(R_p - [x^2 + (b-y)^2 + z^2]^{1/2})\theta(R_T - (x^2 + y^2)^{1/2}). \quad (2)$$

Second, in the Glauber model (method b) N_{part} is calculated by

$$N_{part}^{p}(b) = \int d^2s T_A(\vec{b} - \vec{s})[1 - \exp(-\sigma_{in}T_B(\vec{s}))] + \int d^2s T_B(\vec{s})[1 - \exp(-\sigma_{in}T_A(\vec{b} - \vec{s}))], \quad (3)$$

where $\sigma_{in} \approx 40$ mb is the inelastic nn cross section at RHIC and T_A refers to the nuclear thickness function of nucleus A [7]. The third method is to count the participant or the spectator nucleons in the simulation for nuclear collisions and then to average over simulated events. However, there is multifarious in simulating programs and even in the definition of the participants and spectators. FRITIOF 7.02 (method d) [17] was popularly employed in the past. In FRITIOF the wounded nucleons, i.e., nucleons which suffer at least one inelastic collision, are counted and identified as $< N_{part} >$. In JPCIAE simulations we have devised three counting methods: First, the leading nucleons involved in at least one inelastic nucleon-nucleon collision with string excitation are counted and identified as $< N_{part} >$. This is called method c and the JPCIAE results in Fig. 1 were calculated by $< N_{part} >$ from this method. Second, in method e, spectator nucleons are counted at the final state of JPCIAE simulations without rescattering (i.e., only nucleon-nucleon collisions are included), then $< N_{part} >$ is calculated through

$$< N_{part} >= (A + B) - < N_{spec} >, \quad (4)$$

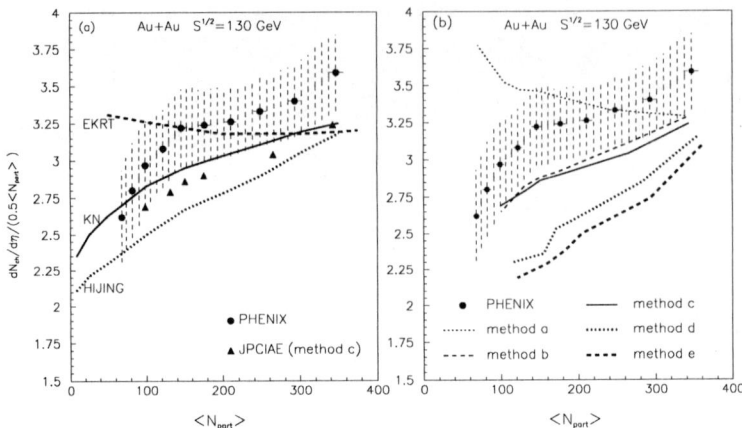

FIGURE 4. The charged particle pseudorapidity density per participant pair at mid-pseudorapidity in $Au + Au$ collisions at $\sqrt{s_{nn}}$=130 GeV as a function of the number of participant nucleons, $< N_{part} >$.

where A and B are the atomic numbers of the target and projectile nuclei. Third, method f is the same as method e but JPCIAE simulations are with rescattering.

Fig.2(a) gives $< N_{part} >$ calculated by different methods as a function of percentage of the geometrical (total) cross section and compares them with results of PHENIX (solid circles with error bar) [2] and of PHOBOS (open circles with error bar) [11]. The thin dotted, dashed, and solid curves in this panel are the results of method a (geometry), b (Glauber model, taken from [7]), and c, respectively. Thick dotted, dashed, and solid curves are, respectively, the results of method d (FRITIOF), e, and f. One knows from Fig. 2 (a) that except method f $< N_{part} >$ from different methods are close to each other for most central collisions but the discrepancies among them increase with decreasing centrality in general. In Fig.2(b) the ratios of $< N_{part} >$ from methods a, b, c, d, e and f to the corresponding results of PHENIX are given.

The charged particle pseudorapidity density at mid-pseudorapidity in $Au + Au$ collisions at $\sqrt{s_{nn}}$=130 GeV as a function of the percentage of geometrical cross section is given in Fig. 3. In this figure the full and open circles with error bar are the PHENIX [2] and PHOBOS [11] data, respectively. The full and open triangles, respectively, are the results of JPCIAE model with rescattering and without rescattering. The abscissa of the data point in this figure is set at the middle of the corresponding bin of percentage of the geometrical cross section. Globally speaking, the experimental data were reproduced fairly well by the JPCIAE model. However, the agreement between the experimental data and the JPCIAE model becomes less satisfied for peripheral collisions.

In panel (a) of Fig. 4 we compare the PHENIX data of the charged particle mid-pseudorapidity density per participant pair (full circles with shaded area of systematic errors) [2] with JPCIAE model (full triangles, $< N_{part} >$ from method c) and with HIJING model (dotted curve), KN model (solid curve), and EKRT model (dashed curve). One can see that except EKRT, all the other three models predict an increase of $(dn_{ch}/d\eta|_{\eta=0})/(0.5 < N_{part} >)$ as a function of $< N_{part} >$ though the theoretical results seem to underestimate the PHENIX data. Fig. 4 (b) compares the PHENIX data to the JPCIAE results of $< N_{part} >$ calculated by method a (thin dotted curve), b (thin dashed curve), c (thin solid curve), d (thick dotted curve), and e (thick dashed curve), respectively. One sees from this panel that starting from a single result of charged particle mid-pseudorapidity density from the JPCIAE model but using $< N_{part} >$ from different methods, it is possible to lead $(dn_{ch}/d\eta|_{\eta=0})/(0.5 < N_{part} >)$ to either increase or decrease with the increase of $< N_{part} >$. Although $< N_{part} >$ from method a are closest to the PHENIX results (cf. Fig. 2 (b)) the $(dn_{ch}/d\eta|_{\eta=0})/(0.5 < N_{part} >)$ from JPCIAE have actually centrality dependence opposite to the PHENIX result (cf. Fig.4 (b)) because the $(dn_{ch}/d\eta|_{\eta=0})$ from JPCIAE is higher than the PHENIX result for peripheral collisions (cf. Fig. 3). On the other hand, even though the discrepancy between $< N_{part} >$ from method c and PHENIX slightly increases with decrease of centrality (cf. Fig. 2 (b)) the $(dn_{ch}/d\eta|_{\eta=0})/(0.5 < N_{part} >)$ from JPCIAE with method c is close to the PHENIX data. If $dn_{ch}/d\eta|_{\eta=0}$ from EKRT model was not normalized by $< N_{part} >$ from method b, as did in [7], but by $< N_{part} >$ from method d the results of $(dn_{ch}/d\eta|_{\eta=0})/(0.5 < N_{part} >)$ might have somewhat similar centrality dependence as in PHENIX data. Therefore one learns here that it is hard using $(dn_{ch}/d\eta|_{\eta=0})/(0.5 < N_{part} >)$ as a function of $< N_{part} >$ to distinguish various theoretical models for particle production since $< N_{part} >$ is not a well defined physical variable.

REFERENCES

1. B. B. Back, et al., PHOBOS collab., Phys. Rev. Lett. 85, 3100 (2000).
2. K. Adcox, et al.,, PHENIX collab., Phys. Rev. Lett. 86, 3500 (2001).
3. L. van Hove, Phys. Lett. B 118, 138 (1982).
4. M. I. Gorenstein, Phys. Lett. B 281, (1992).
5. J. I. Kapusta and A. P. Vischer, Phys. Rev. C 52, 2725 (1995).
6. Xin-Nian Wang and M. Gyulassy, Phys. Rev. Lett. 86, 3496 (2001).
7. K. J. Eskola, K. Kajantie and K. Tuominen, Phys. Lett. B 497, 39 (2001).
8. D. Kharzeev and M. Nardi, nucl-th/0012025.
9. A. Capella and D. Sousa, nucl-th/0106066.
10. Ben-Hao Sa, An Tai, Hui Wang and Feng-He Liu, Phys. Rev. C 59, 2728 (1999);
    Ben-Hao Sa and An Tai, Phys. Rev. C 62, 044905 (2000).
11. B. B. Back, et al., PHOBOS collab., nucl-th/0105011.
12. T. Sjöstrand, Comp. Phys. Comm. 82, 74 (1994).
13. J. Cugnon, T. Mizutani, and J. Vandermeulen, Nucl. Phys. A 352, 505 (1981).
14. An Tai and Ben-Hao Sa, Comp. Phys. Comm. 116, 353 (1999).
15. L. Ahle, et al., E802 Collab., Phys. Rev. C 59, 2173 (1999).
16. Ben-Hao Sa, Yu-Ming Zheng, and Xiao-Ze Zhang, Phys. Rev. C 40, 2680 (1989).
17. Hong Pi, Z. Phys. C 57, 485 (1993).

Bound states of anti-nucleons in finite nuclei

G. Mao*, H. Stöcker* and W. Greiner*

*Institut für Theoretische Physik der J. W. Goethe-Universität
Postfach 11 19 32, D-60054 Frankfurt am Main, Germany

Abstract. We study the bound states of anti-nucleons emerging from the lower continuum in finite nuclei within the relativistic Hartree approach including the contributions of the Dirac sea to
the source terms of the meson fields. The Dirac equation is reduced to two Schrödinger-equivalent
equations for the nucleon and the anti-nucleon respectively. These two equations are solved simultaneously in an iteration procedure. Numerical results show that the bound levels of anti-nucleons
vary drastically when the vacuum contributions are taken into account.

In spite of the great successes of the relativistic mean field (RMF) theory [1, 2, 3, 4]
and the relativistic Hartree approach (RHA) [5, 6] in describing the ground states of
nuclei, the arguments of introduction of strong Lorentz scalar (S) and time-component
Lorentz vector (V) potential in the Dirac equation are largely indirect. So far, no evidence
from experiments ensures the physical necessity. One usually compares the theoretical
predictions only with the experimental data of the nucleon sector (i.e., the shell-model
states), which is subject to a relatively small quantity stemming from the cancellation of
two potentials $S+V$ (V is positive, S is negative.). While the dynamical content of the
Dirac picture certainly lies with both the nucleon and the anti-nucleon sector, the study
of the anti-nucleon sector enjoys an additional bonus: it provides us with a chance to
determine the individual S and V! Due to the *G-parity*, the vector potential changes its
sign in the anti-nucleon sector. The bound states of anti-nucleons are sensitive to the sum
of the scalar and vector field $S-V$. Combining with the information from the nucleon
sector, one may fix the individual values of the scalar and vector field.

The study of the anti-nucleon bound states is extremely interesting for modern nuclear
physics. If the potential of anti-nucleons is much weaker than what one expects or
predicts by means of the RMF/RHA models, that is, the strong scalar and vector field
are not necessary, one may question the validity of the models since some important
physical ingredients, such as quantum corrections, correlation effects, three-body forces
et al., are still missing in these phenomenological approaches. One may think about
constituting a more elaborate model. Alternatively, if a deep potential of anti-nucleons is
indeed observed, that is, the strong scalar and vector potential are realistic, an interesting
phenomenon is that at certain density the energy of anti-nucleons may turn out to be
larger than the free nucleon mass, the system becomes unstable with respect to the
nucleon–anti-nucleon pair creation [7]. On the other hand, as pointed out in Ref. [8],
in high-energy relativistic heavy-ion collisions, the nucleons may be emitted from the
deep bound states emerging from the Dirac sea due to dynamics. These can create a
great number of anti-nucleons in bound states. Such collective creation processes of
anti-matter clusters have a large probability for the production of anti-nuclei, – and

CP597, *Nonequilibrium and Nonlinear Dynamics in Nuclear and Other Finite Systems*,
edited by Z. Li et al.

analogously also for multi-Λ, multi-$\bar{\lambda}$ nuclei. These open two fascinating directions to extend the periodic system, i.e., to extend into the anti-nucleon sector and into the multi-strangeness dimension, in addition to the islands of super-heavy nuclei. In order to reach the quantitative study of the above theoretical conjecture, a prerequisite is to know the exact potential depth of the bound states of anti-nucleons. Up to now, no answers from experimental side or theoretical side are available.

This is the aim of the present work. We study the problem within the relativistic Hartree approach including the vacuum contributions. The starting point is the following effective Lagrangian for nucleons interacting through the exchange of mesons [1, 2, 3]

$$
\begin{aligned}
\mathcal{L} = & \bar{\psi}[i\gamma_\mu \partial^\mu - M_N]\psi + \frac{1}{2}\partial_\mu \sigma \partial^\mu \sigma - U(\sigma) - \frac{1}{4}\omega_{\mu\nu}\omega^{\mu\nu} \\
& + \frac{1}{2}m_\omega^2 \omega_\mu \omega^\mu - \frac{1}{4}\mathbf{R}_{\mu\nu}\mathbf{R}^{\mu\nu} + \frac{1}{2}m_\rho^2 \mathbf{R}_\mu \cdot \mathbf{R}^\mu - \frac{1}{4}A_{\mu\nu}A^{\mu\nu} \\
& + g_\sigma \bar{\psi}\psi\sigma - g_\omega \bar{\psi}\gamma_\mu \psi \omega^\mu - \frac{1}{2}g_\rho \bar{\psi}\gamma_\mu \tau \cdot \psi \mathbf{R}^\mu - \frac{1}{2}e\bar{\psi}(1+\tau_0)\gamma_\mu \psi A^\mu,
\end{aligned}
\tag{1}
$$

here $U(\sigma)$ is the self-interaction part of the scalar field [9]

$$
U(\sigma) = \frac{1}{2}m_\sigma^2 \sigma^2 + \frac{1}{3!}b\sigma^3 + \frac{1}{4!}c\sigma^4.
\tag{2}
$$

Based on this Lagrangian, we have developed a relativistic model describing the bound states of both nucleons and anti-nucleons in finite nuclei [10]. Instead of directly searching for two solutions of the Dirac equation in a finite many-body system, we reduce the Dirac equation to two Schrödinger-equivalent equations for the nucleon and the anti-nucleon respectively. These two equations can be solved simultaneously with the numerical technique of the relativistic mean-field theory for finite nuclei. The contributions of the Dirac sea to the source terms of the meson fields are evaluated by means of the derivative expansion [11] up to the leading derivative order for the one-meson loop and one-nucleon loop. Thus, the wave functions of anti-nucleons, which are used to calculate the single-particle energies, are not involved in evaluating the vacuum contributions to the scalar and baryon density which are, in turn, expressed by means of the scalar and vector field as well as their derivative terms [10]. The Schrödinger-equivalent equation of the nucleon and the equations of motion of mesons (containing the densities contributed from the vacuum) are solved within a self-consistent iteration procedure [2]. Then, the equation of the anti-nucleon is solved with the known mean fields to obtain the wave functions and the single-particle energies of anti-nucleons. The space of anti-nucleons are truncated by the specified principal and angular quantum numbers n and j with the guarantee that the calculated single-particle energies of anti-nucleons are converged when the truncated space is extended. We find that the results are insensitive to the exact values of n and j provided large enough numbers are given. We have used $n = 4$, $j = 9$ for ^{16}O; $n = 5$, $j = 11$ for ^{40}Ca; and $n = 9$, $j = 19$ for ^{208}Pb.

In the previous RHA calculations for the bound states of nucleons [5, 6], the parameters of the model are fitted to the saturation properties of nuclear matter as well as the *rms* charge radius in ^{40}Ca. The *best-fit* routine within the RHA to the properties of spherical nuclei has not been performed yet. Thus, we first fit the parameters of Eqs.

TABLE 1. Parameters of the RMF and the RHA models as well as the corresponding saturation properties. The results of shell fluctuation and the χ^2 values of different sets of parameters are also presented.

	RMF		RHA	
	LIN	NL1	RHA0	RHA1
M_N (MeV)	938.000	938.000	938.000	938.000
m_σ (MeV)	615.000	492.250	615.000	458.000
m_ω (MeV)	1008.00	795.359	916.502	816.508
m_ρ (MeV)	763.000	763.000	763.000	763.000
g_σ	12.3342	10.1377	9.9362	7.1031
g_ω	17.6188	13.2846	11.8188	8.8496
g_ρ	10.3782	9.9514	10.0254	10.2070
b (fm^{-1})	0.0	24.3448	0.0	24.0870
c	0.0	−217.5876	0.0	−15.9936
ρ_0 (fm^{-3})	0.1525	0.1518	0.1513	0.1524
E/A (MeV)	−17.03	−16.43	−17.39	−16.98
m^*/M_N	0.533	0.572	0.725	0.788
K (MeV)	580	212	480	294
a_4 (MeV)	46.8	43.6	40.4	40.4
$\delta\rho$ in ^{208}Pb (fm^{-3})	−0.0075	−0.0070	−0.0016	−0.0030
χ^2	1773	66	1040	812

(1) and (2) within the RHA to the empirical data of binding energy, surface thickness and diffraction radius of eight spherical nuclei ^{16}O, ^{40}Ca, ^{48}Ca, ^{58}Ni, ^{90}Zr, ^{116}Sn, ^{124}Sn, and ^{208}Pb as has been done in Ref. [2] for the RMF model. We distinguish two different cases with (RHA1) and without (RHA0) nonlinear self-interaction of the scalar field. The obtained parameters and the corresponding saturation properties are given in Table I. For the sake of comparison, two sets of the linear (LIN) and nonlinear (NL1) RMF parameters from Ref. [2] are also presented. One can see that the RHA gives a larger effective nucleon mass than the RMF does, which is mainly caused by the feedback of the vacuum to the meson fields, as can be seen from Eqs. (71) $\sim$ (74) of Ref. [10]. When the effective nucleon mass decreases, the scalar density originated from the Dirac sea ρ_S^{sea} increases. It is negative and cancels part of the scalar density contributed from the valence nucleons ρ_S^{val}, which causes the effective nucleon mass to increase again. At the end, it reaches a balance value. In the fitting procedure, we have tried different initial values giving smaller effective nucleon mass. After running the code many times, all of them slowly converge to a large m^*. The larger effective nucleon mass explains why a larger χ^2 value is obtained for the RHA1 compared to the NL1. If one uses the current nonlinear RMF/RHA models to fit the ground-state properties of spherical nuclei, an effective nucleon mass around 0.6 is preferred. The situation, however, might be changed when other physical ingredients, e.g., tensor couplings, correlation effects, three-body forces, are taken into account, which warrants further investigation. On the other hand, in the case of linear model, the RHA0 gives a better fit than the LIN does. This is mainly due to the vacuum contributions which improve the theoretical results of the surface

TABLE 2. The single-particle energies of both protons and anti-protons as well as the binding energy per nucleon and the *rms* charge radius in ^{16}O, ^{40}Ca and ^{208}Pb.

	RMF		RHA		
	LIN	NL1	RHA0	RHA1	EXP.
^{16}O					
E/A (MeV)	7.80	8.00	8.01	8.00	7.98
r_{ch} (fm)	2.59	2.73	2.62	2.66	2.74
PROTONS					
$1s_{1/2}$ (MeV)	42.99	36.18	32.21	30.68	40±8
$1p_{3/2}$ (MeV)	20.71	17.31	16.09	15.23	18.4
$1p_{1/2}$ (MeV)	10.85	11.32	12.98	13.24	12.1
ANTI-PRO.					
$1\bar{s}_{1/2}$ (MeV)	821.30	674.11	413.62	299.42	
$1\bar{p}_{3/2}$ (MeV)	754.62	604.70	369.78	258.40	
$1\bar{p}_{1/2}$ (MeV)	755.43	605.77	370.36	258.93	
^{40}Ca					
E/A (MeV)	8.38	8.58	8.65	8.73	8.55
r_{ch} (fm)	3.36	3.48	3.39	3.42	3.45
PROTONS					
$1s_{1/2}$ (MeV)	51.21	46.86	38.64	36.58	50±11
$1p_{3/2}$ (MeV)	35.05	30.15	27.11	25.32	
$1p_{1/2}$ (MeV)	29.25	25.11	25.17	24.03	34±6
ANTI-PRO.					
$1\bar{s}_{1/2}$ (MeV)	840.76	796.09	456.58	339.83	
$1\bar{p}_{3/2}$ (MeV)	792.36	706.36	424.85	309.24	
$1\bar{p}_{1/2}$ (MeV)	792.75	707.86	425.14	309.52	
^{208}Pb					
E/A (MeV)	7.83	7.89	7.96	7.93	7.87
r_{ch} (fm)	5.34	5.52	5.43	5.49	5.50
PROTONS					
$1s_{1/2}$ (MeV)	58.71	50.41	44.43	40.80	
$1p_{3/2}$ (MeV)	52.74	44.45	39.87	36.45	
$1p_{1/2}$ (MeV)	51.83	43.75	39.49	36.21	
ANTI-PRO.					
$1\bar{s}_{1/2}$ (MeV)	830.16	717.01	476.61	354.18	
$1\bar{p}_{3/2}$ (MeV)	819.15	705.20	466.08	344.48	
$1\bar{p}_{1/2}$ (MeV)	819.22	705.28	466.13	344.52	

thickness substantially, and finally improve the total χ^2 value. An interesting quantity is the shell fluctuation which can be best expressed via the charge density in ^{208}Pb as

$$\delta\rho = \rho_C(1.8 \text{ fm}) - \rho_C(0.0 \text{ fm}). \qquad (3)$$

The empirical value is -0.0023 fm^{-3} [2], which is nicely reproduced in the RHA (see Table I) while the RMF overestimates $\delta\rho$ by a factor of 3, sharing the same disease with the non-relativistic mean field theory [12].

115

In Table II we present the results of both the proton and the anti-proton spectra of ^{16}O, ^{40}Ca and ^{208}Pb. The binding energy per nucleon and the *rms* charge radius are given too. The numerical calculations are performed within two frameworks, i.e., the RHA including the contributions of the Dirac sea to the source terms of the meson fields and the RMF taking into account only the valence nucleons as the meson-field sources. The experimental data are taken from Ref. [13]. From the table one can see that all four sets of parameters can reproduce the empirical values of the binding energies, the *rms* charge radii and the single-particle energies of protons fairly well. For the E/A and the r_{ch}, the agreement between the theoretical predictions and the experimental data are improved from the LIN to the RHA0, RHA1 and NL1 set of parameters. For the spectra of protons, due to large error bars, it seems to be difficult to queue up the different sets of parameters. However, because of the large effective nucleon mass, the RHA has a smaller spin-orbit splitting (see $1p_{1/2}$ and $1p_{3/2}$ state) compared to the RMF. This situation can be improved through introducing a tensor coupling for the ω meson [2] which will be investigated in the future studies. For the anti-nucleon sector, no experimental data are available. In all four cases, the potential of anti-protons is much deeper than the potential of protons. On the other hand, one can notice the drastic difference between the RHA and the RMF calculations – the single-particle energies of anti-protons calculated from the RHA are about half of that from the RMF, exhibiting the importance of taking into account the Dirac sea effects. It demonstrates that the anti-nucleon spectra deserve a sensitive probe to the effective interactions. The spin-orbit splitting of the anti-nucleon sector is so small that one nearly can not distinguish the $1\bar{p}_{1/2}$ and the $1\bar{p}_{3/2}$ state. This is because the spin-orbit potential is related to $d(S+V)/dr$ in the anti-nucleon sector and two fields cancel each other to a large extent. Nevertheless, the space between the $1\bar{s}$ and the $1\bar{p}$ state is still evident, especially for lighter nuclei. This might be helpful to separate the process of knocking out a $1\bar{s}_{1/2}$ nucleon from the background – a promising way to measure the potential of the anti-nucleon in laboratory.

In summary, we have proposed to study the bound states of anti-nucleons in finite nuclei which will provide us with a chance to judge the physical necessity of introducing strong scalar and vector potential in the Dirac picture. Due to the feedback of the vacuum to the meson fields, the scalar and vector fields decrease in the RHA. Numerical calculations show that the single-particle energies of anti-nucleons change drastically in the RMF and the RHA approach for different sets of parameters, while the single-particle energies of nucleons remain in a reasonable range. It is very important to have experimental data to check the theoretical predicted bound levels of anti-nucleons. If the Dirac picture with the large potentials is valid for nucleon-nucleus interactions, a fascinating direction of future studies is to investigate the vacuum correlation and the collective production of anti-nuclei in relativistic heavy-ion collisions. Experimental efforts in this direction are presently underway [14].

ACKNOWLEDGMENTS

The authors thank P.-G. Reinhard, Zhongzhou Ren, J. Schaffner-Bielich and C. Beck-mann for stimulating discussions. G. Mao gratefully acknowledges the Alexander von

Humboldt-Stiftung for financial support and the people at the Institut für Theoretische Physik der J. W. Goethe Universität for their hospitality. This work was supported by DFG-Graduiertenkolleg Theoretische und Experimentelle Schwerionenphysik, GSI, BMBF, DFG and A.v.Humboldt-Stiftung.

REFERENCES

1. B. D. Serot and J. D. Walecka, Adv. Nucl. Phys. **16**, 1 (1986).
2. P.-G. Reinhard, M. Rufa, J. Maruhn, W. Greiner, J. Friedrich, Z. Phys. **A323**, 13 (1986); M. Rufa, P.-G. Reinhard, J.A. Maruhn, W. Greiner, M.R. Strayer, Phys. Rev. **C38**, 390 (1988).
3. Y.K. Gambhir, P. Ring and A. Thimet, Ann. Phys. **198**, 132 (1990).
4. Zhongzhou Ren, Z.Y. Zhu, Y.H. Cai, Gongou Xu, Phys. Lett. **B380**, 241 (1996).
5. C.J. Horowitz and B.D. Serot, Phys. Lett. **B140**, 181 (1984).
6. R.J. Furnstahl and C.E. Price, Phys. Rev. **C40**, 1398 (1989); Phys. Rev. **C41**, 1792 (1990).
7. I.N. Mishustin, L.M. Satarov, J. Schaffner, H. Stöcker and W. Greiner, J. Phys. G: Nucl. Part. Phys. **19**, 1303 (1993).
8. W. Greiner, Heavy Ion Physics **2**, 23 (1995).
9. J. Boguta and H. Stöcker, Phys, Lett. **B120**, 289 (1983).
10. G. Mao, H. Stöcker, and W. Greiner, Int. J. Mod. Phys. E8, 389 (1999).
11. I.J.R. Aitchison and C.M. Fraser, Phys. Lett. **B146**, 63 (1984); O. Cheyette, Phys. Rev. Lett. **55**, 2394 (1985); C.M. Fraser, Z. Phys. **C28**, 101 (1985); L.H. Chan, Phys. Rev. Lett. **54**, 1222 (1985).
12. J. Friedrich and P.-G. Reinhard, Phys. Rev. **C33**, 335 (1986).
13. J.H.E. Mattauch, W. Thiele, and A.H. Wapstra, Nucl. Phys. **67**, 1 (1965); D. Vautherin and D.M. Brink, Phys. Rev. **C5**, 626 (1972); H. de Vries, C.W. de Jager, and C. de Vries. At. Data Nucl. Data Tables **36**, 495 (1987).
14. R. Arsenescu and the NA52 collaboration, J. Phys. **G25**, 225 (1999).

The possibility of a nonzero mean π-field and the self-consistency of relativistic mean field theory for nuclear matter[1]

Qi-Ren Zhang, Chun-Yuan Gao, and Yue-Wei Wu [2]

Department of Technical Physics, Peking University, Beijing 100871, China

SUCCESS OF RMFT AND ITS UNDERLYING PHILOSOPHY

Relativistic mean field theory (RMFT) is now extensively used in nuclear theory, especially for extrapolating the nuclear equation of state to extreme conditions, to research the physics of, for examples, relativistic heavy ion collisions and superdense stars. It is successful in reproducing fundamental data of nuclear matter, average properties of finite nuclei, spin-orbital interactions and shell effects for nuclei. Therefore, people hope to develop it into an effective theory of nuclei at the hadron level. This hope is strengthened by the consideration that its underlying philosophy is quite reasonable. At normal or higher nuclear densities, the physical nucleons are touching each other, the meson clouds should be shared by all nucleons in nuclei rather than adhered to one or another nucleon. The main part of mesons in nuclei may form coherent fields as that assumed in the RMFT. To promote this theory further, its self-consistency becomes more important. Of course, we have to truncate the hadron spectrum at some upper bound in RMFT. But, within the bound, we have to treat all hadrons on the same footing. That is to substitute all meson field operators by their mean values, and quantize all baryon fields in these mean meson fields. The mean values of various meson fields should be determined by the theory self-consistently, for example, by minimizing the energy density of the system at a given baryon density. It is to say we have no right to set the mean value of any of these meson fields to be zero.

THE POSSIBLE NONZERO MEAN π−FIELD, AND THE SELF-CONSISTENCY PROBLEM OF RMFT

An interesting phenomenon is that,in most work on RMFT for nuclear matter, low mass mesons are considered except the lightest meson, the π-meson. The argument is that the expectation value of π-field is zero until the nuclear density becomes rather high. This

[1] Work supported by the Nature Science Foundation of China under grant number 19775004

[2] The present address: China Institute of Atomic Energy, Beijing 102413, China

argument has, of course, to be checked according to the same philosophy of relativistic mean field theory. In the following we show that this argument is not true for some commonly used models, including the Walecka model. The expectation value of π-field is nonzero even for a low density nuclear matter.[1]

Let us consider a system consists of nucleons, scalar mesons, ω mesons and π mesons. The Lagrangian density is

$$\mathcal{L} = -\bar{\Psi}[\gamma_\mu(\partial_\mu - ig_\omega V_\mu) + m - g(\Phi + i\gamma_5 \sum_{k=1}^{3} \tau_k \Phi_k)]\Psi$$
$$-\frac{1}{4}F_{\mu\nu}F_{\mu\nu} - \frac{1}{2}[\partial_\mu\Phi\partial_\mu\Phi + m_s^2\Phi^2]$$
$$+\sum_{k=1}^{3}(\partial_\mu\Phi_k\partial_\mu\Phi_k + m_\pi^2\Phi_k^2) + m_\omega^2 V_\mu V_\mu] . \tag{1}$$

Φ_k and m_π are π field and pion mass respectively, other symbols are defined in the usual way, like those in [2]. This is a minimum generalization of the Walecka model with the π-field being included, and is renormalizable. The π meson-nucleon coupling constant before the renormalization of pseudovector current is assumed to equal the scalar meson-nucleon coupling constant g, like that in the σ model. It is a representation of partial chiral symmetry. But the potential energy densities of π and scalar meson fields are assumed to be the simple quadratic forms. It is well known that the σ model potential energy density could not fit the normal nuclear properties, unless a hardcore of the nucleon is assumed. From (1) we obtain the energy density of the ground state nuclear matter

$$\mathcal{E} = \mathcal{E}' + \mathcal{U} + 2\pi\alpha_\omega \frac{n^2}{m_\omega^2} , \tag{2}$$

in which

$$\mathcal{E}' = \langle | \Psi^\dagger[\vec{\alpha}\cdot(-i\nabla) + \beta(m - g\Phi) - g\rho_2 \sum_{k=1}^{3} \tau_k\Phi_k]\Psi |\rangle \tag{3}$$

is the energy density of nucleons interacting with scalar and π meson fields in the ground state $|\rangle$ of nuclear matter;

$$\mathcal{U} = \frac{1}{2}\left\{\left(\frac{\partial\Phi}{\partial x^0}\right)^2 + (\nabla\Phi)^2 + m_s^2\Phi^2 + \sum_{k=1}^{3}\left[\left(\frac{\partial\Phi_k}{\partial x^0}\right)^2 + (\nabla\Phi_k)^2 + m_\pi^2\Phi_k^2\right]\right\} \tag{4}$$

is the energy density of the scalar and the π meson fields; $\alpha_\omega = g_\omega^2/4\pi$ is the ω- fine structure constant. Suppose the expectation values of the scalar and π fields are non-zero. Substituting these classical values for the corresponding operators in (3) and (4), we see that the nucleon energy density $\mathcal{E}'$ becomes a sum of single nucleon energies. Now the nucleons move in the non-zero scalar and π mean fields. Let us search the solution of the form [3]-[5]

$$\Phi = m/g - \mathcal{A}\cos\theta, \tag{5}$$

$$\Phi_1 = \mathcal{A}\sin\theta\cos(\mathbf{k}\cdot\mathbf{r}), \quad \Phi_2 = \mathcal{A}\sin\theta\sin(\mathbf{k}\cdot\mathbf{r}), \quad \Phi_3 = 0. \tag{6}$$

It may be thought that we are solving the problem by the variational method, with (5) and (6) being the trial field function for scalar and π mesons, $\mathcal{A}, \theta$ and k are variational parameters. To diagonalize the energy operator for the nucleons in (3), we need to solve the Dirac equation

$$[\vec{\alpha}\cdot(-i\nabla) + \beta g\mathcal{A}\cos\theta - \rho_2 g\mathcal{A}\sin\theta(\tau_1\cos\mathbf{k}\cdot\mathbf{r} + \tau_2\sin\mathbf{k}\cdot\mathbf{r})]\psi = \varepsilon\psi. \tag{7}$$

By successive isospin and chiral rotations [3]-[5], we transform

$$\psi = \exp[-i\tau_3(\mathbf{k}\cdot\mathbf{r} - \pi/2)/2]\exp(-i\rho_1\tau_2\theta/2)\psi_e, \tag{8}$$

and the equation becomes

$$\left[\vec{\alpha}\cdot\left(-i\nabla - \frac{\tau_3}{2}\mathbf{k}'\right) + \beta m' + \frac{f'_\pi}{m_\pi}\vec{\Sigma}\cdot\mathbf{k}'\tau_1\phi\right]\psi_e = \varepsilon\psi_e, \tag{9}$$

with

$$\mathbf{k}' = \mathbf{k}\cos\theta, \quad m' = g\mathcal{A}, \tag{10}$$

$$f'_\pi = \frac{g_R m_\pi}{2m'}, \quad \text{and} \quad \phi = \mathcal{A}\tan\theta. \tag{11}$$

$g_R = g_a g$ is the renormalized pion-nucleon coupling constant, $g_a = 1.26$ is the renormalization constant for the pseudovector current. Defining

$$\chi = m'/m = g\mathcal{A}/m \tag{12}$$

as usual, we obtain from (11),

$$\mathbf{k}' = \mathbf{k}/\mathcal{B}', \quad \text{with} \tag{13}$$

$$\mathcal{B}' = \sqrt{1 + \left(\frac{g_R\phi}{g_a m\chi}\right)^2}. \tag{14}$$

(9) has a typical form of the Dirac equation for a nucleon of mass m' interacting with the π meson field

$$\phi_1 = \phi\cos(\mathbf{k}'\cdot\mathbf{r}), \quad \phi_2 = \phi\sin(\mathbf{k}'\cdot\mathbf{r}), \quad \text{and} \quad \phi_3 = 0, \tag{15}$$

by pseudovector coupling with coupling constant f'_π, and has been solved [4]-[7]. Since the operator on the left side of (9) is no longer dependent on space coordinates, we have

$$\psi_e = \exp(i\mathbf{p}\cdot\mathbf{r})U, \tag{16}$$

U is a constant spinor. By further transformations

$$U = S_1 S_2 u, \quad \text{with} \tag{17}$$

$$S_1 = \sqrt{\frac{\mathcal{B}+1}{2\mathcal{B}}} + i\frac{f'\phi}{m_\pi}\sqrt{\frac{2}{\mathcal{B}(\mathcal{B}+1)}}\,\tau_2\rho_1, \tag{18}$$

$$S_2 = \sqrt{\frac{A+\mathbf{p}\cdot\mathbf{k}''}{2A}} + i\frac{g'\phi\tau_2\rho_3\vec{\Sigma}\cdot\mathbf{k}''}{\sqrt{2A(A+\mathbf{p}\cdot\mathbf{k}'')}}, \tag{19}$$

$$A = \sqrt{(\mathbf{p}\cdot\mathbf{k}'')^2 + (g'\phi k'')^2}, \tag{20}$$

$$\mathcal{B} = \sqrt{1 + \left(\frac{2f'\phi}{m_\pi}\right)^2}, \tag{21}$$

$$g' = \frac{2m'f'}{\mathcal{B}m_\pi}, \quad \text{and} \tag{22}$$

$$\mathbf{k}'' = \mathcal{B}\mathbf{k}', \tag{23}$$

we obtain the Dirac equation for a pseudo-free particle

$$(\vec{\alpha}\cdot\mathbf{Q} + m''\beta)u = \varepsilon u, \quad \text{with} \tag{24}$$

$$\mathbf{Q} = \mathbf{p} + \left(\frac{g'^2\phi^2}{A+\mathbf{p}\cdot\mathbf{k}''} - \frac{\tau_3}{2}\right)\mathbf{k}'', \tag{25}$$

$$m'' = m'/\mathcal{B}. \tag{26}$$

The solution is well known:

$$\varepsilon = \sqrt{Q^2 + m''^2}, \tag{27}$$

$$u = (\vec{\alpha}\cdot\mathbf{Q} + m''\beta + \varepsilon)u_0, \tag{28}$$

u_0 is an arbitrary constant spinor both in Minkowski space-time and in the isospin space.

The nucleon number density n and the energy density $\mathcal{E}$ may be calculated by summing up the single nucleon contributions. We minimize the energy density $\mathcal{E}$ under fixed nucleon number density n. After a slightly complex but still elementary derivation we arrive at[1]

$$[(k^2 + m_\pi^2)/\mathcal{B}'^4 - \Pi(\chi, k, \phi, Q_f)]\phi = 0, \tag{29}$$

with

$$\Pi(\chi, k, \phi, Q_f) = \Pi'(\chi, k, \phi, Q_f) - \frac{2g^2}{3\pi^2\alpha\mathcal{B}'^3}\left(\frac{1}{\chi} - \frac{1}{\mathcal{B}'}\right), \tag{30}$$

$$\Pi'(\chi, k, \phi, Q_f) = \frac{g'^2 m'' k'}{2\pi^2 \mathcal{B}}$$

$$\times \begin{cases} 0 & \text{if } Q_f \leq g'\phi - \frac{k''}{2} \\ W(\frac{k''}{2m''}, \frac{g'\phi}{m''}, \frac{Q_f}{m''}) - W(\frac{k''}{2m''}, \frac{g'\phi}{m''}, -\frac{Q_f}{m''}) & \text{if } g'\phi - \frac{k''}{2} < Q_f \leq \frac{k''}{2} - g'\phi \\ W(\frac{k''}{2m''}, \frac{g'\phi}{m''}, \frac{Q_f}{m''}) & \text{if } g'\phi - \frac{k''}{2} < Q_f \leq \frac{k''}{2} + g'\phi \\ \quad - \sqrt{1 + \left(\frac{Q_f}{m''}\right)^2}\left[1 + \frac{C}{2}\left(\frac{g'\phi}{m''}\right)^2\right]\ln\frac{g'\phi}{m''} & \wedge Q_f > \frac{k''}{2} - g'\phi \\ W(\frac{k''}{2m''}, \frac{g'\phi}{m''}, \frac{Q_f}{m''}) - W(-\frac{k''}{2m''}, \frac{g'\phi}{m''}, \frac{Q_f}{m''}) & \text{if } Q_f > g'\phi + \frac{k''}{2}, \end{cases} \tag{31}$$

$$W(a,b,x) = \sqrt{1+x^2}\left\{\left(1+\frac{b^2C}{2}\right)\ln[a+x+\sqrt{(a+x)^2-b^2}]\right.$$
$$\left.+\frac{x-a}{2}C\sqrt{(a+x)^2-b^2}\right\}-G(a,b,x), \tag{32}$$

$$G(a,b,x) = \int_{b-a}^{x}\sqrt{\frac{1+\xi^2}{(\xi+a)^2-b^2}}[1+C(a\xi+\xi^2)]d\xi, \quad\text{and} \tag{33}$$

$$C = \frac{g_a^2-1}{g_a^2+b^2}. \tag{34}$$

(29) is the equation of motion for the π-field in nuclear matter. $\phi=0$ is its trivial solution. The sufficient condition ensuring this solution corresponding to a minimum of $\mathcal{E}$ is

$$\frac{d^2\mathcal{E}}{d\phi^2}\bigg|_{\phi=0} = k^2+m_\pi^2-\Pi(\chi,k,0,Q_f) > 0. \tag{35}$$

If, in contrary,

$$k^2+m_\pi^2-\Pi(\chi,k,0,Q_f) < 0, \tag{36}$$

the solution $\phi=0$ corresponds to a maximum of $\mathcal{E}$, therefore is unstable. A finite ϕ may develop to decrease the energy. This is the π-condensation. Its critical condition is

$$k^2+m_\pi^2-\Pi(\chi,k,0,Q_f) = 0. \tag{37}$$

This condition determines a critical nucleon density of the π-condensation for given parameters χ and k. Varying these parameters we may find the lowest critical density. That is the true critical density for the π-condensation.

Another solution of the field equation (29) is a finite ϕ satisfying

$$(k^2+m_\pi^2)/\mathcal{B}'^4-\Pi(\chi,k,\phi,Q_f) = 0. \tag{38}$$

In the case of (36), this finite solution corresponds to a minimum of the energy density $\mathcal{E}$ for a given nucleon density, and for given parameters χ and k. By varying parameters χ and k, we may further minimize the energy density for a fixed nucleon density. Then we can get the nuclear equation of state $\mathcal{E}(n)$ with the nonvanishing mean π-field being taken into account.

Now let us fix the renormalized π-nucleon coupling constant g_R to be its empirical value: $g_R^2/4\pi = 14.6$. The two remaining parameters α_s and α_ω are then determined by fitting empirical data. They are the equilibrium density $n_0 = (4\pi 1.175^3/3)^{-1}\text{fm}^{-3}$ and the binding energy per-nucleon 15.986MeV in nuclear matter [8]. The determined values are

$$\alpha_s = 24.5 \quad\text{and}\quad \alpha_\omega = 15.4. \tag{39}$$

The numerical result obtained by using these parameters is shown in Fig. 1. We see that a finite mean π-field indeed lowers the energy per-nucleon when the nucleon

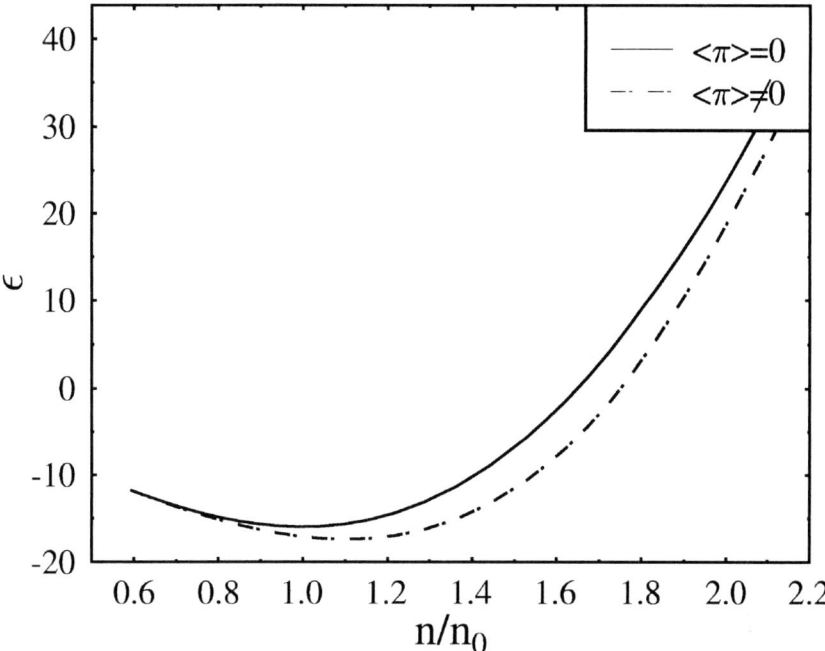

FIGURE 1. Influence of a finite mean π-field on energy per-nucleon in the nuclear matter.

density becomes higher than the critical density $n_c = 0.56n_0$. The amount of further minimization of the energy per-nucleon by the finite π-field may be more than 5MeV. Comparing this amount with the empirical binding energy $\sim$16MeV per-nucleon in normal nuclear matter we see that the effect of a finite mean π-field is not negligible. The equilibrium density is also 20 percent shifted to the higher side. These results show the necessity of readjusting the model parameters to fit the empirical data when taking the mean π field into account. Fig. 2 shows the numerical results obtained by using such a new readjusted set of parameters

$$\alpha_s = 26.6 \quad \text{and} \quad \alpha_\omega = 17.0. \tag{40}$$

On the left of the figure, we see (the solid line) that the empirical values of the equilibrium density and the energy per-nucleon are recovered. The dashed line in it is a copy of the solid line in Fig. 1, for comparison. On the right we see the density dependence of χ for the nucleon, and of **k** and ϕ for the condensed π field. They are sizable, and therefore may have sizable effects in the hadron-nucleus and nucleus-nucleus collisions.

CONCLUSION

It is found[9], that the above example is not a special case. There are also some other RMFT models, in which a nonzero mean π-field develops near the normal nuclear

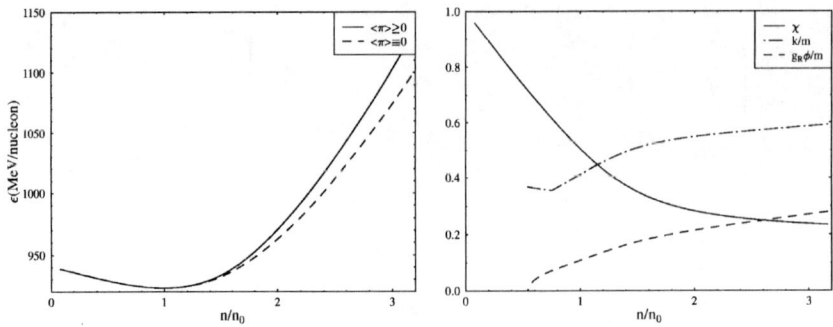

FIGURE 2. The left: Comparison of nuclear equations of state in the models with and without a finite mean π-field. The right: Density dependence of χ for the nucleon, and of **k** (measured by m) and ϕ (measured by m/g_R) for the mean π-field.

density. It shows the necessity of considering the possibility of a nonzero mean π-field whenever a RMFT model for nuclear matter is proposed, if we require our theory to be self-consistent.

REFERENCES

1.   Zhang Q-R and Greiner W 1995 Mod. Phys. Lett. A **10** 2809
2.   Zhang Q-R, Ma B-Q and Greiner W 1992 J. Phys. G **18** 2051
3.   Campbel D K, Dashen R F and Manassah J T 1975 Phys. Rev. D**12** 979
4.   Zhang Q-R 1981 Phys. Ener. Fort. Phys Nucl. **5** 314 (in Chinese)
5.   Zhang Q-R 1989 Nuclear equation of state ptA ed W Greiner and H Stöcker (New York:Plenum) p.625
6.   Zhang Q-R 1981 Phys. Lett. **104B** 347
7.   Zhang Q-R 1981 Phys. Ener. Fort. Phys Nucl. **5** 15 (in Chinese)
8.   Myers W D and Swiatecki W J 1974 Ann. Phys. **84** 186
9.   Gao C-Y, Guo H, Zhang Q-R 1999 Int. J. Mod.Phys. E **8** 39

Meson effects on the chiral symmetry breaking and restoration

Mei Huang*, Pengfei Zhuang* and Weiqin Chao†

*Physics Department, Tsinghua University, Beijing 100084, China
†IHEP, Chinese Academy of Sciences, Beijing 100039, China

Abstract. Based on the self-consistent chiral symmetric scheme of the SU(2) Nambu-Jona-Lasinio model, the meson effects have been investigated on the chiral symmetry breaking vacuum and chiral symmetry restoration by exactly calculating the next-to-leading (NLO) order Feynman diagrams in the $1/N_c$ expansion.

INTRODUCTION

The Nambu-Jona-Lasinio (NJL) model has been regarded as the cornerstone of understanding the chiral symmetry breaking mechanism, and has been widely used in the mean-field approximation. Even though there is no confinement and non-renormalizability, it is still an attractive tool to investigate the chiral phase transiton. However, the thermodynamical system described by the NJL model in the mean-field approximation is dominated by unconfined quark degrees of freedom even at low temperature and density, while the physical meson degrees of freedom, especially the pions should play important roles in this case. Furthermore, the mean-field approximation corresponds to the leading order (LO) in the $1/N_c$ expansion, since $N_c = 3$ in the real world, the next-to-leading order's contribution will be very important, which should not be neglected. With the above and other reasons, most interests had been paid on the NJL model beyond mean-field approximation. This talk is based on our recent two papers [1][2].

MESON FLUCTUATIONS AND CORRECTIONS ON THE QUARK CONDENSATE

The two-flavor NJL model is defined through the Lagrangian density,

$$\mathcal{L} = \bar{\psi}(i\gamma^\mu\partial_\mu - m_0)\psi + G[(\bar{\psi}\psi)^2 + (\bar{\psi}i\gamma_5\vec{\tau}\psi)^2], \tag{1}$$

where G is the effective coupling constant of dimension GeV^{-2}, m_0 is the current quark mass, assuming isospin degeneracy of u and d quarks, ψ, $\bar{\psi}$ are quark fields with flavor, color, and spinor indices suppressed.

CP597, *Nonequilibrium and Nonlinear Dynamics in Nuclear and Other Finite Systems,*
edited by Z. Li et al.

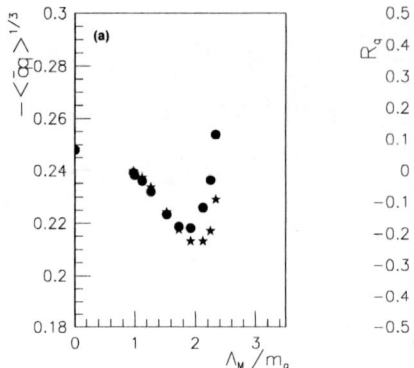

FIGURE 1. Quark condensate (1.a) and the magnitude of mesonic fluctuations and correction on chiral condensate (1.b) as a function Λ_M/m_q. The solid circles and stars in (1.a) correspond to $-<\bar{q}q>^{1/3}$ and $-<\bar{q}q>_H^{1/3}$ respectively, and in (1.b) correspond to the mesonic corrections R_q^{cr} and meson fluctuations R_q^{fl} on quark condensate respectively.

In the chiral symmetric self-consistent scheme [3], the gap equation for the quark mass can be expressed as

$$m_q = m_0 + \Sigma_H + \Sigma^{fl}, \tag{2}$$

where Σ_H and Σ^{fl} are the leading and subleading order of the quark self-enery in the $1/N_c$ expansion. There is a simple relation between the quark condensate and the constituent quark mass m_q,

$$-<\bar{q}q> = -(<\bar{q}q>_H + <\bar{q}q>^{fl}) = \frac{\Sigma_H + \Sigma^{fl}}{4G}. \tag{3}$$

The magnitude of the mesonic quantum fluctuations on the chiral condensate is defined as the ratio of the subleading order contributions over the leading order contributions

$$R_q^{fl} = \Sigma^{fl}/\Sigma_H. \tag{4}$$

The ratio of mesonic correctionsis defined as the chiral condensate including mesonic fluctuation over the mean-field value

$$R_q^{cr} = (<\bar{q}q> - <\bar{q}q>^0)/<\bar{q}q>^0, \tag{5}$$

where $<\bar{q}q>^0$ is the quark condensate calculated in the mean-field approxiamtion.

The total (solid circles) and LO (stars) quark condensate, $-<\bar{q}q>^{1/3}$ and $-<\bar{q}q>_H^{1/3}$, are plotted as a function of Λ_M/m_q in Fig. 1.a, the value at $\Lambda_M/m_q = 0$ corresponds to the mean-field value $-(<\bar{q}q>^0)^{1/3}$. In Fig.1.b, the magnitude of mesonic fluctuations R_q^{fl} (stars) and mesonic corrections R_q^{cr} (solid circles) to quark condensate

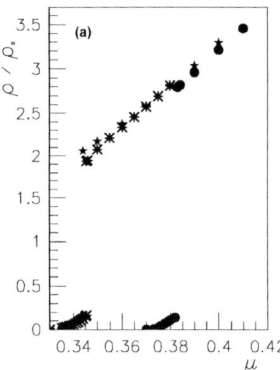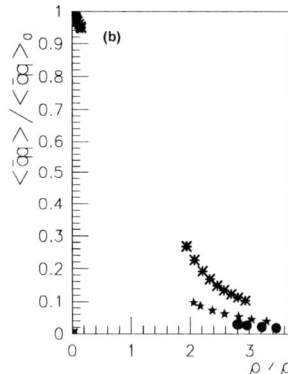

FIGURE 2. The scaled baryon density as a function of the chemical potential μ in (2.a), and the scaled quark condensate in the whole density region in (2.b), the stars correspond to the results in the mean-field approximation, the pentagrams and the solid circles correspond to the results beyond mean-field approximation with the first and second group parameters, respectively.

are plotted as a function of Λ_M/m_q. We can see that the amplitude of meson fluctuations, i.e., the pure meson-loop contributions, can be characterized by the meson-momentum cut-off Λ_M, the larger Λ_M the larger meson-loop contibutions. While the meson corrections, i.e., the difference between in and beyond mean-field approximation, do not vary monotonously with Λ_M, they reach the largest values 30% $\sim$ 40% at about $\Lambda_M/m_q = 2$, then decrease fast. This is because the meson corrections to the mean-field approximation include two parts, the pure meson-loop contributions and the deviation from the one-quark-loop part due to the change of quark mass, and the pure meson-loop contributions are contrary to the deviation from the contributions of the one-quark-loop part due to changing of quark mass. The mesonic corrections to the mean-field approximation are in appropriate scale 10% $\sim$ 40% in the reasonable meson momentum cut-off $\Lambda_M < 1000MeV$.

MESON EFFECTS ON THE CHIRAL PHASE TRANSITION

The obtained model parameters in the mean field approximation are $\Lambda_F = 637.7MeV$, $G\Lambda_F^2 = 2.16$, and $- <\bar{q}q>^{1/3} = 248.0MeV$ with quark mass $m_q = 330MeV$. Going beyond mean-field approximation, we choose two series of NJL parameters: $\Lambda_F = 562.3MeV$, $G\Lambda_F^2 = 2.31$, and $- <\bar{q}q>^{1/3} = 223.2MeV$ with $\Lambda_M/m_q = 1.5$ and $m_q = 330MeV$, and $\Lambda_F = 515.7MeV$, $G\Lambda_F^2 = 2.34$, $- <\bar{q}q>^{1/3} = 218.2MeV$ with $\Lambda_M/m_q = 2$ and $m_q = 370MeV$. Our results of investigating the chiral phase transition by using the above parameters are shown as a function of the scaled density $n_b = \rho/\rho_0$, where $\rho_0 = 0.17fm^{-3}$ is the normal nuclear matter density,

The scaled baryon density as a function of the chemical potential μ is shown in Fig. 2.a, the stars correspond to the results in mean-field approximation, the pentagrams and the solid circles correspond to the results beyond mean-field approximation with

127

the first and second group of parameters, respectively. The jumps in Fig. 2.a show that chiral phase transition at finite density is of the first order both in and beyond mean-field approximation. In both groups of parameters going beyond mean-field approximation, we can see that the meson corrections speed up the chiral phase transition, and the meson corrections extend the mixed phase of the first-order chiral transtion, while the degree of the change depends on the magnitude of the meson fluctuations.

We show the quark condensate scaled by its vacuum value $< \bar{q}q > / < \bar{q}q >_0$ in Fig. 2.b as a function of the scaled density $n_b = \rho/\rho_0$. It is found that in the mean-field approximation, the quark condensate decreases to 30% of its vacuum value after the chiral phase transition, while it decreases to 10% of the vacuum value when meson corrections are included for the first group of parameters, and even decreases to nearly zero for the second group of parameters. This means that the meson corrections can help to melt the chiral condensate more completely.

Acknowledgements: This work was supported in part by the NSFC under Grant No. 19925519 and by the Major State Basic Research Development Program under Contract No. G2000077407.

REFERENCES

1. Mei Huang, P.F. Zhuang, Weiqin Chao, hep-ph/0102305,accepted by Phys.Lett.**B**.
2. Mei Huang, P.F. Zhuang, Weiqin Chao, hep-ph/0011010.
3. V.Dmitrasinovic, H.-J.Schulze, R.Tegen, Ann.Phys.(NY) **238**(1995), 2394.

Density Dependence of Some Quark Condensates in Nuclear Matter in an Effective Model of QCD[1]

Yu-xin Liu[*,†], Dong-feng Gao[*] and Hua Guo[*]

[*]Department of Physics, Peking University, Beijing 100871, China
[†]Institute of Theoretical Physics, Academia Sinica, Beijing 100080, China

Abstract. With the global color symmetry model (GCM) being extended to finite chemical potential, the density dependence of the local and nonlocal two- and four-quark condensates in nuclear matter are investigated. The calculated results indicate that, the quark condensates decrease gradually with the increasing of the nuclear matter density. Nevertheless no sudden change emerges. As a maximal nuclear matter density is reached, the quark condensates vanish. It indicates that the chiral symmetry is restored. Furthermore the nonlocal quark condensates vanish at a smaller nuclear matter density, if the quarks are separated farther away from each other.

INTRODUCTION

It has been known that quark condensates characterize the nonperturbative structure of the QCD vacuum, and hence play essential roles in describing the hadron structure, and further the properties of nuclear matter and finite nuclei[1, 2, 3, 4, 5]. Then a large body of works on the quark condensates in both free space and hot and dense nuclear matter have been existed (see for example Refs.[1, 2, 3, 4, 5, 6, 7, 8, 9, 10, 11, 12, 13, 14, 15]. In the aspect of describing hadron structure in QCD, not only the local quark condensates $<: \bar{q}(0)q(0) :>$, $<: \bar{q}(0)\Gamma q(0)\bar{q}(0)\Gamma q(0) :>$, but also the nonlocal condensates $<: \bar{q}(x)q(0) :>$, $<: \bar{q}(x)\Gamma q(0)\bar{q}(0)\Gamma q(0) :>$ are required, The later are more important to figure the non-point particle property of hadrons[5, 11, 16, 17, 18]. The investigations are accomplished mainly with the QCD sum rule approach. With the Hellmann-Feynman theorm being implemented, one can also evaluated the quark condensates at hadron level[19]. Almost all the approaches can give the descent feature in nuclear matter at low density. However an "upturn", which is believed to be nonphysical, emerges at higher density in the linear Walecka model[20], Dirac-Brueckner method[21], Schwinger-Dyson formalism[22]. Even though the "upturn" can be eliminated with the Brown-Rho scaling[23] being included in the Walecka model[24], the quark condensate vanishes at a density $\rho \approx 3.5\rho_0$. Such a density for the quark condensate to vanish seems to be not high enough.

Recently, the global color symmetry model (GCM), an effective field theory of QCD, has been shown to be quite successful in describing hadron properties[25, 26, 27, 28, 29].

---

[1] Work supported by the National Natural Science Foundation of China and the Foundation for University Key Teacher by the Ministry of Education

CP597, *Nonequilibrium and Nonlinear Dynamics in Nuclear and Other Finite Systems,*
edited by Z. Li et al.

With the Dyson-Schwinger equation (DSE) approach[30, 31] of QCD being extended to finite temperature, the deconfinement and chiral symmetry restoration, the π and ρ meson properties, and a part of the baryon properties in nuclear matter have been investigated[32, 33, 34, 35, 36]. More recently, with the DSE approach and the GCM, the quark condensates in vacuum[14, 15] and those at finite temperature[5] have also been evaluated. However, the condensates at finite density nuclear matter have not yet been analyzed in the DSE scheme and has not in the GCM either. With the global color symmetry model being extended to finite chemical potential μ, we investigate the density dependence of the local and nonlocal two- and four-quark condensates in nuclear matter.

FORMALISM

Extending the GCM in free space to finite chemical potential μ, we have the action

$$S = \int d^4x d^4y \bar{q}(x) \left[\gamma \cdot \partial \delta(x-y) - \mu\gamma_4 + \Lambda^\theta B^\theta(x,y) \right] q(y) + \int d^4x d^4y \frac{B^\theta(x,y)B^\theta(y,x)}{2g^2D(x-y)} ,$$

where Λ^θ are matrices of the Fierz transformation among the spin, color, and flavor spaces of the quarks, $D(x-y)$ is the effective gluon propagator and $B^\theta(x,y)$ is a bilocal field, which can be generally written as

$$B^\theta(x,y) = B^\theta_0(x,y) + \sum_i \frac{B^\theta_0(x,y)}{f_i} \phi^\theta_i(\frac{x+y}{2}), \tag{1}$$

where $B^\theta_0(x,y) = B^\theta_0(x-y)$ is the vacuum configuration of the bilocal field. In the lowest order approximation with only the Goldstone boson being taken into account, the ϕ^θ_i includes σ and π mesons, and f_i $(i = \sigma, \pi)$ stands for the decay constant of σ, π mesons, respectively. The vacuum configuration can be determined by the saddle-point condition $\frac{\partial S}{\partial B^\theta_0} = 0$, After some derivation, one can obtain the vacuum configuration in momentum space by solving the rainbow Dyson-Schwinger equations

$$[A(\tilde{p}) - 1]\tilde{p}^2 = \frac{8}{3} \int \frac{d^4q}{(2\pi)^4} g^2 D(p-q) \frac{A(\tilde{q})\tilde{q}\cdot\tilde{p}}{A^2(\tilde{q})\tilde{q}^2 + B^2(\tilde{q})}, \tag{2}$$

$$B(\tilde{p}) = \frac{16}{3} \int \frac{d^4q}{(2\pi)^4} g^2 D(p-q) \frac{B(\tilde{q})}{A^2(\tilde{q})\tilde{q}^2 + B^2(\tilde{q})}, \tag{3}$$

where $\tilde{p} = p + i\mu$. Basing on the solution of the Dyson-Schwinger equations, one can determine the bilocal field and obtain the GCM action.

In the way proposed in Refs.[[13, 14, 15], the quark condensates can be obtained under the mean-field approximation. For instance, the local two-quark and four-quark condensates can be given as

$$<: \bar{q}q :> = <: \bar{q}(0)q(0) :> = -\frac{12}{(2\pi)^4} \int_{\mu^2}^\infty s' ds' \frac{B(s')}{A^2(s')s' + B^2(s')}, \tag{4}$$

130

$$<: \bar{q}(0)\gamma_\mu \frac{\lambda_C^a}{2} q(0)\bar{q}(0)\gamma_\mu \frac{\lambda_C^a}{2} q(0) :>= -\frac{4}{9}[<: \bar{q}q :>]^2 \tag{5}$$

where $s' = \tilde{p}^2$. Meanwhile, the nonlocal quark condensates are given as

$$<: \bar{q}(x)q(0) :>= -Tr_{\gamma C}[G_0(x,0)]$$

$$= -\frac{12}{16\pi^2} \int_{\mu^2}^{\infty} s'ds' \frac{B(s')}{s'A^2(s')+B^2(s')} \left[\frac{2J_1(\sqrt{s'x^2})}{\sqrt{s'x^2}}\right] \tag{6}$$

$$<: \bar{q}(x)\gamma_\mu \frac{\lambda_C^a}{2} q(x)\bar{q}(0)\gamma_\mu \frac{\lambda_C^a}{2} q(0) :>$$

$$= -\frac{4}{9}[<: \bar{q}(x)q(0) :>]^2 + \frac{32}{(2\pi)^4} \left[\int_{\mu^2}^{\infty} \frac{A(s')}{s'A^2(s')+B^2(s')} \frac{I_2(i\sqrt{s'x^2})}{ix} s'ds'\right]^2 \tag{7}$$

It is definite that the presently obtained result is consistent with that given in the naive vacuum saturation hypothesis if the second term in Eq. (7) is neglected.

In the practical calculation, since the knowledge about the exact behavior of g^2 and $D(p - q)$ in low energy region is still lacking, one has to take some approximations or phenomenological form to solve the Dyson-Schwinger equations. For simplicity, we adopt the infrared dominative form[37, 25]

$$g^2 D(p - q) = \frac{3}{16}\eta^2\delta(p - q), \tag{8}$$

where η is an energy-scale parameter and can be fixed by experiment data of hadrons. We have then

$$A(\tilde{p}) = 2, \qquad\qquad B(\tilde{p}) = (\eta^2 - 4\tilde{p}^2)^{1/2}, \quad \text{for } \tilde{p}^2 < \frac{\eta^2}{4}, \tag{9a}$$

$$A(\tilde{p}) = \frac{1}{2}\left[1 + \left(1 + \frac{2\eta^2}{\tilde{p}^2}\right)^{1/2}\right], \quad B(\tilde{p}) = 0, \qquad\qquad \text{for } \tilde{p}^2 > \frac{\eta^2}{4}. \tag{9b}$$

In order to investigate the dependence of the quark condensates on nuclear matter density ρ explicitly, we must transfer the above obtained μ-dependence to that of the ρ-dependence. Because of the fermionic properties of quarks, the bags can be approximately regarded as a Fermi-Dirac systems[38], and the relation between the nuclear matter density and the chemical potential is given as

$$\rho_B = \frac{n_q}{3} = \frac{2}{3\pi^2}\mu^3. \tag{10}$$

Combining Eqs. (4-7), (9) and (10), we can obtain the dependence of the quark condensates on nuclear matter density ρ_B.

CALCULATION AND RESULTS

By calibrating the nucleon mass $M_0 = 939$ MeV and radius $R_0 = 0.8$ fm in free space (i.e., $\mu = 0$, $\rho = 0$), we get the energy-scale $\eta = 1.220$ GeV, $Z_0 = 3.303$[29]. Such a best fitted energy-scale η fits well the value 1.37 GeV, which was fixed by a good description of π and ρ meson masses[33, 34], and is much more close to the Bjorken-scale 1.0 GeV (see Ref.[38] and the references therein). The obtained Z_0 is larger than the originally fitted value 1.84[39]. However what we refer to here includes all the effects but not only the zero-point energy. Meanwhile other investigations (see for example Ref.[40]) have shown that the zero-point energy parameter can be larger than 1.84, even though the other effects are taken into account separately. With the above parameters η, Z_0 and Eq.(4), we get at first the local quark condensate in free space $< qq >_0 = -(132 \text{ MeV})^3$. Meanwhile the bag constant of a nucleon in free space is obtained as $\mathcal{B}_0 = (172 \text{ MeV})^4$. It is evident that these values of $< \bar{q}q >_0$ and B_0 are quite close to the results given in Refs.[13, 14] and Ref.[41], respectively.

By varying the chemical potential μ, we get the variation behavior of the ratio of the two- and four-quark condensates (denoted as R_2, R_4, respectively) in nuclear matter to the corresponding values in free space against the nuclear matter density. The results are illustrated in Fig. 1.

Looking over the figure, one may easily realize that, as the density of nuclear matter increases, the quark condensates decrease monotonously. When the nuclear matter density reaches a value larger than 12 times the normal nuclear matter density (referred to as ρ_0), the local quark condensates vanish. Moreover the non-local quark condensates vanish in the less dense nuclear matter. On the other hand, the quark condensate $< qq >$ has been regarded as a manifestation to identify the chiral symmetry breaking. The gradual decrease of the quark condensate indicates that the chiral symmetry is restored gradually as the nuclear matter density increases. It has been shown[29] that nucleons can no longer exist as bags consisting of quarks at the density $\rho_B \geq 12\rho_0$, i.e., the nuclear matter becomes quark matter. The present calculation shows that, when the nuclear matter becomes quark matter, the chiral symmetry is restored completely. It indicates that the quark decoupling phase transition and the chiral phase transition may happen at the same nuclear matter density.

The Fig. 1 indicates also that, with the increasing of the time-space distance, the condensates at a definite density of nuclear matter decrease. To show this point more explicitly, we display the changing feature of the ratio of two-quark and four-quark condensates in nuclear matter to the corresponding values in free space against the time-space distance in Fig. 2 (denoted as $G_2(x^2, \rho)$, $G_4(x^2, \rho)$), respectively. The figure manifests that the condensates decrease with the increasing of the time-space distance among the quarks definitely. Especially, the condensates in nuclear matter with the normal density of nuclear matter vanish completely as the quarks are separated about 1.8 fm ($x^2 \sim 80$ GeV^{-2}) far away from each other. It means that the condensates disappear as the time-space distance among the quarks is larger than the empirical diameter of a nucleon. Meanwhile, the higher the nuclear matter density, the smaller the distance for the condensates exist. Such variation characteristics indicate that separating the time-space distance among the quarks can accelerate the chiral symmetry restoration process.

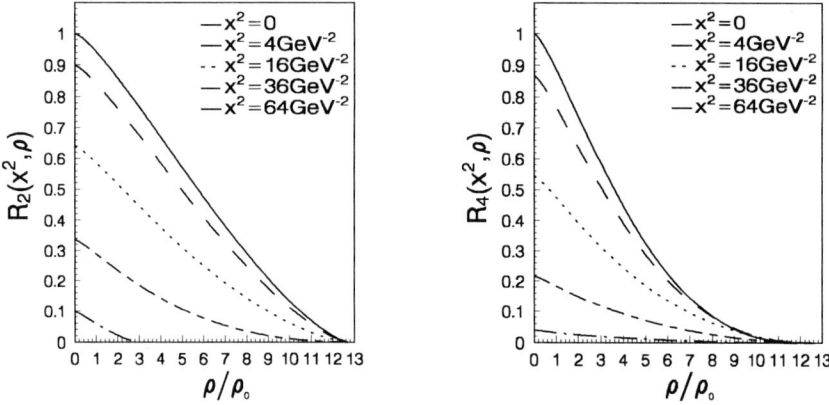

FIGURE 1. Calculated ratio of the two-quark and four-quark condensates in nuclear matter to those in free space at several fixed time-space distance.

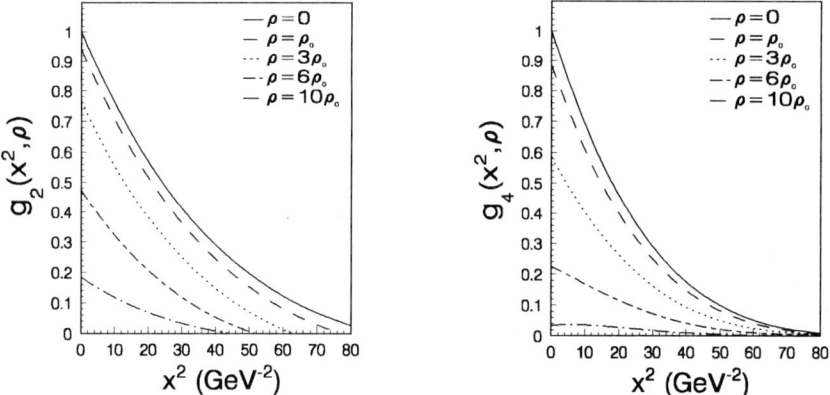

FIGURE 2. Calculated ratio of the non-local two-quark and four-quark condensates in nuclear matter to those in free space at several nuclear matter density.

SUMMARY AND REMARKS

In summary we have investigated the density dependence of the local and non-local two- and four-quark condensates in nuclear matter in an effective field theory model of QCD, namely the global color symmetry model. A maximal density of nuclear matter, which is larger than 12 times the normal nucleon density, for the existence of quark condensates is obtained. The calculated results indicate that the quark condensates decrease with the increasing of the nuclear matter density before the maximal density is reached. As the maximal density is reached, the chiral symmetry is restored. Furthermore the nonlocal quark condensates descend, too, and they vanish at a smaller nuclear matter density, if the quarks are separated farther away from each other. It provides a clue that the chiral restoration process can be enhanced if the time-space distance among the quarks

133

is enlarged.

In the present calculation, the $g^2 D(p-q)$ is taken to be proportional to a $\delta-$function. However, the detailed effects of the running coupling constant, the gluon propagator $D(p-q)$ and the other degrees of freedom on the changing feature have not yet been included. Meanwhile, the relation between the chemical potential and the nuclear matter density was handled with a simple correspondence in statistical mechanics. It means that the present calculation is an approximation of the scheme. Then a more sophisticated investigation is under progress.

REFERENCES

1. Reinders, L. L., Rubenstein, H., and Yazaki, S., Phys. Rep., **127**, 1 (1985).
2. Drukarev, E. G., and Levin, E. M., Nucl. Phys., A **511**, 679 (1990); Prog. Part. Nucl. Phys., **27**, 77 (1991).
3. Henley, E. M., and Pasupathy, J., Nucl. Phys., A **556**, 467 (1993).
4. Jin, X., Phys. Rev., C **51**, 2260 (1995); C **52**, 2964 (1995).
5. Johnson, M. B., and Kisslinger, L. S., Phys. Rev., C **61**, 074014 (2000).
6. Shifman, et al., Nucl. Phys., B **147**, 385 (1979); B **147**, 448 (1979).
7. Ioffe, B. L., Nucl. Phys., B **188**, 317 (1981); B **191**, 591 (1981).
8. Bochkarev, A. I., and Shaposhikov, M. E., Nucl. Phys., B **268**, 220 (1986).
9. Cohen, T. D., Furnstahl, R. J., and Griegel, D. K., Phys. Rev. Lett., **67**, 961 (1991); Furnstahl, R. J., Griegel, D. K., and Cohen, T. D., Phys. Rev., C **46**, 1507 (1992).
10. Celenza, L. S., Shakin, C. M., Sun, W. D., and Sun, X. Q., Phys. Rev., C **48**, 159, (1993).
11. Jung, H., and Kisslinger, L. S., Nucl. Phys., A **586**, 682 (1995).
12. Klingl, F., Kim, S., Lee, S. H., Morath, P., and Weise, W., Phys. Rev. Lett., **82**, 3396 (1999).
13. Meissner, T., Phys. Lett., B **405**, 8 (1997).
14. Kisslinger, L. S., and Meissner, T., Phys. Rev., C **57**, 1528 (1998).
15. Zong, H. S., Lü, X. F., Gu, J. Z., Chang, C. H., and Zhao, E. G., Phys. Rev., C **60** (1999) 055208.
16. Bakulev, A. P., and Radyushkin, A. V., Phys. Lett., B **271**, 223 (1991).
17. Mikhailov, S. V., and Radyushkin, A. V., Phys. Rev., D **45**, 1754 (1992).
18. Wang, Z. G., Wan, S. L., and Wang, K. L., Phys. Lett., B **498**, 195 (2001).
19. Cohen, T. D., Furnstahl, R. J., and Griegel, D. K., Phys. Rev., C **45**, 1881 (1992).
20. Malheiro, M., Dey, M., Delfino, A., and Dey, J., Phys. Rev., C **55**,521(1997).
21. Li, G. Q., and Ko, C. M., Phys. Lett., B **338**, 118 (1994).
22. Mitsumori, T., Noda, N., Kouno, H., Hasegawa, A., and Nakano, M., Phys. Rev., C **55**, 1577 (1997).
23. Brown, G. E., and Rho, M., Phys. Rev. Lett., **66**, 2720 (1991).
24. Guo, H., Yang, S., and Liu, Y. X., Sci. in China, A **44**, 1042 (2001).
25. Cahill, R. T., and Roberts, C. D., Phys. Rev., D **32** (1985) 2419; Cahill, R. T., and Roberts, C. D., and Praschifka, J., Ann. Phys. (N.Y.), **188**, 20 (1988); Roberts, C. D., Cahill, R. T., Sevior, M. E., and Iannella, N., Phys. Rev., D **49**, 125 (1994).
26. Frank, M. R., and Tandy, P. C., Phys. Rev., C **46**, 338 (1992); Johnson, C. W., and Fai, G., Phys. Rev., C **56**, 3353 (1997).
27. Tandy, P. C., Prog. Part. Nucl. Phys., **39**, 177 (1997).
28. Lü, X. F., Liu, Y. X., Zong, H. S., and Zhao, E. G., Phys. Rev., C **58**, 1195 (1998).
29. Liu, Y. X., Gao, D. F., and Guo, Hua, Nucl. Phys., A **695**, 362 (2001).
30. Itzyson, C., and Zuber, J-B., Quantum Field Theory (McGraw-Hill, New York, 1985).
31. Roberts, C. D., and Williams, A. G., Prog. Part. Nucl. Phys., **33**,477(1994).
32. Bender, A., Blaschke, D., Kalinovsky, Yu L., and Roberts, C. D., Phys. Rev. Lett., **77**, 3724 (1996); Bender, A., Roberts, C. D., and Smekal, L. V., Phys. Lett., B **380**, 7 (1996); Bender, A., Poulies, G. I., Roberts, C. D., and Schmidt, S., and Thomas, A. W., Phys. Lett., B **431**, 263 (1998).
33. Blaschke, D., Roberts, C. D., Schmidt, S., Phys. Lett., B **425**, 232 (1998).
34. Maris, P., Roberts, C. D., and Schmidt, S., Phys. Rev., C **57**, 2821 (1998).

35.   Blaschke, D., Grigorian, H., Poghosyan, G., Phys. Lett., B **450**, 207 (1999).
36.   Harada, M., and Shibata, A., Phys. Rev., D **59**, 014010 (1999).
37.   Munczek, H. J., and Nemirovsky, A. M., Phys. Rev., D **28**, 3081 (1983).
38.   Collins, J. C., and Perry, M. J., Phys. Rev. Lett., **34**, 1353 (1975).
39.   Hasenfratz, P., and Kuti, J., Phys. Rep., **40**, 75 (1978).
40.   Miskinis, P., and Karlikauskas, G., Nucl. Phys., A **683**, 339 (2001).
41.   Jin X., and Jennings B. K., Phys. Lett., B **374**, 13 (1996); Phys. Rev., C **54**, 1427 (1996); C **55**, 1567 (1997).

Analysis of multi-particle production by two-source statistical model

Zhong-Dao Lu[*,†,**], Amand Faessler[†], C. Fuchs[†] and E.E. Zabrodin[†,‡]

*China Institute of Atomic Energy, P.O.Box 275(18), Beijing 102413, China
†Institute for Theoretical Physics, University of Tüebingen, D-72076 Tuebingen, Germany
**The China Center of Advance Science and Technology (CCAST), Beijing 100080, China
‡Institute for Nuclear Physics, Moscow State University, RU-119899 Moscow, Russia

Abstract. A two-source statistical model of an ideal hadron gas is developed and used to analyse the experimental data on hadron yields and ratios in 158A GeV Pb+Pb collisions. The model results fit to the experimental data very well and provide a reasonable and clear physical picture. The whole particle source is composed of a hot small inner source and a bigger and cooler outer source. Baryons are mostly distributed in the outer source, while almost all anti-baryons are contained in the inner source. The strangeness density is found to be non-zero having opposite signs in the two sources.

INTRODUCTION

Searching for the quark-gluon plasma (QGP) is one of the major objective in the study of the relativistic heavy ion collisions. The principal question is whether the strongly interacting nuclear, or rather parton matter, reaches the stage of chemical and thermal equilibrium. One possible approach is to study the equilibration process within microscopic models [1, 2]. An alternative way is to fit macroscopic observables obtained in experiments to the statistical model (SM). Here the hadron gas is assumed to form a fireball which is fully equilibrated, i.e., the fireball is a single source with an exclusive temperature. The simplicity of the SM has led to an abundant literature (see [3-8]). However, some imperfections of this simple model are found in explaining the experimental data. Firstly, it cannot well fit to both the particle yields and their ratios simultaneously. For instance, the predicted yield of ϕ meson is twice as high as the measurement. In order to reduce its production, a strangeness suppression factor has been introduced [5]. Secondly, the particle density and the energy density predicted by the model are too large. To resolve this problem the effect of a repulsive hard core of the particles is taken into account, which leads to a Van der Waals type equation of state (EOS) of the hadron gas [3, 4]. In the study of equilibration, however, the microscopic models reveal that at SPS energies the global equilibrium is not reached even at late stage of the reaction [1], and that the baryon distribution has a dip around midrapidity [2]. This finding indicates that the SM based on a single source should be modified. Inspired by this finding, as well as by the experimental observation of low net baryon density in the midrapidity range, we develop a two-source statistical model (TSM). The model divides the whole particle source into two regions, that are assumed to be in local equilibrium and are al-

CP597, *Nonequilibrium and Nonlinear Dynamics in Nuclear and Other Finite Systems,*
edited by Z. Li et al.
© 2001 American Institute of Physics 0-7354-0041-5/01/$18.00

lowed to have different thermodynamic quantities. Especially, the strangeness density is no longer kept zero everywhere as in the single source case.

THE MODEL

The statistical model formulas based on a grand canonical partition function can be found in Refs. [3-8]. For the ideal hadron gas composed of Fermions and Bosons the particle density of species i is

$$n_i = \frac{g_i}{2\pi^2} \int_0^\infty dq \frac{q^2}{e^{(\varepsilon_i - \mu_i)/T} \pm 1} , \qquad (1)$$

where g_i and m_i are the degeneracy and mass of species i, and ε_i, q, and μ_i are its energy, momentum, and chemical potential, respectively. The μ_i can be expressed in terms of baryon chemical potential μ_B and strangeness chemical potential μ_S as $\mu_i = b_i \mu_B + s_i \mu_S$, where b_i and s_i are the baryon charge and strangeness charge of hadron i.

In the two-source statistical model the whole particle source is divided into two regions: the outer region (source 1 or S1) and the inner region (source 2 or S2). The sources are in local thermal and chemical equilibrium, but may possess different temperatures, particle and energy densities, etc. The strangeness densities in the two sources may also have different values. Because of the net strangeness conservation, their signs should be opposite.

The characteristics of a single fireball can be described in terms of four independent parameters. Since the total strangeness number is conserved and equals zero, the number of free parameters in the SM is reduced to three. We choose temperature T, volume V and net baryon density ρ_B as the free parameters. It is worth to mention that the condition of baryon number conservation is not used here, since the number of measured participants is an averaged number. In the TSM the number of free parameters increases to seven. Our choice for them is as follows: Besides temperature, volume and net baryon density of the two sources, i.e. T_1, V_1, ρ_{B1} and T_2, V_2, ρ_{B2}, the strangeness density in source 1, ρ_{S1}, is taken as a free parameter.

RESULTS AND DISCUSSIONS

The yields and ratios of hadrons in 158 A GeV Pb+Pb collisions are listed in Table 1. All hadrons including their resonances and possible anti-particles with masses less than 2 GeV are included in the fitting procedure. No additional constraints such as strangeness suppression or excluded volume are assumed except the feeding-back effect from resonance decay. The relative deviations of the TSM predictions from the fitting data are quite small (8% and less). This results in a high quality of the least-squares fit, $\chi^2/DOF = 4.0/3$, whereas the least-squares fit of the SM is poor, $\chi^2/DOF = 41/7$. Compared to the ideal SM, the TSM significantly improves the agreement with the experimental data.

TABLE 1. Hadron yields and ratios in 158 A GeV Pb+Pb collisions.

	Data	SM	TSM	$\frac{\text{TSM}-\text{Data}}{\text{Data}}\%$	Reference
N_B	372±10	363	375	0.81	[9]
h^-	680±50	606	638	-6.2	[10]
K_s^0	68±10	61.6	65.0	-4.5	[10]
ϕ	7.6±1.1	13.4	8.17	7.5	[11]
$p-\bar{p}$	155±20	125	148	-4.4	[10]
K^+/K^-	1.8± 0.1	1.99	1.71	-5.3	[12]
$\bar{p}/p$	0.07± 0.01	0.065	0.069	-1.1	[13]
$\bar{\Xi}/\Xi$	0.249± 0.019	0.220	0.230	-7.5	[14]
$\bar{\Omega}/\Omega$	0.383± 0.081	0.411	0.409	6.9	[14]
$\bar{\Lambda}/\Lambda$	0.128± 0.012	0.123	0.138	7.8	[15]
χ^2/DOF		41/7	4.0/3		

It is already known that the ideal SM (without strangeness suppression and excluded volume effects) underpredicts the number of negatively charged hadrons h^- and overestimates the yield of ϕ mesons. The problem can be cured by increasing the volume V and decreasing the temperature T of the source [4]. However, the antibaryon to baryon ratios become completely wrong. The TSM enables one to get both the correct multiplicities of h^- and ϕ, and the correct antibaryon/baryon ratios. In the SM, in order to reduce the production of ϕ meson, the strangeness suppression is introduced. In the TSM the yield of ϕ meson is not enhanced, which means that there is no need of strangeness suppression. In fact, according to [15], particles with more strangeness have a larger increase rate and, therefore, the strangeness production should be enhanced instead of being suppressed. This dilemmatic problem may be caused by the assumption that the strangeness is uniformly distributed and its net density equals zero everywhere. The TSM proposes an original solution of this question. The strangeness density in the TSM is non-zero and has different values with opposite signs in the two sources, negative in S2 and positive in S1. The non-zero and non-uniform distribution of the strangeness density results in the good fit for both ϕ and multistrange particle production.

The thermodynamic quantities obtained from fitting the experimental data are shown in Table 2. One can easily find out the sizable differences in temperature and volume in the two sources. The temperature of 149 MeV in source 2 is significantly higher than that of 116 MeV in source 1. The volume of the outer source is almost four times larger than that of the inner source. This gives us a clear and reasonable picture that the two-source object is composed of a hot, relatively small core surrounded by a cooler and larger halo. The size of the halo is consistent with experimental estimates from pion interferometry [16]. The net baryon density in the halo is found to be four times as large as that in the core. Hence, the major part of baryons is contained in the outer source, while the inner source contains almost all antibaryons. This indicates that the outer source characterizes the projectile-like and target-like body and the inner source characterizes the central reaction zone. The total net baryon charge equals, however, the initial number of participants.

The energy density in the TSM is much lower than that in the SM. In the standard SM the energy density and the particle density are too high to treat the system at chemical

TABLE 2. Thermodynamic characteristics of sources

	TSM (S1)	TSM (S2)	SM
$T\,[MeV]$	116	149	157
$V\,[fm^3]$	12862(5.36V_0)	2714(1.13V_0)	4141
$\rho_B\,[fm^{-3}]$	0.028	0.0068	0.088
$\rho_S\,[fm^{-3}]$	0.0005	-0.0023	0.0
$\varepsilon\,[MeV/fm^3]$	74.9	232	423
$P\,[MeV/fm^3]$	11.3	39.6	65.0
N_B	357.2	53.7	396
$N_{\bar{B}}$	0.56	35.4	33.2
$N_B - N_{\bar{B}}$	356.6	18.3	363
N_S	52.4	61.6	112
$N_{\bar{S}}$	46.0	68.0	112
$N_S - N_{\bar{S}}$	6.4	-6.4	0
$\mu_B\,[MeV]$	395	31.7	217
$\mu_S\,[MeV]$	58.1	1.21	49.1

freeze-out as a gas of point-like particles. To resolve this problem the introduction of a repulsive hard-core volume for hadrons might be important [3, 4]. In the TSM neither the energy density nor the particle density is so large and the incorporation of excluded volume effects (at least for the halo) becomes less important. The total volume of the TSM is about four times that of the SM volume. The energy density in S2 is about three times larger than that in S1. Such a low energy density in the outer source corresponds to the energy density at thermal freeze-out rather than at chemical freeze-out.

Acknowledgements. We are grateful to L. Bravina for fruitful discussions and acknowledge the support from NSFC (No. 19975075) of China and from DFG and BMBF (No. 06TÜ986) of Germany.

REFERENCES

1. L.V. Bravina et al., Phys. Rev. C 60 (1999) 024904; *ibid.* 62 (2000) 064906; J. Phys. G 25 (1998) 351.
2. J. Sollfrank, U. Heinz, H. Sorge, N. Xu, Phys. Rev. C 59 (1999) 1637.
3. P. Braun-Munzinger et al., Phys. Lett. B 365 (1996) 1; Phys. Lett. B 465 (1999) 15.
4. G.D. Yen et al., Phys. Rev. C 56 (1997) 2210; G.D. Yen and M.I. Gorenstein, Phys. Rev. C 59 (1999) 2788.
5. J. Rafelski, Phys. Lett. B 62 (1991) 333.
6. J. Cleymans and K. Redlich, Phys. Rev. C 60 (1999) 054908.
7. Z.D. Lu et al., High Energy Phys. Nucl. Phys. 22 (1998) 910.
8. Z.Y. Zhang, Y.W. Yu, C.R. Ching, T.H. Ho, Z.D. Lu, Phys. Rev. C 61 (2000) 065204.
9. I. Huang, NA49 Collab., Ph.D. thesis, University of California at Davis (1997).
10. S.V. Afanasjev et al., NA49 Collab., Nucl. Phys. A 610 (1996) 188c.
11. S.V. Afanasjev et al., NA49 Collab., Phys. Lett. B 491 (2000) 59.
12. C. Borman et al., NA49 Collab., J. Phys. G 23 (1997) 1817.
13. I.G. Bearden et al., J. Phys. G 23 (1997) 1865.
14. R. Caliandro et al., WA97 Collab., J. Phys. G 25 (1999) 171.
15. E. Andersen et al., WA97 Collab., Nucl. Phys. A 638 (1998) 115c; Phys. Lett. B 433 (1998) 209.
16. H. Appelshäuser et al., NA49 Collab., Eur. Phys. J. C 2 (1998) 611.

PART III

TRANSPORT THEORY APPROACH
TO INTERMEDIATE AND
HIGH ENERGY
HEAVY ION COLLISIONS

The production of K^+ mesons in heavy ion reactions

Christoph Hartnack and Jörg Aichelin

SUBATECH, UMR Université, Ecole des Mines, IN2P3/CNRS F-44072 Nantes, France

Abstract. We report on the recent results obtained in QMD calculations of the K^+ production in heavy ion reactions focusing on the physical conclusions one can draw comparing the simulations with the recent data of the KAOS and FOPI collaboration.

INTRODUCTION

The production of K^+ mesons in nuclear reactions is a topic which faces increasing interest from experimentalists as from theoreticians since the first exploratory experiments have been performed about 20 years ago. This is a consequence of the many facets of this production process which allows to study a variety of physical questions. At the beginning the interest was concentrated on the production mechanism and already the first experiments at subthreshold energies have shown that the K^+ yield can be only understood if the two step process with an intermediate resonance is the dominant production channel. The relative momentum in a nucleus-nucleus collision at subthreshold energies at and below 1 AGeV is rarely sufficient to create this meson.

The intermediate resonance carries in form of its mass already a larger fraction of the energy needed to create a K^+ and therefore the relative momentum can be smaller. This dependence on the intermediate resonance allows to use the observed K^+ production cross section to investigate the properties of the resonance in the nuclear environment. Since the probability that the resonance collides with a nucleon is given by $v\tau/\lambda$, where τ is the resonance lifetime and λ is its mean free path in the nuclear environment, the K^+ production depends also on the nuclear density which is achieved in a heavy ion collision. The density itself depends on the equation of state of nuclear matter and therefore there may be the hope that kaons probe its compressibility. In recent times K^+ gained renewed interest because it became evident that the high precision data, available in the mean time, cannot be explained assuming that the properties of the kaons remain unchanged in the nuclear environment. Therefore turning this observation into an advantage there may be the chance to extract from experimental data the properties of the kaons in matter which is compressed to 2 times normal nuclear matter density. This is not only an interesting topic in itself but its knowledge is also a condition for the understanding of how hadronic matter approaches the chiral and confinement phase transition towards a quark gluon plasma. In this contribution we limit ourselves to three of the above mentioned topics. We will first of all study whether the kaon yield tells us something about the nuclear equation of state and how the change of the in

CP597, *Nonequilibrium and Nonlinear Dynamics in Nuclear and Other Finite Systems*,
edited by Z. Li et al.
© 2001 American Institute of Physics 0-7354-0041-5/01/$18.00

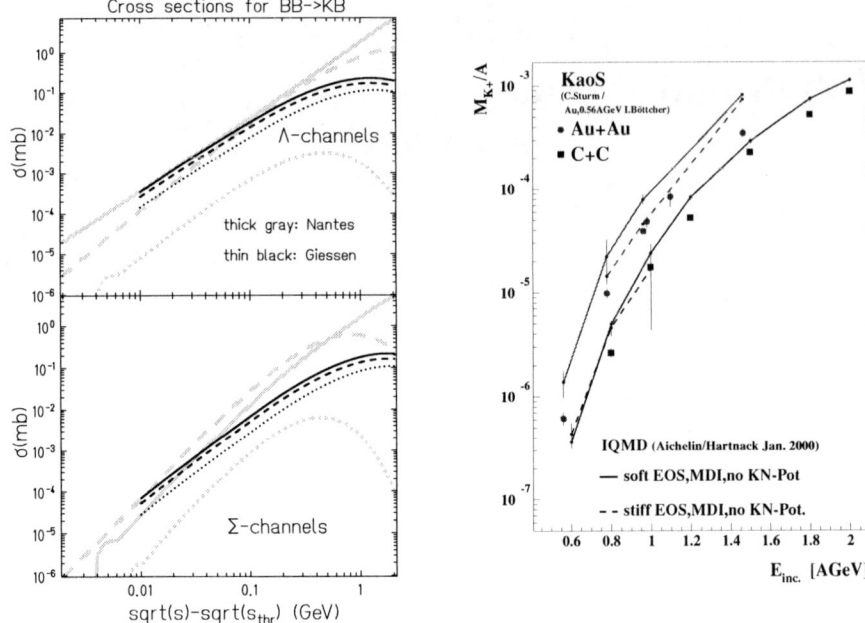

FIGURE 1. Excitation function of K^+ production par participating nucleon for the systems C+C and Au+Au for a soft and hard equation of state, respectively, with momentum dependent interactions. The theoretical curves are compared with experiment. The kaons are assumed to have their on shell mass

medium properties of the K^+'s influences the results on the excitation function of the K^+ production. Then we will concentrate on a physical understanding of the production cross section. Finally we will discuss the in-plane flow of the K^+'s. We will refrain from presenting calculations of the K^- production, which are equally interesting but due to the presence of the $\Lambda(1405)$ resonance difficult to describe. Coupled channel calculations which show how this resonance changes in the medium are presently under way (see the contribution of Lutz in these proceedings) and will allow in near future to simulate the K^- production as well. For the moment the uncertainties are still too large to draw firm conclusions from the comparison of these calculations with data.

KAONS IN THE NUCLEAR ENVIRONMENT

In heavy ion reactions the K^+'s are created by hadron hadron collisions which are sufficiently energetic. There are contributions from meson-baryon, baryon-baryon, resonance-baryon and resonance-resonance reactions. Some of the cross sections are known experimentally. But especially the resonance-baryon and resonance-resonance cross sections are unknown and one has to rely on theoretical calculations which differ quite a bit. Depending on which cross section is used the output of the numerical

144

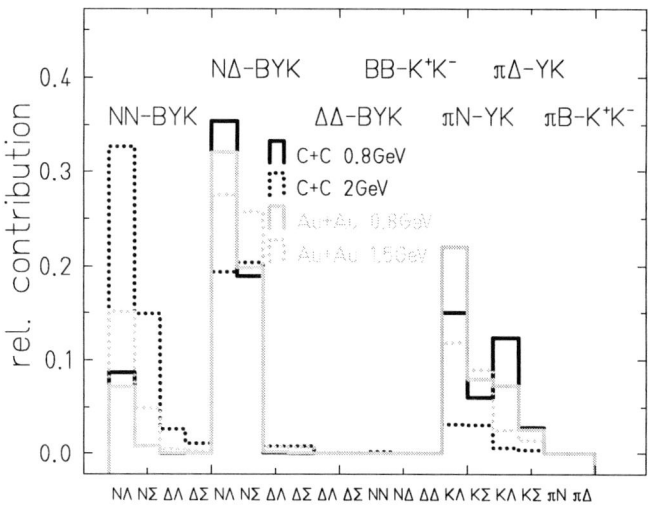

FIGURE 2. Relative contribution of the different production channels for different beam energies and different systems

simulation differ of course. In fig. 1 we display the baryonic cross sections used in two different simulations, in that of the Giessen group[1] and that of the Nantes group[2]. At higher energies the difference in the NN channel comes from different exit states which are included, but this difference is of minor importance at the beam energy considered here. We will discuss later how these differences in the cross section influence the final result. Using these cross sections we performed simulations with the standard IQMD program. For details we refer to ref.[3]. We use 2 different assumptions about the interaction among the nucleons. Both are obtained by using a potential which reproduces the optical potential but they differ in the assumption about the compressibility of nuclear matter. The equation of state (EOS) which is called soft assumes a compressibility of nuclear matter of 200 MeV whereas the hard EOS has a compressibility of 380 MeV. This is about the range one expects from the analysis of experiments (giant resonances) and from nuclear matter calculations. Fig. 2 shows the excitation function of the kaon production yield per participating nucleon for Au+Au and C+C as compared with the data of the KAOS collaboration[4]. It is assumed in this calculation that the kaons are created with their free mass. First of all we see a large difference between the two systems. This indicates that there are strong collective effects (in contradistinction to the pions which are almost proportional to the number of participants). We see as well that for the light system the EOS dependence is negligible. This is expected because the system is too small to become really compressed. Rather projectile and target passes each other with a moderate change in momentum. For the heavy system the situation is different. At lower energies the yield for the different EOS's differs by almost a factor of 2. This is a consequence of the higher compression obtained for a soft EOS which

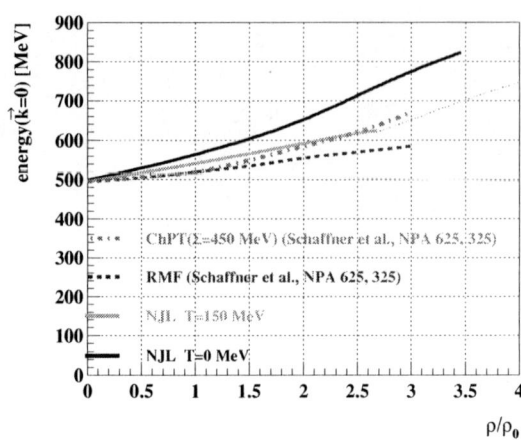

FIGURE 3. Energy of the K^+ for zero momentum as a function of the baryon density obtained in three different models, a mean field approach, chiral perturbation theory(ChPT) and the Nambu Jona Lasinio (NJL) model at two different temperatures of the matter

yields a shorter mean free path. Above threshold kaons can also be produced in first nn collisions and hence the EOS dependence of the cross section is weakened. The IQMD results overpredict systematically the observed kaon yield. We would like to stress that in contradistinction to earlier results we have included here that the optical potential contributes in the average about 40 MeV to the energy which is available for the kaon production because the baryons are slower after the production and therefore their optical potential is more attractive. In addition it is assumed that the nuclear potential of the hyperons is 2/3 of that of the nucleons.

That this excitation function is all but trivial shows fig.3 where the relative contributions of the different channels are displayed. We see that this contributions change by factors of up to 4 from the lowest energy point where the nucleon-Δ and π-baryon channel dominate to the largest energy where the NN collisions contribute most.

In reality the kaons are not produced as free on shell particles but their properties are changed in the medium. The measured scattering length of the K^+-n reaction allows to calculate the increase of the K^+ mass at low nuclear densities. How the kaon masses increase towards higher energies has been calculated in quite different approaches[6, 7], all of them giving slightly different results. In fig. 4 we display the increase of this mass as predicted by several approaches. If we implement these in medium properties of the K^+'s in the simulation we obtain the excitation function as seen in fig.5. The yield has decreased as compared to fig.2 but in addition the sensitivity to the nuclear EOS has almost disappeared. Both is easy to understand. An increasing mass reduces the production because the threshold is higher. The larger density and hence the smaller mean free path for a soft EOS as compared to the hard EOS is now counterbalanced

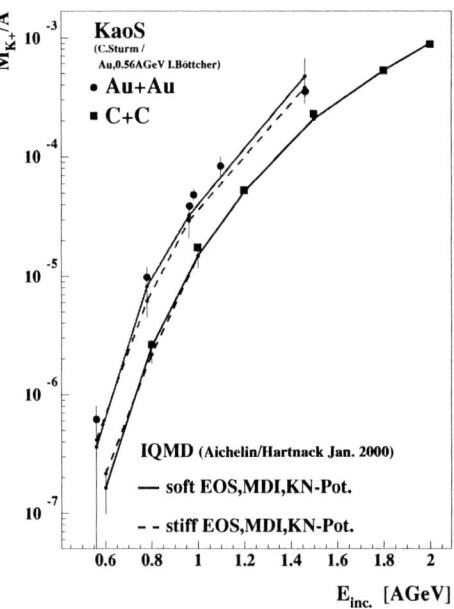

FIGURE 4. Excitation function of K^+ production par participating nucleon for the systems C+C and Au+Au for a soft resp hard equation of state with momentum dependent interactions. The theoretical curves are compared with experiment. The kaons are assumed to have a mean field potential as proposed by Schaffner

by the increase of the kaon mass with the density. With a soft momentum dependent interaction the excitation function is well reproduced. One should stress, however, that the uncertainties of the cross sections are still too large for using this agreement as a confirmation of the softness of the EOS.

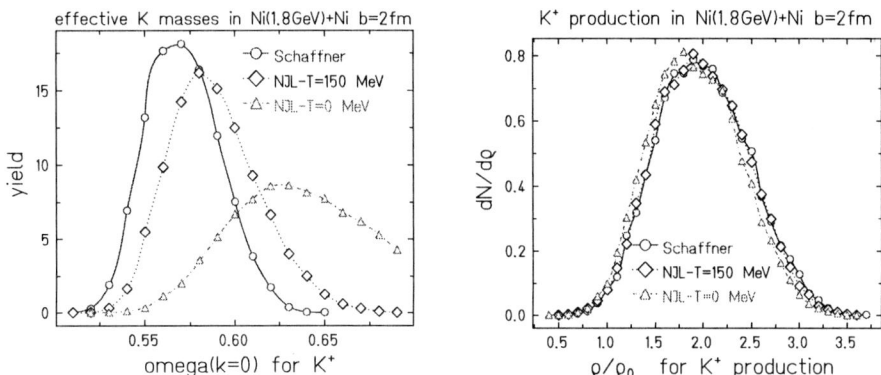

FIGURE 5. Distribution of baryonic density during the K^+ production for three different models (see text) in a IQMD calculation Ni(1.8 AGeV) + Ni

147

Another facet of the in medium modifications of the kaons is shown in fig.6 which displays $\omega(k = 0)$, the kaon energy at zero momentum, at production. We see that this distribution deviates strongly from the free value of 498 MeV. On the average the kaon has in the mean field approach of Schaffner an energy of close to 570 MeV. This reflects the fact that the kaons are created at high density, around twice nuclear matter density, as seen in fig. 7. Hence the K^+ production in heavy ion collisions is sensitive to the K^+ properties at twice normal nuclear matter density.

COMPARISON WITH THE EXPERIMENTAL SPECTRA

The comparison of our calculation for the systems C+C [5] and Au+Au [4] at different energies with the experimental spectra is displayed in fig.8. We display the numerical results for the 2 equations of state and including (Pot) resp. excluding (NoPot) the mean field potential of kaons. The different symbols refer to different laboratory angles. We

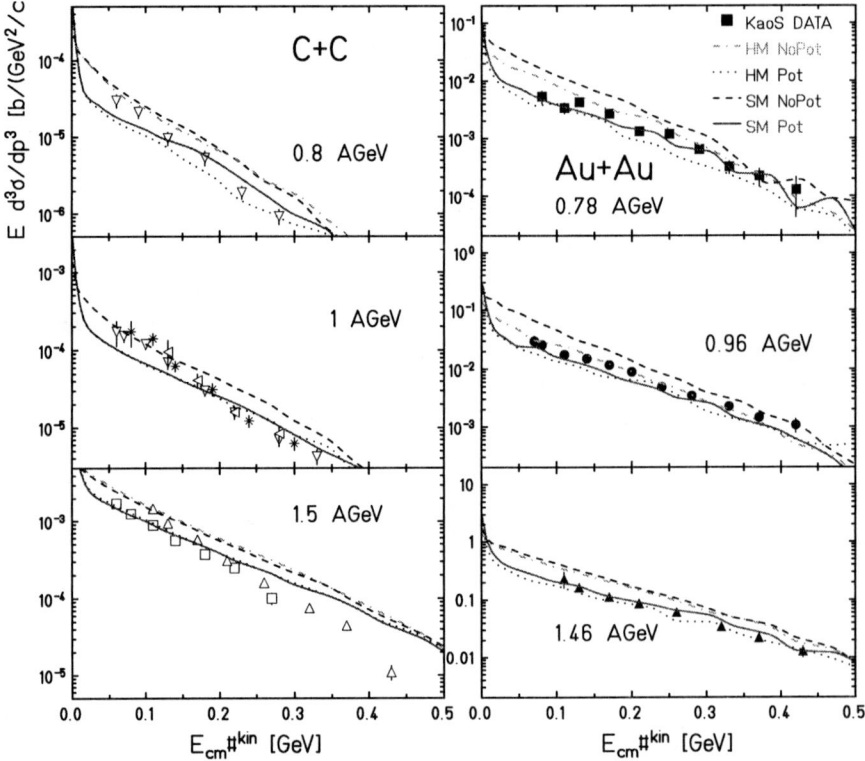

FIGURE 6. $E^{cm} \frac{d\sigma}{dp^3_{cm}}$ of K^+ for the systems C+C and Au + Au at three different energies. We compare the calculations for 2 nuclear equations of state with and without kaon mean field potential with the preliminary experimental data of the KAOS collaboration

see that the spectra are reasonable well reproduced and that the influence of the nuclear equation of state is very small if the kaon mean field potential is taken into account. The rapidity distribution observed in the reaction Ni(1.93 AGeV)+Ni is displayed in fig.9. We display these data for different kaon potentials. These calculations are compared

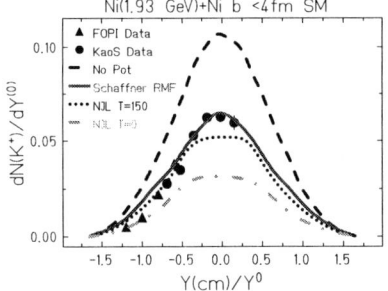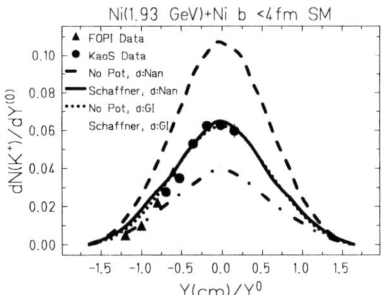

FIGURE 7. The rapidity distribution of K^+'s for Ni(1.93 AGeV) + Ni, central collisions, measured by the FOPI and KaoS collaboration as compared with the calculations using different production cross sections

with the data obtained by the FOPI collaboration [8] and those of the KaoS collaboration [9]. We see the noticeable influence of the kaon potential in nuclear matter on the yield, a very promising result. Because the uncertainties of the cross sections influence also the pion production, there is hope that the open questions concerning the cross section can be solved without using the kaon production. This may allow then to use the kaon production yield for a determination of the kaon in medium properties. How presently the uncertainties of the cross section influence the kaon yield is demonstrated in fig. 10. Here we have used (σ GI) the cross section of the Giessen group in the otherwise unchanged time evolution of the IQMD approach. The kaon potential of the Giessen group is not identical with our parametrisation but both are very similar. We see a change of about 30% of the dN/dy at midrapidity, well below the change due to the mean field potential. The almost identical result is obtained by the Giessen collaboration (fig. 5.5 of ref.[1]) although both calculations have only the kaon production cross section in common whereas the whole time evolution of the system is treated differently. This shows that the difference between both results is fully explained by the different cross sections a very remarkable and encouraging result which points towards the fact that the time evolution of heavy ion reactions at these energies is well under control.

THE IN-PLANE FLOW OF KAONS

The in plane flow of kaons [8] has gained in the past a lot of interest because it was claimed that its observation allows a distinction between a vector potential and a scalar potential of the kaons[10]. It is, however, not evident what one can learn from separating vector and scalar potential because only both together describe the behavior of kaons in the medium. The interesting quantity to look at is the comparison between the in-plane flow with and without interaction of the K^+ with the nuclear medium. As seen in fig.

11 this effect is quite small. The sources, in which the kaons are produced have almost

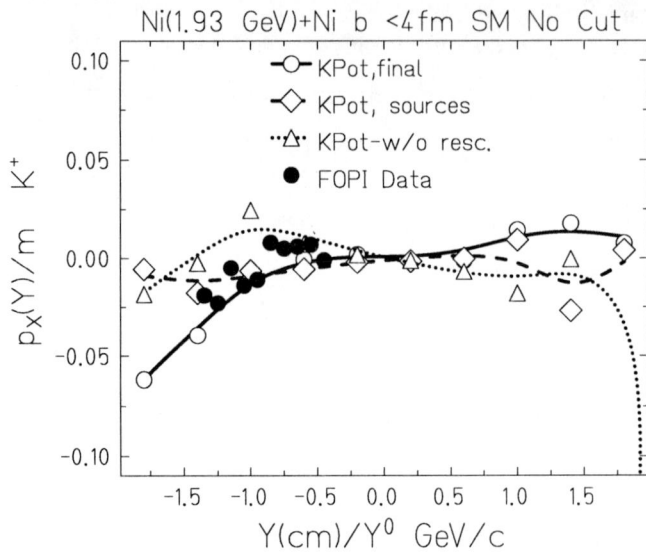

FIGURE 8. The in-plane flow of the kaons at production (sources) and asymptotically (with and without K^+-n rescattering.as compared to data

no flow. This has been explained in ref.[11]. The potential interaction of the kaons with the medium (KPot-w/o resc.) creates some flow, of the order of 15 MeV/c at backward rapidity. But if one includes the rescattering of the kaons with the nucleons most of the in-plan flow disappears and we have only a very small flow of the order of 5 MeV/c. The dramatic effect shown in ref.[10] is due to the fact that the total kaon potential is the difference of two large numbers, the scalar and the vector potential. Each of them gives a quite large in-plane flow but taking either vector or scalar potential only has no physical significance. In view of the smallness of the kaon flow and its large error bars it is not evident that it can serve for drawing conclusion about the potentials.

In conclusion, we have seen that the production cross section of K^+'s in nuclear reactions depends strongly on the in medium properties of the K^+. The kaons are produced around twice nuclear matter density. Their sources carry no in-plane flow. The KN potential creates some flow, KN collisions reduce it again. Finally the in-plane flow at production and at the end is almost identical. If we include the change of the K^+ mass in the nuclear environment in the calculation the dependence of the kaon yield on the nuclear equation of state is strongly reduced and probably too small to serve for distinguishing experimentally between the different EOS's.

We have further shown that for the same input, two very different numerical programs predict the same K^+ yield. Hence the complicated dynamics which governs the production process is numerically well under control.

REFERENCES

1. W. Cassing and E. Bratkovskaja, *Phys. Rep* **308** (1999) 65
2. C. Hartnack and J. Aichelin, *to be published*
3. C. Hartnack et al, *Eur. Phys. J.* **A 1** (1998) 151
4. C. Sturm et al, *Phys. Rev. Lett.* **86** (2001) 39 and C. Sturm, PhD thesis.
5. F. Laue, C. Sturm et al, *Phys. Rev. Lett* **82** (1999) 1640
6. J. Schaffner et al. , *Nucl. Phys.* **A625** (1997) 325
7. R. Nebauer, *PhD thesis and to be published*
8. J.L. Ritman et al., *Z. Phys.* **A352** (1995) 355
9. M. Menzel et al, *Phys. Lett.* **B495** (2000) 86
10. G.Q.Li and C.M. Ko, *Nucl. Phys.* **A594** 460
11. C. David et al, *Nucl. Phys.* **A650** (1999) 358

Probing Equilibrium in Intermediate Energy Heavy Ion Collisions [1]

Qingfeng Li* and Zhuxia Li*,†,**

*China Institute of Atomic Energy, P. O. Box 275(18), Beijing 102413, P. R. China
†Institute of Theoretical Physics, Academia Sinica, P. O. Box 2735, Beijing 100080, China
**Center of Theoretical Nuclear Physics, National Laboratory of Lanzhou Heavy Ion Accelerator, Lanzhou 730000, P. R. China

Abstract. By using the Isospin Dependent Quantum Molecular Dynamics Model (IQMD), we show the influence of initial N/Z, isospin symmetry potential and medium correction of two-body cross sections on equilibration in momentum space in intermediate energy heavy ion collisions. By investigating the isospin distribution in emitted nucleons, light charged particles and intermediate mass fragmentations for four mass symmetric and isospin asymmetric 96+96 systems Ru+Ru, Ru+Zr, Zr+Ru, and Zr+Zr at 400 AMeV and 100 AMeV we further study the degree of equilibration with respect to isospin degree of freedom in intermediate energy heavy ion collisions. We also propose the sensitive probes to the degree of equilibration in intermediate energy heavy ion collisions.

INTRODUCTION

The study of whether equilibrium is reached or not in a colliding system is a prerequisite for extraction of valid information about EOS and the reaction mechanism. This problem has been studied extensively both theoretically and experimentally[1]-[6], mostly based on one-component description of nuclear system. Following the establishment of radioactive beam facilities in many laboratory it becomes possible to study neutron (or proton) rich nuclear collisions at intermediate energies. Therefore it is worth to re-study the degree of equilibration in isospin asymmetric colliding systems based on two-component description of nuclear system. We will first talk about the influence of initial N/Z ratio, isospin dependent part of nuclear potential and isospin dependence of medium correction of two-body cross sections on the degree of equilibration in momentum space. Then we will talk about testing the non-equilibrium effect by means of isospin dependent probes stimulated by the 'mixing experiments' by using four mass 96+96 systems Ru+Ru, Zr+Zr, Ru+Zr, Zr+Ru at 400 AMeV[7, 8] performed by FOPI collaboration. We study the isospin distribution of emitted nucleons, light charged particles(LCP) and intermediate mass fragments(IMF) in the same collision systems at 400 AMeV as well as 100 AMeV to explore the non-equilibrium effect with respect to isospin degree of

[1] Supported by National Natural Science Foundation of China under No. 19975073, Science Foundation of Nuclear Industry of China and Major State Basic Reseach Development Program in China under contract No. G20000774

CP597, *Nonequilibrium and Nonlinear Dynamics in Nuclear and Other Finite Systems*, edited by Z. Li et al.

freedom.

THE MODEL

The Isospin Dependent Quantum Molecular Dynamics(QMD) model[9, 10] is used in the calculations. The following modifications in QMD model are introduced: Firstly, the isospin dependent part of the nuclear potential is taken into account in addition to the Coulomb interaction. The symmetry potential enegy per nucleon is taken as the following form,

$$V_{sym}(\rho,\delta) = \frac{C_S}{2}(\frac{\rho}{\rho_0})\delta^2,$$ (1)

where

$$\delta = \frac{\rho_n - \rho_p}{\rho_n + \rho_p},$$ (2)

In this work C_S is taken to be 35MeV and the corresponding symmetry energy is about 29MeV. Secondly, the isospin dependent binary elastic scattering cross section is used. It is well known that up to hundreds MeV the free elastic proton-neutron cross section is about 2-3 times larger than that of proton-proton (neutron-neutron)'s. Finally, in the treatment of the Pauli blocking, we firstly distinguish protons and neutrons, and then we use the following two criteria:

$$\frac{4\pi}{3}r_{ij}^3 \cdot \frac{4\pi}{3}p_{ij}^3 \geq \frac{h^3}{4},$$ (3)

and

$$P_{block} = 1 - (1 - f_i)(1 - f_j),$$ (4)

where f_i is the distribution function in phase space for particle i and reads as

$$f_i(\vec{r},\vec{p},t) = \frac{1}{\pi\hbar^3}\exp(-(\vec{r} - \vec{r}_i(t))^2/2L^2)\exp(-(\vec{p} - \vec{p}_i(t))^2 2L^2/\hbar^2).$$ (5)

Where L is a parameter which represents the spatial spread of wave packet, $\vec{r}_i(t)$ and $\vec{p}_i(t)$ denote the center of the wave packet in coordinate and momentum space respectively. The first condition gives the criterion for the uncertainty relation of the centroids of Gaussion wave packets of two particles. The second one is the probability of the Pauli blocking effect for the scattering of two particles, which is especially useful for collisions of heavy nuclei.

THE NUCLEAR STOPPING

As we expect that the degree of equilibrium is influenced by the mean field, medium correction of two-body cross section as well as the size of colliding system. To consider-ing the influence of the isospin dependence of the medium correction of two-body cross section, we use 4 different forms of in-medium two-body cross sections, they are: 1) σ_0, the free nucleon-nucleon cross section which is isospin dependent[11], 2) σ_0^*, which reads as [12]

$$\sigma_0^* = [1 + \alpha(\rho/\rho_0)]\sigma_0, \tag{6}$$

Here $\alpha = -0.2$. 3) σ_1, the Dirac-Brueckner Hartree-Fock calculation results of Li and Machleidt [13] where the medium correction of ρ meson mass was not taken into account. For σ_1, the in-medium cross sections first decreased as density and then increased slightly when density is higher than normal nuclear density and energy is higher than about 125 MeV(in Lab. system). 4)σ_2, in which the medium correction is obtained based on the results of [14]. In [14] the in-medium binary scattering cross section is derived based on QHD-II type Lagrangian within the framework of the microscopic transport theory where the medium correction of ρ meson mass was taken into account[15] and it was found that because of the medium correction of ρ meson mass the medium correction of nucleon-nucleon cross section was also isospin dependent, i.e., σ_{np}^* depended on the baryon density weakly while $\sigma_{nn(pp)}^*$ depended on the baryon density obviously and a slightly increasing as energy was also shown when density was higher than normal density and energy was higher than about 200 MeV(in Lab. system). The influence of the mean field is considered by taking EOS to be soft EOS(K=200MeV) and hard EOS(K=380MeV) as well as with and without isospin dependent part of nuclear potential. In order to study the dependence of the degree of equilibrium on the nuclear size and the initial isospin asymmetry, we calculate two groups of colliding systems 1) $^{58}Ni + ^{58}Ni$ and its neighbor nuclei with A=60 Z=24-32 and 2)$^{120}Sn + ^{120}Sn$ and its neighbor nuclei with A=120 Z=47-55. We study the degree of equlibrium in momentum space by investigating the momentum quadrupole Q_{ZZ} which usually is called nuclear stopping. Q_{ZZ} is defined as

$$Q_{ZZ} = \Sigma_i[2 * P_z(i)^2 - P_x(i)^2 - P_y(i)^2], \tag{7}$$

Here summation runs over all nucleons in projectile and target. First of all, we investigate the effect of symmetry potential and isospin dependence of the medium correction of two body cross section as well as the isospin asymmetry of initial system on Q_{ZZ}. Fig.1 shows average Q_{ZZ} as function of Z for projectile and target taken to be a) $A = 60$ and Z ranging from 24 to 32 at 50 AMeV, 150 AMeV and 400 AMeV with $C_s = 0$ MeV and 35 MeV and two body cross section taking to be σ_0 and σ_2, respectively. As a whole, we can find that the influence of isospin dependent part of nuclear potential and the initial isospin asymmetry on average Q_{ZZ} is weak for both σ_0 (no medium correction) and σ_2(with isospin dependent medium correction) cases. But we still can find the influence of the isospin dependent part of nuclear potential on Q_{ZZ} for neutron rich nuclear system at 150 AMeV case, i.e., the difference of average Q_{ZZ} between cases with $C_s = 0$ and 35 MeV slightly increases as isospin asymmetry increases. This effect disapears at energy 400 AMeV case.

In Figs. 2 a) and b) we present the exitation function of average Q_{ZZ} with different EOS ('Soft' and 'Hard'), different cross sections (σ_0, σ_0^*, σ_1 and σ_2) for a) $^{58}Ni + ^{58}Ni$ and b) $^{120}Sn + ^{120}Sn$, respectively. It is shown that average Q_{ZZ} first increases as beam energy then is saturated at beam energy at about 150 AMeV - 300 AMeV depending on the form of in-medium cross section used. Different forms of medium correction of two-body cross sections significantly affect the behavior of excitation function of average Q_{ZZ}. From comparison between Fig.2 a) and b) we find that for $^{58}Ni + ^{58}Ni$ case the influence of different EOS on average Q_{ZZ} is much weaker than that of different forms of in-medium cross sections. While for $^{120}Sn + ^{120}Sn$ case, the influence of different EOS and different forms of in-medium cross sections is comparable. And it is also shown that the degree of equilibrium in momentum space depends on the system size, the larger the system is the higher the degree of equilibrium is reached. From this figure, it is implied that the behavior of excitation function of average Q_{ZZ} for medium size systems like $^{58}Ni + ^{58}Ni$ can provide more clear information of isospin dependence of the medium correction of cross sections than that for heaver systems like $^{120}Sn + ^{120}Sn$ can.

EQUILIBRIUM WITH RESPECT TO ISOSPIN DEGREE OF FREEDOM

First we study the normalizing proton counting defined [7] by

$$R_Z = \frac{2*Z - Z^{Zr} - Z^{Ru}}{Z^{Zr} - Z^{Ru}} \qquad (8)$$

By definition, $R_Z=1$ for Zr+Zr, $R_Z=-1$ for Ru+Ru. For asymmetric reaction Ru+Zr, Zr+Ru, if the reaction is completely mixed, R_Z should be equal to 0.0 at all rapidity and if the reaction is full transparent, R_Z should be along the diagonal line from (-1,+1) to (1,-1) for Ru+Zr and from (-1,-1) to (+1,+1) for Zr+Ru. Fig.3 shows R_Z as function of rapidity at beam energy 100 AMeV, impact parameter b=0 fm and 400 AMeV, b=0 fm and b=5 fm. The experimental data (at 400 AMeV) is also given in the figure. From this figure, one can easily find that the absolute R_Z value goes from zero to about 0.5 for reactions

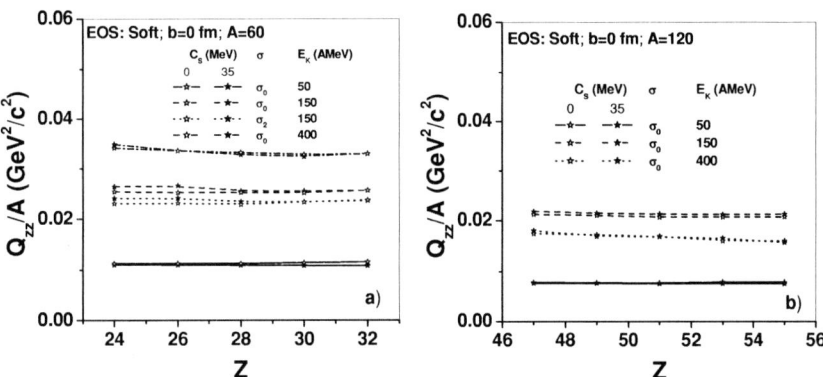

FIGURE 1. The average Q_{ZZ} as function of Z for projectile and target taken to be a) $A = 60$ and Z ranging from 24 to 32 and b) $A = 120$, Z ranging from 47 to 55 at beam energy 50 AMeV, 150 AMeV, 400 AMeV with $C_S = 0$ and 35MeV and two-body cross section taking to be σ_0 and σ_2(see text), respectively.

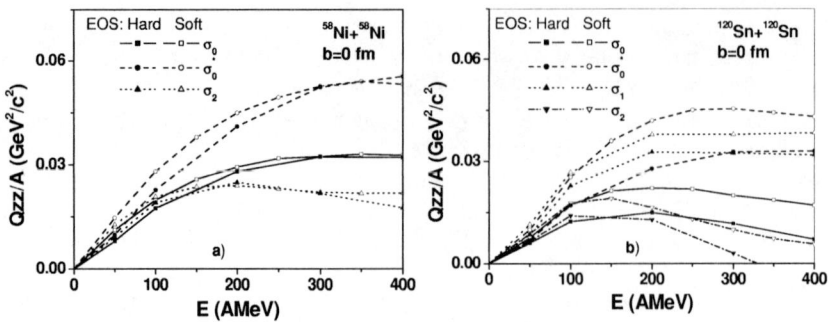

FIGURE 2. The average Q_{ZZ} as function of beam energy for a) $^{58}Ni + ^{58}Ni$ and b) $^{120}Sn + ^{120}Sn$ at b=0 fm with different EOS and two-body cross sections σ(see text).

Zr+Ru and Ru+Zr at energy 100 AMeV and 400 AMeV b=0 fm, and about 0.75 for the same reactions at beam energy 400 AMeV, b=5 fm. Our calculation is in reasonable agreement with experiments data and consequently, the same conclusion concerning the non-equilibrium effect can be drawn for 400 AMeV case. The results for b=0 fm and b=5 fm show the non-equilibrium effect strongly depends on the impact parameter. However, the results of R_Z for 400 AMeV and 100 AMeV at b=0 fm are undistinguishable and it leads to the conclusion that the protons are produced in a non-equilibirium source at both 400 AMeV and 100 AMeV.

It would be more desirable to study the isovector density as function of rapidity for isospin asymmetric nuclear systems. Therefore we introduce the neutron-proton differential rapidity distribution. Fig. 4a), 4b) and 4c) show the neutron-proton differential

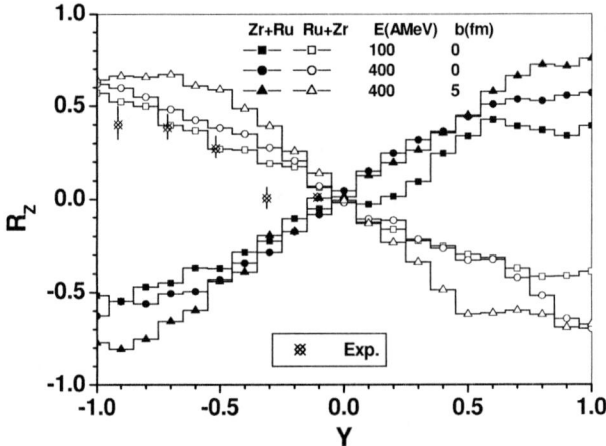

FIGURE 3. The proton counting number R_z as function of rapidity for $^{96}Ru + ^{96}Ru$, $^{96}Ru + ^{96}Zr$, $^{96}Zr + ^{96}Ru$, $^{96}Zr + ^{96}Zr$ at E=100 AMeV b=0 fm, E=400 AMeV b=0 fm and b=5 fm. The experimental data for 400AMeV are also given in the figure.

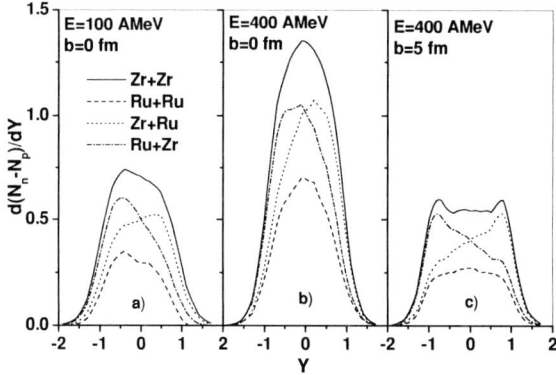

FIGURE 4. The neutron-proton differential rapidity distribution for the same reactions as Fig.1 a) at 100 AMeV, b=0 fm and b) at 400 AMeV, b=0 fm and c) at 400 AMeV, b=5 fm.

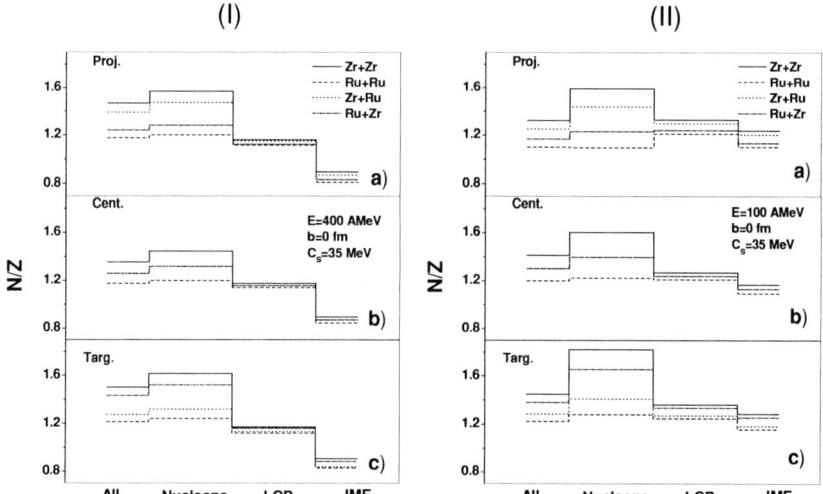

FIGURE 5. (I) The average N/Z ratio of emitted nucleons, light charged particles and intermediate mass fragments at a) projectile rapidity region, b)central rapidity region, c)target rapidity region for the same reactions as Fig.1 at E=400 AMeV, b=0 fm. (II)The same as (I) but at E=100 AMeV, b=0 fm.

rapidity distribution for $^{96}Ru +^{96} Ru$, $^{96}Zr +^{96} Zr$, $^{96}Zr +^{96} Ru$, $^{96}Ru +^{96} Zr$ at a)100 AMeV, b=0 fm, b)400 AMeV, b=0 fm and c)400 AMeV, b=5 fm. First, for all three cases a), b) and c), the centroids of neutron-proton differential rapidity distribution for $^{96}Ru +^{96} Zr$, $^{96}Zr +^{96} Ru$ are located at the side of Zr (as target or porjectile) and strongly deviate from $Y = 0$. The centroid of distribution should be at $Y = 0$ if a system is in equilibrium. The deviation of the centroid of neutron-proton differential rapidity distribution from $Y = 0$ means that there is non-equilibrium effect. The larger the deviation from $Y = 0$ the stronger the non-equilibrium effect is. The deviation of the centroid of neutron-proton differential rapidity distribution from $Y = 0$ for b=5 fm case is much larger than that for b=0 fm case. This is of course quite understandable. Further more, one can find that the neutron-proton differential rapidity distribution of symmetric reactions $^{96}Ru +^{96} Ru$, $^{96}Zr +^{96} Zr$ at 100 AMeV deviats from the Gaussion shape more strongly than that at 400 AMeV. It implys that there exist obvious non-equilibrium effect in the emitting nucleon process. Therefore we can conclude that the neutron-proton differential rapidity distribution is a good probe to explore the degree of equilibration for an isospin asymmetric system. We may generalize the neutron-proton differential rapidity distribution by introducing t-^3He differential rapidity distribution to probe equilibration in isospin asymmetric colliding systems.

However, the sources of nucleon emission are complicated therefore we further study the isospin distribution in LCP and IMF in addition to nucleons. In Fig.5 (I) and (II), we show the average N/Z ratio in emission of nucleons, LCP and IMF at projectile (a)), central (b)) and target (c)) rapidity region in the four systems at 400 AMeV, and 100 AMeV, b=0 fm, respectively. The projectile rapidity region is defined by $1.5 \geq Y \geq 0.5$, the target rapidity region $-0.5 \geq Y \geq -1.5$ and the central rapidity region $0.5 \geq Y \geq -0.5$. The figure firstly tells us a basic feature that the difference between the average N/Z ratios of emitted nucleons of 4 colliding systems with different isospin asymmetry is much larger than that between the average ratios of LCP and IMF of 4 systems at three rapidity regions. The more neutron(proton) rich systems emit more neutrons(protons) while the average N/Z ratios of LCP and IMF for these four systems are relatively close. This behaviour is stronger at 100 AMeV case. The experimental measurements at tens AMeV energy region [16] found that the more asymmetric the system is the stronger the system will be breaking up into still more neutron rich (deficient) light fragments while the N/Z ratio of heavier fragments remains relatively insensitive. Our calculations show similar tendency, only because of the energy difference, here the N/Z ratio of emitted nucleons, LCP and IMF are compared instead of comparing the N/Z ratio for LCP and IMF in [16] there the energy was relatively low.

It is more meaningful to investigate whether the N/Z ratio of IMF for mixing reaction converges or not as far as the degree of equilibrium is concerned because IMF produce at late stage of reaction. When we attend to the N/Z ratio at target and projectile rapidity region, we find that not only the N/Z ratio of emitted nucleons for mixing reactions $^{96}Zr+^{96}Ru$ $^{96}Ru+^{96}Zr$ but also that of LCP and IMF does not converge. The N/Z ratio of emitted nucleons as well as that of LCP and IMF for two mixing reactions separates each other and is more close to Zr+Zr or Ru+Ru at respective rapidity region. Although the differnce of average N/Z ratio of IMF for reactions Zr+Ru and Ru+Zr is weaker than emitted nucleons we can still find there exist difference and the difference is larger at 100 AMeV than at 400 AMeV. It means that not only the nucleons but also IMF are not emitted from an completely equilibrium source. Furthermore, we find by comparing Fig.5 (I) and (II) that the N/Z ratio decreases as energy increases from 100 AMeV to 400 AMeV for all 4 systems. By study the isotope yield of the most abundant IMF we find that for 100 AMeV case the most abundant isotopes are always located at most stable one and the relative yield of neutron rich isotopes is always higher than at 400 AMeV case. And the relative yields of neutron rich isotopes increase as N/Z ratio of initial reaction system increases for 100 AMeV case.

SUMMARY

In conclusion we have investigated the degree of equilibrium in momentum space in isospin asymmetry colliding systems. Our study shows that the isospin dependent part of nuclear potential and the isospin asymmetry of colliding system influence the nuclear stopping weakly. The behavior of the excitation function of nuclear stopping depends on both EOS and in-medium cross section but for medium size system like $^{58}Ni +^{58}Ni$ the dependence on EOS is much weaker than that on in-medium cross section. Therefore, the behavior of excitation function of average Q_{ZZ} for medium size systems like $^{58}Ni +^{58}Ni$ can provide more clear information of isospin dependence of the medium correction of cross sections than that for heaver systems like $^{120}Sn +^{120}Sn$ can. Then we test non-equilibrium effect by taking the probes with isospin degree of freedom. We study the normalizing proton counting as well as the neutron-proton differential rapidity distribution for 4 mass symmetry and isospin asymmetry systems $^{96}Ru +^{96}Ru$, $^{96}Zr +^{96}Zr$, $^{96}Zr +^{96}Ru$, $^{96}Ru +^{96}Zr$ at 400AMeV and 100 AMeV we can conclude that the nucleons are produced fron a non-equilirium source at both energies. Then we further study the average N/Z ratios of LCP and IMF for 4 systems and we find that even for emission of IMF at projectile and target region, there exist non-equilibrium effect.

REFERENCES

1. J. Cugnon, T. Mizutani and J. Vandermeulen, Nucl. Phys. A359, 345c (1981).
2. J. Randrup, Nucl. Phys. A 314, 429 (1979).
3. M. Cubero, M. Schonhofen, B. L. Frimen and W. Norenberg, Nucl. Phys. A340, 385 (1991).
4. A. Lang, B. Blattel, V. Koch, K. Weber, W. Cassing and U. Mosel, Z.Phys.A340, 287 (1991).
5. Li Zhuxia, Zhuo Yizhong, Gu Yingqi, et.al., Nucl. Phys. A559, 603(1993).
6. Li Zhuxia, Zhuo Yizhong, Wu Xizhen, M. Sano, J. Phys. G: Nucl.Part.Phys. 20, 1829 (1994).
7. F. Rami et.al. Phys.Rev.Lett. 84 (2000)1120,
8. W.Reisdorf, FOPI collabration "Multifragmentation", edited by H.Fedmeier, J.Knoll, W. Nörenberg, J.Wambach,1999.
9. J. Aichelin, Phys. Rep. 202, 233 (1991) and references therein.
10. Ch. Hartnack, Li Zhuxia, et.al., Nucl. Phys. 495, 303 (1989).
11. K. Chen, Z. Fraenkel, G. Friedlander, et.al. Phys. Rev. 166, 949(1968).
12. D.Klakow, G. Wilke and W. Bauer, Phys. Rev. C48, 1982 (1993).
13. G.Q. Li and R. Machleidt, Phys. Rev. C 48, 1702 (1993) and Phys. Rev. C 49, 566 (1994).
14. Q.F. Li and Z.X. Li, Phys. Rev. C 62, 014606 (2000).
15. G.J. Lolos, et.al., Phys. Rev. Lett. 80, 241 (1998).
16. S.J. Yennello, M.Veslesky, R.Laforest, et.al., Nucl. Phys. 681, 317c(2001).

Universality in Fragment Inclusive Yields from Au + Au Collisions

A. Insolia[1], C. Tuvè[1], S. Albergo[1], F. Bieser[2], F. P. Brady[5], Z. Caccia[1], D. Cebra[2,4], A. D. Chacon[6], J. L. Chance[5], Y. Choi[4] [1], S. Costa[1], J. B. Elliott[4], M. Gilkes[4], J. A. Hauger[4], A. S. Hirsch[4], E. L. Hjort[4], M. Justice[3], D. Keane[3], J. Kintner[5], M. Lisa[2], H. S. Matis[2], M. McMahan[2], C. McParland[2], D. L. Olson[2], M. D. Partlan[5], N. T. Porile[4], R. Potenza[1], G. Rai[2], J. Rasmussen[2], H. G. Ritter[2], J. L. Romero[5], G. V. Russo[1], R. Scharenberg[4], A. Scott[3], Y. Shao[3], B. K. Srivastava[4], T. J. M. Symons[2], M. L. Tincknell[4] [2], S. Wang[3] [3], P. G. Warren[4], H. H. Wieman[2], K. L. Wolf[6]

(EOS Collaboration)

[1]*Università di Catania & INFN, 95129 Catania, Italy*
[2]*Nuclear Science Division, LBNL, Berkeley, California, 94720*
[3]*Kent State University, Kent, Ohio, 44242*
[4]*Purdue University, West Lafayette, Indiana, 47907–1396*
[5]*University of California, Davis, California, 95616*
[6]*Texas A&M University, College Station, Texas, 77843*

Abstract. The inclusive light fragment ($Z \leq 7$) yield data in Au+Au reactions, measured by the EOS Collaboration at the LBNL Bevalac, are presented and discussed. For peripheral collisions the measured charge distributions develop progressively according to a power law which can be fitted by a single τ exponent independently of the bombarding energy in the range 250 - 1200 A MeV. In addition to this universal feature, we observe that the location of the maximum in the individual yields of different charged fragments shift towards lower multiplicity as the fragment charge increases from $Z = 3$ to $Z = 7$. This trend is common to all six measured beam energies. Moments of charge distributions and correlations among different monents are reported. Finally, the $T_{He,DT}$ thermometer has been constructed for central and peripheral collisions using the double yield ratios of He and D,T projectile fragments. The *measured* nuclear temperatures are in agreement with experimental findings in other fragmentation reactions.

[1] † Current address: Sung Kwun Kwan University, Suwon 440-746, Republic of Korea
[2] ⋆ Current address: Lincoln Laboratory, S1-257, 244 Wood Street, Lexington, MA 02173-9108
[3] ‡ Current address: Harbin Institute of Technology, Harbin 150001, People's Republic of China

CP597, *Nonequilibrium and Nonlinear Dynamics in Nuclear and Other Finite Systems,*
edited by Z. Li et al.
© 2001 American Institute of Physics 0-7354-0041-5/01/$18.00

INTRODUCTION

The pioneering work of the Purdue group [1], with the recognition of a power law dependence in the fragment yield in $p + Xe$ and $p + Kr$ reactions, was recognized early on as one of the possible signatures that, under proper conditions, nuclear matted could undergo a phase transitions[1]. Since then a lot of scientific effort has been devoted to the subject [2], along with the experimental discovery of universal features in nucleus - nucleus collisions in the energy range 100 A - 1000 A MeV. In particular, the EOS Collaboration, in an event-by-event analysis of the Au + C reaction at 1 A GeV, has been able to characterize the experimental data in terms of critical exponents [3] thus bringing the first clear evidence for a second order liquid - gas phase transition in multifragmentation reaction between heavy ions. The purpose of the present analysis of Au + Au collisions, in the energy range $250A - 1200AMeV$ is to show that the inclusive yields produced by the fragmentation of the projectile spectator carry most of the universal features, already recognized by different groups as signatures for a critical behaviour of nuclear matter [2, 3, 4]. The fingerprints of the *liquid - gas* phase transition will be therefore found in the power law trend observed in the inclusive yields for peripheral collisions, in the moments of charge distributions. Most of the results presented in this talk have been recently published [6].

INCLUSIVE YIELDS AND CONDITIONAL MOMENTS

The EOS Collaboration has recently measured fragment production in Au + Au collisions at the LBNL Bevalac beam energies of 0.25, 0.4, 0.6, 0.8, 1.0 and 1.15 A GeV. The experiment was conducted with the EOS Time Projection Chamber (TPC), a multiple sampling ionization chamber (MUSIC) as well as a TOF wall [7] and a neutron detector (MUFFINS) [8, 9]. Only data from the TPC detector will be used in the present analysis. The TPC [7, 10] afforded almost complete coverage in the forward hemisphere. The EOS TPC resolution allows one to identify particles up to $Z = 7$, measuring their momentum with virtually no p_t cut. The multiplicity scale has been defined according to the Plastic Ball analysis prescription [11]. Thus, most central collisions will correspond to the highest multiplicity bins.

The measured inclusive rapidity distribution, see fig. 1, allow to define proper cuts to select the fragmentation of the projectile spectators. We have used a sharp cut near beam rapidity, $y' \geq 0.60$. It is seen, in fig. 1, that rapidity spectra for $Z = 1, 2$ exhibit a large prompt component at lower rapidity. For larger Z values the fragments are produced with a rapidity close to the beam rapidity with an almost simmetric distribution. The problem posed by the presence of prompt particles has been discussed widely in the literature [7, 12, 13]. In the present analysis the prompt component is strongly suppressed by the sharp cut close to beam rapidity that we use to select projectile spectator fragmentation. In addition, one should recall the result of ref. [7], where it has been shown, for the Au + C reaction at $1AGeV$ that the total multiplicity is proportional to the second stage multiplicity. This fact makes less crucial the problem of a complete suppression of the prompt component. The measured inclusive yields for the six beam energies are reported

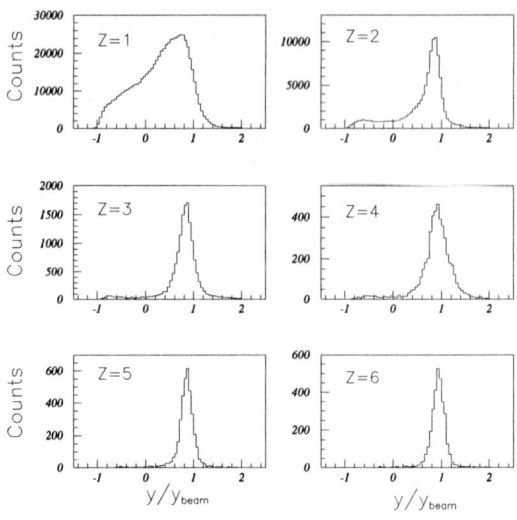

FIGURE 1. Typical inclusive particle rapidity distributions for Au + Au, where $y' = y/y_{beam}$, at $E_{lab} = 1000\,A\,MeV$, with the proper selection for peripheral collisions ($M_{bin} = 3, 4, 5$).

in fig. 2 versus the fragment charge Z.

We see that for central collisions (multiplicity bins 8 or 9, see fig.2) an almost a linear dependence (in semilog scale) is observed ($Y = Y_0 e^{-1.58*Z}$), while approaching the semi-peripheral domain a power law starts to develop. This is nicely seen, in the figure, for the multiplicity bins $M_{bins} = 3, 4$ and 5, where the yields fall one on top of the other. For these bins, the experimental dependence on Z can well be fitted with a power law of the type $Y = q_0 Z^{-2.1}$. Moreover, the observed power law is independent of the beam energy in the range of the experiment. This shows that the main mechanism for fragment production from the projectile remnant is very much the same in this energy range. The numerical values of q_0 and Y_0 are the same for all the beam energies.

This power-law-like distribution has been already reported for Au + Au reactions by the ALADiN Collaboration [14]. This feature of the multifragmentation phenomenon has been, on the other hand, interpreted as a possible signature for a continuous liquid-gas phase transition [1, 2]. Actually, if a liquid - gas phase transition takes place for a given value of excitation energy transferred to the system, one has to expect a typical power law dependence of the relevant physical quantity which, in our case, is just the fragment distribution. Different model calculations have reported very clearly this behaviour. M. Baldo et al. [15] find that deterministic chaos inside the spinodal zone is associated with the multifragmentation and the predicted fragment distribution shows a power law trend with an exponent τ very close to what was previously reported in ref.[2, 3]. Furthermore, A. Bonasera et al. [16], in the frame of a simple purely classical molecular dynamic model, have found that a system of 100 particles exhibits an expanding scenario that for a given initial temperature produces multifragmentation.

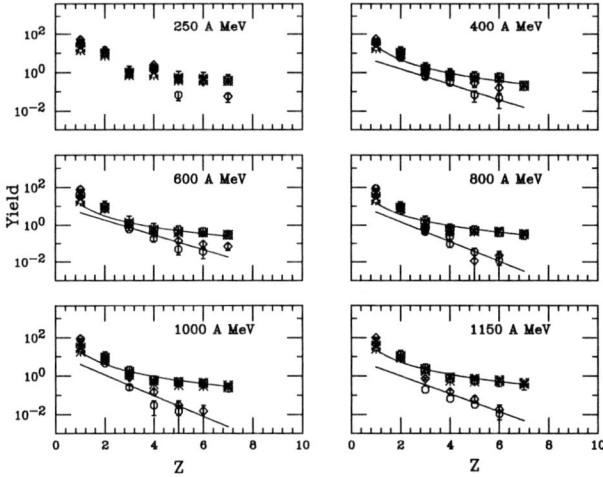

FIGURE 2. Charge Yields for different multiplicity bins. Data points are represented by different symbols for different multiplicity bins. However, data corresponding to peripheral collisions (multiplicity bins 3,4 and 5) are almost undistinguishable for they fall one on top of the other. Only one multiplicity bin corresponding to central collisions ($M_{bin} = 9$ at 250 and 400 MeV, or $M_{bin} = 8$ at the remaining energies) is reported. The latter shows an almost exponential decrease of the yield.

The mass distribution displays a power law behaviour $Y(A) \simeq A^{-\tau}$ with $\tau = 2.23$ [16], in agreement with the value predicted by the Fisher droplet model[17, 18].

The second moment of the charge distribution (for a beam energies) are given in figs. 3(a,b,c,d). The correlations between different moments are shown in fig. 3(e) and 3(f). One can easily recognize that a maximum appears at $M_{bin} = 3$. This is the first multiplicity bin for which we observe the universal trend in the yield versus Z. In particular, as expected, the second moment is characterized by a maximum which tends to soften when a mass cutoff is applied [20]. In our case this comes in a natural way due to the detector cut at charge $Z \geq 8$. A correlation among different moments, third versus second one, fig. 3(e), and fifth versus second one, in fig. 3(f), is nicely seen, in agreement with the percolation model calculations [19, 18, 20]. The slopes are consistent with the τ exponent which best fits the yields for $M_{bin} = 3,4,5$ in fig. 2.

INTERMEDIATE MASS FRAGMENTS

The integrated intermediate mass ($Z = 3 - 7$) fragment yield is almost energy independent for peripheral collisions, while it is suppressed as the beam energy increases for central collisions, as seen in fig. 4.

It is quite interesting to look at the individual fragment yields versus multiplicity. The data are reported in fig. 5, where only two typical values ($Z = 3$ and $Z = 6$) are considered for simplicity. All six energies show similar features. Indeed, increasing multiplicity

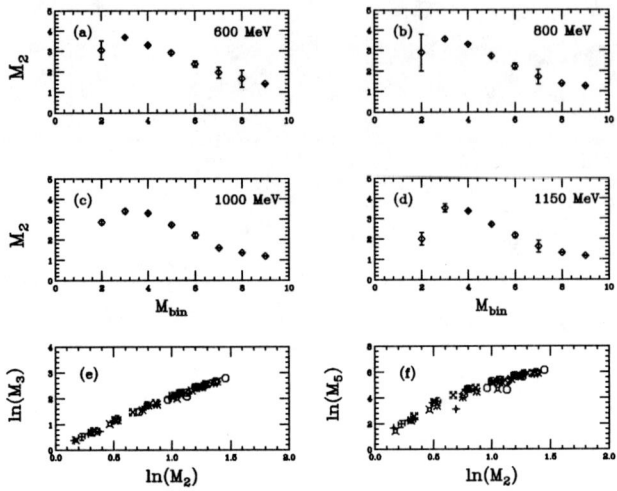

FIGURE 3. (a,b,c,d) Second moment of the charge distribution for the indicated beam energies.. The maximum is observed at $M_{bin} = 3$. (e) Correlation between the third and the second moment. (f) Correlation between the fifth and the second moment.

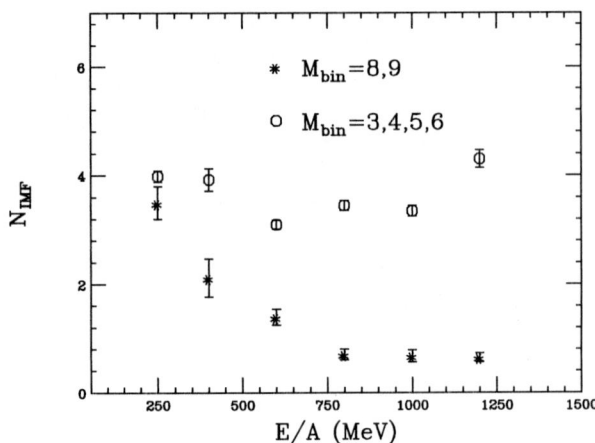

FIGURE 4. Energy depedence of the IMF (Z=3-7) production. The peripheral integrated yields are averaged over the multiplicity bins 3,4 and 5. For central collisions a mean between the bins 7 and 8 has been reported.

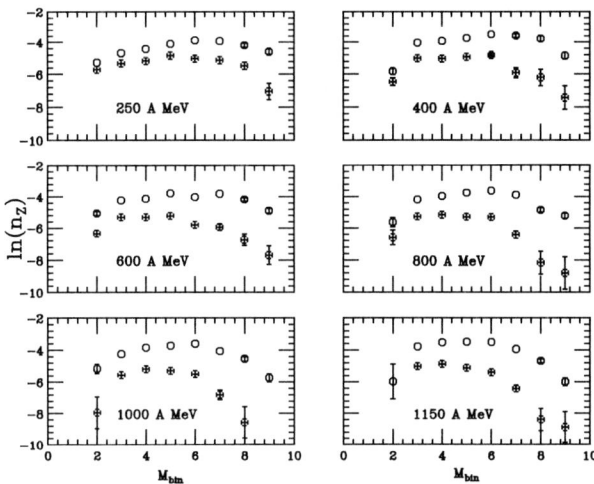

FIGURE 5. Individual yields $ln(n_Z)$ versus multiplicity. The number of fragment for each species has been divided by $Z = 79$ (the nuclear charge of gold).

corresponds to events with greater excitation energy of the projectile remnant. This produces the rise of the number of each species of IMF as the multiplicity increases from the lower M_{bin} values. We do observe a strict ordering: lighter species is always more abundant than the heavier. As noticed in [21], all IMF yields decrease at the highest multiplicity. For those M_{bin} values, the excitation energy is so large that only the lightest fragments can survive. Finally we observe that, in spite of the unavoidable smearing produced by the fact that we average the individual yields over many events within the multiplicity bins, the location of the maximum still shifts towards lower multiplicity for the larger Z values. This is a very important feature for it allowed to extract for the first time the critical exponent σ in the Au + C analysis at 1 *A GeV* [21]. In Au + Au inclusive individual yields, with which this paper is concerned, the shift of the maximum location appears to be, again, independent of the beam energy along with the expected and well known *rise and fall* of the fragment production versus the centrality of the collision.

NUCLEAR TEMPERATURE

A lot of interest has been raised by the possibility to characterize the remnant by determining its nuclear temperature. Different hadronic thermometers have been used so far [4] (slope parameter, isotopic composition, relative population of states). Different thermometers usually produce different temperatures. We will present our own results with the widely used double yield ratio [5] thermometer. In this case one has to assume that the nuclear system finds itself at low density and in chemical and thermal equilibrium. Under those conditions it has been shown [5] that the temperature can be obtained

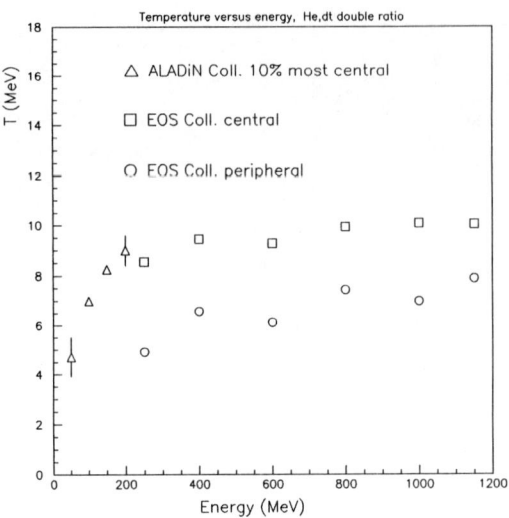

FIGURE 6. $T_{He,DT}$ thermometer built from the double yield ratios versus beam energy for peripheral (multiplicity bins $M = 3, 4, 5$) and central collisions (multiplicity bins corresponding to the $10\% - 15\%$ most central collisions). The ALADiN data are from ref. [22].

through the double yield ratio of two isotope pairs. differing by the same number of neutrons and/or protons. We will consider in the present report only the $T_{He,DT}$ thermometer.

The $T_{He,DT}$ could be partially affected by uncertainty in the prompt particles separation (mainly for Z=1 and Z=2) from the corresponding $Z = 1, 2$ fragments emitted by the remnant decay in the second stage [7], while side feeding effects[4, 7, 2] should be less important. Our results are reported in fig. 6 for peripheral and central collisions. The results for central collisions are compared with experimental data obtained by the ALADiN Collaboration in Au + Au reaction at lower beam energy [22]. For central collisions our data show a trend consistent with the available ALADiN data. In addition our data suggest a sort of saturation versus bombarding energy for central collisions, when looking at the projectile spectator fragmentation. For peripheral reactions the temperature is still slowly increasing with the energy. The statistical error is within the size of the data points.

SUMMARY

In conclusion, we have shown that EOS data for the inclusive yield of fragments ($Z = 1 - 7$) in Au + Au, from $E_{beam} = 250\ A\ MeV$ up to $E_{beam} = 1200\ A\ MeV$, show a typical power law distribution when triggering on peripheral collisions ($M_{bin} = 3, 4, 5$) in agreement with previous model calculations as well as with other experimental data. The second moment of the charge distribution shows a maximum at the $M_{bin} = 3$ for all considered energies. In addition to this universal feature, we have found that the location of the maximum in the individual yields of different charged fragments shift towards lower multiplicity as the fragment charge increases from $Z = 3$ to $Z = 7$. This trend is common to all six energies. Finally, preliminary data have been reported on the $T_{He,DT}$ thermometer built from double yield ratios for central and peripheral collisions versus beam energy. We conclude that the present data are consistent with previous EOS Collaboration data and bring additional experimental evidence of a liquid - gas phase transitions in nuclei, as previously reported by the EOS Collaboration for the Au + C reaction.

REFERENCES

1. J. E. Finn et al., Phys. Rev. Lett. **49** (1982) 1321
2. *Critical Phenomena and Collective Observables*, Proc. of the Int. Conf. CRIS'96, Acicastello (Italy, 1996); Ed.s S. Costa, S. Albergo, A. Insolia and C. Tuvé, World Scientific (1996)
3. M. Gilkes et al. (EOS Collaboration), Phys. Rev. Lett. **73** (1994) 1590
4. J. Pochodzalla , Prog. Part. Nucl. Phys. **39** (1997) 443
5. S. Albergo, S. Costa, E. Costanzo and A. Rubbino, Il Nuovo Cim. A**89** (1985) 1
6. A. Insolia, C. Tuvè et al., (EOS Collaboration), Phys. Rev. C**61** (2000) 044902
7. J. A. Hauger, P. Warren et al. (EOS Collaboration), Phys. Rev. C**57** (1998) 764
8. C. Tuvè et al. (Transport Collaboration), Phys. Rev. C**56** (1997) 1057; Phys. Rev. C**59** (1999) 233
9. S. Albergo et al., Nucl. Instr. Meth. A**311** (1992) 280;
   J. Engelage et al. (Transport Collaboration), Radiat. Measurements **27** (1997) 549
10. G. Rai et al., IEEE Transactions on Nuclear Science **37** (1990) 56
11. H. H. Gutbrod, A. M. Poskanzer and H. G. Ritter, Rep. Prog. Phys. **52** (1989) 1267
12. W. Bauer and W. A. Friedman, Phys. Rev. Lett. **75** (1995) 767; W. Bauer and A. Botvina, Phys. Rev. C**75** (1995) R1760
13. M.L. Gilkes et al., Phys. Rev. Lett. **75** (1995) 768
14. G.J. Kunde et al., Phys. Rev. Lett. **74** (1995) 38
15. M. Baldo, G.F. Burgio and A. Rapisarda, Phys. Rev. C**51** (1995) 198;
16. V. Latora, M. Belkacem and A. Bonasera, Phys. Rev. Lett. **73** (1994) 1765;   M. Belkacem, V. Latora and A. Bonasera, Phys. Rev. C**52** (1995) 271;
17. M. E. Fisher, Rep. Prog. Phys. **30** (1987) 615
18. D. Stauffer and A. Aharony, *Introduction to Percolation Theory* (Taylor & Francis, London, 1994).
19. X. Campi, J. Phys. A **19** (1986) L917; Phys. Lett. B**208** (1988) 351
20. W. Bauer, Phys. Rev. C**38** (1988) 1297
21. J.B. Elliott et al. (EOS Collaboration), Phys. Lett. B**381** (1996) 35
22. V. Serfling et al., (ALADiN Collaboration), Phys. Rev. Lett. **80** (1998) 3928

Isospin Effects on Nuclear Collective Flow in Heavy-Ion Collisions at Intermediate Energies

Lie-Wen Chen,[1,2] Feng-Shou Zhang,[2,3] Gen-Ming Jin,[2,3] and Zhi-Yuan Zhu[2,4]

1 Institute of Theoretical Physics, Shanghai Jiao Tong University, Shanghai 200030, China
2 Center of Theoretical Nuclear Physics, National Laboratory of Heavy Ion Accelerator, Lanzhou 730000, China
3 Institute of Modern Physics, Academia Sinica, P. O. Box 31, Lanzhou 730000, China
4 Shanghai Institute of Nuclear Research, Academia Sinica, Shanghai 201800, China

Abstract. Within the framework of an isospin-dependent quantum molecular dynamics (IDQMD) model, we study systematically the isospin dependence of four main categories of collective motions in intermediate energy heavy ion collisions, i.e., the in-plane directed flow, in-plane rotational flow, out-of-plane squeeze-out flow, and radial expansion flow. In particular, the influence of symmetry energy and isospin-dependent nucleon-nucleon cross sections on the isospin dependence of the collective flows is studied.

INTRODUCTION

Collective flow phenomena observed in heavy ion collisions (HIC's) from the Fermi energy range to (ultra-) relativistic energies have become a subject of intensive theoretical and experimental studies during the last decade[1,2]. At (ultra-) relativistic energies, nuclear collective flow may carry some important information about the quark gluon plasma (QGP) phase transition[3]. At a few hundreds of MeV/nucleon, many experimental and theoretical studies have revealed several collective flow phenomena, such as the directed flow[4,5], the squeeze-out flow[6], and the radial flow[7,8]. At a few tens of MeV/nucleon, there exist the so-called rotational collective flow[9-11]. The main goals of studying collective flow are to determine the nuclear equation of state (EOS) and the in-medium nucleon-nucleon (N-N) cross section[5,12-16]. The recent advance in radioactive nuclear beam (RNB) physics provides people a unique opportunity to investigate isospin effects in HIC's[17,18]. The investigation on collective flow in HIC's induced by the radioactive nuclei may provide people some important information about the isospin-dependent nuclear EOS, particularly the symmetry energy, and the isospin-dependent in-medium N-N cross section.

THE IDQMD MODEL

The IDQMD model includes the symmetry energy, Coulomb interaction, isospin-dependent experimental N-N cross sections, and particularly the isospin-dependent

CP597, Nonequilibrium and Nonlinear Dynamics in Nuclear and Other Finite Systems,
edited by Z. Li et al.

Pauli blocking[19,20]. In the initialization process of the IDQMD model, the neutron and proton are distinguished from each other and meanwhile the nonphysical rotations in the initialized nuclei have been removed[21]. The nuclear mean field in the IDQMD model can be parameterized by

$$U(\rho, \tau_z) = \alpha(\frac{\rho}{\rho_0}) + \beta(\frac{\rho}{\rho_0})^\gamma + \frac{1}{2}(1-\tau_z)V_c + C\frac{\rho_n - \rho_p}{\rho}\tau_z + U^{Yuk} \qquad (1)$$

with ρ_0 the normal nuclear matter density (here is 0.16 fm^{-3}); ρ, ρ_n and ρ_p are the total, neutron, and proton interaction densities, respectively; τ_z is the z-th component of the isospin degree of freedom, which equals 1 or -1 for neutrons or protons, respectively; C is the symmetry strength; V_c is the Coulomb potential; and U^{Yuk} is the finite range Yukawa (surface) potential which will vanish for infinite nuclear matter. The forms and parameters of Eq. (1) can be found in Ref. 20. The IDQMD is different from the so-called IQMD (Isospin-QMD) model[22-25] by the Pauli blocking, the initialization process, and construction of fragment. This model has been used recently to explain successfully several phenomena in HIC's at intermediate energies, which depend on the isospin of the reaction system[19-21,26-28]. In the present calculations, the so-called soft EOS with an incompressibility of K=200 MeV is used and the symmetry strength C=32 MeV without particular consideration.

RESULTS AND DISCUSSIONS

Using IDQMD model, we studied the directed flow and its isospin effects in reactions ^{58}Fe + ^{58}Fe and ^{58}Ni + ^{58}Ni, which have the same mass number but different neutron/proton ratios. The calculated results indicate that the neutron-rich system (^{58}Fe + ^{58}Fe) displays stronger negative directed flow at energy of 55 MeV/nucleon and thus

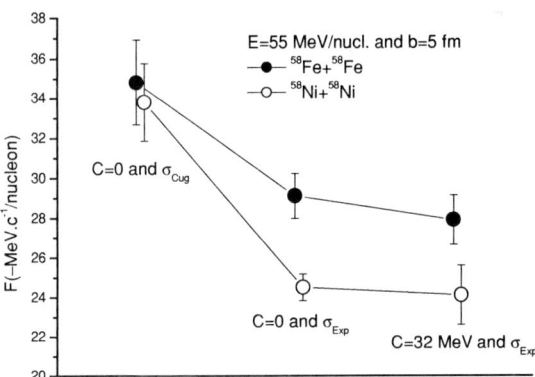

FIGURE 1. The flow parameters using different symmetry energy strength C and parameterizations of N-N cross sections for ^{58}Fe + ^{58}Fe and ^{58}Ni + ^{58}Ni at beam energy of 55 MeV/nucleon and impact parameter b=5 fm.

has a higher balance energy, which could be qualitatively in agreement with the experimental data[20,29-31]. Using different symmetry energy strength C and parameterizations of N-N cross sections, we displays in Fig. 1 the flow parameters for $^{58}\text{Fe}+^{58}\text{Fe}$ and $^{58}\text{Ni}+^{58}\text{Ni}$ at 55 MeV/nucleon and impact parameter b=5 fm. σ_{Cug} represents the Cugnon's parameterization of cross sections, which is isospin-independent. σ_{Exp} is the experimental parameterization of cross sections, which is isospin-dependent. It is indicated from the difference of flow parameters between the two systems that the isospin dependence of directed flow depends on both the isospin-dependent N-N cross section and the symmetry energy, but more sensitive to the former.

The isospin dependence of the azimuthal distribution from reactions of $^{58}\text{Fe} + ^{58}\text{Fe}$ and $^{58}\text{Ni} + ^{58}\text{Ni}$ at 40 MeV/nucleon for different impact parameters is studied in the IDQMD model. It is found that the more neutron-rich system ($^{58}\text{Fe} + ^{58}\text{Fe}$) displays stronger rotational collective flow than the system $^{58}\text{Ni} + ^{58}\text{Ni}$. This phenomenon is more obvious in the semi-peripheral collisions[21]. The azimuthal distribution can be fit by the Legendre polynomial up to the second order, i.e.,

$$\frac{dN}{d\phi} = c(1 + a_1 \cos(\phi) + a_2 \cos(2\phi)). \tag{2}$$

In Eq. (2) the coefficient a_1 represents the strength of the first moment, namely, the in-plane directed flow with preferred emission at $\phi = \pm 180°$ (i.e., $a_1 < 0$) for low rapidity values (in the backward hemisphere) and $\phi = 0°$ (i.e., $a_1 > 0$) for high rapidity values (in the forward hemisphere) if the incident energy is above the balance energy. The coefficient a_2 reflects the strength of the second moment, i.e., it represents a flattening of the ellipsoid. A negative value of a_2 (i.e., the azimuthal distribution peaks at $\phi = \pm 90°$, simultaneously) reflects the squeeze-out effects and a positive value (i.e., the azimuthal distribution peaks at $\phi = 0°$ and $\pm 180°$, simultaneously) the rotational

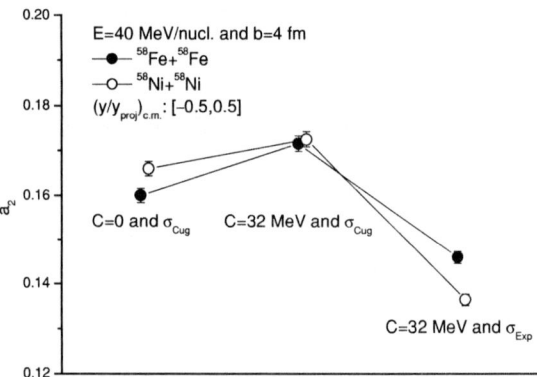

FIGURE 2. The values of a_2 from $^{58}\text{Fe} + ^{58}\text{Fe}$ (solid circles) and $^{58}\text{Ni} + ^{58}\text{Ni}$ (open circles) at 40 MeV/nucleon and b=4 fm for mid-rapidity nucleons using different symmetry energy strength C and parameterizations of N-N cross sections.

collective motion. Using different symmetry energy strength C and parameterizations of N-N cross sections, we show in Fig. 2 the values of a_2 from ^{58}Fe + ^{58}Fe (solid circles) and ^{58}Ni + ^{58}Ni (open circles) at 40 MeV/nucleon and $b=4$ fm for mid-rapidity nucleons. The difference of a_2 between the two systems indicates that the isospin dependence of rotational flow depends on both the isospin-dependent N-N cross section and the symmetry energy, but the symmetry energy seems to enhance the a_2 while the σ_{Exp} reduces it more strongly.

The squeeze-out flow is of special interest since it comes from the only direction where the nuclear matter might escape without being hindered by the presence of the cold spectator remnants and thus a less disturbed information on the matter of high density and temperature is expected[6]. Using the IDQMD model, we studied the out-of-plane squeeze-out flow in reactions of ^{124}Sn + ^{124}Sn and ^{124}Ba + ^{124}Ba, which have the same mass number but different neutron/proton ratios of 1.48 and 1.21, respectively. It is found that the more neutron-rich system (^{124}Sn + ^{124}Sn) exhibits weaker squeeze-out flow. Meanwhile, it is indicated that the squeeze-out flow depends strongly on the impact parameter and incident energy[28]. For the mid-rapidity azimuthal distribution, the ratio

$$R_N = \frac{\frac{dN}{d\phi}(90°) + \frac{dN}{d\phi}(-90°)}{\frac{dN}{d\phi}(0°) + \frac{dN}{d\phi}(180°)} = \frac{1-a_2}{1+a_2} \tag{3}$$

measures the strength of the squeeze-out flow in a quantitative way. A value $R_N>1$ corresponds to a preferential out-of-plane emission. Using different symmetry energy strength C and parameterizations of N-N cross sections, we show in Fig. 3 the values of ratio R_N from ^{124}Sn + ^{124}Sn (solid circles) and ^{124}Ba + ^{124}Ba (open circles) at 350 MeV/nucleon and $b=6$ fm for mid-rapidity nucleons. It is indicated that the isospin dependence of the squeeze-out flow depends on both the isospin-dependent N-N cross section and the symmetry energy.

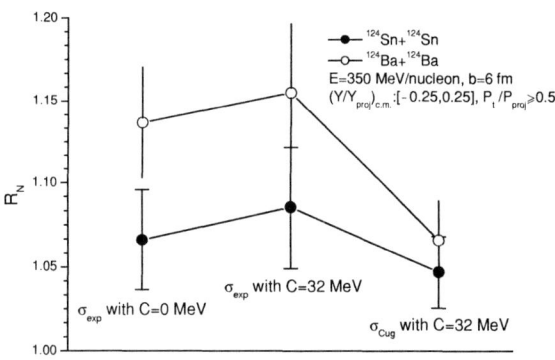

FIGURE 3. The values of ratio R_N from ^{124}Sn + ^{124}Sn (solid circles) and ^{124}Ba + ^{124}Ba (open circles) at 350 MeV/nucleon and $b=6$ fm for mid-rapidity nucleons using different symmetry energy strength C and parameterizations of N-N cross sections.

The radial flow was first found in central collisions of heavy nuclei which are particularly interesting since, for them, stopping, compression, and thermal equilibration are expected to be largest and all the initial kinetic energy is deposited into "one source" located at mid-rapidity which greatly simplifies the collision dynamics, reduces surface effects, and enhances volume effects[7,8]. The experimental and theoretical studies have shown that the radial flow shows no dependence on the nuclear incompressibility and thus it is very different from the in-plane transverse collective flow[7,8,26]. Fig. 4 displays the IDQMD model predicted radial expansion flow from central collisions ^{124}Sn + ^{124}Sn (solid circles) and ^{124}Ba + ^{124}Ba (open circles) at different energies. Meanwhile, the difference of the radial flow between the two systems is also included (up-triangles). From Fig. 4 one can find that the isospin dependence of radial flow is more and more pronounced with the increment of the energies considered here. This isospin dependence is shown to mainly result from the isospin dependence of N-N cross sections and be almost independent of the nuclear symmetry energy[26]. Therefore, a novel experimental recipe is proposed by which detailed comparisons between experimental data and model predictions in the future will shed light on the isospin-dependent in-medium N-N cross sections.

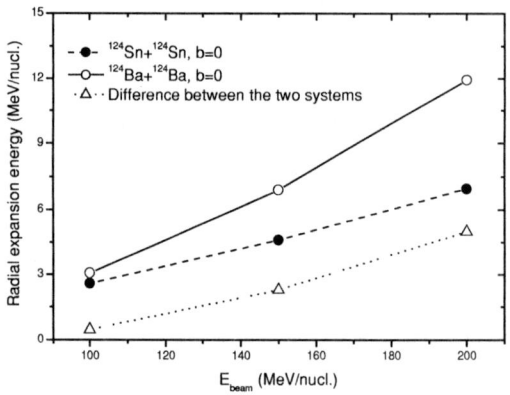

FIGURE 4. The IDQMD model predicted radial expansion flow from central collisions ^{124}Sn + ^{124}Sn (solid circles) and ^{124}Ba + ^{124}Ba (open circles) at different energies. The difference between the two systems is also included (up-triangles).

SUMMARY

Within the framework of an isospin-dependent quantum molecular dynamics (IDQMD) model, we study systematically the isospin dependence of four main categories of collective motions in intermediate energy heavy ion collisions, i.e., the in-plane directed flow, in-plane rotational flow, out-of-plane squeeze-out flow, and radial expansion flow. In particular, the influence of symmetry energy and isospin-dependent nucleon-nucleon cross sections on the isospin dependence of the collective flows is studied. From the calculated results, it is found that the isospin dependence of in-plane directed flow, in-plane rotational flow and out-of-plane squeeze-out flow is determined mainly by both the symmetry energy and isospin-dependent nucleon-

nucleon cross sections. However, the isospin dependence of radial expansion flow has been shown to be determined mainly by only the isospin-dependent nucleon-nucleon cross sections and it is insensitive to the nuclear symmetry energy. Our study proposes that one can investigate the isospin-dependent reaction dynamics by studying the isospin effects on the nuclear collective flows and suggests that the isospin dependence of the radial expansion flow could be as a probe of the isospin-dependent in-medium N-N cross section.

ACKNOWLEDGMENTS

We thank J. Randrup for his kindly providing us the code WIX. We also thank W. Q. Shen and B. A. Li for interesting discussions. This work was supported by the National Natural Science Foundation of China under Grant NOs. 19875068 and 19847002, the Major State Basic Research Development Program under Contract NO. G2000077407, and the Foundation of the Chinese Academy of Sciences.

REFERENCES

1. W. Reisdorf and H. G. Ritter, Annu. Rev. Nucl. Part. Sci. **47**, 663 (1997).
2. J. Y. Ollitrault, *Nucl. Phys.*, 1998, A638: 195c
3. N. S. Amelin, L. P. Csernai, E. F. Staubo, and D. Strottman, Nucl. Phys. **A544**, 463 (1992)
4. H. Stocker, J. A. Maruhn, and W. Greiner, *Phys. Rev. Lett.* **44**, 725 (1980).
5. P. Danielewicz and G. Godyniec, *Phys. Lett.* B **157**, 146 (1985).
6. H. H. Gutbrod et al., *Phys. Lett.* B **216**, 267 (1989).
7. S. C. Jeong et al., *Phys. Rev. Lett.* **72**, 3468 (1994).
8. M. A. Lisa et al., *Phys. Rev. Lett.* **75**, 2662 (1995).
9. W. K. Wilson et al., *Phys. Rev.* C**41**, R1881 (1990).
10. W. K. Wilson, R, Lacey, C. A. Ogilvie, and G. D. Westfall, *Phys. Rev.* C**45**, 738 (1992).
11. W. Q. Shen et al., *Nucl. Phys* .**A551**, 333 (1993).
12. G. D. Westfall et al., *Phys. Rev. Lett.* **71**, 1986 (1993).
13. C. Gale et al., *Phys.Rev.* C**35**, 1666 (1987).
14. H. M. Xu, *Phys. Rev. Lett.* **67**, 2769 (1991).
15. H. B. Zhou, Z. X. Li, Y. Z. Zhuo, G. J. Mao, *Nucl. Phys.* **A580**, 627 (1994).
16. G. J. Mao, Z. X. Li, Y. Z. Zhuo, *Phys. Lett.* B**327**, 183 (1994).
17. B. A Li , C. M. Ko, W. Bauer, *Inte. J. Mod. Phys.* E**7**, 147 (1998).
18. M. Di Toro et al., Prog. Nucl. Part. Phys. **42**, 125 (1999).
19. L. W. Chen, L. X. Ge, X. D. Zhang, and F. S. Zhang, J. Phys. G **23**, 211 (1997).
20. L. W. Chen, F. S. Zhang, G. M. Jing, *Phys. Rev.* C**58**, 2283 (1998).
21. L. W. Chen, F. S. Zhang, Z. Y. Zhu, *Phys. Revt.* C**61**, 067601 (2000).
22. J. Aichelin, Phys. Rep. **202**, 233 (1991).
23. C. Hartnack, Z. X. Li, L. Neise et al., Nucl. Phys. **A495**, 303c (1989).
24. S. A. Bass, C Hartnack, H. Stöcker, and W. Greiner, Phys. Rev. C **51**, 3343 (1995).
25. C. Hartnack, R. K. Puri, J. Aichelin et al., Eur. Phys. J. A **1**, 151 (1998).
26. L. W. Chen, F. S. Zhang, G. M. Jin, and Z. Y. Zhu, *Phys. Lett.* B **459**, 21 (1999).
27. F. S. Zhang, L. W. Chen, Z. Y. Ming, and Z. Y. Zhu, *Phys. Rev.* C**60**, 064604 (1999).
28. F. S. Zhang, L. W. Chen, W. F. Li, Z. Y. Zhu, *Eur. Phys. J.* A **9** (short notes), 149 (2000).
29. B. A. Li, Z. Z. Ren, C. M. Ko, and S. J. Yennello, *Phys. Rev. Lett.* **76**, 4492 (1996).
30. R. Pak, W. Benenson, O. Bjarki et al., *Phys. Rev. Lett.* **78**, 1022 (1997).
31. R. Pak, B. A. Li, W. Benenson et al., *Phys. Rev. Lett.* **78**, 1026 (1997).

AMD Study of Nuclear Clustering

H. Horiuchi

Department of Physics, Kyoto University, Kyoto 606-8502, Japan

Abstract. Study of nuclear clustering by the use of antisymmetrized molecular dynamics (AMD) is discussed for both nuclear structure problems and nuclear collision problems. AMD enables us to describe both the mean field dynamics and the cluster formation. In the study of nuclear structure of light nuclei up to the neutron dripline, it presents us with a unique theoretical method which can describe both shell model strucure and the cluster structure and the structure-change between the two. Capability of treating both the mean field dynamics and the dynamical formation of clusters is especially important in the study of heavy ion collisions at intermediate energies because the reaction process is dominated by various types of fragmentation mechanisms.

INTRODUCTION

The formation of clusters is a fundamental aspect of nuclear many body dynamics together with the formation of mean field [1]. It is largely due to the saturation property of nuclear binding energy, which implies that nucleons and nuclei easily assemble and disassemble without large input or output of energy. In light stable nuclei it is well known that the clustering structure is of basic importance [2, 3]. We will show in this talk from AMD studies that the clustering is also of basic importance in light unstable nuclei. It will be shown from AMD studies that the quantum mechanical description is decisive in many respects for studying the cluster production in nuclear collisions. One example is the description of the shell effect for the dynamical production of alpha particles. Another example is the channel branching due to the quantum mechanical fluctuation which is indispensable for the multifragmentation. The quantum mechanical channel branching ensures the quantum statistical property of the AMD theory.

CLUSTERING IN THE STRUCTURE OF STABLE NUCLEI

Clustering is known to be of basic importance in light stable nuclei [2, 3]. Recently an ab initio calculation with realistic nuclear force by the method of Green's function Monte Carlo [4] has given the result that ^{8}Be has the t wo-alpha dumbbell cluster structure where the inter-alpha distance is about 4 fm. This result gives an explicit theoretical verification of the existence of the cluster structure which has the spatial localization of constituent clusters and cannot be reduced to a special shell model state with a specific spatial symmetry.

AMD [6, 5, 7] is an ab initio calculation with effective nuclear force. AMD can describe formation and dissolution of clusters without any model assumptions.

CP597, *Nonequilibrium and Nonlinear Dynamics in Nuclear and Other Finite Systems,*
edited by Z. Li et al.
© 2001 American Institute of Physics 0-7354-0041-5/01/$18.00

A good example which shows clearly that AMD gives a good description of the formation and dissolution of clusters is a study of ^{12}C made by Kanada-En'yo [8]. It is shown that the calculation reproduces the coexistence of mean field structure of the ground band states and the 3α cluster states near and above 3α breakup threshold. Intra and inter band E2 transitions between the ground band and the cluster band are well reproduced and beta decay strengths from the 1^+ ground state of ^{12}N to the ground band and cluster band states are also well reproduced even though these two bands are very different from each other in structure.

CLUSTERING IN THE STRUCTURE OF NEUTRON-RICH NUCLEI

The AMD studies of Be isotopes [9, 10, 11] present us a very good example of the importance of clustering in neutron-rich nuclei. AMD calculations show that in all Be isotopes from ^{8}Be up to the neutron dripline nucleus ^{14}Be there exists the α-α core. The motion of neutrons is well understood by the concept of the molecular orbits [12] around α-α core. Long time ago Seya, Kohno, and Nagata made the structure study of Be and B isotopes by the model of molecular orbit around α-α core up to the neutron dripline [13]. The ab initio model AMD now confirmed theoretically the formation of the $\alpha - \alpha$ core in Be isotopes up to the neutron dripline. An important valence orbit for neutrons in Be isotopes is the so-called σ orbit coming down from the sd-shell due to the clustering deformation.

Especially in ^{10}Be AMD reproduces well the energy spectra of the observed states which are classified into four rotational bands; $K^\pi = 0_2^+$ and $K^\pi = 0_1^-$ bands have the neutron configurations with two neutrons and one neutron in σ orbit, respectively while $K^\pi = 0_1^+$ and $K^\pi = 2_1^+$ bands have neutron configurations with no neutrons in σ orbit. The Gamow-Teller transition strengths from the ^{10}B ground state to $K^\pi = 0_1^+$ and $K^\pi = 2_1^+$ band states are well reproduced by AMD in addition to the good reproduction of the electric transitions in ^{10}Be [10].

Recent AMD study of ^{12}Be [11] shows us further exciting features of clustering in Be isotopes. The ground rotational band with $K^\pi = 0^+$ has a 2p-2h neutron configuration, which means the vanishing of the N=8 neutron magic number. The second $K^\pi = 0^+$ band has a closed shell structure of neutrons. What is interesting is the third $K^\pi = 0^+$ band which again has a 2p-2h neutron configuration. Its band-head 0^+ state is located around 10 MeV excitation energy and is close to the ^{6}He + ^{6}He threshold at 10.17 MeV. The third $K^\pi = 0^+$ band has a moment of inertia larger than the ground band and its decay width into the ^{6}He + ^{6}He channel is larger than that into the ^{8}He + ^{4}He channel. The experimentally observed excited rotational band with $K^\pi = 0^+$ firstly at RIKEN [14] and recently at GANIL [15] in more detail has a good correspondence with this theoretical third $K^\pi = 0^+$ band. This band has a character of molecular band of ^{6}He + ^{6}He structure.

According to the AMD study of B isotopes [16], B isotopes near neutron dripline nucleus ^{19}B have prominent di-cluster density distribution. Here five protons are divided spatially into two groups with two and three protons, respectively, which are surrounded by neutrons. Reliability of such AMD results is assured by the good reproduction of data for ^{13}B $\sim$ ^{19}B which include binding energies, radii, electric quadrupole moments, and magnetic moments.

SHELL EFFECT AND ALPHA PRODUCTION

AMD which is an antisymmetrized version of molecular dynamics of wave packets assures of course the Fermi statistics of nucleons and hence it can describe the shell effect. However, on account of the antisymmetrization the centroids of nucleon wave packets do not necessarily represent the physical positions and momenta of nucleons when the overlap between wave packets is not small. In order to treat the the two nucleon collision process we have to find the physical positions and momenta of nucleons. In AMD we construct the physical position and momentum coordinates of nucleons by borrowing the ideas of the physical nucleon coordinate [17] of the time-dependent cluster model [18].

The shell effect in the dynamical production of fragments shows up in many kind of reaction processes. A typical example is the alpha production of ^{12}C + ^{12}C collision around the incident energy of 30 MeV/nucleon [6]. AMD calculations have reproduced within a factor 2 the observed production cross sections of various nuclides, namely protons, neutrons, isotopes of He, Li, Be, B, C, and N. What is remarkable is the re-production of the large production cross sections of alpha particles which is comparable to the production cross sections of nucleons. We have studied in more detail the reason why the production cross sections of alpha particles is so large, and have found that the large part of alpha particle production comes through the dynamical process that ^{12}C is first excited to the excited states in the excitaion energy region from 7 MeV to about 15 MeV and then the excited ^{12}C is broken up into 3 alpha particles [19]. The excitaion energy region from 7 MeV to about 15 MeV is the region where many excited states have 3-α cluster structure [2].

Recently, Takemoto et al. studied the fragmentation reaction of B isotopes using AMD [20]. According to them, the clustering structure of B isotopes near the neutron dripline is reflected in their fragmentation reaction. Coincidence cross sections of observed Li and He fragments are predicted to be large compared to those in the case of nonclustering isotope ^{13}B. We expect that the fragmentation experiments will be useful for the study of clustering features of unstable nuclei.

WAVE PACKET SPLITTING (QUANTUM FLUCTUATION) AND CHANNEL BRANCHING

One of the most important advantages of the molecular dynamics theory over the mean field theory in the description of nuclear collisions is the capability of the channel branching. The event calculation of the molecular dynamics enables the description of the channel branching. A basic ingredient of the molecular dynamics which causes the channel branching is the two nucleon collision process. However, theoretical ingredients of the molecular dynamics theory which bring about the channel branching need not to be unique. The aim with which we incorporate into the molecular dynamics theory a new ingredient of the channel branching is that we improve the theory so that it has more quantum mechanical characters.

The equation of motion of the usual molecular dynamics describes the time-development of the centroids of nucleon wave packets. The formation and dissolution of clusters are treated only through the dynamics of the centroids of nucleon wave packets. This feature of the usual molecular dynamics theory of wave packets is clearly unsatisfactory. A wave packet of a nucleon implies that both position and momentum of the nucleon have spread distributions around the centroid values of the wave packet. Hence the the molecular dynamics theory should properly take into account effects of the spread distributions of nucleon position and momentum. The following two cases ellucidate the importance of the proper treatment of the spread distributions of position and momentum around the centroid values.

Let us consider a wave packet approaching the edge of the mean field potential. This wave packet ϕ contains high and low momentum components, $\phi(p > p_0)$ and $\phi(p < p_0)$. The value p_0 is the border momentum value implying that a nucleon whose momentum is higher (lower) than p_0 can (can not) go out of the potential region. According to the quantum mechanics, the wave packet function ϕ predicts that the nucleon goes out of the potential region with the probability $|\alpha|^2$ while the nucleon is reflected back to the potential region with the probability $|\beta|^2$. However according to the usual molecular dynamics, the nucleon is either to go out of the potential region or to be reflected back to the potential region with the 100 % probability depending upon whether the centroid momentum is larger or smaller than p_0. This insufficient character of the usual molecular dynamics of wave packet makes unreasonably the nucleon emission difficult and the nucleus temperature (internal energy) higher. This example shows the necessity of the incorporation of wave packet splitting and its resultant channel branching process into the molecular dynamics theory.

Another example which shows the importance of the proper treatment of the position spread of the nucleon wave packet is the formation process of a cluster from two neighboring wave packets. Let us consider a case that the distance between two wave packet centroids is slightly too far to gain sufficient potential energy to form a cluster in the usual molecular dymamics theory. Even if the usual molecular dynamics does not allow the cluster formation, the quantum mechanics tells us that there should be a finite probability of the cluster formation in this case. It is because there is a finite probability of two nucleons to be close enough to each other through the spread distribution of position even if two wave packet centroids is far apart from each other. This example again shows the necessity of the incorporation of wave packet splitting and its resultant channel branching process into the molecular dynamics theory.

The above-mentioned second example tells us that it is dangerous to allow the width parameter of the nucleon wave packet to be time-dependent without wave packet split-ting. It is because if we allow the time development of the width parameter of the wave packet, the wave packet will become swollen with time and the attraction between wave packets will become too weak resulting in no formation of clusters. We thus realize that if the time development of the width parameter of the wave packet is incorporated into the molecular dynamics theory, the wave packet splitting should also be incorporated into the theory simultaneously.

AMD with wave packet splitting was first successfully applied to the system of ^{40}Ca + ^{40}Ca at the incident energy of 35 MeV/nucleon [21]. Transverse energy spectra of many kinds of fragments including p, α, Li, B, C, O, Ne are well reproduced. The calculated production cross sections of fragments were found to depend on the choice of the effective nuclear force; by using the Gogny force with soft incompressibility we can reproduce well the observed data while if we use the SKG2 force with stiff incompressibility the reproduction of the data is poor.

AU + AU CENTRAL COLLISIONS AND CLUSTERING IN DILUTE NUCLEAR MATTER

AMD with wave packet splitting was applied also to the study of fragmentation in radial expansion in the system of ^{197}Au + ^{197}Au at the incident energy of 150 and 250 MeV/nucleon [22]. Good reproduction of the observed linearity of the fragment kinetic energy as a funtion of fragment mass was obtained. Again here it was found that the calculated production cross sections of fragments depend on the choice of the effective nuclear force; by using the Gogny force with soft incompressibility we can reproduce well the observed data while if we use the SKG2 force with stiff incompressibility the reproduction of the data is poor. In these AMD calculations a new channel branching process is utilized and below we explain this new channel branching process.

In central collisions of Au + Au above about 100 MeV/nucleon on the basis of FOPI collaboration, it is reported that even at 1 GeV/nucleon, about 50 % of protons are contained in clusters. Clusters are regarded as being formed in radially expanding dilute nuclear matter [23]. These experiments can be said to present us with the first observation of the clustering in dilute nuclear matter.

Cluster formation in expanding dilute nuclear matter is made through the coalescence process. Ono pointed out that in order to make proper description of the coalescence process in dilute nuclear matter by using the molecular dynamics of nucleon wave packets, we need to introduce a new channel branching process. The reason why we need the introduction of a new channel branching process is again due to the inherent nature of the wave packet theory. Let us consider proton and neutron wave packets which are very close to each other and have very small relative velocity. We expand the product wave function of two wave packets by the energy eigen-states of the proton-neutron system:

$$\phi_p \phi_n = C_d \Phi_d + \int d^3k C_{\mathbf{k}} \Phi_{\mathbf{k}}. \tag{1}$$

Here Φ_d is the deuteron wave function and $\Phi_{\mathbf{k}}$ is the continuum wave function of the proton-neutron system with asymptotic relative momentum $\mathbf{k}$. According to quantum mechanics, the proton and neutron form a deuteron with a probability $|C_d|^2$ and this deuteron no more disintegrates so far as it does not interact with other particles in expanding dilute nuclear matter. However in the molecular dynamics, if the expectation value of the proton-neutron relative energy of $\phi_p \phi_n$ is positive, we do not have a deuteron cluster.

CONCLUSION

Clustering dynamics is of fundamental importance in nuclear many-body dynamics together with mean field dynamics. They show up both in nuclear structure problems and in nuclear collision problems. AMD which is a quantum molecular dynamics of nucleon wave packets can describe simultaneously the mean field dynamics and the clustering dynamics both in nuclear structure problems and in nuclear collision problems. In the study of nuclear structure of light nuclei up to the neutron dripline, it predicts the existence of variety of clustering states and the predictions are supported by the successful reproduction of various kinds of experimental data. In the study of nuclear fragmentation collisions where quantum mechanical effects are decisively important, AMD has proved to be very successful if we incorporate a few channel branching prescriptions which take into account properly important quantum mechanical fluctuations. The quantum mechanical channel branching ensures the quantum statistical property of the theory.

ACKNOWLEDGMENTS

This talk is based on collaboration works with A. Ono, Y. Kanada-En'yo, and H. Takemoto.

REFERENCES

1. H.Horiuchi, *Summary Talk*, Proc. 7th Int. Conf. on Clustering Aspects in Nuclear Structure and Dynamics, Rab (1999), edited by M. Korolija, Z. Basrak, and R. Čaplar, (World Scientific, Singapore), p.405.
2. H.Horiuchi and K.Ikeda, *Cluster Model of the Nucleus*, International Review of Nuclear Physics, Vol.4 (ed. T.T.S.Kuo and E.Osnes, World Scientific, Singapore, 1986), pp.1 ~ 258.
3. K. Wildermuth and Y. C. Tang, *Unified Theory of the Nucleus*, (Vieweg, Braunschweig, Germany, 1977)
4. R. B. Wiringa, Steven C. Pieper, J. Carlson, and V. R. Pandharipande, Phys. Rev. **C62**, 014001 (2000).
5. H.Horiuchi, Proc. Int. Symp. in Honor of Akito Arima; Nuclear Physics in 1990's, Santa Fe (1990), edited by D.H.Feng, J.N.Ginocchio, T.Otsuka and D.Strottman, Nucl.Phys. **A522**, 257c (1991).
6. H. Horiuchi, Toshiki Maruyama, A. Ohnishi, and S. Yamaguchi, *Proc. Int. Conf. on Nuclear and Atomic Clusters*, Turku (1991), eds. M. Brenner, T. Lönnroth, and F. B. Malik, (Springer, 1992), p.512; A. Ono, H. Horiuchi, Toshiki Maruyama, and A. Ohnishi, Phys. Rev. Letters **68**, 2898 (1992); Prog. Theor. Phys. **87**, 1185 (1992);
7. H.Horiuchi, *Clustering in Nuclear Structure and Collisions*, Proc. NATO Advanced Study Institute on Correlations and Clustering Phenomena in Subatomic Physics, Dronten (1996), edited by M. N. Harakeh, J. H. Koch, and O. Scholten, NATO ASI Series B: Physics Vol.359 (Plenum Press, 1997), p.29.
8. Y. Kanada-En'yo, Phys. Rev. Letters **81**, 5291 (1998).
9. Y. Kanada-En'yo, H. Horiuchi, and A. Ono, Phys. Rev. **C52**, 628 (1995); A. Doté, H. Horiuchi, and Y. Kanada-En'yo, Phys. Rev. **C56**, 1844 (1997).
10. Y. Kanada-En'yo, H. Horiuchi, and A. Doté, Phys. Rev. **C60** 064304 (1999).
11. Y. Kanada-En'yo, H. Horiuchi, and A. Doté, Proc. 7th Int. Topical Conf. on Giant Resonances, Osaka (2000), edited by M. Fujiwara, H. Ejiri, M. Nomachi, and T. Wakasa, Nucl. Phys. **A687**, 146c (2001); Y. Kanada-En'yo, Proc. Int. Workshop on Physics with Radioactive Nuclear Beams, Puri

(1998), edited by B. Sinha, R. Shyam, and I. J. Thompson, J. Phys. G: Nucl. Part. Phys. **24**, 1499 (1998).

12. W. von Oertzen, Proc. 7th Int. Conf. on Clustering Aspects in Nuclear Structure and Dynamics, Rab (1999), edited by M. Korolija, Z. Basrak, and R. Čaplar, (World Scientific, Singapore), p.27.

13. M. Seya, M. Kohno, and S. Nagata, Prog. Theor. Phys. **65**, 204 (1981).

14. A. A. Korscheninnikov et al., Phys. Letters **B343**, 53 (1995).

15. M. Freer et al., Phys. Rev. Letters **82**, 1383 (1999).

16. Y. Kanada-En'yo and H: Horiuchi, Phys. Rev. **C52**, 647 (1995).

17. M. Saraceno, P. Kramer and F. Fernandez, Nucl. Phys. **A405** (1983) 88.

18. S. Drożdż, J. Okolowcz, and M. Ploszajczak, Phys. Lett. **109B** (1982), 145; E. Caurier, B. Grammaticos, and T. Sami, Phys. Lett. **109B** (1982), 150; W. Bauhoff, E. Caurier, B. Grammaticos and M. Ploszajczak, Phys. Rev. **C32** (1985), 1915.

19. H. Takemoto, H. Horiuchi, A. Engel, and A. Ono, Phys. Rev. **C54**, 266 (1996); H. Takemoto, H. Horiuchi, and A. Ono, Phys. Rev. **C57**, 811 (1998).

20. H. Takemoto, H. Horiuchi, and A. Ono, Prog. Theor. Phys. **101**, 101 (1999); Phys. Rev. **C63**, 034615 (2001).

21. A. Ono and H. Horiuchi, Phys. Rev. **C53**, 2958 (1996); A. Ono, Phys. Rev. **C59**, 853 (1999); A. Ono, H.Horiuchi, H. Takemoto, and R. Wada, *Proc. Sixth Int. Conf. on Nucleus-Nucleus Collisions*, Gatlingburg (1997), edited by M. Thoennessen, F. E. Bertrand, J. D. Garrett, and C. K. Gelbke, Nucl. Phys. **A630**, 148c (1998).

22. A. Ono and H.Horiuchi, *XVII RCNP Int. Symp. on Innovative Computational Methods in Nuclear Many-Body Problems*, Osaka (1997), edited by H. Horiuchi, M. Kamimura, H. Toki, Y. Fujiwara, M. Matsuo, and Y. Sakuragi, (World Scientific, Singapore, 1998), p.443; A. Ono, Proc. 7th Int. Conf. on Clustering Aspects in Nuclear Structure and Dynamics, Rab (1999), edited by M. Korolija, Z. Basrak, and R. Čaplar, (World Scientific, Singapore), p.294.

23. W. Reisdorf et al, Nucl. Phys. **A612**, 493 (1997); W. Reisdorf, Proc. 7th Int. Conf. on Clustering Aspects in Nuclear Structure and Dynamics, Rab (1999), edited by M. Korolija, Z. Basrak, and R. Čaplar, (World Scientific, Singapore), p.323.

Nonlinear Response from Classical Transport Theory and Quantum Field Theory

M.E. Carrington[*,†], Defu Hou[*,†,**] and R. Kobes[†,‡]

[*]*Department of Physics, Brandon University, Brandon, Manitoba, R7A 6A9 Canada*
[†]*Winnipeg Institute for Theoretical Physics, Winnipeg, Manitoba*
[**]*Institute of Particle Physics, Huazhong Normal University, 430070 Wuhan, China*
[‡]*University of Winnipeg, Winnipeg, Manitoba, R3B 2E9 Canada*

Abstract. We study nonlinear response in weakly coupled nonequilibrium ϕ^4 theory in the context of both classical transport theory and real time quantum field theory, based on a generalized Kubo formula which we derive. A novel connection between these two approaches is established which provides a diagrammatic interpretation of the Boltzmann equation.

Fluctuations occur in a system perturbed slightly away from equilibrium. The responses to these fluctuations are described by transport coefficients [1]. The investigation of transport coefficients in high temperature gauge theories is important in cosmological applications such as electroweak baryogenesis and in the context of heavy ion collisions [2]. There are two basic methods to calculate transport coefficients: transport theory and linear response theory [3, 4, 5, 6]. As is typical in finite temperature field theory, it is not sufficient to calculate perturbatively in the coupling constant: there are certain infinite sets of diagrams that contribute at the same order in perturbation theory and have to be resummed [7, 8]. To date, most calculations of transport coefficients have been done to the order of linear response. In many physical situations however nonlinear response can be important [9, 10, 11, 12]. Here we study nonlinear response using transport theory and using quantum field theory, and explain the connection between these approaches. We perform a Chapman-Enskog expansion of the Boltzmann equation keeping up to quadratic contributions. We obtain a generalized nonlinear Kubo formula, and a set of integral equations which resum ladder and extended ladder diagrams. We show that these two equations have exactly the same structure, and thus provide a diagrammatic interpretation of the Chapman-Enskog expansion of the Boltzmann equation, up to quadratic order[13].

We start from the definition of shear viscosity. In a system that is out of equilibrium, the existence of gradients in thermodynamic parameters like the temperature and the four dimensional velocity field give rise to thermodynamic forces which lead to deviations from the equilibrium expectation value of the viscous shear stress:

$$\delta \langle \pi_{\mu\nu} \rangle = \eta^{(1)} H_{\mu\nu} + \eta^{(2)} H_{\mu\nu}^{T2} + \cdots, \quad H_{\mu\nu} = \partial_\mu u_\nu + \partial_\nu u_\mu - \frac{2}{3} \Delta_{\mu\nu} \Delta_{\rho\sigma} \partial^\rho u^\sigma$$

$$H_{\mu\nu}^{T2} := H_{\mu\rho} H^\rho_{\;\nu} - \frac{1}{3} \Delta_{\mu\nu} H_{\rho\sigma} H^{\rho\sigma}$$

CP597, *Nonequilibrium and Nonlinear Dynamics in Nuclear and Other Finite Systems*,
edited by Z. Li et al.

where $u_\mu(x)$ is the four dimensional four-velocity field which satisfies $u^\mu(x)u_\mu(x) = 1$. $\eta^{(1)}$ and $\eta^{(2)}$ are the coefficients of the terms that are linear and quadratic respectively in the gradient of the four-velocity. The first coefficient is the usual shear viscosity. The second has has not been widely discussed in the literature – we will call it the quadratic shear viscous response.

The Boltzmann equation can be used to calculate transport properties for weak coupling $\lambda\phi^4$ theory with zero chemical potential[6]. We introduce a phase space distribution function $f(x,\underline{k})$ (the underlined momenta are on shell). The form of $f(x,\underline{k})$ in local equilibrium is,

$$f^{(0)} = \frac{1}{e^{\beta(x)u_\mu(x)\underline{k}^\mu} - 1} := n_k; \quad N_k := 1 + 2n_k. \tag{1}$$

We expand f around f^0 using a gradient expansion in the local rest frame where $\vec{u}(x) = 0$. We keep only linear terms that contain one power of $H_{\mu\nu}$ and quadratic terms that contain two powers of $H_{\mu\nu}$, since these are the only terms that contribute to the viscosity coefficients we are trying to calculate. We obtain [13],

$$\delta\langle\pi_{ij}\rangle = -\frac{\beta}{15}\int\frac{d^3k}{(2\pi)^3 2\omega_k}n_k(1+n_k)k^2 B(\underline{k})H_{ij} +$$
$$\frac{2\beta^2}{105}\int\frac{d^3k}{(2\pi)^3 2\omega_k}[n_k(1+n_k)N_k]k^2 C(\underline{k})H_{ij}^{T2}.$$

Comparing with (1) we have,

$$\eta^{(1)} = \frac{\beta}{15}\int\frac{d^3k}{(2\pi)^3 2\omega_k}n_k(1+n_k)k^2 B(\underline{k}) \tag{2}$$

$$\eta^{(2)} = \frac{2\beta^2}{105}\int\frac{d^3k}{(2\pi)^3 2\omega_k}[n_k(1+n_k)N_k]k^2 C(\underline{k}). \tag{3}$$

Next we will show that $B(\underline{k})$ and $C(\underline{k})$ can be obtained from the first two equations in the hierarchy of equations obtained from the gradient expansion of the Boltzmann equation, which has the form:

$$\underline{k}_\mu\partial^\mu f(x,\underline{k}) = C[f]. \tag{4}$$

The collision term is $C[f]$

Enskog expansion leads to the first order equation [6, 14],

$$I_{ij}(k) = \frac{1}{2}\int_{123}d\Gamma_{12\leftrightarrow 3k}d_n[B_{ij}(\underline{p}_1) + B_{ij}(\underline{p}_2) - B_{ij}(\underline{k}) - B_{ij}(\underline{p}_3)] \tag{5}$$

where $d_n = (1+n_1)(1+n_2)n_3/(1+n_k)$. The second order equation turns out to be[13],

$$N_k I_{ij}(k)B_{lm}(\underline{k}) = \frac{1}{2}\int_{123}d\Gamma_{12\leftrightarrow 3k}d_n\{[N_1 C_{ijlm}(\underline{p}_1) + N_2 C_{ijlm}(\underline{p}_2) - N_k C_{ijlm}(\underline{k})$$
$$-N_3 C_{ijlm}(\underline{p}_3)] + \frac{1}{2}[N_{12}B_{ij}(\underline{p}_1)B_{lm}(\underline{p}_2) - N_{k3}B_{ij}(\underline{p}_3)B_{lm}(\underline{k}) + \tilde{N}_{31}B_{ij}(\underline{p}_1)B_{lm}(\underline{p}_3)$$
$$+\tilde{N}_{k1}B_{ij}(\underline{p}_1)B_{lm}(\underline{k}) + \tilde{N}_{32}B_{ij}(\underline{p}_3)B_{lm}(\underline{p}_2) + \tilde{N}_{k2}B_{ij}(\underline{k})B_{lm}(\underline{p}_2)]\} \tag{6}$$

where we used $N_{ij} = N_i + N_j$, $\tilde{N}_{ij} = N_i - N_j$ $(i, j = 1, 2, 3, k)$. This equation can be solved self consistently for the quantity $C_{ijlm}(\underline{k})$ using the result for $B_{ij}(\underline{k})$ from (5).

Now we turn to response calculation [5]. We work with the density matrix in the Heisenberg representation which satisfies, $\frac{\partial \rho}{\partial t} = 0$ and write $\rho = e^{-A+B}/\text{Tr}e^{-A+B}$ Here A is the equilibrium part of the Hamiltonian and B is a perturbative contribution. We expand the density matrix up to the quadratic term of B. The linear response approximation gives [13],

$$\delta\langle\pi_{\mu\nu}\rangle^l = \frac{H_{\mu\nu}}{10}\int d^3x' \int_{-\infty}^{t} dt' e^{\varepsilon(t'-t)} \int_{-\infty}^{t'} dt'' D_R(x,t;x',t'')$$

where $D_R(x,t;x',t'') = -i\theta(t-t'')\langle[\pi(x,t),\pi(x',t'')]\rangle$. Extracting the shear viscosity using the definition (1) we obtain in momentum space

$$\eta^{(1)} = \frac{1}{10}\frac{d}{dq_0}\text{Im}[\lim_{\vec{q}\to 0} D_R(Q)]|_{q_0=0}. \tag{7}$$

This is the well known Kubo formula [4, 5].

Now we consider corrections to the linear response approximation [10, 11]. After a lengthy calculation we find that the result can be written as a retarded three-point correlator [13]:

$$\delta\langle\pi_{\mu\nu}(x,t)\rangle^q = \eta^{(2)}H_{\mu\nu}^{T2} \tag{8}$$

$$\eta^{(2)} = \frac{3}{70}\frac{d}{dq_0}\frac{d}{dq_0'}\text{Re}\left[\lim_{\vec{q}\to 0} G_{R1}(-Q-Q',Q,Q')\right]|_{q_0=q_0'=0}$$

with

$$\begin{aligned} G_{R1}(x,y,z) =\ & \theta(t_x-t_y)\theta(t_y-t_z)\langle[[\pi(x),\pi(y)],\pi(z)]\rangle \\ & + \theta(t_x-t_z)\theta(t_z-t_y)\langle[[\pi(x),\pi(z)],\pi(y)]\rangle. \end{aligned}$$

We have obtained a type of nonlinear Kubo formula that allows us to obtain the quadratic shear viscous response from a retarded three-point function using equilibrium quantum field theory.

Next we obtain a perturbative expansion for the correlation functions of composite operators $D_R(x,y)$ and $G_{R1}(x,y,z)$ which appear in (7) and (8). We use the CTP formulation of finite temperature field theory, and work in the Keldysh representation [15, 16]. We define the vertices Γ_{ij} and M_{ijlm} by truncating external legs from the following connected vertices:

$$\Gamma_{ij}^C = \langle T_c\pi_{ij}(x)\phi(y)\phi(z)\rangle, \quad M_{ijlm}^C = \langle T_c\pi_{ij}(x)\pi_{lm}(y)\phi(z)\phi(w)\rangle \tag{9}$$

where $\pi_{ij}(x) = \partial_i\phi(x)\partial_j\phi(x) - \frac{1}{3}\delta_{ij}(\partial_m\phi(x))(\partial_m\phi(x))$, and T_c is the time ordering operator on the CTP contour. These definitions allow us to write the two- and three-point correlation functions as integrals of the form depicted in Fig. [1].

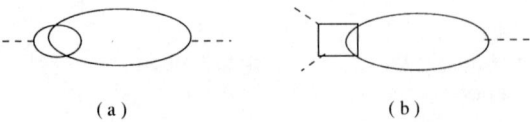

(a)   (b)

FIGURE 1. (a) Two-point function for shear viscosity from linear response; (b) Three-point function for quadratic shear viscous response.

After performing the sum over Keldysh indices using the Mathematica program described in [17] we obtain,

$$D_R(Q) = i \int dK (N_{k+q} - N_k) \Gamma_{R2}^{ij}(K,Q,-K-Q) D_A(K) D_R(K+Q) I_{ji}(k) \quad (10)$$

$$G_{R1}(-Q-Q',Q,Q') = -4 \int dK (\bar{M}_F)_{ikkj}(K,Q,Q') D_R(K)$$
$$\times D_A(K+Q+Q') I_{ji}(k) \quad (11)$$

where $\bar{M}_F = M_F + N_1 M_{R4} + N_4 M_{R1}$ is a particular combination of four-point vertices [18]. Rotational invariance leads to $M_{ijlm} := \hat{I}_{ij}\hat{I}_{lm}M$; $\Gamma_{ij} := \hat{I}_{ij}\Gamma$. We regulate the pinching singularity with the imaginary part of the HTL self energy Σ_k and obtain [14],

$$D_R(K)D_A(K+Q) \rightarrow -\frac{\rho_k}{2 \text{Im} \Sigma_k} \quad (12)$$

with $\rho_k = i(D_R(K) - D_A(K))$. Now we expand in q_0 and q'_0. In (10) and (11) we keep terms proportional to q_0 and $q_0 q'_0$ respectively, since these terms are the only ones that contribute to (7) and (8). Substituting (10) and (11) into (7) and (8) we obtain,

$$\eta^{(1)} = \frac{\beta}{15} \int dK \, k^2 \rho_k n_k (1 + n_k) \left[\frac{\text{Re}\Gamma_{R2}(K)}{\text{Im}\Sigma_k} \right] \quad (13)$$

$$\eta^{(2)} = -\frac{2\beta^2}{105} \int dK \, k^2 \rho_k n_k (1 + n_k) N_k \left[\frac{\text{Re}M_{R1}(K)}{\text{Im}\Sigma_k} \right]. \quad (14)$$

Comparing with (2) and (3) we see that the results are identical if we identify

$$B(\underline{k}) = \frac{\text{Re}\Gamma_{R2}(\underline{k})}{\text{Im}\Sigma_k}, \quad C(\underline{k}) = -\frac{\text{Re}M_{R1}(\underline{k})}{\text{Im}\Sigma_k} \quad (15)$$

with the momentum K on the shifted mass shell: $\delta(K^2 - m_{th}^2)$ where $m_{th}^2 = m^2 + \text{Re}\Sigma_K$.

It is well known that ladder diagrams give the dominate contributions to the vertex Γ_{ij}. They contribute to the viscosity to the same order in perturbation theory as the bare one loop graph and thus need to be included in a resummation. The integral equation that one obtains from resumming ladder contributions to the three-point vertex (see Fig.[2]) has exactly the same form as the equation obtained from the linearized Boltzmann equation (5) with a shifted mass shell describing effective thermal excitations [6, 14].

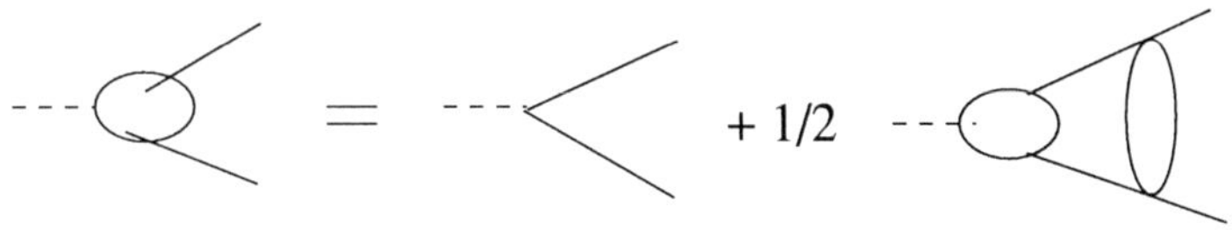

FIGURE 2. Integral equation for the ladder resummation.

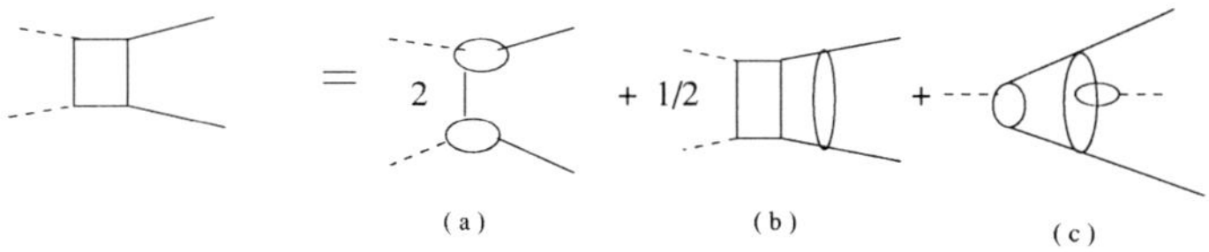

FIGURE 3. Integral equation for an extended-ladder resummation.

Following the pinch effect argument[6, 13], one can show that an infinite set of ladder graphs and some other contributions which we will call extended ladder graph contribute to the same order to vertex M_{ijlm} as the tree diagram. Therefore, for consistent calculation, we consider an integral equation which resums all of these diagrams, as shown in Fig. [3].

We keep only the pinching contributions, using (12) to regulate, and expand in q_0 and q_0' keeping only the term proportional to $q_0 q_0'$, since that is the only term that will contribute to the quadratic shear viscous response coefficient. We obtain [13]:

$$N_k M_{R1}^{ijlm}(K) = -N_k I_{ij} \frac{\Gamma_{R2}^{lm}(K)}{\text{Im}\Sigma_k} \frac{\lambda^2}{4} \int dP \, dR \, d_n \rho_p \rho_p' \rho_r \left[-\frac{N_p M_{R1}^{ijlm}(P)}{\text{Im}\Sigma_p} \right.$$
$$\left. + \frac{1}{2} \frac{\Gamma_{R2}^{ij}(P)}{\text{Im}\Sigma_p} \frac{\Gamma_{R2}^{lm}(R)}{\text{Im}\Sigma_r} \tilde{N}_{rp} + \frac{1}{2} \frac{\Gamma_{R2}^{ij}(P)}{\text{Im}\Sigma_p} \frac{\Gamma_{R2}^{lm}(K)}{\text{Im}\Sigma_k} \tilde{N}_{pk} \right].$$

We introduce the symmetric notation: $P := P_1$; $P' := P_2$; $R := P_3$ and rewrite the equation above after symmetrizing on the integration variables. We obtain,

$$N_k I_{ij} \frac{\Gamma_{R2}^{lm}(K)}{\text{Im}\Sigma_k} = \frac{\lambda^2}{12} \int dP_1 dP_2 dP_3 (2\pi)^4 \delta_{(P_1+P_2-P_3-K)}^4 d_n \rho_1 \rho_2 \rho_3 \left[\frac{N_{p3} M_{R1}^{ijlm}(P_3)}{\text{Im}\Sigma_{p3}} \right.$$
$$+ \frac{N_k M_{R1}^{ijlm}(K)}{\text{Im}\Sigma_k} - \frac{N_{p1} M_{R1}^{ijlm}(P_1)}{\text{Im}\Sigma_{p1}} - \frac{N_{p2} M_{R1}^{ijlm}(P_2)}{\text{Im}\Sigma_{p2}} + \frac{1}{2} \{ N_{12} \frac{\Gamma_{R2}^{ij}(P_1)}{\text{Im}\Sigma_{p1}} \frac{\Gamma_{R2}^{lm}(P_2)}{\text{Im}\Sigma_{p2}}$$
$$- N_{k3} \frac{\Gamma_{R2}^{ij}(K)}{\text{Im}\Sigma_k} \frac{\Gamma_{R2}^{lm}(P_3)}{\text{Im}\Sigma_{p3}} + \tilde{N}_{31} \frac{\Gamma_{R2}^{ij}(P_1)}{\text{Im}\Sigma_{p1}} \frac{\Gamma_{R2}^{lm}(P_3)}{\text{Im}\Sigma_{p3}}$$
$$+ \tilde{N}_{k1} \frac{\Gamma_{R2}^{ij}(P_1)}{\text{Im}\Sigma_{p1}} \frac{\Gamma_{R2}^{lm}(K)}{\text{Im}\Sigma_k} \tilde{N}_{32} \frac{\Gamma_{R2}^{ij}(P_3)}{\text{Im}\Sigma_{p3}} \frac{\Gamma_{R2}^{lm}(P_2)}{\text{Im}\Sigma_{p2}} + \tilde{N}_{k2} \frac{\Gamma_{R2}^{ij}(K)}{\text{Im}\Sigma_k} \frac{\Gamma_{R2}^{lm}(P_2)}{\text{Im}\Sigma_{p2}} \} \right]. \quad (16)$$

Note that once again we have obtained an integral equation that is decoupled: it only involves M_{R1} and Γ_{R2}. With Γ_{R2} determined from the integral equation for the ladder

185

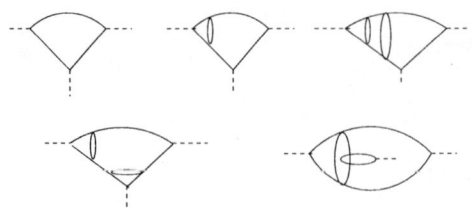

FIGURE 4. Some of the ladder and extended ladder diagrams that contribute to quadratic shear viscous response.

resummation, (16) can be solved to obtain M_{R1}. Finally, comparing (14) and (16) with (3) and (6) we see that calculating the quadratic shear viscous response using transport theory describing effective thermal excitations and keeping terms that are quadratic in the gradient of the four-velocity field in the expansion of the Boltzmann equation is equivalent to calculating the same response coefficient from quantum field theory at finite temperature using the next-to-linear response Kubo formula with a vertex given by a specific integral equation. This integral equation shows that the complete set of diagrams that need to be resummed includes the standard ladder graphs, and an additional set of extended ladder graphs. Some of the diagrams that contribute to the viscosity are shown in Fig. [4].

This result provides a diagrammatic interpretation of the Chapman-Enskog expansion of Boltzmann equation, up to quadratic order.

REFERENCES

1. S.R. de Groot, W.A. van Leeuwen, and Ch.G. van Weert, *Relativistic Kinetic Theory*, (North-Holland Publishing, 1980).
2. See, for example, D. Teaney and E. V. Shuryak, Phys. Rev. Lett. **83**, 4951 (1999).
3. G. Baym, et al , Phys. Rev. Lett. **64**, 1867 (1990).
4. D.N. Zubarev, *Nonequilibrium Statistical Thermodynamics*, (Plenum, New York, 1974).
5. A. Hosoya, M. Sakagami, and M. Takao, Ann. of Phys. (NY) **154**, 229 (1984), and references therein.
6. S. Jeon, Phys. Rev. **D52**, 3591 (1995); S. Jeon and L. Yaffe, Phys. Rev. D **53**, 5799 (1996).
7. R.D. Pisarski, Phys. Rev. Lett. **63**, 1129 (1989); E. Braaten and R.D. Pisarski, Nucl. Phys. B **337**, 569 (1990).
8. V.V. Lebedev and A.V. Smilga, Physica A **181**, 187 (1992); M.E. Carrington, Phys. Rev. **D48**, 3836 (1993).
9. D. Bodeker, Phys. Lett. **B426**, 351 (1998); D. F. Litim and C. Manuel, Phys. Rev. Lett. **82**, 4981 (1999).
10. R. Jackiw and V.P. Nair,Phys. Rev. bf D48, 4991 (1993).
11. J -P Blaizot and E. Iancu, hep-ph/0101103 .
12. P. Arnold, D. T. Son and L. G. Yaffe, Phys. Rev. **D59**, 105020 (1999).
13. M.E. Carrington, Hou Defu and R. Kobes, Phys. Rev. **D65**, 025001 (2001).
14. M.E. Carrington, Hou Defu and R. Kobes, Phys. Rev. **D62**, 025010 (2000).
15. L.V. Keldysh, Zh. Eksp. Teor. Fiz. **47**, 1515 (1964).
16. K.-C. Chou, Z.-B. Su, B.-L. Hao, and L. Yu, Phys. Rep. **118**, 1 (1985).
17. M.E. Carrington, Hou Defu, A. Hachkowski, D. Pickering and J. C. Sowiak, Phys. Rev. **D61**, 25011 (2000).
18. M.E. Carrington, Hou Defu and J.C. Sowiak, Phys. Rev. **D62**, 065003, (2000).

Stochastic One-Body Transport and Coherent Collision Term

Sakir Ayik

Tennessee Technological University, Cookeville, TN 38505, USA

Abstract. The stochastic one-body transport theory in non-relativistic framework is briefly reviewed, and a coherent damping mechanism arising from the coupling between the mean-field fluctuations and the single-particle motion is discussed. For small amplitude fluctuations, the coherent mechanism appears as a particle-phonon collision term in the transport equation for the average density matrix, which play a dominant role in damping of giant resonance excitations and nuclear collective motion at low energies.

INTRODUCTION

In transport theory, we consider a reduced one-body description of dynamics of a many-body system in terms of single-particle density matrix. The simplest form of such a description is provided by the time-dependent Hartree-Fock (TDHF) theory, in which dynamics is treated in the mean-field approximation by neglecting coupling to two-body correlations. Over last two decades, much work has been done to improve the TDHF theory beyond the mean-field approximation. In, so called extended TDHF theory, two-body correlations are incorporated into the equation of motion by truncating the BBGKY hierarchy at the second level by retaining the leading order terms [1-2]. Most applications have been carried out in the semi-classical limit with a BUU form of a collision term. We refer this standard description as the mean-field transport theory, which provides a good description for the average behavior of collision dynamics. However, the mean-field transport theory is totally inadequate for describing dynamics of density fluctuations, in which higher order correlations play a dominant role. In the stochastic transport theory, higher order correlations are incorporated into the equation of motion by a stochastic mechanism according to the generalized Langevin description of Mori, and it has emerged as a promising microscopic approach for dynamics of density fluctuations [3-6].

In the mean-field transport theory, dissipation is described by an incoherent collision term due to coupling of the single-particle motion with 2p-2h excitations. Such an incoherent damping mechanism is very important at relatively high-energy heavy-ion collisions to convert collective energy of the relative motion into incoherent excitations and thermalize the system. However, at low energies and also in giant resonance excitations, the incoherent mechanism is not very effective due to relatively long nucleon mean-free-path [7-9]. Therefore, for a proper description of the damping mechanism, the coherence between p-h pairs should be taken into account. One

CP597, Nonequilibrium and Nonlinear Dynamics in Nuclear and Other Finite Systems,
edited by Z. Li et al.
© 2001 American Institute of Physics 0-7354-0041-5/01/$18.00

possibility for this purpose is provided by the time-dependent density matrix formalism [10]. On the other hand, such a coherent damping mechanism is naturally included in the stochastic transport description as a coupling of single-particle motion with the mean-field fluctuations [11-12]. In section 2, we briefly describe the stochastic one-body transport theory. In section 3, we present a derivation of the coherent collision term in the form of particle-phonon coupling. In section 4, we discuss self-energies of single-particle excitations and collective vibrations due to coherent mechanism. Some conclusions are given in section 5.

STOCHASTIC TRANSPORT EQUATION

Temporal evolution of the reduced single-particle density matrix $\hat{\rho}(t)$ is determined by the first equation of the BBGKY hierarchy,

$$i\frac{\partial}{\partial t}\hat{\rho}(t) - [h(\hat{\rho}), \hat{\rho}(t)] = Tr_2[v, \hat{\sigma}_{12}(t)] \tag{1}$$

where $h(\hat{\rho})$ is the effective mean-field Hamiltonian and v denotes the effective residual interactions [13]. The quantity on the right-hand-side is usually referred to as the collision term, which is specified by the correlated part of the two-particle density matrix $\hat{\sigma}_{12}(t) = \hat{\rho}_{12}(t) - \hat{\rho}_1(t) \cdot \hat{\rho}_2(t)$, where $\hat{\rho}_1(t) \cdot \hat{\rho}_2(t)$ represents the anti symmetric product of single-particle density matrices. At low energies, we can neglect the coupling to two-body correlations entirely, that gives rise to the TDHF theory in which nucleons move in the common mean-field without seeing each other. At intermediate energies both the mean-field and binary collisions are important, therefore we need to incorporate two-body correlations into the equation of motion. The correlated part of two-particle density matrix is determined by the second equation of the BBGKY hierarchy, which depends on the three-body correlations, and so on. At sufficiently low energies for determining propagation of two-body correlations over short time intervals, we truncate the hierarchy at the second level by keeping the leading order terms in the equation of motion [14-15],

$$i\frac{\partial}{\partial t}\hat{\sigma}_{12}(t) - [h(\hat{\rho}), \hat{\sigma}_{12}(t)] = F_{12}(\hat{\rho}) \tag{2}$$

where $F_{12}(\hat{\rho}) = (1 - \hat{\rho}_1(t))(1 - \hat{\rho}_2(t))v\hat{\rho}_1(t) \cdot \hat{\rho}_2(t) - \hat{\rho}_1(t) \cdot \hat{\rho}_2(t)v(1 - \hat{\rho}_1(t))(1 - \hat{\rho}_2(t))$ is the source term. Solving this equation formally, we can express development of correlations over a time interval from an initial time t_0 to time t as,

$$\hat{\sigma}_{12}(t) = -i\int_{t_0}^{t}dt'\hat{G}(t,t')F_{12}(\hat{\rho}(t'))\hat{G}^+(t,t') + \delta\hat{\sigma}_{12}(t) \tag{3}$$

where $\hat{G}(t,t') = T \cdot \exp\left[-i\int dsh(\hat{\rho}(s))\right]$ denotes the mean-field propagator. In this expression, the first term represents correlations developed by the residual interactions during the time interval and the second term describes propagation of the initial correlations $\hat{\sigma}_{12}(t_0)$ from t_0 to t,

$$\delta\hat{\sigma}_{12}(t) = \hat{G}(t,t_0)\hat{\sigma}_{12}(t_0)\hat{G}^+(t,t_0). \tag{4}$$

The time interval cannot be taken arbitrary large, but should be sufficiently short to justify truncation of BBGFY hierarchy at the second level by retaining the leading order terms. However, dominant effects of correlations is still accounted for by the initial correlation term $\hat{\sigma}_{12}(t_0)$, which, in principle, contains all order correlations that accumulated from remote past up to time t_0.

We consider an ensemble of identical systems that are prepared with slightly different initial conditions at remote past. Then, the exact two-body correlations accumulated until t_0 exhibit nearly random fluctuations. In the standard treatment, one considers the average evolution taken over such an ensemble, and the average value of initial correlations in (3) is assumed to vanish, $\overline{\hat{\sigma}_{12}(t_0)} = 0$. This assumption is known as the "molecular chaos assumption" in semi-classical transport theory and it corresponds to factorization of two-particle phase-space density before each binary collisions [14-15]. As a result, evolution of the average single-particle density matrix $\rho(t) = \overline{\rho(t)}$ is determined by a mean-field transport equation,

$$i\frac{\partial}{\partial t}\rho(t) - [h(\rho), \rho(t)] = -i\int_{t_0}^{t} dt'\, Tr_2[v, G(t,t')F_{12}(t')G^+(t,t')]. \quad (5)$$

The collision term essentially involves two different characteristic times: the correlation time τ_{cor} of matrix elements of the residual interactions and the relaxation time τ_{rel} of occupation numbers of the natural states, that corresponds to the average duration time of binary collisions and the mean-free-time between collisions in the semi-classical limit, respectively. The collision term in (5) is valid for a sufficiently dilute system when the binary collisions are well separated in time, $\tau_{cor} \ll \tau_{rel}$, which is referred to as the weak-coupling limit. In this case, the decay time of the collision kernel is determined by the correlation time, and the memory effects associated with occupation numbers over this time maybe neglected. Employing the approximation $G(t,t')\rho(t')G^+(t,t') \approx \rho(t)$, we can make the following substitution in (5),

$$G(t,t')F_{12}(t')G^+(t,t') = (1-\rho_1)(1-\rho_2)v(t,t')\rho_1 \cdot \rho_2 - \rho_1 \cdot \rho_2 v(t,t')(1-\rho_1)(1-\rho_2) \quad (6)$$

where $v(t,t') = G(t,t')vG^+(t,t')$, and all density matrices are evaluated at time t. Numerical applications of the mean-field transport theory have been carried out mostly in the semi-classical approximation with a BUU form collision term. The mean-field transport theory provides a good description for average nuclear static properties and also for the average behavior of the nuclear collision dynamics, however it is inadequate for description of density fluctuations. Correlations between binary collisions are neglected by incorporating the molecular chaos assumption. As a result, independent collisions drive system along the average trajectory without breaking the initial symmetries.

In order to describe dynamics of density fluctuations, correlations should be incorporated into the description in some manner. According to the generalized Langevin description of the relevant variables developed by Mori [16], the correlations due to coupling with degrees of freedom that are not considered explicitly have two different but intimately connected effects: (i) dissipation of energy associated with the relevant variables leading to thermalization of the system, which is described by the friction or the collision term in the equation of motion, and (ii) dynamical fluctuations of the relevant variables, which are described by the random force term originating from initial correlations. Therefore, temporal evolution of the reduced single-particle density matrix should have a stochastic nature. The associated "random force" in the equation of motion should originate from initial correlations and the statistical properties of this term should be specified in accordance with the fluctuation-dissipation relation. In the stochastic description, the initial correlation term is retained, but it is treated as a random quantity specified by a Gaussian distribution: in any representation each matrix element has a Gaussian distribution determined by a zero mean and a second moment. It is convenient to employ the natural representation, $\rho(t) = \sum_i |\phi_i(t)\rangle n_i(t)\langle\phi_i(t)|$, which diagonalizes the average density matrix, where $n_i(t)$ denotes occupation numbers [17-18]. Employing closure approximation, it is possible to determine the second moment of the initial correlation term (4) in the weak-coupling limit. As a result, in the natural representation, the second moment of the each matrix element of the initial correlation term is given by,

$$\overline{<ij|\delta\hat{\sigma}_{12}(t)|kl><k'l'|\delta\hat{\sigma}_{12}(t)|i'j'>} = \frac{1}{2}S_{ij;i'j'}S_{kl;k'l'}N_{ijkl}(t) \tag{7}$$

where $S_{ij;i'j'} = \delta_{ii'}\delta_{kk'} - \delta_{ij'}\delta_{ji'}$ and $N_{ijkl}(t) = [(1-n_i)(1-n_j)n_k n_l + (1-n_k)(1-n_l)n_i n_j]_t$. In the initial correlation term, the initial time t_0 is not relevant, at any time $\delta\hat{\sigma}_{12}(t)$ is a Gaussian random quantity with its second moment specified in terms of one-body quantities according to (7). Substituting the expression (3) for the two-particle correlations into (1) yields a transport equation for the fluctuating density matrix,

$$i\frac{\partial}{\partial t}\hat{\rho}(t) - [h(\hat{\rho}), \hat{\rho}(t)] = -i\int_{t_0}^{t} dt'\, Tr_2[v, \hat{G}(t,t')\hat{F}_{12}(t')\hat{G}^+(t,t')] + \delta\hat{K}(t). \tag{8}$$

Here, the first term on the left-hand-side is a binary collision term, which has the same form as in (5) but expressed in terms of fluctuating quantities and the second term arises from the initial correlations, $\delta\hat{K}(t) = Tr_2[v, \delta\hat{\sigma}_{12}(t)]$, and it describes the stochastic part of collision term. In analogy with the generalized Langevin description of Mori, this equation is regarded as a stochastic transport equation for the fluctuating density matrix, in which the stochastic part of the collision term $\delta\hat{K}(t)$ acts as a random noise. According to the stochastic properties of the initial correlation term, the random noise also has a Gaussian distribution with zero mean and second moment determined by a correlation function,

$$C_{ij;kl}(t,t') = \overline{<i|\delta\hat{K}(t)|j><k|\delta\hat{K}(t')|l>} \tag{9}$$

which can be explicitly determined in terms of one-body properties by substituting the expression (7) for the second moment of the initial correlation term.

PARTICLE-PHONON COUPLING

In the stochastic transport theory, higher order correlations beyond the mean-field are incorporated into the dynamical evolution in an approximate manner by a stochastic mechanism. Dynamical evolution of the system is described by constructing an ensemble of solutions of the stochastic transport equation (8). In this manner, the stochastic theory provides a basis for describing the average evolution, as well as, dynamics of density fluctuations. Furthermore, the stochastic evolution involves, in addition to the incoherent damping mechanism due to coupling with 2p-2h excitations, a coherent damping mechanism arising from the coupling between the single-particle motion and randomly excited mean-field fluctuations [9,14]. When the amplitude of mean-field fluctuations is small, this mechanism appears as a coupling between the single-particle motion and the time dependent RPA phonons around the mean trajectory. In order to illustrate this mechanism, we consider average evolution of the density matrix $\rho(t) = \overline{\hat{\rho}(t)}$, taken over the ensemble generated by the stochastic transport equation. We calculate the ensemble average of (8), by expressing the mean-field and the density matrix as, $h(\hat{\rho}) = h(\rho) + \delta\hat{h}(t)$ and $\hat{\rho}(t) = \rho(t) + \delta\hat{\rho}(t)$, where $\delta\hat{\rho}(t)$ and $\delta\hat{h}(t) = (\partial h / \partial \rho) \cdot \delta\hat{\rho}(t)$ represent the fluctuating parts of the density matrix and the mean-field respectively. Noting that the ensemble average of the random noise vanishes, $\overline{\delta K(t)} = 0$, the evolution of the average density matrix is governed by the transport equation,

$$i\frac{\partial}{\partial t}\rho(t) - [h(\rho), \rho(t)] = K_C(\rho) + K(\rho). \tag{10}$$

Here, $K(\rho)$ represents the same incoherent collision term that appears in (5), but there is an additional term arising from correlations between the mean-field fluctuations and the density fluctuations,

$$K_C(\rho) = \overline{[\delta\hat{h}(t), \delta\hat{\rho}(t)]}, \tag{11}$$

which is referred to as the coherent collision term.

In order to calculate the coherent collision term, we consider that the amplitude of density fluctuations is small, which can be described by the linearized transport equation around the average evolution $\rho(t)$,

$$i\frac{\partial}{\partial t}\delta\hat{\rho} - [\delta\hat{h}, \rho] - [h(\rho), \delta\hat{\rho}] = -i\int_{t_0}^{t}dt'Tr_2[v, \delta\{\hat{G}\hat{F}_{12}\hat{G}^+\}] + \delta\hat{K}. \tag{12}$$

where the first term on the right-hand-side is the linearized form of the non-Markovian incoherent collision term. We analyze the small density fluctuations in a time-dependent RPA approach by introducing the expansion,

$$\delta\hat{\rho}(t) = \sum \delta z_\lambda(t)\rho_\lambda^+(t) + \delta z_\lambda^*(t)\rho_\lambda(t) \tag{13}$$

where $\rho_\lambda^+(t)$ and $\rho_\lambda(t)$ denote the time-dependent RPA functions, and $\delta z_\lambda(t)$ and $\delta z_\lambda^*(t)$ are the stochastic amplitudes associated with these modes. The time-dependent RPA functions describe the correlated p-h excitations around the average trajectory and their time evolutions are determined by,

$$i\frac{\partial}{\partial t}\rho_\lambda^+(t) - [h(\rho), \rho_\lambda^+(t)] - [h_\lambda^+, \rho(t)] = 0. \tag{14}$$

We, also, introduce dual wave functions, $Q_\lambda(t)$ and $Q_\lambda^+(t)$ associated with the RPA modes according to,

$$i\frac{\partial}{\partial t}Q_\lambda(t)-[h(\rho),Q_\lambda(t)]+\tilde{h}_\lambda(t)=0. \tag{15}$$

In these expressions, the fluctuating part of the mean field and its conjugate are defined as $h_\lambda^+(t)=(\partial h/\partial\rho)\cdot\rho_\lambda^+(t)$ and $\tilde{h}_\lambda(t)=-[Q_\lambda(t),\rho(t)]\cdot(\partial h/\partial\rho)$. It is possible to show that [14], if the RPA functions and their dual functions form a bi-orthonormal system at the initial time, they remain orthonormal in time according to, $TrQ_\lambda(t)\rho_\mu^+(t)=\delta_{\lambda\mu}$ and $TrQ_\lambda(t)\rho_\mu(t)=0$. Projecting transport equation (12) on the RPA modes, we can deduce stochastic equations for the random amplitudes,

$$i\frac{d}{dt}\delta z_\lambda(t)=\int_{t_0}^{t}dt'\Sigma_\lambda(t,t')\delta z_\lambda(t')+F_\lambda(t). \tag{16}$$

where $\Sigma_\lambda(t,t')$ is the incoherent self-energy of the RPA modes arising from the collision term and $F_\lambda(t)$ denotes the projected noise. Since, these amplitudes follow linear Langevin equations, they have Gaussian distributions determined by a zero mean and a second moment, $|\delta z_\lambda(t)|^2=N_\lambda(t)+1/2$. It is possible to show that [17], near equilibrium, the quantity $N_\lambda(t)$ is nothing but the finite temperature phonon occupation factor, $N_\lambda^0=1/[\exp(\omega_\lambda/T)-1]$. Following this property, we regard $N_\lambda(t)$ as the time-dependent occupation factors for the RPA modes.

We calculate the ensemble average in the coherent collision term (11) by expanding density fluctuations and mean-field fluctuations in terms of the time-dependent RPA functions; $\delta\hat{n}(t)=\sum\delta z_\lambda(t)h_\lambda^+(t)+\delta z_\lambda^*(t)h_\lambda(t)$, and $\delta\hat{\rho}(t)$ is given by (13). The collision term involves, in addition to the diagonal terms $\overline{\delta z_\lambda\delta z_\lambda^*}$, the off-diagonal terms $\overline{\delta z_\lambda\delta z_\mu^*}$ arising from the coupling between different RPA modes through the incoherent collision term and its stochastic part. This coupling is neglected in eq. (16) for the random amplitudes, which may not be important between the collective RPA modes. However, as a result of the collisional coupling, the non-collective RPA modes are strongly mixed up and loose their bosonic character. Therefore, the non-collective modes should be excluded from the coherent collision term, since their effects can be included into the incoherent collision term by renormalizing the residual interactions. The diagonal contribution of the collective modes to the coherent collision term are given by,

$$[\delta\hat{n}(t),\delta\hat{\rho}(t)]_{coll}=\sum\{(N_\lambda+1/2)[h_\lambda^+,\rho Q_\lambda(1-\rho)]-(N_\lambda+1/2)[h_\lambda^+,(1-\rho)Q_\lambda\rho]\}-h.c. \tag{17}$$

By inspection, it can be seen that the first and second terms in this expression correspond to absorption and excitation of RPA phonons. These rates should be proportional to N_λ and $N_\lambda+1$ respectively, but the average value $N_\lambda+1/2$ appears in both rates. There are other contributions in $K_C(\rho)$ arising from the cross-correlations between the collective and non-collective modes. In schematic models, it is possible to show that these cross-correlations give rise to additional contributions to the collision term, so that the excitation and absorption rates become proportional to $N_\lambda+1$ and N_λ, as it should be. However, in the RPA analysis it is difficult to extract such contributions. For the time being, we replace the excitation and absorption factors in (17) by $N_\lambda + 1$ and N_λ, and express the coherent collision term as,

$$K_C(\rho) = \sum_{coll} \left\{ N_\lambda \left[h_\lambda^+, \rho Q_\lambda (1-\rho) \right] - (N_\lambda + 1)\left[h_\lambda^+, (1-\rho)Q_\lambda \rho \right] \right\} - h.c. \quad (18)$$

where all the quantities are evaluated at time t. Incoherent binary collisions are the dominant mechanism for damping of collective energy in nuclear reactions at intermediate energies. However, at low energy nuclear dynamics and also in giant resonance excitations, the coherent mechanism dominates the damping of collective motion into more complex configurations. Therefore, by retaining only the coherent collision term in (10), we obtain a new extended TDHF theory in which the evolution is self-consistently determined by the coupled transport equations (10), (14) and (15) for the single-particle density matrix and for the time-dependent RPA functions.

FLUCTUATIONS AROUND EQUILIBRIUM

Here, we consider small amplitude density fluctuations around a finite temperature equilibrium state ρ_0, and deduce expressions for coherent damping widths of the single-particle excitations and the collective vibrations. Since the small amplitude density fluctuations are harmonic, the time-dependent RPA equations (14) are reduced to the ordinary finite temperature RPA equations,

$$(\omega_\lambda + i\eta)\rho_\lambda^+ - \left[h_0, \rho_\lambda^+ \right] - \left[h_\lambda^+, \rho_0 \right] = 0 \quad (19)$$

As a result, the equilibrium mean-field fluctuations can be expressed in terms of static RPA modes and their occupation factors N_λ^0.

Damping of single-particle excitations

For describing damping of single-particle excitations, we replace the mean-field Hamiltonian in (10) by its static value, $h(\rho) \to h_0$, and express the density matrix in the Hartree-Fock representation, $\rho(t) = \sum_i |\phi_i(t)> n_i(t) < \phi_i(t)|$. Then, from transport equation (10), we can deduce a master equation for the occupation numbers,

$$\frac{d}{dt} n_i(t) = -\Gamma_i^>(t)n_i(t) + \Gamma_i^<(t)(1 - n_i(t)) \quad (20)$$

where the first and second terms represent loss and gain terms respectively. The loss term is determined by the imaginary part of the self-energy, $\Gamma_i^>(t) = -2 \operatorname{Im}\Sigma_i^>(t)$ with,

$$\Sigma_i^>(t) = \sum (1 - n_j(t)) \left\{ \frac{|< i | h_\lambda^+ | j >|^2}{\varepsilon_i - \varepsilon_j - \omega_\lambda + i\eta} \right\} (N_\lambda^0 + 1) + \left\{ \frac{|< i | h_\lambda | j >|^2}{\varepsilon_i - \varepsilon_j + \omega_\lambda + i\eta} N_\lambda^0 \right\}. \quad (21)$$

Here, the first and second contributions describe decay of the single-particle state by excitation and absorption a phonon respectively. The expression $\Gamma_i^<(t) = -2\operatorname{Im}\Sigma_i^<(t)$ in the gain term is obtained from $\Sigma_i^>(t)$ by making the substitutions $N_\lambda^0 + 1 \to N_\lambda^0$, $N_\lambda^0 \to N_\lambda^0 + 1$ and $1 - n_j(t) \to n_j(t)$. Similar expressions for the single-particle self-energies have been derived in literature by following different approaches [12,19].

Damping of collective vibrations

In order to describe collective vibrations, we linearize the transport equation (10) around a finite temperature equilibrium state ρ_0. Here, we retain only the coherent collision term. The small amplitude vibrations $\delta\rho(t) = \rho(t) - \rho_0$ are determined by,

$$i\frac{\partial}{\partial t}\delta\rho(t) - [h_0, \delta\rho(t)] - [\delta h(t), \rho_0] = \delta K_c(t) \qquad (22)$$

where the quantity $\delta K_c(t)$ represents the linearized form of the coherent collision term. We can analyze density vibrations in the RPA framework similar to the treatment employed in the fluctuation analysis, and expand density vibrations in terms of static RPA functions as, $\delta\rho(t) = \sum z_\lambda(t)\rho_\lambda^+(t) + z_\lambda^*(t)\rho_\lambda(t)$. Here, $z_\lambda(t)$ and $z_\lambda^*(t)$ are deterministic amplitudes of the modes, and the RPA functions are related to dual functions according to $\rho_\lambda^+ = [Q_\lambda^+, \rho_0]$. By projecting (22) on the RPA modes, we find that the associated amplitudes are determined by [12,17,20],

$$i\frac{d}{dt}z_\mu(t) - \omega_\mu z_\mu(t) = \int_{t_0}^{t} dt' \Sigma_\mu(t - t') z_\mu(t') \qquad (23)$$

Here, $\Sigma_\mu(t - t')$ denotes the self-energy of the mode due to the coherent damping mechanism, and its Fourier transform is given by,

$$\Sigma_\mu(\omega) = \sum \left\{ \frac{|< i|[Q_\mu, h_\lambda^+]| j >|^2}{\omega - \omega_\lambda - \varepsilon_j + \varepsilon_i + i\eta} M_{ij}^\lambda - \frac{|< i|[Q_\mu, h_\lambda]| j >|^2}{\omega + \omega_\lambda - \varepsilon_j + \varepsilon_i + i\eta} M_{ji}^\lambda \right\} \qquad (24)$$

where $M_{ij}^\lambda = (N_\lambda + 1)(1 - n_j^0)n_i^0 - N_\lambda n_j^0(1 - n_i^0)$. The first term in the self-energy describes damping of the collective vibration by excitation of a phonon and a p-h pair. At finite temperature, the reverse process is also possible, which decreases the damping. There is another contribution to the self-energy given by the second term in this expression. It describes absorption of a phonon accompanied with excitation of a p-h pair, that is possible only at finite temperature. The self-energy of collective modes has also been investigated by employing the Matsubara formalism, which gives essentially the same result presented here [21-22].

CONCLUSIONS

Development of one-body transport descriptions may lead to novel theoretical tools for understanding the nuclear dynamics at low energies, as well as, the complex reaction mechanism in heavy-ion collisions at intermediate energies. In the stochastic description, the one-body transport theory is further improved by incorporating higher order correlations beyond the mean-field in an approximate manner by a stochastic mechanism that is consistent with the fluctuation-dissipation theorem of non-equilibrium statistical mechanics. As a result, in the stochastic theory, it is possible to address the average evolution as well as the dynamics of density fluctuations in a manner similar to the generalized Langevin description of relevant variables. Furthermore, the stochastic dynamics contains, in addition to incoherent binary collisions, a coherent damping mechanism resulting from the coupling from the mean-field fluctuations and the single-particle motion. It is possible to analyze small amplitude density fluctuations in the time-dependent RPA approach. Then, the coherent damping appears as a particle-phonon collision term in the transport equation for the average density matrix, which play a dominant role in the damping of giant resonance excitations and the collective motion at low energies. While the present results are very encouraging, more work remains to be done for a consistent description of the coherent and the incoherent damping mechanism in connection with dynamics of nuclear collective motion.

ACKNOWLEDGMENTS

Author gratefully acknowledges useful discussions with Y. Abe, Ph. Chomaz and D. Lacroix. This work is supported in part by the US DOE grant DE-FG05-89ER40530.

REFERENCES

1. Bertsch, G. F., and Das Gupta, S., Phys. Rep. **160**, 190 (1988).
2. Cassing, W., and Mosel, U., Prog. in Part. And Nucl. Phys. **25**, 235 (1990).
3. Ayik, S., and Gregoire, C. Phys. Lett. **B212**, 269 (1988); Nucl. Phys. **A513**, 187 (1990).
4. Randrup, J., and Remaud, B., Nucl. Phys. **A514**, 339 (1990).
5. Burgio, G. F., Chomaz, Ph., and Randrup, J., Phys. Rev. Lett. **69**, 885 (1992).
6. Abe, Y., Ayik, S., Reinhard, P. –G., and Suraud, E., Phys. Rep. **275**, 49 (1996).
7. Bertsch, G. F., Bortignon, P. F., and Broglia, R. A., Rev. of Mod. Phys. **55**, 287 (1983).
8. Bertsch, G. F., and Broglia, R. A., Oscillations in Finite Quantum Systems, Cambridge, 1994.
9. Yilmaz, O., Gokalp, A., Yildirim, S., and Ayik, S., Phys. Lett. **B472**, 258 (2000).
10. Bortignon, P. F., Bracco, A., and Broglia, R. A., Giant Resonances and Nuclear Structure at Finite Temperature, Harwood Academic, 1998.
11. Ayik, S., Z. Phys. **A350**, 45 (1994).
12. Ayik, S., Phys. Lett. **B493**, 47 (2000).
13. Ring, R., and Schuck, P., The Nuclear Many-Body Problem, New York: Springer, 1980.
14. Kadanoff, L. P., and Baym, G., Quantum Statistical Mechanics, New York: Benjamin, 1962.
15. Daniellewicz, P., Ann. Phys. **152**, 239 (1984); **197**, 154 (1990).
16. Mori, H., Theoret. Phys. **33**, 423 (1965).
17. Ayik, S., and Abe, Y., Phys. Rev. **C64**, 024609 (2001).
18. Gardiner, C. W., Quantum Noise, Berlin: Springer, 1991.
19. Danielewicz, P., and Schuck, P., Nucl. Phys. **A567**, 78 (1994).
20. Lacroix, D., Ayik, S., and Chomaz, Ph., Phys. Rev. **C63**, 064305 (2001).
21. Bortignon, P., Broglia, R. A., Bertsch, G. F., and Pacheco, J., Nucl. Phys. **A460**, 149 (1986).
22. DeBlasio, F. et al., Phys. Rev. Lett. **68**, 1663 (1992).

PART IV

LARGE AMPLITUDE
COLLECTIVE MOTION
IN FINITE SYSTEMS

Breakdown of Local Equilibrium and the Equation of State in Non-Equilibrium Systems

Dimitri Kusnezov* and Kenichiro Aoki†

*Center for Theoretical Physics, Sloane Physics Lab, Yale University, New Haven, Connecticut
06520-8120 USA
†Department of Physics, Keio University, 4-1-1 Hiyoshi, Kouhoku-ku, Yokohama 223-8521, Japan

Abstract. Classical lattice field theories are studied in non-equilibrium steady states from first principles by placing them in strong thermal gradients. The breakdown of local equilibrium as well as general observables is quantified. We find that linear response predictions are violated at the same rate as the breakdown of local equilibrium, and that there is no threshold for this behavior. Further, the equation of state develops non-local contributions even in theories with bulk behavior in the non-equilibrium steady state.

CLASSICAL LATTICE MODELS

The statistical mechanics of non-equilibrium systems is a field which dates back to the inception of equilibrium statistical mechanics and J.W.Gibbs[1]. However, in spite of over a century of work, the field still lacks a firm theoretical foundation and very little is known analytically beyond the linear response regime. Often, to circumvent to lack of formal techniques, one assumes that linear response provides a good definition of transport coefficients or that there exists a well defined local equilibrium state[2]. Lattice models are a convenient starting point to understand these types of problems since they allow a first principles approach to the non-equilibrium steady state through the use of thermal boundary conditions[3]. In this study we consider classical lattice ϕ^4 theory in $1-3$ dimensions, as well as the Fermi-Pasta-Ulam (FPU) β–model in $1-3$ dimensions. One important distinction between these models is that the former has a well defined bulk limit in $1-d$, while the latter does not[4]. We place these models in a thermal gradient in the x-direction by adding thermostats at temperatures T_1 and T_2 at $x = 0$ and L. For $d > 1$, we impose periodic boundary conditions in the y and z directions.

We have computed the thermal conductivity for these models. In the ϕ^4 theory in $d = 1 - 3$, the conductivity displays a simple power law behavior:

$$\kappa(T) = \frac{c}{T^\gamma} \qquad \begin{cases} c = 2.72(4) & \gamma = 1.38(2) & d = 1 \\ c = 8.8(4) & \gamma = 1.64(8) & d = 3 \end{cases} \qquad (1)$$

This has a well defined bulk behavior in $d = 1 - 3$. In contrast, since the FPU β-model has no bulk limit in $d = 1$[4], the measured conductivity depends both on T and L, as indicated in Fig.1[5]. When the thermal conductivity can be expressed in the form (1) over some temperature range $[T_1, T_2]$, it was shown that the temperature profile has the

CP597, *Nonequilibrium and Nonlinear Dynamics in Nuclear and Other Finite Systems,*
edited by Z. Li et al.
© 2001 American Institute of Physics 0-7354-0041-5/01/$18.00

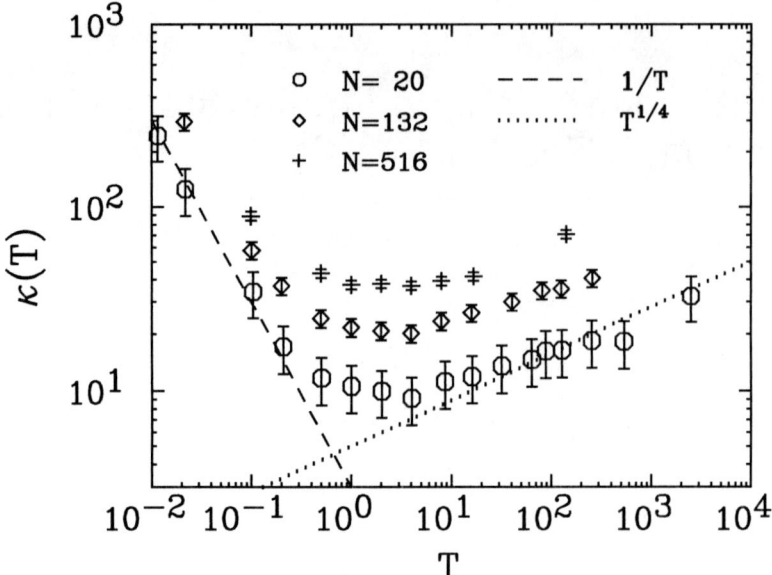

FIGURE 1. Temperature and size dependence of the thermal conductivity in the FPU β−model. The asymptotic power laws are predicted analytically[5]. Even though the behavior is non-trivial, Eq. (2) describes the temperature profile $T(x)$ in non-equilibrium steady states.

form [3, 6]

$$T(x) = T_1 \left(1 - \left(1 - \left[\frac{T_2}{T_1} \right]^{1-\gamma} \right) \frac{x}{L} \right)^{1/(1-\gamma)} \tag{2}$$

While the power law behavior is only asymptotic in Fig. 1, this actually applies very well to the FPU model in any temperature range. This is illustrated in the top panel of Fig. 2. The results of direct measurements are shown by the lines while the symbols indicate the predictions of the formula.

QUANTITATIVE MEASURES

Concepts like local equilibrium are typically invoked to simplify calculations through the use of equilibrium statistical mechanics and thermodynamics. Some efforts have investigated the violation of linear response or local equilibrium[7]-[12], unfortunately there still no quantitative measures of how these quantities begin to breakdown. For systems in a thermal gradient, it is natural to consider how an observable A in the non-equilibrium steady state departs from its equilibrium value, denoted A_{eq}. The normalized deviation from equilibrium is expanded as

$$\delta_A \equiv \frac{\delta A}{A} = \frac{A - A_{eq}}{A_{eq}} = C_A \left[\frac{\nabla T}{T}\right]^2 + C_A' \left[\frac{\nabla T}{T}\right]^4 + \cdots \tag{3}$$

The expansion is expected to be an even power in the temperature gradient since the behavior should not depend on which side of the system the temperatures are applied. This will be explicitly verified below. This is not a priori obvious since in sheared systems analogous expansions have been suggested to have non-analytic behavior[13]. The coefficients and are in principle dependent on T and L, but if the behavior is only local, we would expect the results to depend only on local conditions and not on the overall system size. We will see that this is not the case, and that non-local behavior generally develops.

There is an interesting consequence of the expansion (3). Since the temperature profile (2) is valid quite generally, we can compute the spatial variation of any deviation by computing $\nabla T / T$. The result is generic, given by

$$\frac{\delta A}{A}(x) = C_A \left[\frac{J}{\kappa T}\right]^2 = C_A \frac{1}{[a + bx]^2} \tag{4}$$

The coefficients depend on the power law of the conductivity in the given temperature range and the endpoint temperatures:

$$a = T_1^{1-\gamma} c / J \tag{5}$$
$$b = \gamma - 1 \tag{6}$$
$$\kappa = c T^{-\gamma} \tag{7}$$

While the spatial dependence is generic, the prefactor C_A is model dependent. We now consider several choices of operators A and their properties.

LOCAL EQUILIBRIUM

For **local equilibrium**, let $A = \langle p_x^4 \rangle$ denote the non-equilibrium ensemble average of the fourth moment of momentum measured at a given location x. Then the deviation associated with this observable is $\delta A = \langle p_x^4 \rangle - \langle p_x^4 \rangle_{eq}$. In thermal equilibrium, the fourth moment has the value $\langle p_x^4 \rangle_{eq} = 3 \langle p_x^2 \rangle_{eq}^2$. We use the ideal gas thermometer to identify the local temperature at a position x as $\langle p_x^2 \rangle = T(x)$ both at and away from global thermal equilibrium. Hence the deviation for local equilibrium is $\delta A = \langle p_x^4 \rangle - 3 \langle p_x^2 \rangle^2 = \langle p_x^4 \rangle - 3T(x)^2 = \langle\langle p_x^4 \rangle\rangle$. Thus we define a quantitative measure for the breakdown of local equilibrium as

$$\delta_{LE} = \frac{\langle\langle p_x^4 \rangle\rangle}{3T^2} = C_{LE} \left[\frac{\nabla T}{T}\right]^2. \tag{8}$$

The symbol $\langle\langle \cdots \rangle\rangle$ represents the cumulant. This is sensible since this cumulant should vanish for a Gaussian distribution and hence a non-zero value indicates non-Maxwellian thermal distributions. In the bottom panel of Fig. 2 we show the measurement of δ_{LE}

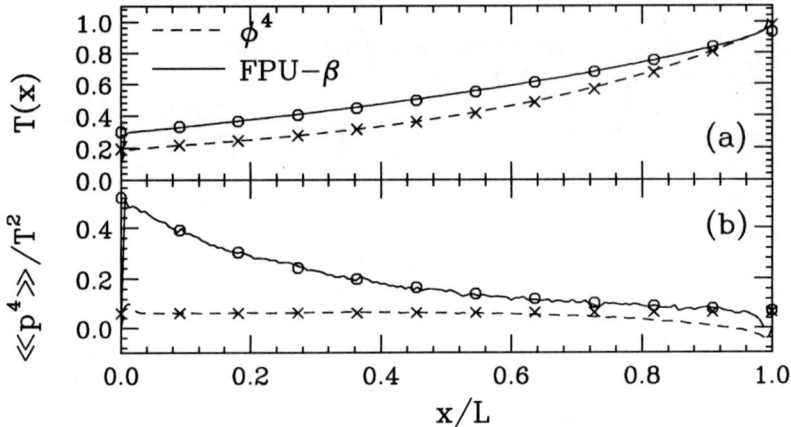

FIGURE 2. Top: Predictions of Eq. (1) (symbols) compared to direct measurements of the temperature profile (lines) in non-equilibrium steady states. One can see that even though the FPU model does not have a global power law behavior, Eq. (1) still describes the behavior. Bottom: Spatial dependence of the breakdown of local equilibrium, as measured through the fourth cumulant. One can see that the direct measures in both models (solid) agrees well with Eq. (4).

as a function of location in the system (lines), compared to Eq. (4) where the coefficients (5)-(6) are fixed (symbols). The analytic predictions are readily confirmed. The x-dependence of (4) suggests further that coarse graining cannot improve the notion of local equilibrium, since the function is positive definite in x. In Fig. 3 we measure δ_{LE} as a function of $\nabla T/T$, verifying explicitly the quadratic dependence. The slopes of these lines are the coefficients C_A, which are tabulated in Table 1 for selected models and temperatures. They are expressed in the form

$$C_A = \mu(T,d)L^{\alpha(T,d)} \tag{9}$$

which demonstrates explicit temperature and dimensional dependence on the model. The best fits are indicated together with the corresponding parameters if the power law is fixed at $\alpha = 1$.

LINEAR RESPONSE

For **linear response**, we compute the difference of the measured heat flux J in the non-equilibrium steady-state and its value in local thermal equilibrium:

$$\delta_{LR} = \frac{J - J_{LR}}{J_{LR}} = C_{LR}\left[\frac{\nabla T}{T}\right]^2 . \tag{10}$$

When Fourier's law holds locally, we know that linear response measurements of transport coefficients through the Green-Kubo autocorrelation functions agree well with direct measurements of heat flow. Hence measuring the breakdown of linear response

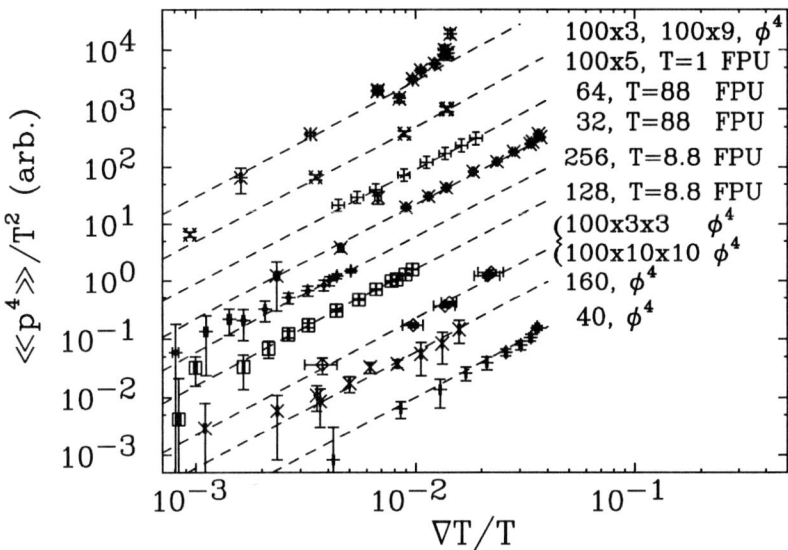

FIGURE 3. Breakdown of local equilibrium δ_{LE} as a function of $\nabla T/T$. The dashed lines indicate quadratic dependence.

TABLE 1. Non-Equilibrium Coefficients for Local Equilibrium

		μ	α
ϕ^4 Theory :			
$d = 1$	$T = 1$	3.3(24)	0.96(15)
		2.7(4)	$\equiv 1$
$d = 1$	$T = 5$	1.6(6)	1.18(9)
		3.1(3)	$\equiv 1$
$d = 2$	$T = 1$	1.9(4)	1.09(5)
$d = 2$	$T = 5$	0.4(2)	1.6(2)
$d = 3$	$T = 1$	4	0.96
$d = 3$	$T = 5$	0.2(5)	1.6(6)
FPU $\beta-$ Model in $d = 1$			
$T = 1$		29(5)	0.87(4)
		16(1)	$\equiv 1$
$T = 8.8$		13(1)	0.99(1)
		12.3(2)	$\equiv 1$
$T = 88$		7.4(4)	1.04(2)
		8.6(2)	$\equiv 1$

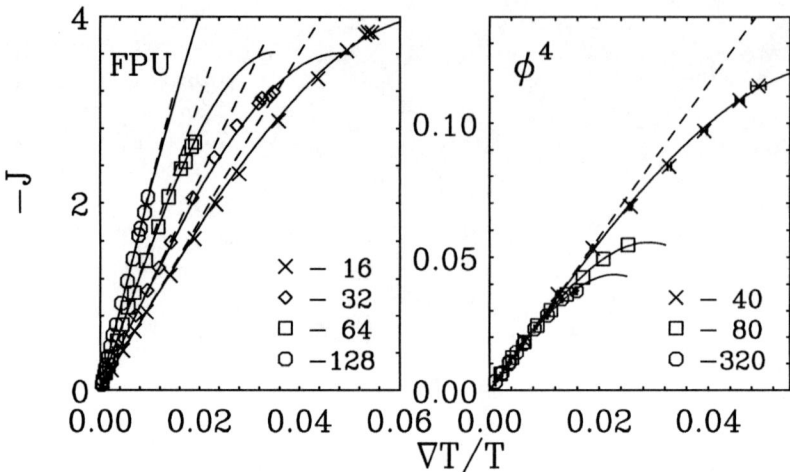

FIGURE 4. Breakdown of linear response in the FPU and ϕ^4 models in $1-d$. The dashed line is the expectation from Fourier's law. As the FPU model has no bulk behavior in $1-d$, there are several dashed lines corresponding to the different system sizes.

can be done through the heat flow. In Fig. 4 we show δ_{LR} as a function of $\nabla T/T$. The curvature in the figure is not due to the curvature in the temperature profile $T(x)$, but indicates how Fourier's law is locally violated. If we consider the heat flow obtained when Fourier's law is satisfied globally, $J_0 = -\kappa((T_1+T_2)/2)(T_2-T_1)/L$, and the linear response heat flow when the temperature profile is linear or curved, J_{LR}, we see that the effects of the non-linear profile are already contained in J_{LR}:

$$\frac{J_{LR}-J_0}{J_0} = \frac{\gamma(\gamma+1+)}{24}L^2\left[\frac{\nabla T}{T}\right]^2 + \cdots \tag{11}$$

This is obtained from Eqs. (1)-(2), using Fourier's law locally and subtracting J_0. We see that while this deviation from Fourier's law being satisfied globally is also quadratic in the temperature, but it depends on size in a specific manner. In (10) we are studying how Fourier's law breaks down locally. The coefficients are displayed in Table 2 for selected systems.

EQUATION OF STATE

For these single component lattice models, the equation of state can be expressed at $P(T)$ where P is the pressure. Beyond global thermal equilibrium, we can explore what happens by choosing $A = P$ in (3). We denote $\delta P = P - P_{eq}$ or equivalently

$$P(T,\nabla T,L) = P_{eq}(T)\left[1+C_P\left(\frac{\nabla T}{T}\right)^2 + \cdots\right]. \tag{12}$$

TABLE 2. Non-Equilibrium Coefficients for Linear Response

	μ	α
ϕ^4 Theory :		
$d = 1 \quad T = 1$	4.0(24)	0.96(26)
	3.6(5)	$\equiv 1$
$d = 1 \quad T = 5$	0.36(16)	1.5(1)
	2.0(5)	$\equiv 1$
$d = 2 \quad T = 5$	0.7(2)	1.2(1)
FPU $\beta-$ Model in $d = 1$		
$T = 8.8$	6.6(8)	0.87(4)
	4.2(2)	$\equiv 1$
$T = 88$	10(5)	0.7(1)
	3.5(4)	$\equiv 1$

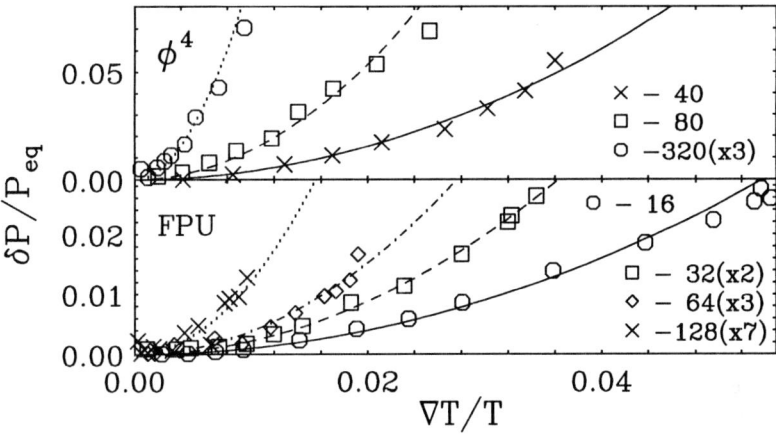

FIGURE 5. Variation of the equation of state $P(T)$ for increasing thermal gradients. A clear system size dependence is seen even though the ϕ^4 model has bulk behavior for these non-equilibrium steady states.

Note that we anticipate new 'state' variables developing in the non-equilibrium state. However, if C_P develops size dependence, the non-equilibrium equation of state becomes non-local. In Fig. 5 we measure the deviation of the pressure and compare it to the functional form (12) to extract the coefficients, tabulated in Table 3.

As the system moves away from equilibrium, the pressure is observed to increase. This should be contrasted with the results from Extended Irreversible Thermodynamics, in which the First Law of Thermodynamics is augmented to include heat flow. In that approach, if one assumes that the concepts of temperature are more or less equivalent (see below), we expect pressure to decrease away from local equilibrium[2]. This

TABLE 3. Non-Equilibrium Coefficients for the Equation of State

	μ	α
ϕ^4 Theory :		
$d = 1 \quad T = 1$	1.5(12)	0.96(17)
	1.2(1)	$\equiv 1$
$d = 1 \quad T = 5$	0.08(2)	1.35(6)
	0.30(5)	$\equiv 1$
$d = 2 \quad T = 1$	9.4(3)	0.61(6)
$d = 2 \quad T = 5$	1.4(4)	0.87(7)
FPU $\beta-$ Model in $d = 1$		
$T = 8.8$	4.1(3)	0.3(2)
$T = 88$	1.8	0.44

TABLE 4. Non-Equilibrium Coefficients for the Energy Density

	μ	α
ϕ^4 Theory :		
$d = 1 \quad T = 1$	0.45(28)	0.91(15)
	0.31(4)	$\equiv 1$
$d = 1 \quad T = 5$	0.03(1)	1.2(1)
	0.056(5)	$\equiv 1$
$d = 2 \quad T = 1$	0.4(1)	1.07(4)
FPU $\beta-$ Model in $d = 1$		
$T = 8.8$	1.7(7)	0.3(1)
$T = 88$	2.2	0.15

apparent contradiction would be interesting to reconcile. These types of lattice models provide a good testing ground since their first principles results can be used to address the construction of non-equilibrium thermodynamic states. It is also possible to make the same analysis for the energy density. If we choose $A = E$ we obtain similar results to the pressure, which are summarized in Table 4.

OBSERVATIONS

One immediate observation is that linear response and local equilibrium break down simultaneously:

$$\delta_{LE} \simeq \delta_{LR}. \tag{13}$$

This is independent of the model or the dimension. They are violated at the same rate, and there is no threshold - violations develop immediately as a system moves from global thermal equilibrium in accordance to Eq. (3).

It is worth making some remarks on Tables 1-4. For the ϕ^4 theory in $d = 1$, all four types of coefficients are consistent with a linear size dependence, while the prefactor still retains T dependence:

$$C^{\phi, d=1} = \mu(T) \times L. \tag{14}$$

For $T = 1$, all the coefficients are compatible with each other for all dimensions. In contrast, the FPU model in $d = 1$, the coefficients for the deviations of linear response and local equilibrium are consistent with a linear L dependence, while the pressure and energy density are similar to $L^{0.3}$.

For local equilibrium, the $\mu(T)$ has a reasonably simply behavior for both models:

$$C_{LE}(L, T) = a \times T^b \times L \qquad \begin{cases} a = 16.1(4) & b = -0.14(1) & FPU \\ a = 2.7 & b = 0.01 & \phi^4 \end{cases} \tag{15}$$

TEMPERATURE RENORMALIZATION

Away from global thermal equilibrium, the definition of temperature can become unclear. Many definitions have been suggested[2, 14]. However, to the order we discuss in this analysis, our results are invariant under redefinition of temperature as follows. If we redefine the temperature in the non-equilibrium case

$$T = \tilde{T} + v \left(\frac{\nabla T}{T} \right)^2 \tag{16}$$

then the redefinition of the coefficients

$$\tilde{C}_A = C_A + v \frac{dA}{dT} \tag{17}$$

will leave the deviations unchanged. The additional term in (16) is due to the temperature dependence of A in (3).

CONCLUSIONS

We have quantified the behavior of observables both near and far from global thermal equilibrium for classical lattice models in 1-3 dimensions. In doing so, we find that a generic spatial dependence emerges for all observables which has been explicitly verified in simulations. By choosing different observables, we have quantified the breakdown of local equilibrium, the goodness of linear response predictions for transport coefficients far from equilibrium, as well as the evolution of the equation of state as one moves to strong thermal gradients. One surprising result is that observables become non-local, developing explicit system size dependence in spite of the existence of a bulk limit. Finally, the notion of non-equilibrium temperature leads one to question the generality of these results. But a suitable redefinition of the coefficients yields all the physical results invariant

ACKNOWLEDGMENTS

We acknowledge support from grants at Keio University, DOE grant DE-FG02-91ER40608 and the Office of Naval Research.

REFERENCES

1. J.W. Gibbs, unpublished notes circa 1895, Yale University.
2. See, *e.g.*, S.R. de Groot, P. Mazur, *"Non-equilibrium Thermodynamics"*, North-Holland, Amsterdam (1962); D. Jou, G. Lebon, J. Casas-Vazques, *"Extended Irreversible Thermodynamics"* (Springer, Berlin, 1996).
3. K. Aoki, D. Kusnezov, *Phys. Lett.***B477** (2000) 348; hep-ph/0002160.
4. H. Kaburaki, M. Machida,*Phys. Lett.* **A181** (1993) 85, S. Lepri, R. Livi, A. Politi, *Phys. Rev. Lett.* **78** (1997) 1896; *Europhys. Lett.* **43** (1998) 271, A. Maeda, T. Munakata, *Phys. Rev.***E52** (1995) 234.
5. K. Aoki, D. Kusnezov, *Phys. Rev. Lett.* (2001) 86, 4029.
6. K. Aoki, D. Kusnezov, *Phys. Lett.***A265** (2000) 250
7. A. Tenenbaum, G. Ciccotti, R. Gallico, *Phys. Rev.* **A25** (1982) 2778.
8. G. Ciccotti, A. Tenenbaum, *J. Stat. Phys.* **23** (1980) 767; C. Trozzi, G. Ciccotti, *Phys. Rev.* **A29** (1984) 916.
9. B. Hafskjold, S.K. Ratkje, *J. Stat. Phys.***78** (1995) 463
10. W. Loose, G. Ciccotti, *Phys. Rev.* **A45** (1992) 3859; M. Mareschal, E. Kestemont, F. Baras, E. Clementi, G. Nicolis, *Phys. Rev.* **A35** (1987) 3883; A. Tenenbaum, *Phys. Rev.* **A28** (1983) 3132.
11. A. Dhar, D. Dhar, *Phys. Rev. Lett.***82** (1999) 480
12. S. Takesue, *Phys. Rev. Lett.* **64** (1990) 252
13. G. Marcelli, B.D.Todd, R. Sadus, *Phys. Rev.* **E63** (2001) 021204; J.P.Ryckaert, A.Bellemans, G.Ciccotti, G.V.Paolini, *Phys. Rev. Lett.* **60** (1988) 128; S. Rastogi, N. Wagner, S. Lustig, *J. Chem. Phys.***104** (1996) 9234; D.Evans, H.J.M.Hanley,*Phys. Lett.* **80A** (1980) 175.
14. J. Casas-Vazquez, D.Jou, *Phys. Rev.* **E49** (1994) 1040 and references therein; J. Keizer, *"Statistical Thermodynamics of Nonequilibrium Processes"* (Springer, New York, 1987).

Fluctuation-Dissipation Dynamics in Heavy-Ion Fusion and Synthesis of the Superheavy Elements

Y. Abe[*], C.W. Shen[†,‡,§] and G. Kosenko[**,§]

*Yukawa Institute for Theoretical Physics, Kyoto Univ., Kyoto 606-8502, Japan
†China Institute of Atomic Energy, P.O.Box 275(18), Beijing 102413, China
**Department of Theoretical Physics, Omsk State Univ. Omsk 644077, Russia
‡Center of Theoretical Nuclear Physics, National Laboratory of Lanzhou Heavy Ion Accelerator,
Lanzhou 730000, China
§RIKEN, Wako-shi, 351-0198, Japan

Abstract. A dynamical framework for synthesis of the superheavy elements is presented, where fusion process is treated by fluctuation-dissipation dynamics with Langevin equations in two steps; approaching phase and shape evolution. Preliminary results are given for residue cross sections of Z=114 with $^{48}Ca+^{244}Pu$ entrance channel, combining the statistical decay calculations.

INTRODUCTION

Theoretical predictions on the existence of an extraordinary stable superheavy elements (SHE) in the nuclear chart (superheavy island) have been made and refined by Strutinski's shell correction method, using phenomenological single particle models and microscopic mean-field calculations[1]. And experimental efforts to reach the superheavy island have been made, but more or less in empirical ways[2]. This is due to the fact that there was no reliable theoretical framework in describing reaction processes starting from an encounter of projectiles and targets through quasi-stable residues of superheavy nucleides, especially fusion processes of massive systems. One of the present authors (Y. A.), together with Konan Univ. group, initiated to construct a dynamical framework which enables us to predict residues cross sections[3]. In the present talk, a framework based on the assumption of the compound nucleus formation is presented, although whole processes might have to be treated dynamically. According to the compound nucleus theory of reaction, residue cross sections are given by a product of fusion probability P_{fus} and survival probability P_{surv} which are independent with each other except conserved quantities such as energy, total angular momentum, etc.;

$$\sigma_{res} = \pi \lambda^2 \Sigma_J (2J+1) \cdot P_{fus}^J(E_{c.m.}) \cdot P_{surv}^J(E_{ex}), \qquad (1)$$

where the excitation energy of the compound nucleus E_{ex} is equal to the *c.m.* incident energy $E_{c.m.}$ plus the Q-value of fusion reaction. In section 2, the survival probability P_{surv} is briefly reminded with an emphasis of special aspects of SHEs. A fusion mechanism is proposed to be of two steps, i.e., an approaching up to the sticking of projectile and

CP597, *Nonequilibrium and Nonlinear Dynamics in Nuclear and Other Finite Systems*,
edited by Z. Li et al.

target and a shape evolution from the sticking configuration to the ground state shape of the superheavy nucleides in section 3. Preliminary results are also given for ^{48}Ca+^{244}Pu system where P_{surv} is calculated with the code HIVAP[4]. The results are to be compared with the recent experiments at Dubna. Remarks on remaining problems are made in section 4.

BRIEF REMINDER OF STATISTICAL DECAYS OF SHE

Since decay modes of the compound nuclei of SHE are mainly fission and neutron emission, the probability P_{surv}^J is given as

$$P_{surv}^J(E_{ex}) = \frac{\Gamma_n^J(E_{ex})}{\Gamma_n^J(E_{ex}) + \Gamma_f^J(E_{ex})}, \qquad (2)$$

where Γ_f and Γ_n are the decay widths of fission and neutron emission and are given by the statistical theory i.e., by Bohr-Wheeler formula or Kramers formula, and Weisskopf formula, respectively[5]. Then, dependences on the excitation energy can be understood by approximate expressions with nuclear temperature, $\Gamma_n \sim \exp[-B_n/T]$, and $\Gamma_f \sim \exp[-B_f/T]$, where B_f and B_n are fission barrier height and neutron separation energy, respectively. (Explicit specifications of angular momentum dependences are suppressed in physical quantities, and should be properly taken into account.) The temperature T is related to the excitation energy by the expression $E_{ex} = a \cdot T^2$ with so-called level density parameter a. The special aspect here in SHE is that the barrier B_f is approximately given by the minus of the shell correction energy, ΔE_{shell}, because Liquid Drop Model (LDM) fission barrier is almost equal to zero in accord with the fissility parameter x being nearly equal to 1. Thus, $\Gamma_f \gg \Gamma_n$ in most superheavy compound nuclei, and then, $P_{surv} \simeq \Gamma_n/\Gamma_f \sim \exp[-(B_n - B_f)/T]$. And the energy ΔE_{shell} naturally depends on the excitation energy, which is readily understood by considering the situation in high excitation where many particle-hole excitations diminish the shell effect. Its dependence is parameterized with the shell-damping energy E_d, by $\Delta E_{shell}(E_{ex}) = \Delta E_{shell}(0) \cdot exp[-E_{ex}/E_d]$ where $\Delta E_{shell}(0)$ denotes the shell correction energy of the ground state and E_d is about 18MeV theoretically[6]. This means that compound nuclei with rather high excitation comparable to E_d have very small fission barriers and decay mostly through fission. Thus, the probabilities calculated with Eq. (2) are very small. If the excitation energy is large enough for multiple emissions of neutrons, the expression of the r.h.s. of Eq. (2) has to be used repeatedly (This is properly taken into account in HIVAP code[4]. and then, the final survival probability would be extremely small. This is the reason why experiments have been done so as for compound nuclei formed to be as low as possible in excitation, especially in so-called cold-fusion path.

It is worth to make semi-quantitative remarks here. The shell correction energies ΔE_{shell} of SHEs are not known precisely, though there are refined theoretical estimates available. If there is an ambiguity of 1 MeV in ΔE_{shell}, accordingly in fission barrier B_f, then the ambiguity in the survival probability is about one order of magnitude, using a typical temperature of 0.5 MeV. Similarly, if one can form an initial neutron-rich compound nucleus with the neutron separation energy B_n being 1 MeV less than a usual value, then one can expect a larger survival probability by about one order of magnitude. This is especially crucial in hot fusion path, because this factor is repeated.

FUSION PROCESSES OF MASSIVE SYSTEMS

In lighter mass region, P^J_{fusion} is equal to the transmission coefficient $T^J(E_{\text{c.m.}})$. In massive systems, however, as is well known, fusion does not occur even if the incident energy is well above the barrier height[7], which means that P^J_{fusion} is much smaller than T^J. This is called the fusion-hindrance. There are two possible interpretations for the hindrance. One is due to effects of frictions in the entrance channels, which cause a dissipation of the incident kinetic energy and thus reduce probability for the system to overcome the barrier. An example is the surface friction model (SFM) proposed by Gross which was successful in explaining characteristic features of deep-inelastic collisions and was applied to the fusion hindrance with fair success in less massive systems, though it appears to hinder not enough in very massive systems[8]. The other one is due to effects of energy dissipation during shape evolutions toward the spherical compound nucleus, starting from the dumb-bell or pear-shaped configuration formed by the sticking of the two nuclei of the entrance channel. Important here is that there is a conditional saddle point (more precisely ridge line or ridge surface in a multi-dimensional deformation space) between the compound nucleus configuration and the sticked dinucleus configuration, which has to be overcome for fusion. The latter mechanism was proposed by Swiatecki and was fairly successful in explaining experimental features of the hindrance observed in many massive systems, though not so good in less massive systems[9]. It would be reasonable to consider both of them to exist, but which one is dominating is not clarified yet. That would depend on systems. Therefore, we propose that the fusion probability is given by the two factors, the sticking and the formation probabilities,

$$P_{\text{fus}} = P_{\text{stick}} \cdot P_{\text{form}}, \tag{3}$$

which is reasonable from the dynamical viewpoint of the reaction process. The two factors in the r.h.s. of Eq. (3) are not independent. As will be discussed below, mechanisms for passing-over the Coulomb barrier not only determine the probability P_{stick}, but also provide initial conditions for the next stage of dynamics of shape evolutions, i.e., for calculations of P_{form}. Below, we will discuss the approaching phase with the SFM, remind the Langevin treatment of shape evolutions, and preliminary results of residue cross sections obtained by the use of HIVAP for P_{surv}.

Approaching phase to the sticking over the Coulomb barrier

It is important to know how much flux reaches the contact point of the two nuclear surface, i.e., the relative distance being the sum of the radii of the projectile and the target. For, from there dynamical evolutions of shapes of the compound system start toward the spherical shape. In order to calculate probabilities of passing over the Coulomb barrier to the contact and to analyse how contacted di-nucleus systems are, or more precisely how much the sticking limit is reached among them, we employ the SFM, i.e., a classical trajectory model in the approaching phase of colliding ions. Here, we simply recapitulate the equations, together with the Langevin forces added consistently with the dissipation-fluctuation theorem,

$$
\begin{aligned}
\frac{dr}{dt} &= \frac{1}{\mu}p \\
\frac{dp}{dt} &= -\frac{dV}{dr} - K_r\frac{p}{\mu} + R_r(t) \\
\frac{d\varphi}{dt} &= \frac{l}{\mu r^2} \\
\frac{dl}{dt} &= -K_\varphi\frac{(l-l_s)}{\mu} + R_\varphi(t),
\end{aligned}
\tag{4}
$$

where the notations are standard, and Langevin forces are assumed to be Gaussian and to satisfy the relation

$$
\langle R_i(t)R_j(t') \rangle = \delta_{ij} \cdot \delta(t-t') \cdot 2 \cdot K_i \cdot T
\tag{5}
$$

with i specifying the coordinate r or φ.

The potential is a sum of the Coulomb potential V_C and the folding potential V_N employed in Ref. 8, and the friction form factors are

$$
K_i = K_i^o \cdot \left(\frac{dV_N}{dr}\right)^2.
\tag{6}
$$

The values of K_i^o are those of Ref. 8,

$$
\begin{aligned}
K_r^o &= 3.5 \times 10^{-23}[\text{s/MeV}] \\
K_\varphi^o &= 0.01 \times 10^{-23}[\text{s/MeV}]
\end{aligned}
\tag{7}
$$

The sticking limit angular momentum denoted by l_s is given by

$$
l_s = l_o \times \frac{\mu r_c^2}{\mu r_c^2 + I_1 + I_2},
\tag{8}
$$

where l_o is the incident angular momentum, r_c is the contact point and I_1, I_2 are the rigid body moments of inertia of the incident ions 1 and 2, respectively.

Preliminary results on ^{48}Ca+^{244}Pu collisions are given in Fig. 1. It is readily seen that an extra energy necessary for the probability to be 1/2 is about 15 MeV, which could be compared with an experimental extra-push energy if the later-on process is neglected, i.e., P_{form} is assumed to be equal to 1. From the results shown in Fig. 1, it can be considered that the hindrance in the approaching phase is not so strong effectively. Quantum-tunneling effects which surely exist but are neglected in the present study

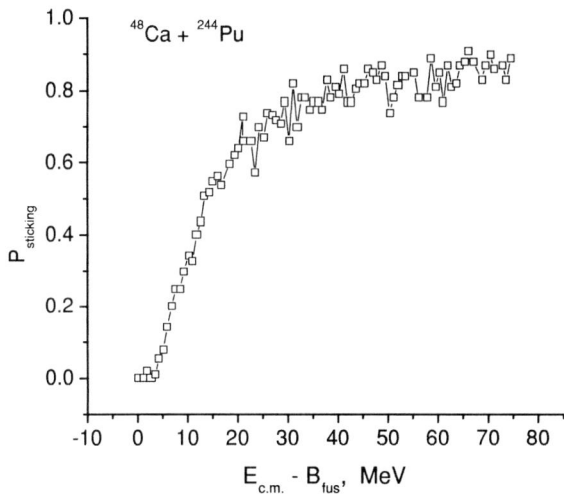

FIGURE 1. Calculated probability for the system to reach the contact point of two nuclear surface. As discussed in the text, this corresponds to the sticking probability. The quantum tunneling would enhance it slightly.

would make it even weaker. On the other hand, results on ^{86}Kr+^{208}Pb system indicate extremely strong hindrance, the extra-push energy being 50 MeV, although the present model does not take into account effects of the shell structures of the projectile and the target. It should be noted here that in both cases the radial momentum at the contact is that of the thermal equilibrium.

As for the angular momentum dissipation, we have calculated orbital angular momenta of the trajectories as a function of the radial distance with $E_{c.m.}$ being 10MeV above the barrier, and found that the average angular momenta reach the sticking limit at the contact point, up to the initial angular momentum of 30 which is the maximum of the calculated examples. This means that the systems averagely form the sticking configuration when they reach the contact point. In the present model, so-called rolling friction is neglected, while tangential friction is included. Analyses with a complete framework are now being made.

The results given above indicate that the SFM is rather strong in the dissipations, although resultant effects depend on systems. There is another model for the approaching phase, i.e., the proximity friction model[10]. Analyses with the model are now being made for comparisons.

213

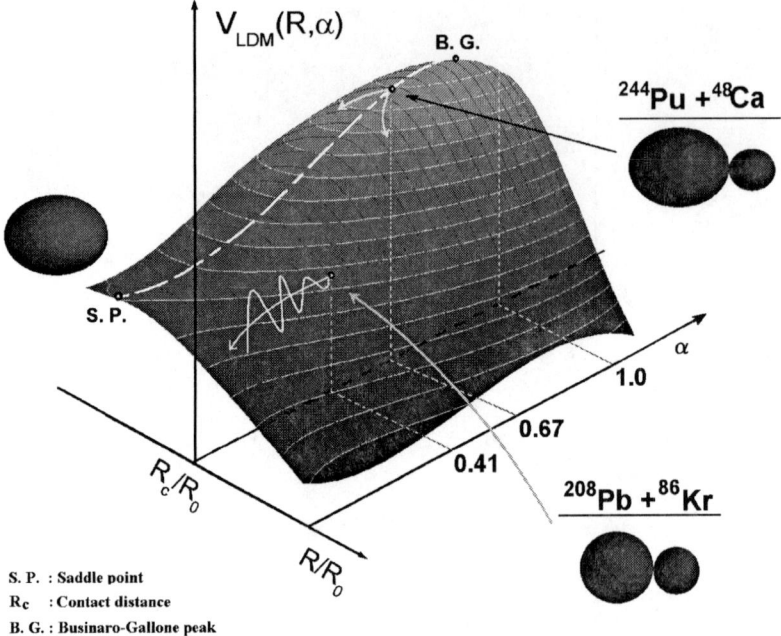

FIGURE 2. LDM energy surface as a function of the distance between two fragments and the mass asymmetry $\alpha = (A_1 - A_2)/(A_1 + A_2)$. As for the former, R_c is the distance at the contact with R_o being the c.m. distance of one-center limit, i.e., the c.m. distance between two semi-spheres. The dashed line is connecting the conditional saddle points, i.e., denotes the ridge line. *S.P.* is the true saddle point for fission. Note that the figure is schematic, because the neck degree of freedom etc. are freezed.

Shape evolutions to the ground state over the ridge line

In the previous subsection, it is shown that a projectile and a target reach the sticking limit after overcoming the Coulomb barrier. The compound system formed is of a pear shape or of a two-center shape which corresponds to that of a connected projectile and target. This configuration does not always fall into the spherical ground state, but rather mostly undergoes reseparation into two fragments. This is readily understood qualitatively by LDM energy surface, which is shown schematically in Fig. 2. The important observation is that in the heavy mass region the contact point is generally outside of the ridge line, and thus the compound system initially formed has to overcome a conditional saddle point or generally the ridge line. The incident kinetic energy or radial momentum, however, is already mostly dissipated until the incident system reaches the contact as is discussed in the previous subsection. Therefore, collective degrees of freedom which describe shape evolutions are subjected to the fluctuation-dissipation dynamics with the heat bath of nucleons inside. Since we need at least a few number of the collective coordinates to describe shape evolutions from the pear-shape to the compact shape or to the reseparation, again we have to employ a multi-dimensional Langevin equation,

$$\frac{dq_i}{dt} = (m^{-1})_{ij} \cdot p_j$$

$$\frac{dp_i}{dt} = -\frac{\partial V}{\partial q_i} - \frac{1}{2}\frac{\partial}{\partial q_i}(m^{-1})_{jk}p_jp_k - \gamma_{ij}(m^{-1})_{jk} \cdot p_k + g_{ij}R_j(t), \qquad (9)$$

where V is the LDM potential energy, m_{ij} and γ_{ij} are the shape-dependent collective inertia mass and friction tensors, respectively. The normalized random force $R_i(t)$ is assumed to be a white noise as usual, i.e., $\langle R_i(t) \rangle = 0$ and $\langle R_i(t_1)R_j(t_2) \rangle = 2 \cdot \delta_{ij} \cdot \delta(t_1 - t_2)$. The strength of the random force g_{ij} is given by $\gamma_{ij} \cdot T = g_{ik} \cdot g_{jk}$. Since compound systems are well excited, Hydrodynamical inertia tensor is adopted with Werner-Wheeler approximation[11]. As for the friction tensor, two models are available; Hydrodynamical viscosity[12] and the wall-and-window one-body models[13]. In order to obtain the formation probability P_{form}, we have to solve many Langevin trajectories with the initial condition of the thermal equilibrium of the radial momentum, most of which undergo reseparation and very small fraction of which falls into the spherical shape with a very shallow LDM pocket. If cooling due to neutron evaporations is quick enough, then the system can have an additional binding due to a restoration of the shell correction energy which makes the pocket deeper. Apparently in Fig 2., mass asymmetric systems like ^{48}Ca+^{244}Pu are favourable, while more mass symmetric systems like ^{86}Kr+^{208}Pb are unfavourable for formation of the spherical compound nuclei. For, in the former there is almost no conditional saddle height, while in the latters finite height saddle points have to be overcome.

Preliminary results of residue cross sections for ^{48}Ca+^{244}Pu system

Since the fusion hindrance originating from the approaching phase appears rather weak in ^{48}Ca+^{244}Pu system, we neglect it for the moment, i.e., we employ the transmission coefficient calculated with the proximity potential, instead of P_{stick} (possible ambiguity would be one-order of magnitude), but for P_{form}, we calculate it as is described in the previous subsection, with the collective coordinates being the relative distance of the fragments, the mass asymmetry, and the deformation of ^{244}Pu, i.e., we solve classical trajectories in the three-dimensional collective coordinate space. As the friction tensor the Hydrodynamical viscosity is employed. The statistical decay part, i.e., the survival probability P_{surv} is calculated with the part of the code HIVAP[4]. Unknown masses are taken from the table predicted by Liran[14]. Residue cross sections calculated for xn reactions are shown in Fig. 3. Apparently, 3n and 4n are to be the favourable channels for obtaining Z=114 element and the magnitude is comparable to the experiments at Dubna[15]. It should be noted here that Liran's mass table gives an extremely small shell correction energy for Z=114, because Z=120 is adopted to be the next shell closure in his model. This is very unfavourable in P_{surv} (probably a few order of magnitude, compared, say with the case of P. Möller et al mass prediction), while if the one-body wall-and-window formula is used as the friction tensor in the Langevin calculations,

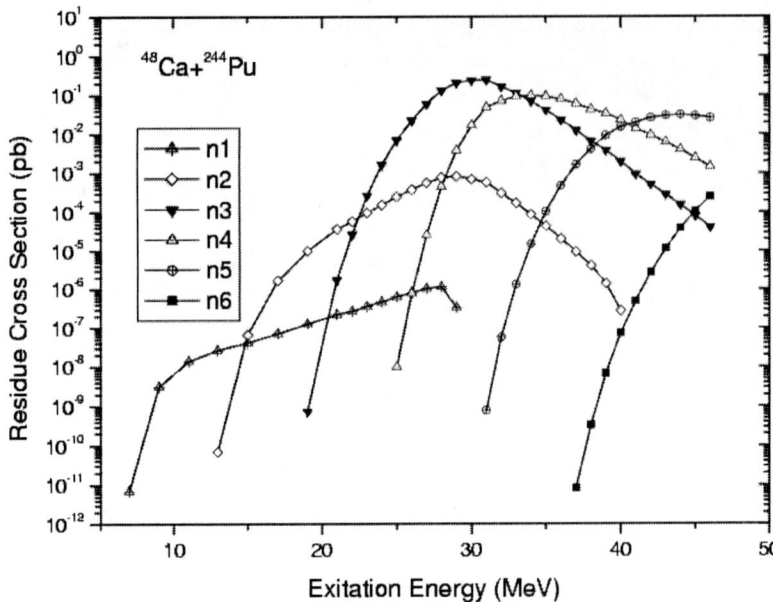

FIGURE 3. Calculated residue cross sections are given for xn reactions of $^{48}Ca+^{244}Pu$ system. Details are discussed in the text.

P_{form} would be reduced dramatically, because the one-body friction is much stronger than the two-body viscosity model.

REMARKS

The preliminary results of the dynamical calculations are in good agreement with Dubna experiments, which is extremely encouraging for the present theoretical framework. Of course, this does not mean that so-called hot fusion path is more favourable than the cold fusion path. The latter is now being under investigation, but it can be stated at least that the former path is favourable in fusion probability and unfavourable in survival probability, while the latter is inverse, and that the final residue cross section is a product of the two probabilities as is given in Eq. (1).

As is mentioned at the end of the previous section, strength of the friction tensor plays a crucial role in determining P_{form} quantitatively, but unfortunately there is a wide variety of theoretical predictions as well as of experimental indications which are obtained by analyses of particle multiplicity prior fission etc.[16] Furthermore, its temperature dependence is not well established[17], which is decisive to quantitative predictions for the cold fusion. Another point to be explored is quantum effects, especially, so-called dissipative tunneling in the Coulomb barrier as well as the conditional saddle. In belief, a theoretical framework for predictions of residue cross sections of SHEs is now available, but there remain many ambiguities in physical parameters which should be fixed, consistent with other phenomena and observables.

ACKNOWLEDGMENTS

The authors acknowledge the supports provided by the theory project of RIKEN Accelerator Research Facility (RARF) and two of them (C.W.C. and G.K) appreciate those by Yukawa Institute for Theoretical Physics (YITP), Kyoto Univ., without which the collaborations on SHE are not possible for the development of the theoretical framework and for the numerical feasibility study of the multi-dimensional Langevin equation. They thank Drs. S. Yamaji and W. Reisdorf for providing them with two center parameterization of the LDM and the statistical decay part of the code HIVAP. At last but not at least they also thank many experimentalists at GSI, Dubna, GANIL and RIKEN for useful discussions and information, especially Drs. S. Hofmann and F.P. Hessberger for the communications of their experiences in the analyses.

REFERENCES

1. W.D. Myers and W.J. Swiatecki, *Nucl. Phys.* **81** (1966) 1.
2. S. Hofmann, *Rep. Prog. Phys.* **61** (1999) 639.
3. Y. Abe et al., *J. Phys* **G23** (1997) 1275.
   Y. Aritomo et al., *Phys. Rev.* **59** (1999) 796.
   T. Wada et al., *Proc. 4th Intern. Conf. on Dynamical Aspects of Nuclear Fission*, Slovakia, Oct. 1998 (World Sceintific 2000) p.77.
4. W. Reisdorf, the fusion evaporation code VIHAP.
5. Y. Abe et al., *Phys. Rept.* **275** (1996) Nos. 2 and 3.
6. A.V. Ignatyuk et al., *Sov. J. Nucl. Phys.* **21** (1975) 255.
7. K.-H. Schmidt and W. Morawek, *Rep. Prog. Phys.* **54** (1991) 949.
8. D.H.E. Gross and H. Kalinowski, *Phys. Rept.* **45** (1978) 175.
   P. Fröbrich et al., *Nucl. Phys.* **A406** (1983) 557.
9. W.J. Swiatecki, *Phys. Scripta* **24** (1981) 113.
   S. Bjornholm and W.J. Swiatecki, *Nucl. Phys.* **A391** (1982) 471.
10. J. Randrup, *Ann. Phys.* **112** (1978) 356.
11. K.T.R. Davis et al., *Phys. Rev.* **C13** (1976) 2385.
12. Ref.11. For the comparisons of two models for energy dissipation, see T. Wada et al., *Phys. Rev. Lett.* **70** (1993) 3538.
13. J. Blocki et al., *Ann. Phys.* **113** (1978) 330.
14. Liran's mass table, http://csnwww.in2p3.fr/AMDC/theory/zel_adt2.pdf.
15. Yu. Oganessian et al., *Phys. Rev. Lett.* **83** (1999) 3154.
16. D. Hilscher et al., *Proc. XII Meeting on Physics of Nuclear Fission*, Obninsk, Russia, Sept. 1993.
17. S. Yamaji et al., *Nucl. Phys.* **A612** (1997) 1.

New Results from Three-Dimensional TDHF

J. A. Maruhn*, D. Dean†, M. R. Strayer† and K. J. Roche**

*Institut für Theoretische Physik, Universität Frankfurt, 60054 Frankfurt, Germany, and Joint
Institute of Heavy Ion Research, Oak Ridge National Laboratory, Oak Ridge, TN 37831, USA
†Physics Division, Oak Ridge National Laboratory, Oak Ridge, TN, USA
**Dept. of Computer Science, University of Tennessee, Knoxville, TN, USA

Abstract. In the late 70's the time-dependent Hartree-Fock (TDHF) approximation originally proposed by Bonche, Koonin, and Negele [1] was used in a number of studies of heavy-ion collisions. These calculations yielded a lot of insight into the reaction dynamics, explaining such phenomena as neck formation and neck rapture and the collective excitations of the fragments in a natural way. There were a number of limitations, however, which decreased interest in the method: while the energy losses in deep-inelastic reactions could be understood semiquantitatively, there was insufficient energy loss for central collisions, leading to a window of deep-inelastic reactions at small angular momenta within the fusion region. This raised the question of whether two-body collisions can be neglected. Also, because of computer limitations at that time, most calculations were done in axial symmetry and the spin-orbit coupling could not be included. In later work, a limited exploration of TDHF with spin-orbit coupling seemed to indicate that this may provide sufficient additional dissipation. We believe that present computer capabilities are sufficient to use TDHF in fully three-dimensional geometry and including the spin-orbit force. This will for the first time allow the investigation of collisions of deformed nuclei. The theory will be explained and first results for collisions of deformed light nuclei will be presented.

INTRODUCTION

The simplest way to obtain the TDHF equations is from a time-dependent variational principle. Demanding that

$$\delta \int dt \, \langle \Psi(t) \| i\hbar \frac{\partial}{\partial t} - H \| \Psi(t) \rangle = 0 \tag{1}$$

with H a standard Hamiltonian contaning two-body interactions,

$$H = \sum_k \frac{-\hbar^2 \nabla^2}{2m} + \sum_{k<l} v_{kl}, \tag{2}$$

and restricting the wave function $\Psi(t)$ to a single Slater determinant built out of the single-particle wave functions $\phi_k(t)$ leads to the time-dependent Hartree-Fock equations

$$i\hbar \frac{\partial}{\partial t} \phi_k(\vec{r}, t) = V(\vec{r}, t) \phi_k(\vec{r}, t) \tag{3}$$

$$- \sum_k \phi_k(\vec{r}, t) \int d^3 r' \phi_l(\vec{r}, t) * v(\vec{r}, \vec{r}') \phi_l(\vec{r}, t)$$

CP597, *Nonequilibrium and Nonlinear Dynamics in Nuclear and Other Finite Systems*,
edited by Z. Li et al.

The important ingredient is the mean field V, which in the case of a pure space dependence of the interaction can be written simply as

$$V(\vec{r},t) = \int d^3 r' v(\vec{r},\vec{r}') \sum_k \left(\phi_k(\vec{r}',t) \right)^2, \tag{4}$$

but including a spin or isospin dependence introduces additional densities in the mean field, while a momentum dependence leads to current densities.

Practical applications of TDHF always used zero-range interactions, in whcih case the exchange term can be written as a function of densities and currents as well. Thus previous calculations employed anything between the highly simpliifed BKN force [1] and a full-fledged Skyrme interaction.

The derivation from the time-dependent variational principle makes one feature quite clear: such an approximation can always only try to get as close as possible to the true *time derivative* of the wave function; there is no guarantee that the solution will not deviate arbitrarily much from the true solution as time goes on.

The setup for a nuclear collision is quite straightforward: the nuclear ground-state wave functions are obtained from a static Hartree-Fock calculations; the states localized in each fragment are inserted into the computational grid such that the center of mass is located at the desired point, and then the single-particle wave functions are multiplied by a common plane wave phase factor to set them into motion with a specified velocity. Since computationally the initial distance of the fragments cannot be large, there also has to be some computation for following the initial and final Coulomb trajectories from and to infinite distance.

ACHIEVEMENTS AND FAILURES

Upon the original proposal [1] there were quite exaggerated expectations concerning what might be learnt from this method, as even the first one-dimensional calculations showed a rich variety of collisional behaviour in nuclear collisions. Two- (in axial symmetry) and three-dimensional calculations soon followed in a large number of papers; we refer the reader to the review [2, 3, 4]. Then as the nature of the restriction to a single Slater determinant become better understood, the limitations of the method also became apparent:

1. The crucial feature to understand is the interaction of all wave function configurations in the one mean field generated by the total density. Thus, for example, even though the reseparated fragments in the final state do not have fixed nucleon numbers (since the single-particle wave functions are spread out over both fragments), they are located in a potential that is correct for the *average* nucleon number — if $^{16}O + ^{16}O$ is the average, the contribution for $^{20}Ne + ^{12}C$ is located in ^{16}O potentials, which is obviously energetically unfavorable and leads to a suppression of such a channel. Thus the spread of fragment masses was always found to be quite small.

2. In the same way the uncertainty in the center of mass, scattering angle, and collective kinetic energy is suppressed: TDHF describes essentially classical behaviour.

219

3. Since the TDHF equations are time-reversal invariant, the TDHF solution is reversible. This does not, however, imply that there is no dissipation from the collective degrees of freedom: certainly the randomization of the single-particle degrees of freedom is very important. In this sense TDHF should be compared to a microscopic simulation like molecular simulations in thermodynamics. On the other hand, the lack of two-body collisions will lead to an underestimation of dissipation.

It is thus not surprising that TDHF was most successful in describing the classical properties of heavy-ion reactions epitomized in the Wilczynski plots, where reasonable agreement with experiment was reached, although most calculations were based on axial symmetry and needed the "clutching" approximation, i. e., the rapid relaxation of the colliding nuclei into a joint rotation for noncentral collisions.

Another successful field of application was that of fusion cross sections. Usually in TDHF calculations at large angular momenta there is a rapid reseparation of the nuclei, but for more central collisions a long-lived compound state may be formed. While TDHF does not allow true fusion, in the sense that the compound nucleus cannot settle down by particle or γ-emission, nevertheless it is highly likely that for such angular momenta fusion will occur.

This leads to another problematic feature: there is usually also a window of quite central angular where rapid reseparation occurs, looking like transparency albeit with the highest energy loss values. This "feed-through" window reduces tie fusion cross section considerably and there were no indications of it in experiment. It was thus an open proble whether this effect is due to the highly restrictive symmetry in most calculations — such as the lack of spin-orbit coupling — or to the omission of two-body collisions. On the other hand, it should be clearly seen that the description of fusion cross sections within an order of magnitude without adjustable parameters, based only on Skyrme interactions whose fit does not include any dynamical properties of nuclei, is a remarkable achievement. Interestingly, in detail the fusion properties seem to crucially depend on the exact Skyrme force used.

To summarize, the positive results from TDHF studies were that many qualitative features of heavy-ion reactions can be understood that are difficult to describe otherwise, such as

- the formation and the snapping of the neck,
- the general behaviour of collective dynamics,
- the role of single-particle dissipation
- the gross features of fusion.

One the goals of the present investigations is to check whether a better quantitative description can be achieved by solving the equations under less restrictive assumptions.

THE CODE BASIS

The first and up to now only calculations for TDHF in three dimensions and including the spin-orbit force were done by Umar and Strayer [6]. The inclusion of the spin-orbit

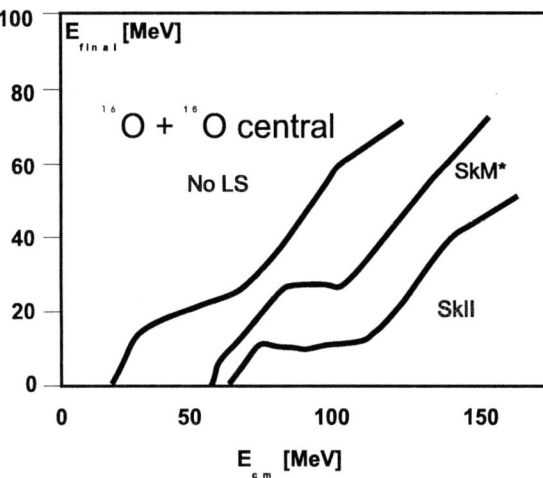

FIGURE 1. Final energy in central $^{16}O+^{16}O$ collisions, both in older (without spin orbit) and the newer calculations, for the latter case with two different Skyrme forces. From [6].

term, of course, implied that for the first time nuclei with a correct shell structure could be simulated in such a way. At that time the most modern Skyrme force the code could deal with was SkM*, but this aspect can easily be generalized for more recent forces.

Because of the computational expense, only a few actual calculations were carried out, both nuclear excitations and heavy-ion reactions. The relaxed symmetry as expected led to a larger energy loss and a much higher threshold for the cental "feed-through", as is illustrated in Fig. 1.

PROBLEMS AND FUTURE DEVELOPMENTS

Code development

The code as developed in [6] represents the wave functions in a three-dimensional grid using B-splines (typically of fifth order) and solving the time-dependent equations with the collocation method. This allows the use of a relatively coarse grid; for example, in the $^{20}Ne + ^{20}Ne$ calculations a grid of 34 by 18 by 18 points was used for almost central collisions, requiring abit more than one million wave function values or about 16 MB for double-precision complex values.

A practicable timestep is of the order of 0.1 fm/c, and a typical collision with fast reseparation lasts of the order of several hundred fm/c. Typical running times on a PC are 10 hours to several days. While this may be tolerable for exploratory calculations such as presented in this talk, using the method systematically will be fraught with problems of computational expense:

- Calculating heavier nuclei means an increase both in the number of wave functions

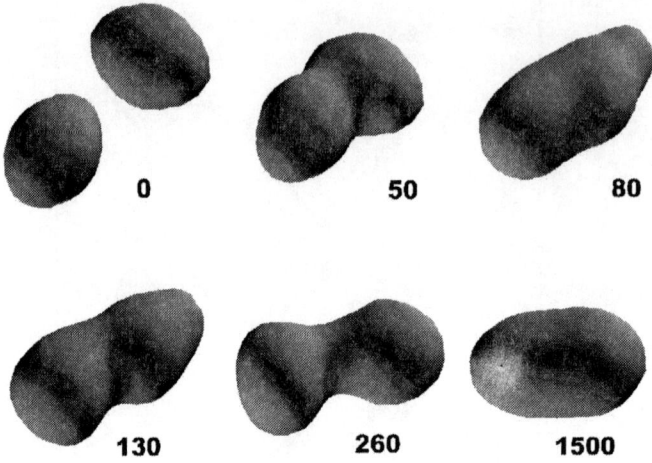

FIGURE 2.   Collision of two deformed Ne nuclei in a "tip-to-belly" configuration at impact parameter 2 fm. The times corresponding to each plot are given in fm/c.

and the grid size required: going from ^{20}Ne $+^{20}$Ne to, e. g., ^{70}Zn $+ ^{208}$Pb, a reaction of interest for the formation of superheavy nuclei, means a factor 6 in the number of wave functions, and one may safely double the number of grid points in each direction, yielding an overall factor of 48.

- Then it must be remembered that for fusion studies a reasonable number of c.m. energies and angular momenta have to be explored.
- Finally, the biggest problem is that of relative orientations. If both nuclei are axially deformed, there are three angular degrees of freedom determining the relative orientation. If one of the nuclei is spherical, there are still two angles left.

From this discussion it is clear that a treatment of heavy-ion collisions will require the use of massively parallel computers, and even then the deformation dependence can probably be studied only in an explorative way. A parallelization fo the code is underway.

Pairing

Traditional time-dependent Hartree-Fock calculations did not include pairing, because the prevalent way of calculating pairing — the BCS approximation with constant gap or constant strength — is based only on the single-particle spectra and cannot be generalized easily to separated fragments (the same holds true of the Strutinsky shell-correction method). Using a state-dependent pairing matrix element, generated, for example, from a delta force, and employing the time-dependent Hartree-Fock-Bogolyubov equations

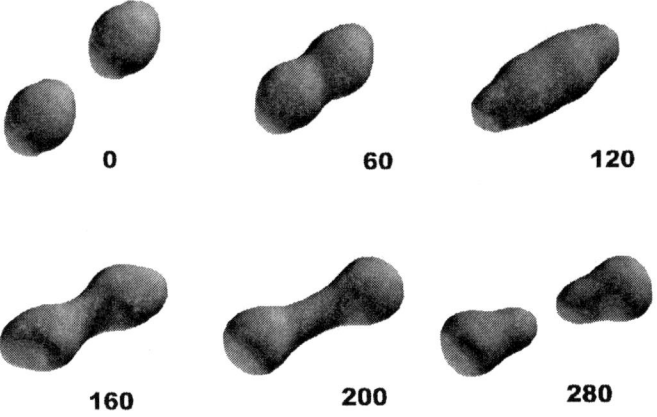

FIGURE 3. Collision of two deformed Ne nuclei in a "tip-to-tip" configuration at impact parameter 2 fm. The times corresponding to each plot are given in fm/c.

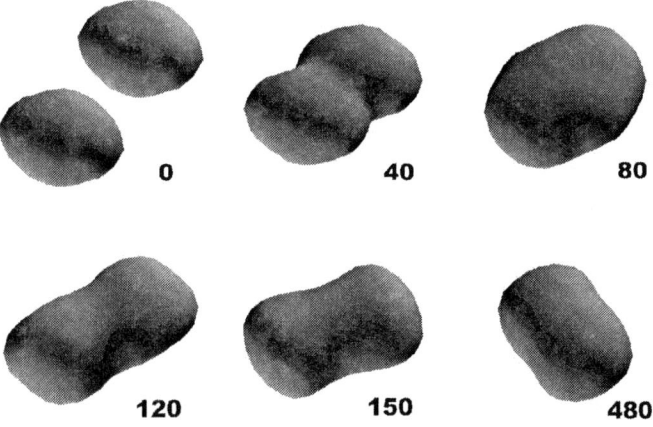

FIGURE 4. Collision of two deformed Ne nuclei in a "belly-to-belly" configuration at impact parameter 2 fm. The times corresponding to each plot are given in fm/c.

would provide an attractive option. The calculations presented here do not include any pairing at all, so that the deformation dependence and the moments of inertia are not quite realistic. This should be borne in mind.

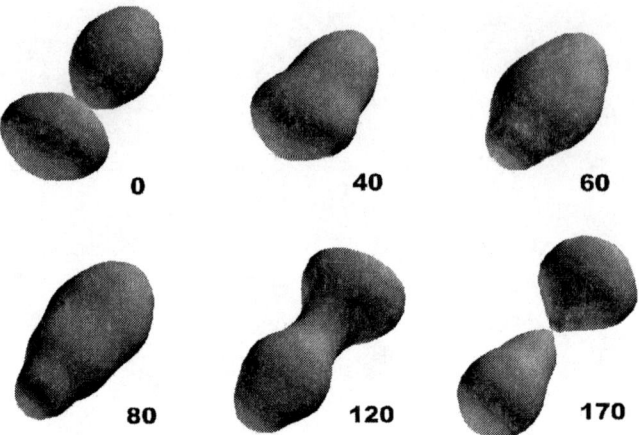

FIGURE 5. Central collision of two deformed Ne nuclei in a "tip-to-belly" configuration. The times corresponding to each plot are given in fm/c.

Initial Condition

The initial conditions for a heavy-ion reaction are, in principle, given at infinite separation distance. Practically, however, the TDHF calculation has to be started at a finite and, for reasonable sizes of the computational box, quite small distance. It is then highly nontrivial to generate the correct initialization at this finite distance, because

- the correct positioning and initial velocity of the nuclei are difficult to calculate for the deformed case, and
- there may be alignment or even Coulomb excitation happening on the way in.

The first problem can be tackled by calculating the kinetic energy of the fragments and the angular momentum from the TDHF wave functions and then correcting their translational velocities in an iterative way. The second problem will have to be ignored for the time being: it should not introduce major errors into the celculation.

The same problem occurs for extrapolating the final state to infinity. Again the interaction between internal excitations and the orbital motion must be neglected, while the pure extrapolation of the Coulomb orbit should be possible using a higher-order multipole expansion of the Coulomb potential.

Both problems will be alleviated if one can afford to use a larger computational grid.

FIRST CALCULATIONS

It is clear from the foregoing sections that in this paper only initial applications of an exploratory character can be presented, which serve to illustrate that collisions between

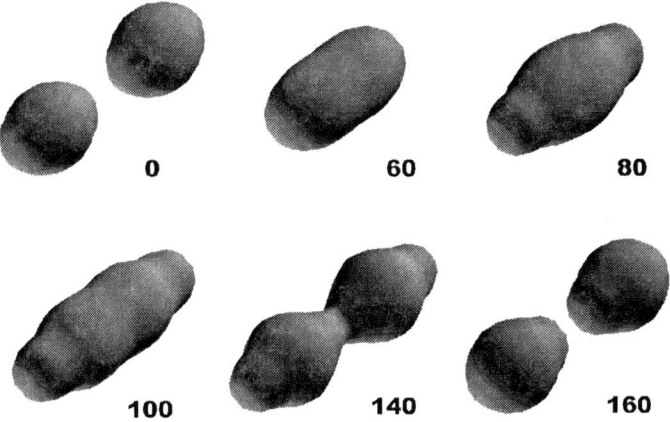

FIGURE 6. Central collision of two deformed Ne nuclei in a "tip-to-tip" configuration. The times corresponding to each plot are given in fm/c.

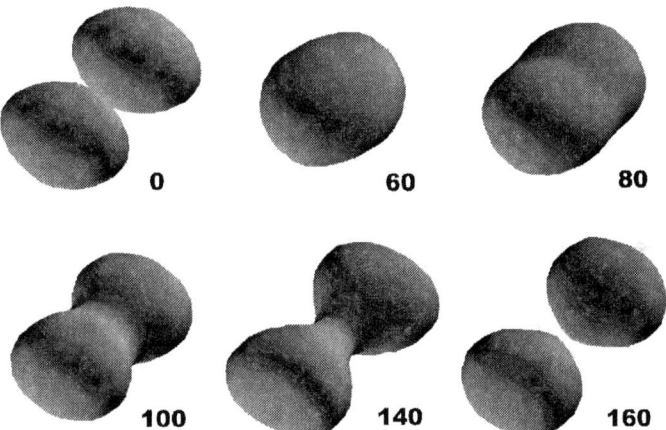

FIGURE 7. Central collision of two deformed Ne nuclei in a "belly-to-belly" configuration. The times corresponding to each plot are given in fm/c.

deformed nuclei are in principle accessible.

Figs. 2 through 4 demonstrate the sensitivity of the collision dynamics to the relative orientation of the nuclei. In each case two deformed ^{20}Ne nuclei interact at a c.m. energy of 78 MeV and with an impact parameter of $b = 2$ fm. The only difference between the collisions shown is the different orientation, which is clearly visible at the initial time. The three-dimensional plots show the isosurface at one quarter of the central density.

The interplay between collective and single-particle motion is particularly intriguing in these results. Thus the "tip-on-belly" configuration at 80 fm/c shows some transparency effect with the "tip" nucleus seeming to try to emerge on the other side; similarly for the "tip-on-tip" configuration at 100 fm/c. The former is the only configuration not leading to fusion in this case, demonstrating that the fusion behaviour is strongly influenced by deformation.

For central collisions (Figs. 5–7) effects are of a similar qualitative nature. In this case the somewhat higher center-of-mass energy of 115 MeV was chosen, leading to rapid reseparation in all cases. For the "tip-to-belly" configuration, the most fascinating result is that the orientation of the reseparated nuclei appears to indicate transparency again, in the sense that the nucleus initially in the "tip" position appears to emerge on the other side. For the "tip-tip" configuration, notice that the octupole deformation reverts just at the moment the neck is snapping. By contrast, the "belly-to-belly" configuration appears relatively uninteresting.

SUMMARY

The world of three-dimensional TDHF has only begun to be exploited. Proceeding along this way with a qualitatively improved computational effectiveness will allow the study of:

- the role of nuclear deformation in heavy-ion reactions,
- fusion reactions with possibly the most important application the formation of superheavy elements,
- collective excitations, especially those of a non-axial structure.

ACKNOWLEDGMENTS

The Joint Institute for Heavy Ion Research has as member institutions the University of Tennessee, Vanderbilt University, and the Oak Ridge National Laboratory; it is supported by the members and by the DOE through contract No. DE–FG05–87ER40361

REFERENCES

1. P. Bonche, S. E. Koonin, and J. W. Negele, Phys. Rev. C13, 1226 (1976).
2. K. T. R. Davies, K. R. S. Devi, S. E. Koonin, and M. R. Strayer, in: *Treatise on Heavy Ion Science*, Vol. 3 (ed. D. A. Bromley), Plenum, New York 1985, p. 3.
3. J. W. Negele, Rev. Mod. Phys. 54, 913 (1982).
4. J. P. Svenne, Adv. Nucl. Phys. 11, 179 (1979).
5. J. A. Maruhn, K. T. R. Davies, and M. R. Strayer, Phys. Rev. C31, 1289 (1985).
6. A. S. Umar, M. R. Strayer, P.-G. Reinhard, Phys. Rev. Lett. 56, 2793 (1986); P.-G. Reinhard, A. S. Umar, K. T. R. Davies, M. R. Strayer, and S. J. Lee, Phys. Rev. C37, 1026 (1988); A. S. Umar and M. R. Strayer, Comp. Phys. Comm. 63, 179 (1991).

The many facets of multinucleon transfer processes in low energy heavy-ion collisions

Lorenzo Corradi

INFN-Laboratori Nazionali di Legnaro, Via Romea 4, I-35020, Legnaro (Padova), Italy

Abstract. Examples of important recent achievements in the study of multinucleon transfer reactions at Coulomb barrier energies are presented. The unambiguos identification of transfer products up to the pick-up of six neutrons and the stripping of six-eight protons allows to investigate into detail the degrees of freedom acting in the transfer process. This is important also for γ spectroscopic studies and for future research to be done with radioactive beams.

INTRODUCTION

Multinucleon transfer reactions at Coulomb barrier energies is an important field of research in low energy heavy-ion physics. Through this mechanism one can in fact investigate nucleon correlation in nuclei [1, 2], the transition from the quasi-elastic to the deep-inelastic regime [3] and channel coupling effects in sub-barrier fusion reactions [4]. This mechanism has been also shown [5] to be a competitive tool for the production of neutron-rich nuclei, in alternative to spallation with energetic proton beams or to asymmetric fission, at least for certain mass regions.

An important and still poorly investigated question is what are the relevant degrees of freedom acting in the transfer process, i.e. single nucleon, pair or even cluster transfer modes. Theoretically, in the quasi-elastic regime (i.e. few transferred nucleons and small energy losses) one or, at most, two nucleon transfer has been treated using DWBA or Coupled Channel codes, while in the deep inelastic regime (i.e. many transferred nucleons and large energy losses) mostly stochastic, friction or diffusion models have been used [6]. Experimentally, nuclei produced in transfer reactions have been identified with good A, Z and Q-value resolutions for medium-light systems, but only in a few cases measurements were sensitive enough to identify channels corresponding to the transfer of more than two nucleons. For medium-heavy systems, especially in the study of deep-inelastic processes, at best nuclear charge identification was possible, without mass or energy resolution.

The "transition" regime, where many nucleons are transferred and where shell effects still play a significant role in the dynamics, represents a window not well studied in its detail and for which both experimental and theoretical advances are in progress. Thanks to the development of high resolution and high efficiency experimental set-up's, one could recently unambiguously detect in mass and charge the nuclei produced in transfer reactions up to the pick-up of six neutrons and the stripping of six protons [7, 8, 9, 10, 11]. Also, with the advent of the last generation large γ-arrays [12], one is

CP597, *Nonequilibrium and Nonlinear Dynamics in Nuclear and Other Finite Systems,*
edited by Z. Li et al.

presently able to perform γ-γ-particle coincidences for nuclei far from stability produced via nucleon transfer or deep-inelastic reactions, especially in the neutron-rich region [13, 14, 15]. These studies are of primary importance also for the reactions to be done with radioactive ion beams [5, 16]. Below, examples are presented of specific items where considerable progress has been recently made.

NUCLEON CORRELATION EFFECTS IN NUCLEI

The tendency of identical nucleons to couple in pair with total angular momentum zero is attributed to the existence of short-range residual forces, pairing in particular, that are responsible for several nuclear effects. Enhanced transition rates in the transfer of two nucleons were observed in light ion reactions, namely (p,t), (t,p), and in collisions between heavy ions [2]. Collisions between heavy-ions offer the possibility of observing multiple transfer of pairs and thus to measure, at least in principle, the pair-density in the nuclear medium.

This is in general a difficult task since the reaction mechanism increases in complexity with the increasing number of transferred particles [3], it is, for instance, not clear if the pairs transfer proceed in a cluster- or sequential-like picture. The reaction is dominated by the Coulomb interaction that favors the excitation of inelastic states and thus pushes the transfer strength in a region of excitation energy where pair correlations are weaker.

Recent high resolution data obtained with momentum and time-of-flight spectrometers allowed to measure transfer cross sections up to three nucleon pairs, opening the road to a more quantitative study of correlation effects in nuclei. Parallel to this experimental achievement, semiclassical theories have been developed so as to properly treat the enormous number of channels observed in the reaction. In particular, a new model has been implemented by the Copenhagen group [17, 18] treating quasi-elastic and deep-inelastic processes on the same ground and allowing a meaningful quantitative comparison with the experimental observables, i.e., mass and charge distributions, cross sections and angular momenta.

In a recent study of the system ^{62}Ni+^{206}Pb [11] multinucleon transfer cross sections have been measured at three bombarding energies up to three nucleon pairs, focusing on pure neutron channels which all have well matched Q-values. The aim of the experiment was to observe an eventual odd-even staggering in the cross sections for pure neutrons indicating a contribution from pair transfer mode. The total kinetic energy loss distributions of Fig.1 show that only the +1n and +2n channels have the main population close to the ground-ground state Q-value (indicated by the down arrows) while the more massive transfer channels display a population towards more negative Q-values with the tail increasing with the number of transferred neutrons. The observation of these long tails in the Q-values spectra suggests that any contribution from "cold" transfers, associated with low excitation energy, is likely hidden by the dominant "warm" sequential transfer process. As a matter of fact the trend of the experimental Q-value integrated differential cross sections as a function of the number of transferred neutrons is almost energy independent and, at least within the experimental accuracy, displays a constant drop without any clear odd-even staggering (see Ref. [11] for details).

228

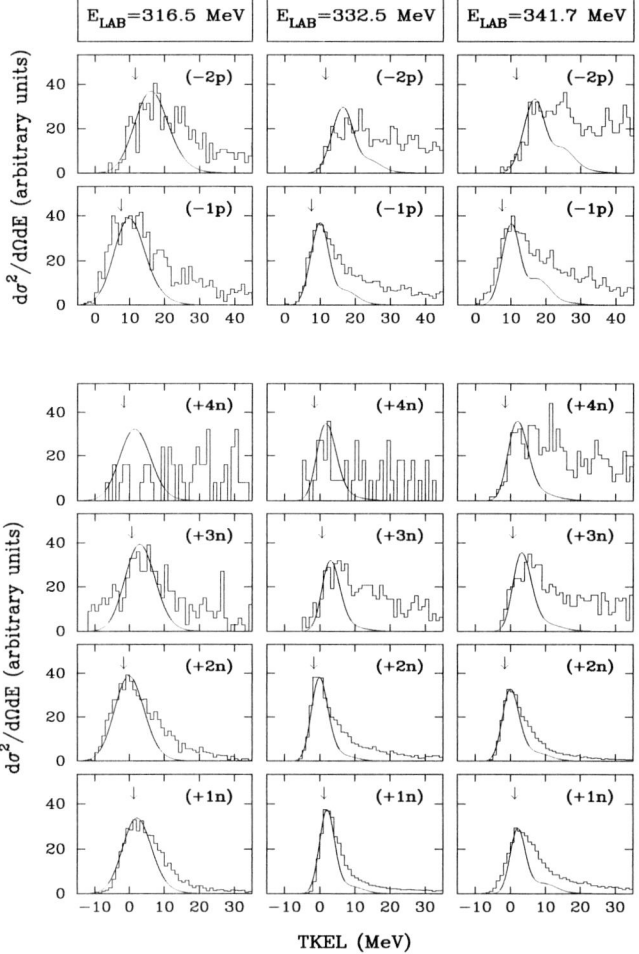

FIGURE 1. Experimental (histograms) and theoretical (lines) total kinetic energy loss distributions for pure neutron pick-up channels in the reaction ^{62}Ni+^{206}Pb (see Ref. [11] for details).

It is presently not understood if and how nucleon correlation effects take place in this time-dependent dynamical process, in particular it is not clear why, for instance, proton pairs should be favoured with respect to neutron pairs [2]. Certainly, however, beyond the present interpretations, detailed studies of this kind performed on other suitable systems and with the being constructed large solid angle spectrometers [19, 20], possibly in conjunction with powerful γ-arrays, able to distinguish the population to individual states, allow to get a deeper insight into interesting and long standing problems in low energy nuclear physics, often linking nuclear and condensed matter phenomena, like the appearance of a Josephson effect or nucleon superconductivity [21, 22].

229

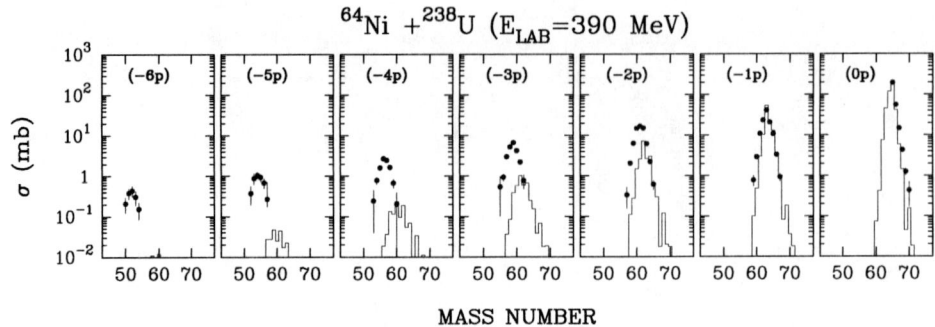

FIGURE 2. Experimental (points) and theoretical (lines) total cross sections for the transfer channels observed in the reaction ^{64}Ni+^{238}U (from Ref. [10]).

MULTINUCLEON TRANSFER : A TOOL FOR THE PRODUCTION OF UNSTABLE NUCLEI

Different systems have been studied at LNL [9, 10, 11, 23], where transfer products have been detected with cross sections down to $\simeq$ 50-100μb/sr. In this way the whole population pattern in the A-Z plane could be followed and compared with state-of-art coupled channels calculations. Fig.2 gives as an example the experimental and theoretical isotope distributions obtained in ^{64}Ni+^{238}U [10]. Taking into account that calculations [17, 18] include in this case only independent single nucleon transfer modes, various important results could be deduced, for instance it has been shown how channels involving neutron pick-up and neutron stripping are populated by two different mechanism. Also it has been shown how, even at these low bombarding energies, large energy losses may be present, generating a significant nucleon (mainly neutron) evaporation from the primary fragments.

More than discussing the origin of the discrepancies between data and theory when more than two protons are involved (see Refs.[9, 10] for details) I just want to remark that the data demonstrate that multinucleon transfer is a competitive tool for the production of neutron-rich (light) nuclei. Also, one can properly test the reliability of the calculations for reactions involving radioactive beams [5]. With this respect, it was shown [16] that in moving from a stable projectile to an unstable neutron-rich one, the transfer flux proceeds from neutron pick-up and proton stripping channels to a "reverse" mode. This trend has been observed e.g. in the case of ^{48}Ca+^{124}Sn [23] and in other γ spectroscopic data [15, 24]. This process, which may be a very effective one using radioactive beams, brings, for instance, to neutron rich heavier partners, which are interesting from the spectroscopic point of view in different areas of the nuclide chart.

Fig.3 gives an example of predictions for the representative cases of ^{x}Sm+^{248}Cm systems, where one sees clearly the larger cross sections in the neutron rich mass region of Cm, Am (+1p) and Pu (+2p) when moving from ^{144}Sm to the radioactive ^{160}Sm. For instance, the mass 254 in the +2p case has a $\sigma \simeq$ 1 mb, which is sufficient for

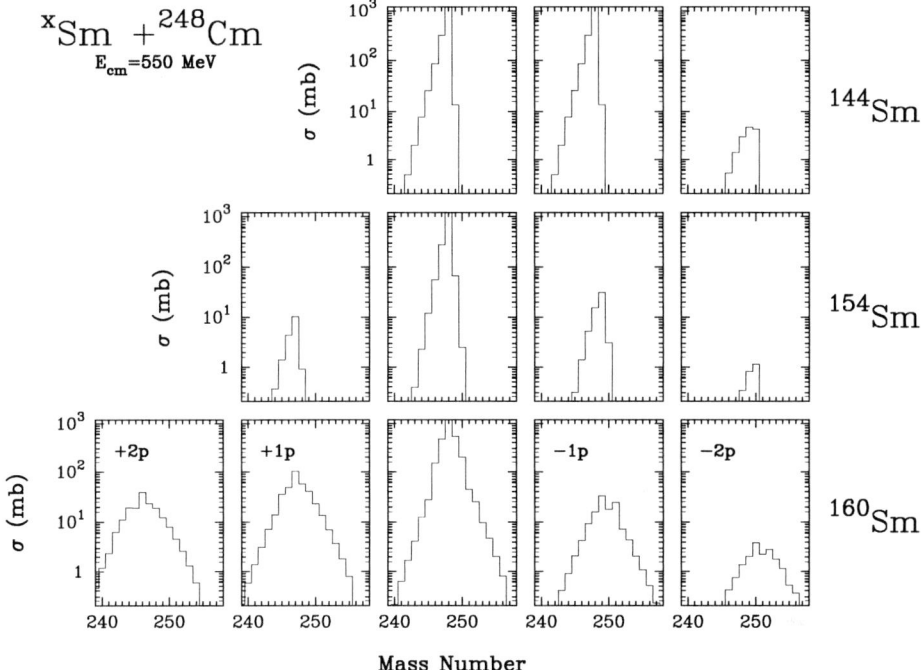

^{x}Sm + ^{248}Cm
E_{cm}=550 MeV

σ (mb)

^{144}Sm

^{154}Sm

+2p +1p −1p −2p

^{160}Sm

Mass Number

FIGURE 3. Total transfer cross sections calculated with the program GRAZING [18] for the reaction ^{x}Sm+^{238}Cm.

performing γ spectroscopy after tagging with a suitable spectrometer. These actinides and transactinides regions are a particular sub-class which is inaccessible with other types of reactions. Of course, an important question is how much of the primary yield of the heavy partner survives against fission, which likely becomes the dominant decay channel. Therefore high resolution measurements of both light and heavy products in multinucleon transfer, for suitable projectile/target combinations which could serve as a guide for future studies in that region, are highly desirable and started already at LNL.

γ SPECTROSCOPY PERFORMED VIA MULTINUCLEON TRANSFER

The neutron rich side of the stability valley is much less investigated than the (easier to reach with stable beams) proton rich side, and there are entire nuclear regions where little or almost nothing is known besides the very lowest levels. Multinucleon transfer or deep-inelastic reactions have been already demonstrated to be a competitive tool to investigate, with stable beams and via proton stripping channels, neutron rich nuclei in the light and heavy mass regions (see e.g. Ref.[25] and references therein). γ-spectrocopic studies in neutron-rich nuclei are important for a wealth of physical phenomena, like for instance the change of shell structure or the interplay between single particle and collective degrees of freedom [26]. Other subjects are interesting for both nuclear srtucture and dynamics, for instance how pair correlations behave as a function of spin and excitation energy [1], or the precise determination of the transfer form factors for the population of specific nuclear levels. These studies can be now performed with high efficiency thanks to the development of large γ-arrays [12]. An important further possibility offered by these devices is that to observe the coincident γ transitions of the light and the heavy partners, which give important informations on the evaporation effects from the primary fragments. Even more efficient and detailed studies will become soon possible with the being constructed high solid angle spectrometers [19, 20], used in conjunction with γ detectors.

As an example of the kind of data quality obtained so far and to have an idea of the agreement between experiment and theory, I mention a recent experiment performed at LNL [27], where multinucleon transfer reactions in ^{40}Ca+^{124}Sn with the GASP Ge-array and a couple of PPAC was studied. One of the purposes of the experiment was to get the differential and total cross sections for the lowest excited levels of the transfer products populated in the reaction. The comparison with theory, at least for the simple one particle transfer process, allows the extraction of the single-particle formfactors that constitute the fundamental ingredient for more elaborate coupled-channels calculations. In Fig.4 the experimental and theoretical differential cross sections for the individual levels of the nuclei populated in the +1n and -1p channels are shown (to obtain the final cross sections the contribution of the feeding from higher-lying states has been subtracted). The agreement between data and theory turns out to be very good, for the transfer channels of both the light and the heavy partners.

The calculations concerning the transfer processes have been performed in the CWKB formalism [28], which is adequate both below and above the Coulomb barrier and which has been successfully applied in previous works to analyze similar reactions (see Ref.[27] for details). This formalism involves the same approximations which were exploited to calculate the absorptive and polarization component of the optical potential and the off-diagonal inelastic couplings. With such a theory one can confidently compute the differential cross sections for the population to specific excited states of the final nuclei produced after the transfer of one particle (in the present case neutron pick-up and proton stripping). The above formalism has been generalized to calculate the more complicated multinucleon transfer channels. With the form factors used to compute these transfer cross sections recently also sub-barrier fusion excitation functions and barrier distributions have been calculated and compared with high accuracy data [29].

SUMMARY AND OUTLOOK

From the examples of new studies presented above one can deduce features of the reaction mechanism not clearly identified in the past, which can be observed only if one is able to detect the transfer of $\simeq$ 4-6 nucleons. The improvement of both experimental accuracy and theoretical models allows to quantitatively investigate the relevant degrees

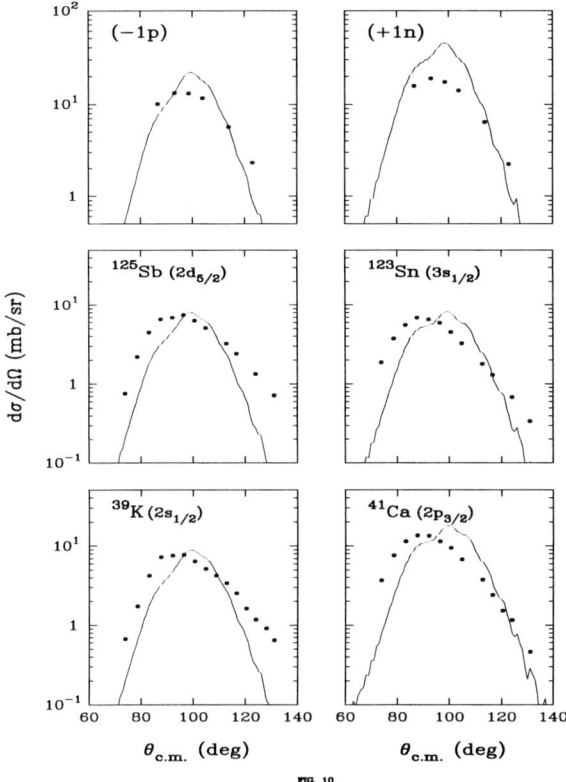

FIG. 10

FIGURE 4. Experimental (points) and theoretical (lines) differential transfer cross sections for specific levels populated in the reaction ^{40}Ca+^{124}Sn at E_L=170 MeV. The experimental angular distributions in the top row are from the inclusive data of Ref.[9].

of freedom acting in the transfer process. This turns out to be important not only for understanding the dynamics of the reaction but also for nuclear structure. Multinucleon transfer studies with stable beams have been shown to be an interesting field also in view of future research with radioactive beams, especially for neutron-rich nuclei.

Even more detailed studies will become soon possible with the being constructed high solid angle spectrometers [19, 20]. With such devices one can make a decisive step in the detection solid angle (i.e. $\simeq$ 80 msr), and good charge, mass and energy resolution will be obtainable even for very heavy ions. These new devices will exploit fully the many possibilities offered by binary reactions in the study of both reaction dynamics and nuclear structure, which are closely interconnected at energies close to the Coulomb barrier. An important field of research will be opened for γ-spectroscopy with the spectrometers used in conjunction with large γ-arrays. Their huge solid angle and broad energy acceptance will in fact allow to identify and tag exotic neutron-rich nuclei populated via multinucleon transfer and deep-inelastic reactions.

ACKNOWLEDGMENTS

I like to acknowledge the following people who contributed to the work done on multi-nucleon transfer reaction studies at LNL : A.M. Stefanini, D. Ackermann, S. Beghini, F. Cerutti, C.J. Lin, G. Montagnoli, G. Pollarolo, F. Scarlassara, M. Trotta, M.A. Vinodkumar, and A. Winther.

REFERENCES

1. C.Y. Wu et al., *Annu. Rev. Nucl. Part. Sci.* **40**, 285 (1990).
2. W. von Oertzen, in *"Heavy Elements and Related New Phenomena"*, edited by W. Greiner and R. K. Gupta, (North-Holland, Amsterdam), 1999.
3. K.E. Rehm, *Annu. Rev. Nucl. Part. Sci.* **41**, 429 (1990).
4. M. Dasgupta et al., *Annu. Rev. Nucl. Part. Sci.* **48**, 401 (1998).
5. G. Pollarolo, *"XXXVII Int. Winter Meeting on Nuclear Physics"*, Bormio, January 25-30, 1999, I.Iori ed., Univ. degli Studi di Milano, Vol.N.114, p.369.
6. W.U. Schröder and J.R. Huizenga, in *"Treatise on Heavy Ion Collisions"*, ed. D.A.Bromley, Vol.2 chapter 3 (1984), Plenum Press, New York.
7. C.L. Jiang et al., *Phys. Lett.* **B337**, 59 (1994).
8. C.L. Jiang et al., *Phys. Rev.* **C57**, 2393 (1998).
9. L. Corradi et al., *Phys. Rev.* **C54**, 201 (1996).
10. L. Corradi et al., *Phys. Rev.* **C59**, 261 (1999).
11. L. Corradi et al., *Phys. Rev.* **C63**, 021601-1 (R) (2001).
12. C.W. Beausang and J. Simpson, *J. Phys.* **G22**, 527 (1996).
13. R. Broda et al., *Phys. Rev. Lett.* **74**, 868 (1995).
14. J.F.C. Cocks et al., *Nucl. Phys.* **A645**, 61 (1999).
15. S.J. Asztalos et al., *Phys. Rev.* **C61**, 014602 (2000).
16. C.H. Dasso et al., *Phys. Rev. Lett.* **73**, 1907 (1994).
17. A. Winther, *Nucl. Phys.* **A572**, 191 (1994); *Nucl. Phys.* **A594**, 203 (1995).
18. A. Winther, *program GRAZING*, unpublished.
19. A.M. Stefanini et al., *"5th Int. Conf. on Radioactive Nuclear Beams, RNB2000"*, Divonne (Switzerland), 3-8 April 2000, Proc. to be published.
20. H. Savajols, Proc. *"Workshop on the Experimental Equipment for an Advanced ISOL Facility"*, Berkeley, July 22-25, 1998, I.Y.Lee ed.
21. A. Bohr and B. Mottelson, *"Nuclear Structure"*, Vol.II, p.392 (1999), World Scientific.
22. F. Iachello, *Nucl. Phys.* **A570**, 145c, (1994).
23. L. Corradi et al., *Phys. Rev.* **C56**, 938 (1997).
24. L. Zhang et al., *Phys. Rev.* **C58**, 156 (1998).
25. L. Corradi et al., *RNB2000*, Divonne.
26. W. Nazarewicz, *Nucl. Phys.* **A654**, 195c (1999).
27. L. Corradi et al., *Phys. Rev.* **C61**, 024609 (2000).
28. E. Vigezzi and A. Winther, *Ann. Phys.* (N.Y.) **192**, 432 (1989).
29. G. Pollarolo and A. Winther, *Phys. Rev.* **C62**, 054611 (2000).

Nonlinear dynamics in metal clusters

P.-G. Reinhard* and E. Suraud†

*Institut für Theoretische Physik, Universität Erlangen, D-91058, Erlangen, Germany
†Laboratoire de Physique Quantique, Université Paul Sabatier, 118 Route de Narbonne, F-31062
Toulouse, cedex, France

Abstract. We discuss the response of metal clusters subject strong electromagnetic fields in the framework of density functional theory. This allows a fully nonadiabatic treatment of both electronic and ionic degrees of freedom. We give a few general properties of metal clusters which can be accessed with this kind of nonadiabatic approaches and we present an example of application of a metal cluster subject to an intense laser field.

INTRODUCTION

The study of the dynamical response of metal clusters to electro-magnetic excitations has been a topic of intense investigations over the past two decades [1, 2]. Depending on the intensity of the exciting perturbation, such studies have given access to various phenomena, ranging from the linear to the semi-linear regime. More recently, experiments with intense laser beams and heavy ion projectiles have given access to the strongly non-linear regime, involving, at different paces, both electronic and ionic degrees of freedom [3, 4, 5, 6]. The recent developments of laser technology, in particular, provide almost instantaneous excitation mechanisms, at a time scale of order of the electronic motion. Such excitations almost instantaneously place the cluster far away from equilibrium. And the ensuing evolution of the system couples electronic to ionic degrees of freedom in a complex manner. In this contribution we shall present a few recent theoretical results devoted to the understanding of these strongly out of equilibrium situations. Details can be found in a recent review paper on this topic [6].

PHYSICAL CONTEXT

Some scales in metal clusters

Before entering the core of our topic we would like to remind a few basics of cluster physics, in order to replace our investigations in a more general context. As compared to atoms, metal clusters are loosely bound structures composed of ions and quasi free valence electrons. The simplest metal clusters, on which we shall focus in the following, are alkaline clusters, with one valence electron per atom; sodium clusters constitute an archetype of such clusters. Sizes of alkaline clusters scale with number of atoms in a way similar to nuclei with radii $R \propto n^{1/3}$ for a n atom cluster. The proportionality coefficient

CP597, *Nonequilibrium and Nonlinear Dynamics in Nuclear and Other Finite Systems*,
edited by Z. Li et al.

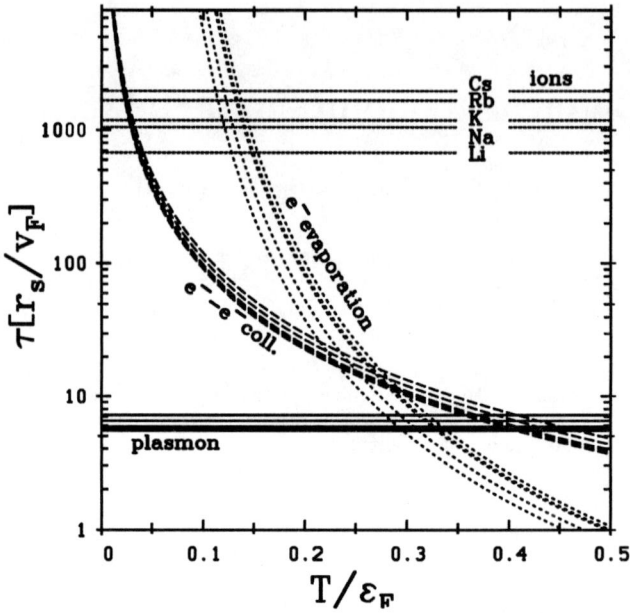

FIGURE 1. Summary of time scales in alkaline metal clusters (see text for details).

r_s is the Wigner Seitz radius which represents the radius of the average sphere occupied by one atom in the bulk and is thus characteristic of the material.

Because electrons move basically independently from each other in such simple metal clusters, they can be superficially seen as a quasi free Fermi gas. The Fermi energy ε_F $\sim$ 2-15 eV, is directly related to r_s. From the corresponding Fermi velocity v_F one can estimate a typical microscopic time scale $\tau_\mu \sim 2\, r_s/v_F \sim$ of order 1 fs = 10^{-15} s. In most dynamical situations the electron response is dominated by the Mie plasmon, namely a collective oscillation of electrons with respect to ions, much similar to the Giant Dipole Resonance in nuclei. The typical energy of the Mie plasmon $\hbar\omega_{Mie}$ is of order $\sim$ 2-10 eV (depending on the material) which corresponds to a period of order $\tau_{Mie} \sim$ 0.5-2 fs. In turn, ions, which are much heavier than electrons, move at a (relatively) slow pace of order 1 ps, at least for ground state dynamics.

In the course of the violent excitations we consider here, various dynamical time scales interfere (see also Figure 2). We briefly summarize here the most relevant time scales involved and their possible dependence on the excitation energy deposited in the cluster, which we characterize here by its electronic temperature. The microscopic time scale typically takes sub fs values in alkalines. The Mie plasmon period also ranges in the fs domain. In strongly dynamical situations one has also to consider the typical time scale associated to electron electron collisions, which scales with $1/T^2$ [7] and the time scale associated to thermal electron emission, following Weisskopf's rule [8]. Typical ionic time scales can finally be estimated from ionic frequencies of the dimer. These ionic times range in the hundreds of fs domain. These various (electronic and ionis) times are plotted for alkalines in Figure 1. It is striking to see from this figure the nice decoupling of time scales, whatever the material.

Some experimental directions

The field of cluster physics is huge and we do not aim here, by far, at exhaustivity. Rather we would like to mention a few directions in which promising phsyics shows up and for which the theoretical developmemts underway seem to be properly suited. With a degree of arbitrariness we have selected three dynamical regimes: a linear, a semi-linear and a nonlinear domain. The linear domain is the realm of the optical response which has already been widely studied in the past years [5, 9]. Photoelectron spectroscopy [4] is also a promising field, with recent renewed interest [10]. Finally the study of ionization potentials and shell closures remains a nontrivial question in nonsimple clusters [11]. Roughly speaking these various "linear domain" observables give information on the structure of clusters, even if they address some dynamics. In the so called "semi linear" domain we consider situations in which the system is already sizeably perturbed with a few electrons emitted and nonnegligible energy deposit. Still, in spite of this excitation, the cluster remains, at least for some time, in one piece. This is typically the situation encountered in the measurements of kinetic energy spectra of emitted electrons from irradiated clusters or in the seek for second harmonic generation [6]. These various directions clearly point towards dynamics of clusters. They can typically not be addressed by linear response theory and the kind of nonadiabatic dynamics presented in the following is ideally adapted to such situations. Finally, the nonlinear domain corresponds to high excitations as attained by ionic collisions [12] or through irradiations by intense lasers [13, 14, 15]. The fate of the excited clusters is strongly affected by dynamical effects and the system usually disappears in a more or less violent way by fission or fragmention. In that case one explore strongly dynamical situations and one may even question the relevance of clusters physics, in the sense that clusters may become more a laboratoty that a true object of investigation. This is typically the case of irradiations by very intense laser fields [16].

From hit to explosion

The typical nonlinear excitations are illustrated in Figure 2. We have considered two classes of rapid, intense, excitations: collisions with highly charged energetic ions [12] and irradiation by intense femtosecond laser pulses [13, 14]. Both mechanisms take place within tens of fs down to below 1 fs, thus in the same time range as electronic times. The upper part of Figure 2 symbolizes the excitation mechanisms, (frequency selective) laser excitation by a weavy line, and (frequency unselective) fast ion collisions by a short peak. The cluster response takes place in different stages. First phases are obviously dominated by electronic effects; Ions, because of their high mass, more "slowly" couple to the whole process, even for significant excitations. The first electronic phases are thus characterized by a direct electron emission and a collective oscillation

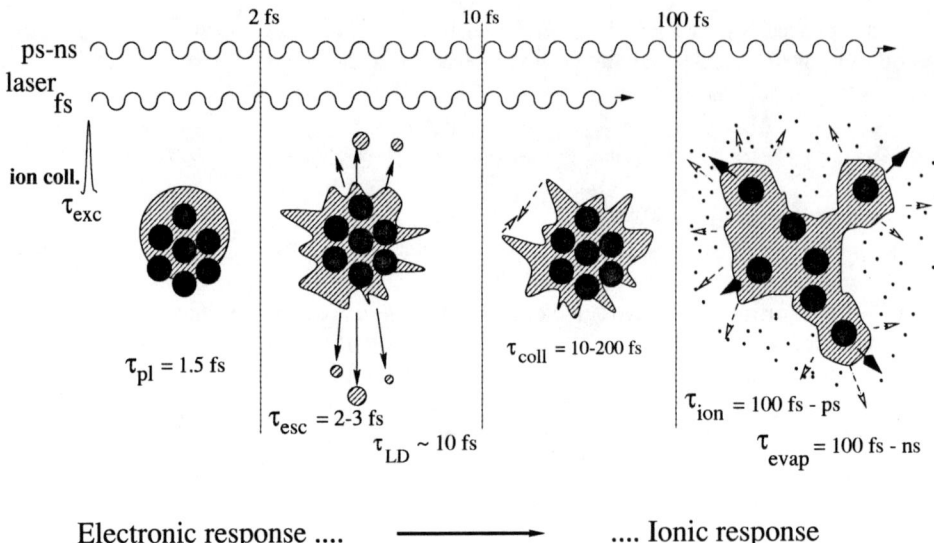

ps-ns
laser
fs

ion coll.

τ_{exc}

$\tau_{pl} = 1.5$ fs

$\tau_{esc} = 2\text{-}3$ fs

$\tau_{LD} \sim 10$ fs

$\tau_{coll} = 10\text{-}200$ fs

$\tau_{ion} = 100$ fs - ps

$\tau_{evap} = 100$ fs - ns

Electronic response .... $\longrightarrow$ .... Ionic response

FIGURE 2. Sketch of the various stages of the typical response of a metal cluster to a short intense laser pulse. Various time scales are indicated (see text for details).

of the Mie plasmon (typically below about 10 fs). The "second" phase is still of purely electronic nature, and corresponds to damping of collective motion via Landau damping and electron-electron collisions. After that, electronic degrees of freedom will slowly couple to ionic motion. Two effect interfere coherently here: i) the net charge of the cluster following ionization, and ii) energy exchanges between the now "hot" electron cloud and the still "cold" ions. This "activatation" of ionic motion may then lead, on a time scale of several hundreds of fs to several ps, to evaporation, fission or fragmentation. Finally, thermal evaporation of electrons usually proceeds on a very long time scale, in most cases slower than ionic processes (monomer evaporation, fragmentation) and can thus be overlooked in many "violent" situations.

How to solve the problem

A complete theoretical description of the whole dynamical process, from details of the excitation phase to the actual fragmentation, represents a formidable task, because one should integrate both electron dynamics (at the sub-fs level) *and* ion dynamics (in the ps time scales). Basically three ways have been proposed to attack such a mutiscale dynamics: i) the ab-initio methods of quantum chemistry and molecular physics have up to now only be applied to very small systems; ii) the adiabatic methods involving a Born Oppenheimer treatment of electrons can be used for large systems on long times, but they do allow only very small electronic excitations; iii) the effective approaches of Density Functional Theory [17, 18] provide the presently unique way to treat such questions for medium size systems, at any level of nonlinearity. The developments presented below will thus be based on Time Dependent DFT [18]. Such methods allow an effective propagation of the most relevant degrees-of-freedom *without* any adiabatic approximation which is crucial in our context. These methods have been used since long in metal clusters in particular for structure and optical response problems [5]. It is only quite recnetly that they have started to be used in truly dynamical problems, beyond the linear regime, and with possible coupling to ions [9, 19, 20]. The many calculations performed with these methods since a few years have shown that the DFT approaches provide an efficient and flexible access to truly nonlinear dynamics of the electrons, if one takes care that the electronic wavefunctions are represented in a large enough Hilbert space [9, 20].

MODEL AND OBSERVABLES

As time-dependent density-functional theory is still in a developing stage, many of the much refined approaches remain too involved to be directly applicable to large scale calculations [18]. We shall thus restrict ourselves to a robust and most efficient scheme, namely Time Dependent Local Density Approximation (TDLDA) for the electrons. The latter TDLDA is coupled to classical Molecular Dynamics (MD) for the ions giving rise to the so called TDLDA-MD. Let us point out here that this coupled ionic and electronic dynamics, constitutes a true TDLDA-MD, which definitively goes beyond the usual Born-Oppenheimer molecular dynamics. The interaction between the various constituents of the system is purely coulombic. While the ion-ion interaction may safely be reduced to a mere point charge interaction the electron-ion interaction requires some care, as only valence electrons do effectively participate in the global binding of the system. Most electrons (core electrons) remain bound to their parent nucleus and have to included in the interaction between ions and valence electrons. This is accounted for in the so-called pseudo-potentials, which properly modelize the screening of the nuclear charge by (frozen) core electrons. In the case of simple metal clusters discussed here, the pseudo-potential may be taken as local and consists, in our case, out of a sum of two error functions. Parameters of the pseudo potential can be adjusted to provide a good description of structure as well as of dynamics [21].

The electron-electron contribution can be splitted into 3 components: a direct and an exchange one, and a contribution steming from correlations. The direct contribution is the standard Hartree term. Most of the binding comes from the "attractive" exchange and correlation contributions, which furthermore are, to a large extent, of local nature. The exchange contribution just stem from the indistinguishability of electrons. The correlation term reflects the fact that the 2-body density matrix which should enter the calculation of the electron-electron potential energy does not reduce to a mere (even antisymmetrized) product of 1-body densities. The electronic TDLDA equations are finally solved numerically on a grid in coordinate space and a Verlet algorithm is used for the simultaneous (classical) ionic propagation. We employ absorbing boundary conditions all over, to treat ionisation correctly [6].

In the following we shall focus on dynamical electronic and ionic observables. A key observable at the side of electrons is the electronic dipole moment, computed in a vicinity of the cluster. From this dipole signal, one can acceed excitation spectra by Fourier transformation into the frequency domain which has become meanwhile a standard technique to evaluate spectra via TDLDA [9, 20, 22]. Ionisation also constitutes a key observable and it is also computed as a function of time, by recording the number of electrons lost through the absorbing boundaries. It should be noted that other more detailed electronic quantities can be extracted from these simulations. A typical example is provided by the kinetic energy spectra of emitted electrons [10]. Ions are treated as classical particles and from their positions and momenta as a function of time we can access the ionic extension and the ionic temperature of the system.

AN EXAMPLE OF APPLICATION

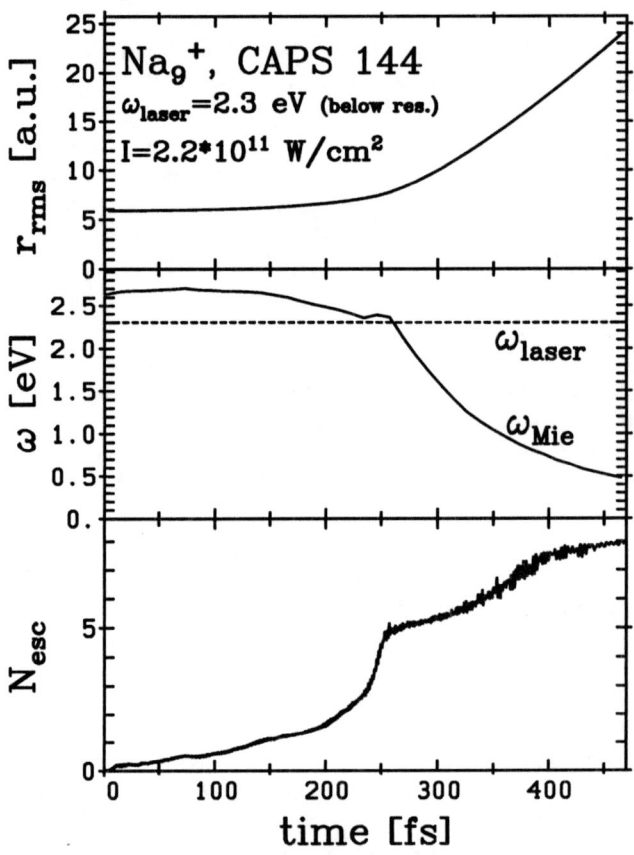

FIGURE 3. Ionic and electronic dynamics of a Na_9^+ cluster irradiated by a continuous laser pulse with parameters as indicated. Lowest panel: number of emitted electrons N_{esc}; middle panel: time evolution of plasmon resonance position (full line) and laser frequency (dotted); upper panel: ionic extension along laser polarisation (z).

As an example of application we consider the case of the irradiation of a simple sodium cluster by an intense laser pulse. We present here a case in which the pulse is long enough so that a direct coupling between electronic and ionic degrees of freedom occurs, in the spirit of the experiments perform on such cases [15]. In the case of laser irradiation one observes, not surprisingly, that the plasmon resonance plays a key role in relation with the laser frequency, in particular in terms of ionisation [6]. For laser frequencies far from the plasmon, the electronic response exhibits a typical off resonant behaviour, directly following the laser pulse. The ionisation is correspondingly small. On the contrary, when the laser attaches the plasmon frequency, one observes a resonant effect: both the dipole response and ionisation become large and the dipole oscillations survive long after the laser has been switched off. For short enough laser pulses (below a few tens of fs in the case of sodium [23]), these effects remain at the pure electronic level. For longer pulses one however observes that ionic degrees of freedom very quickly couple, which may play an important role in the degree of ionisation attained by the cluster, as observed recently in experiments [15].

We concentrate here on the high energy case of a Coulomb explosion of a metal cluster [23], with a setup qualitatively similar to experiments [15]. The laser is tuned far below the initial plasmon resonance of the cluster but strong enough to produce sizeable ionisation; the laser pulse remains switched on for a time span long enough to explore the changes in resonant conditions as induced by ionic motion (basically ionic expansion of the charged cluster). In figure 3 we show a Na_9^+ test case. Initially the laser stays below resonance. But ionisation (here the number of emitted electrons N_{esc}) steadily grows in the course of the irradiation. The thus produced net charge provokes a slow ionic expansion (uppermost panel). As the plasmon frequency $\hbar\omega_{Mie}$ sensitively depends on the radius R of the system $(\hbar\omega_{Mie} \propto R^{-3/2})$ resonant conditions do develop around 250 fs. This resonant beahvior can be traced in terms of a much enlarged dipole response (lowest panel) and a sudden increase of ionisation (middle panel). But the charge increase shifts the resonance position up again, which balances the decrease due to ionic expansion. This makes that resonant conditions can be preserved for several tens of fs, hence their efficiency. Finally, the huge ionisation (more than 50 % ionisation) leads to a violent Coulomb explosion (upper panel).

Experiments performed along a similar line [15] qualitatively show the same behaviour as seen in Figure 3. An enhanced ionisation is observed as soon as the laser is shining long enough to coincide with a stage of resonant conditions. The above example is only one out of many conceivable scenarios, and many aspects of both ionic motion and final fragments may sensitively depend on laser parameters. Even multiple occurrence of resonant conditions can be observed for laser frequencies initially close to the Mie plasmon resonance [23].

CONCLUSIONS AND PERSPECTIVES

In this text we have presented the TDLDA-MD approach which provides a flexible and robust tool of investigation of metal clusters in strong fields. The far off-equilibrium situations considered here indeed require a simultaneous treatment of both ionic and electronic degrees of freedom. TDLDA-MD furthermore allows to study a wide range of dynamical situations ranging from the linear to the strongly nonlinear regime. When available, experimental data are in good agreement with theoretical predictions.

Most of these studies are still in an early exploratory stage and many further developments are conceivable. These developments concern various aspects both in terms of formalism and applications. of the theory and of its applications. From the formal side let us mention the inclusion of SIC corrections to LDA or the extension of the theory to the dissipative regime by inclusion of a collision term [6]. Concerning applications, the semi-linear regime constitutes a promising field of development, for example with more thorough analyses of photoelectrons [10]. The impact on clusters of intense laser fields is also a very attractive domain but the focus is here less on clusters but rather on plasma physics and potential applications [16]. Nevertheless, more systematic investigations with the microscopic TDLDA-MD approach would be welcome here. Finally, it should be noted that applications to more extended systems are also conceivable, as for example defect formation in irradiated materials. Work along these various lines is in progress.

ACKNOWLEDGMENTS

The authors thank Institut Universitaire de France and french german exchange program PROCOPE number 99074 for financial support during the realization of this work.

REFERENCES

1. *Large clusters of atoms and molecules*, NATO ASI **E313**, ed. by T. P. Martin, Kluwer, Dordrecht, 1996
2. *Small particles and inorganic clusters*, Euro. Phys. J. D **9**, 1 (1999)
3. U. Näher et al, Phys. Rep. **285**, 245 (1997)
4. W. de Heer, Rev. Mod. Phys. **65**, 611 (1993)
5. M. Brack, Rev.Mod.Phys. **65**, 677 (1993)
6. F. Calvayrac, P.-G. Reinhard, E. Suraud, C. Ullrich, Phys. Rep., **337**, 493 (2000)
7. A. Domps, P.-G. Reinhard, E. Suraud, Phys. Rev. Lett. **81**, 5524 (1998)
8. V. Weisskopf, Phys. Rev. **52**, 295 (1937)
9. K. Yabana et al Phys. Rev. B **54**, 4484 (1996)
10. A. Pohl, P. G. Reinhard, E. Suraud, Phys. Rev. Lett., **84**, 5090 (2000)
11. T. Diederich et al, Phys. Rev. Lett. **86**, 4807 (2001)
12. F. Chandezon et al, Phys. Rev. Lett. **74**, 3784 (1995)
13. T. Baumert, G. Gerber, Adv. At. Mol. Opt. Phys. **35**, 163 (1995)
14. R. Schlipper et al, Phys. Rev. Lett. **80**, 1194 (1998)
15. L. Köller et al, Phys. Rev. Lett. **82**, 3783 (1999)
16. J. Zweiback et al, Phys. Rev. Lett. **84**, 2634 (2000)
17. R.M. Dreizler, E.K.U. Gross, *Density functional theory*, Springer, Heidelberg, 1990
18. E. K. U. Gross et al, Top.Curr.Chem **181**, 81 (1996)
19. U. Saalmann, R. Schmidt, Z. Phys. D **38**, 153 (1996)
20. F. Calvayrac, P. G. Reinhard, E. Suraud, Phys. Rev. B **52**, R17056 (1995)
21. S. Kümmel et al, Eur.J.Phys. D9, 149 (1999)
22. F. Calvayrac, P.-G. Reinhard, E. Suraud, Ann. Phys. (N.Y.) **255**, 125 (1997)
23. E. Suraud, P.-G. Reinhard, Phys.Rev.Lett. **85**, 2296 (2000)

Fission Mass Division and Topology of Potential Energy Surface

P. Möller*, D. G. Madland*, A.J. Sierk* and A. Iwamoto[†]

*Theoretical Division, Los Alamos National Laboratory, Los Alamos, NM 87545, USA
[†]Department of Materials Science, Japan Atomic Energy Research Institute, Tokai-mura, Ibaraki, 319-1195 Japan

Abstract. We present calculations of fission potential energy surface based on Strutinsky's prescription for realistic 5-parameter shape parameterization of the fissioning nucleus. The determination of the saddle points in this 5-dimensional potential energy surface was performed directly. The calculated fission saddle points present typical feature that most of the actinide nuclei have the lowest two saddle points, one is mass-symmetric and the other is mass-asymmetric, separated by a ridge.

INTRODUCTION

The distribution of fragment masses following a nuclear fission has been observed since just after the discovery of fission by Hahn and Strassmann in 1939. In case of typical actinide nuclei, the mass distribution is asymmetric, i.e.; the most probable mass division produces one heavy fragment and one light fragment. The Radium, however, is known long time to have two components simultaneously, i.e., symmetric and asymmetric mass division coexist. In heavy mass region, for example, some isotope of Fm fissions asymmetrically but another isotope of Fm fissions symmetrically. Therefore, the mechanism of mass division is not so simple and has been a challenging problem for theoreticians but general understanding of it is not yet achieved.

Theoretically, most basic quantity is the potential energy surface of fissioning nucleus as was first introduced in the pioneering work of Bohr-Wheeler. The later development of the calculation was due to Strutinsky, which enabled to take into account the quantum mechanical shell effect. A remaining problem in the calculation of the potential energy surface is to calculate the realistic potential energy surface with at least 5-shape parameters of fine mesh points, which was unable before due to the lack of computer power. Another problem to solve is to determine the saddle location in the 5-dimensional potential energy surface without any approximation. We solved the former problem by utilizing the modern multi-processor system and the latter problem, by introducing the numerical technique that will be described in the next section [1, 2].

CP597, *Nonequilibrium and Nonlinear Dynamics in Nuclear and Other Finite Systems*,
edited by Z. Li et al.
© 2001 American Institute of Physics 0-7354-0041-5/01/$18.00

SHAPE PARAMETRIZATION

We use the Three-Quadratic-Surface parameterization for the shape of the nucleus specified in terms of three smoothly joined portions of quadratic surfaces of revolution as is given by

$$\rho^2 = \begin{cases} a_1{}^2 - \dfrac{a_1{}^2}{c_1{}^2}(z-l_1)^2 \ , & l_1 - c_1 \leq z \leq z_1 \\[2mm] a_2{}^2 - \dfrac{a_2{}^2}{c_2{}^2}(z-l_2)^2 \ , & z_2 \leq z \leq l_2 + c_2 \\[2mm] a_3{}^2 + \dfrac{a_3{}^2}{c_3{}^2}(z-l_3)^2 \ , & z_1 \leq z \leq z_2 \end{cases} \tag{1}$$

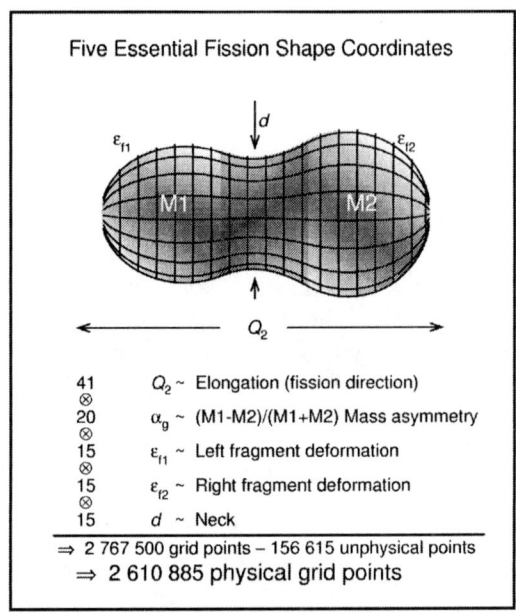

FIGURE 1. Five-dimensional shape parametrization used in our potential energy calculations.

There are 9 numbers of parameters, 3 of which are eliminated by the volume conservation and the continuation of the first derivative at z1 and z2. Finally, the position of the total center of mass is eliminated and we are left with 5 shape parameters. In Fig.1, we show the 5 shape parameters , elongation, mass asymmetry, left fragment deformation, right fragment deformation and the neck. From our experience, it is inevitable to use fine mesh points in order not to miss the complicated structure of the potential energy surface. Therefore, we use the mesh point given in Fig.1, that is, total 2,610,883 physical grid points.

RECIPE FOR SADDELE ASSIGNMENT

There is a common misunderstanding about the multi-dimensional potential energy surface about the method of obtaining the saddle points. In case of two-dimensional potential energy surface, we can assign saddle points together with minima and maxima by eye. When the surface is 3-dimensional or more, however, the assignment of the saddle points is not simple. Often, people project out the full potential energy surface to two-dimensional surface by minimizing with respect to the remaining variables. It is, however, not mathematically justified [1]. Sometimes, such procedure fail to find out important saddles and sometimes, the obtained saddle is not a real saddle. This situation is not at all safer even if we adopt the restricted Hartree Fock method. This method with several restriction variables meets the same difficulty to assign the saddle points.

In order to avoid this difficulty, we need to adopt mathematically correct method to determine the saddle point. An idea to solve this problem in the case of fission potential energy surface was demonstrated by A.Mamdouh et al. [3] and was developed in our previous paper [1]. We will give essential points of this method in the followings:

Flooding method to obtain the saddle height is based on the most intuitive definition of the saddle point. We start from some minimum, e.g., like the second minimum point on the potential energy surface. We fill the second minimum point with an imaginary water up to some level. Mesh points which is lower than this level is defined as "wet". Next, we increase the water level a little and check immediate neighborhood of the surface of wet points. Number of wet points increase and we get a new envelop of wet points. Continuing this procedure till the water level comes to some value at which a further increase of the water level causes a flooding across the barrier. Thus if we are waiting at a mesh point beyond the barrier, the point will be assigned as wet for the next increase of the water lever. This water level coincides with the height of the barrier. In this way, we can determine the height of the barrier in an arbitrary dimensional space.

In order to determine the location of the saddle, we used a technique called "fluctuation method". It is necessary because sometimes the energy of the mesh point is degenerate and if the energies of several points including the saddle point is within the increment of the water level adjustment, all these points are the candidates of the saddle position and we need a method to choose the correct saddle point. In that case, we add random numbers to these points and then perform the "flooding method" for this modified potential energy surface with an increment of water level less than the range of the random numbers. Among mesh points which are chosen as candidates of saddle point in this modified potential energy surface, we seek for the point which also appeared in the original potential energy surface and we define it as the real saddle point. In this way, even if some points accidentally come the the same energy as the saddle point within the accuracy of numerical calculation, we can eliminate it.

Using the technique of "flooding method" and "fluctuation method", we can determine the location and the energy of the lowest saddle point. Next problem is to determine the second and higher saddle points because we expect that there are many saddle points in the multi-dimensional potential energy surface. The method we developed for this purpose is called "dam method". This name comes from the fact that we build a dam at the position of the lowest saddle point to block the water flow. At the beginning, the dam is like a column because only the energy of the lowest saddle point is increased artificially. Then we perform the flooding and fluctuation method for this artificially modified potential energy surface. In this new potential energy surface, what we get for the saddle point is the point just immediate neighborhood of the column (dam element) because the water surely flow out through a point just next to the column. As we know, this newly chosen point is not a real saddle point in the original potential energy surface and thus we build an element of dam on this point. Since this point is next to the original dam, the width of the dam is increased. We perform the flooding and fluctuation methods with this further modified potential energy surface with a dam of increased width. Continuing this procedure, the width of the dam is widening in the multi-dimensional space and at some water level, the water flow out suddenly through the second saddle point that is located at discretely different position from the dam neighborhood. In this way, we can find the second saddle point location and its energy. The third saddle can be found out by blocking the lowest saddle and building the dam in the second saddle point.

CALCULATION FOR ACTINIDE NUCLEI AND DISCUSSIONS

Numerical calculations have been done on the basis of the Strutinsky prescription for the mesh points of more than 2.5 million for 138 even-even actinide nuclei [1, 2]. We obtained the following conclusions:

1. Most of the actinide nuclei have the lowest two saddle points, one is mass symmetric and the other is mass-asymmetric. This characteristic of the potential energy surface seems to be very general for actinide nuclei. This feature has been shown experimentally [4]. Thus the idea of the coexistence of the symmetric and asymmetric saddles is not new. There has been, however, no realistic and systematic calculations which has shown it. We also showed that the valley structure is well developed beyond these two saddle points. Two valleys are separated by the ridge and we can define the symmetric and asymmetric fission path separately.

2. The relative height of the symmetric and asymmetric saddles changes according to the parent nucleus. For nuclei at the beginning of the actinide nuclei, for example ^{228}Ra, the height of two saddles are almost degenerate. This feature correspondsto the triple peak structure of ^{228}Ra fission well known for a long time. For the heavier actinides, the asymmetric saddle becomes lower than the symmetric saddle, which support the asymmetric fission of actinide nuclei observed experimentally. This feature continues until we come to the region of the element Fm. For ^{256}Fm and ^{258}Fm, the height of two saddle points once again becomes almost degenerate resulting in the appearance of the symmetric fission for very heavy nuclei that is observed experimentally.

3. Special attention should be paid to ^{256}Fm and ^{258}Fm. From a precise analysis of the saddle point, we showed that the relative height of two saddle points in these nuclei changes, i.e., the lowest asymmetric and the next symmetric saddles for ^{256}Fm and lowest symmetric and the next asymmetric saddle for ^{258}Fm. This is an explanation of the bimodal fission observed in these nuclei. We could explain the features of the relative kinetic energy between two fragments at the same time from the fact that the symmetric saddle has a compact shape and the asymmetric saddle has an elongated shape. In

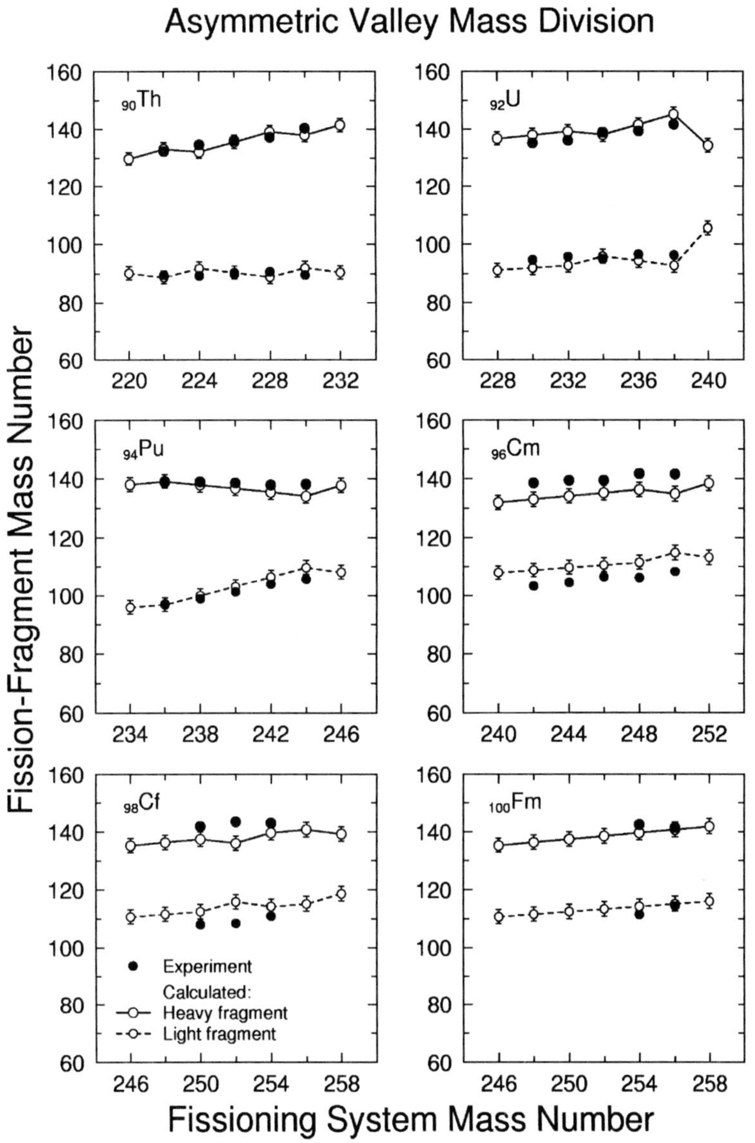

Asymmetric Valley Mass Division

FIGURE 2. Calculated (white circle) and measured (black circle) most probable mass division in asymmetric fission for sequence of even isotopes of Th, U, Cm, Cf and Fm.

demonstrating these features, it was quite necessary to use very fine mesh points with enough number of shape parameters because two saddle points are almost degenerate. The mass asymmetry and the neck variable play important roles. That is the reason why this phenomena has been a mystery for long time. Present day computer power and the correct method of fixing the saddle point were inevitable for this explanation.

4. Systematic analysis of the degree of mass-asymmetry for isotopes of Th, U, Pu, Cm, Cf and Fm have been done. The degree of the asymmetry was calculated at the minimum of valley beyond the mass-asymmetric saddle point at quadrupole moment Q_2=64b. The obtained values are compared with the experimentally observed most probable mass asymmetry, i.e., the mass asymmetry corresponding the the top of the fragment mass distribution. This is shown in fig.2 and we see that the calculated values and the experimental values coincide quite well. A maximum deviation is the order of mass three. From this comparison, we can say that the most probable mass division is explained well by the precise analysis of the static potential energy surface.

SUMMARY

We calculated the potential energy surface with full 5-dimensional grid points of more than 2.5 million for 138 even-even actinide nuclei. Special attention was paid to determine the saddle points without any approximation. We showed that there is a multiple-saddle structure, the most important ones are the lowest and the second lowest saddle points. They are mass-symmetrically and mass-asymmetrically deformed. The relative height of two saddle points well explain the change of the most probable mass division observed experimentally. Especially, the bimodal feature of very heavy nuclei has been explained from the multiple-saddle structure. These multiple-saddle structure is due to the competition between the macroscopic and microscopic energies of the compound and the fragment nuclei.

REFERENCES

1. P. Möller and A. Iwamoto, Phys.Rev.**C61** (2000) 40672-1.
2. P. Möller, D.G. Madland, A.J. Sierk and A. Iwamoto, Nature **406** (2001) 785.
3. A.Mamdouh et al.Nucl.Phys. **A644** (1988) 389.
4. Y. Nagame et al., Radiochim. Acta **78** (1997) 3.
5. E.K. Hulet et al., Phy.Rev.Lett. **56** (1986) 313.

Toward a Dynamics of Evolution of Matter - Past, Present and Future of the Self-consistent Collective Coordinate Method-

Fumihiko Sakata*, Yukio Hashimoto†, Shiwei Yan*, Lu Guo*, Hiroto Imagawa†, Akiyuki Seki* and Shigeyasu Fujiwara*

*Department of Mathematical Sciences, Ibaraki University, Mito, Ibaraki 310-8512
†Institute of Physics, University of Tsukuba, Ibaraki 305-8571

Abstract. Now-a-days, many fields of science are seemed to share the same interests as to explore the dynamics of *evolution of matter*, which intends to make clear what dynamical relations are there between different strata or different levels of the nature. Relations between the "macroscopic" and "microscopic" effects, between the "classical" and "quantum" dynamics, between the dynamics and thermo-dynamics and so on are basic subjects to explain how the nature evolutes from one stratum to the next stratum. As an example, the nucleus provides us with a very nice benchmark field to explore the above stated subjects, because it shows a coexistence of "macroscopic" and "microscopic" effects in association with various "phase transitions", and a mutual relation between "classical" and "quantum" effects related with the macro-level and micro-level variables, respectively. At certain energy region, the nucleus exhibits some statistical aspects which are associated with dissipation phenomena well described by the phenomenological transport equation. In this talk, a few topical subjects undertaken in our group and future problems will be discussed by putting special emphasis on clarifying the dynamics of *evolution of matter*.

INTRODUCTION

Recently, there appeared a review article[1] describing the development of the self-consistent collective coordinate (SCC) method, which was proposed in 1980 by Marumori, Maskawa, Kuriyama and Sakata[2] for aiming at describing the large-amplitude collective motion observed in a finite quantum system as nucleus. Since its development has just coincided with that of the order-to-chaos transition dynamics in both the classical and the quantum systems, with that of the microscopic dynamics underlying the transport phenomenon as well as that of the nuclear states under the various extreme conditions, it has been developed into a variety of forms appropriate for many interesting subjects. In this symposium, its applications to the high-spin and high iso-spin physics[3], to the microscopic dynamics responsible for the dissipation phenomena[4] and to the quantum realization of the classical bifurcation phenomena[5] are reported.

As a further application of the SCC method formulated within the time-dependent Hartree-Fock theory, it is strongly desired to study what actually happens in the heavy-ion deep inelastic collision when it is treated within the mean-field theory. From the viewpoint of the microscopic theory of many-body system, it is a very interesting subject to explore how the dissipation dynamics observed in the deep inelastic collision occurs

CP597, *Nonequilibrium and Nonlinear Dynamics in Nuclear and Other Finite Systems,*
edited by Z. Li et al.
© 2001 American Institute of Physics 0-7354-0041-5/01/$18.00

as a result of the microscopic Hamilton dynamics. From our investigation[4] [5] on how the transport phenomena usually described by the phenomenological Langevin equation is derived as a result of its underlying microscopic dynamics, and on how an exceedingly rich structure of the quantum space of states can be understood like in a case of the classical phase space in the context of the order-to-chaos transition, it is strongly desired to extend our approach to a case of the quantum transport phenomena. It is also desired how the finite system reaches to a statistical state, which is usually expressed as a heat bath. These problems will be discussed in this paper by putting special emphasis on clarifying the dynamics of evolution of matter.

DYNAMICAL CHANGE OF TIME-DEPENDENT MEAN-FIELD

It is one of the main issues in various fields of contemporary science to relate the *macro-level* dynamics with the *micro-level* dynamics. In order to study this inter-stratum relation, the nuclear system provides us with a good testing field, because it shows the collective motion described by a few macro-variables which ought to be expressed by the nucleonic degrees of freedom.

The first step in exploring this subject within the nuclear physics may be provided within the mean-field theory, because the microscopic nucleon degrees of freedom are fully taken into account, and the interplay between the collective degrees of freedom describing the time-dependent variation of the mean-field and the intrinsic degrees of freedom, which adjust their features to self-consistently follow the change of the mean-field, is properly included. On top of that, the time-dependent Hartree-Fock (TDHF) equation is known to satisfy the same formal structure as that of the classical Hamilton equation of motion, as well as the full quantum nature through the boson expansion method. Namely, the TDHF theory provides us with an excellent framework to study an inter-relation between the *quantum*-level dynamics and the *classical*-level dynamics in a very convenient and appropriate way.

Even in the mean-field theory, there still remains a long-standing problem on how to define the collective degrees of freedom in terms of the nucleon degrees of freedom. One dynamical way in defining the collective operator has been well-known to use the RPA, which is defined at a given stable mean field $|\phi_0\rangle$, and is only capable of describing the small-amplitude oscillation around the mean field. The other way known for many years in introducing the collective operator is limited to a case where the collective motion under consideration is related to a kinematical or dynamical symmetry preserved by a system, like the translational or the rotational motions in an isolated system. Within the mean-field theory, for example, the rotational motion related to an angular momentum conservation law is derived from the variation principle

$$\delta\langle\phi(t)|i\frac{\partial}{\partial t}-\hat{H}|\phi(t)\rangle = 0, \quad |\phi(t)\rangle = \exp\{-i\omega t\hat{J}_x\}\exp\{i\hat{F}(\omega)\}|\phi_0\rangle, \quad (1)$$

where $|\phi_0\rangle$ is one of the Hartree-Fock states satisfying $\delta\langle\phi_0|\hat{H}|\phi_0\rangle = 0$, $\hat{J}_x$ the x-component of the total angular momentum operator perpendicular to the de-formation axis, and $\hat{F}(\omega)$ the one-body Hermitian operator to be determined for a given ω or for a given angular momentum with a constraint condition $\langle\phi_0|\exp\{-i\hat{F}(\omega)\}\hat{J}_x\exp\{i\hat{F}(\omega)\}|\phi_0\rangle = J$. Equation (1) is nothing but the usual self-consistent cranked Hartree-Fock equation given as

$$\delta\langle\phi(\omega)|\left\{\omega\hat{J}_x - \hat{H}\right\}|\phi(\omega)\rangle = 0, \quad |\phi(\omega)\rangle \equiv \exp\{i\hat{F}(\omega)\}|\phi_0\rangle, \tag{2}$$

which describes a uniform rotational motion of the whole nuclear system around the x-axis.

An important difference between the collective RPA operator and the angular momentum operator generating the rotational motion rests on the following points: The microscopic structure of the collective RPA operator is fixed by the RPA and is valid only when it operates on the stable mean-field $|\phi_0\rangle$, whereas that of the angular momentum operator changes as its operand $\exp\{i\hat{F}(\omega)\}|\phi_0\rangle$ changes depending on ω, and is always valid regardless of its operand. Here it should be noted that the *macro* information of the rotational motion is substantialized into a concept of *moment of inertia* $\mathcal{J}(\omega)$, which is usually defined by using the RPA (Thouless-Valatin) and has a sense locally around a given constrained Hartree-Fock state $\exp\{i\hat{F}(\omega)\}|\phi_0\rangle$.

The SCC method was firstly proposed to describe how the collective RPA operator undergoes a microscopic structure change, when the collective motion develops into a large-amplitude region. Since the collective motion is known to couple with many non-collective degrees of freedom (defined by RPA at $|\phi_0\rangle$) as its amplitude becomes large, this problem is strongly related to a problem how to dynamically extract the collective-manifold out of the huge-dimensional time-dependent Hartree-Fock manifold, and how to introduce the collective operator which has a sense not only in the small-amplitude region but also in the large-amplitude region.

A set of the basic equations of the SCC method capable of extracting the collective sub-manifold, on which a given collective motion is confined, is given as

$$0 = \delta\langle\phi(\eta,\eta^*)|\left\{i\frac{\partial}{\partial t} - \hat{H}\right\}|\phi(\eta,\eta^*)\rangle, \tag{3}$$

$$\frac{\eta^*}{2} = \langle\phi(\eta,\eta^*)|\hat{O}^\dagger(\eta,\eta^*)|\phi(\eta,\eta^*)\rangle, \tag{4}$$

$$i\dot{\eta} = \frac{\partial\mathcal{H}}{\partial\eta^*}, \quad i\dot{\eta}^* = -\frac{\partial\mathcal{H}}{\partial\eta}, \tag{5}$$

where

$$|\phi(\eta,\eta^*)\rangle \equiv \exp\{i\hat{F}(\eta,\eta^*)|\phi_0\rangle, \tag{6}$$

$$\hat{O}^\dagger(\eta,\eta^*) \equiv \exp\{-i\hat{F}(\eta,\eta^*)\}\frac{\partial}{\partial\eta}\exp\{i\hat{F}(\eta,\eta^*)\}, \tag{7}$$

$$\mathcal{H} \equiv \langle\phi(\eta,\eta^*)|\hat{H}|\phi(\eta,\eta^*)\rangle. \tag{8}$$

Here, it is worthwhile mentioning that the collective operator $\hat{O}^\dagger(\eta,\eta^*)$ defined in Eq. (7) has a meaning irrespective of the amplitude η,η^* of its operand $|\phi(\eta,\eta^*)\rangle$ like the angular momentum operator $\hat{J}_x$ in the case of the cranking Hartree-Fock theory. It should be also mentioned that a number of collective variables η,η^* necessary for describing a given collective motion are also obtained dynamically by applying the stability condition formulated within the SCC method.

With the aid of Eqs. (5) and (7), Eq. (3) is expressed as

$$0 = \delta \langle \phi(\eta, \eta^*) | \left\{ \frac{\partial \mathcal{H}}{\partial \eta^*} \hat{O}^\dagger(\eta, \eta^*) + \frac{\partial \mathcal{H}}{\partial \eta} \hat{O}(\eta, \eta^*) - \hat{H} \right\} | \phi(\eta, \eta^*) \rangle, \tag{9}$$

which has the same structure as the self-consistent cranking Hartree-Fock equation (2). In the case of the cranking equation (2), the coupling between the collective degrees of freedom $\hat{J}_x$ and the rest does not exist, because there holds a relation $[\hat{J}_x, \hat{H}] = 0$. In this case, the auxiliary Hamiltonian $\hat{H} - \omega \hat{J}_x$ gives an intrinsic Hamiltonian in the rotating frame and a term $\omega \hat{J}_x$ is regarded as the Colioris term. On the other hand, the SCC method defines the collective sub-manifold in such a way that a *linear* coupling with the rest space is eliminated dynamically (satisfying the maximally decoupled condition). In this sense, the auxiliary Hamiltonian appearing in the parenthesis $\{\cdots\}$ on the rhs of Eq. (9) is understood to be an approximate intrinsic Hamiltonian in the moving frame, which contains an extended Colioris term. From the above discussion, it may be understood that the SCC method gives a general framework for treating the large-amplitude collective motion, whose small amplitude limit is described by the usual RPA collective mode. More generally, the SCC method is capable of extracting a KAM torus of the regular motion out of the huge-dimensional phase space. It has been shown[1] that an instability point of the collective sub-manifold just corresponds to the bifurcation point and approximately to the level-crossing point of the single-particle orbits in the moving frame, which constitutes a good example of understanding an interrelation between the quantum-level and classical-level dynamics.

At the best of our knowledge, however, the basic equation of the SCC method can be solved perturbatively by starting with the stable mean-field $|\phi_0\rangle$ or some fixed point $|\phi(\eta_0, \eta_0^*)\rangle$ satisfying the stationary condition. Namely, what one may learn from the SCC method is how there appears an essential dynamical change when the system acquires an additional amplitude, i.e., when the system develops from $|\phi(\eta_0, \eta_0^*)\rangle$ to $|\phi(\eta_0 + \Delta\eta, \eta_0^* + \Delta\eta^*)\rangle$. In a case of the cranking model, one may learn how the microscopic structure of the collective rotation operator changes and how the moment of inertia $\mathcal{J}(\omega)$ changes as the nuclear system acquires an additional angular momentum or an additional deformation.

According to the general theory of the nonlinear dynamics, an exceedingly rich structure of the phase space like an appearance of the nonlinear resonance, bifurcation, the main island, the secondary island, the tertiary island structure $\cdots$ of the regular motion and an appearance of the chaotic sea may be explored analytically by applying the perturbation theory also. Since the structure change of the phase space is associated with some singularity which can be detected by applying a proper perturbation theory, the SCC method will be useful in understanding an inexhaustible structure embedded in the time-dependent Hartree-Fock manifold.

In the above stated context, it is very interesting to explore what happens in the *micro-level* dynamics when one applies a mean-field theory to the heavy-ion reaction. In treating the heavy-ion reaction within the mean-field theory, one starts with boosting the static Hartree-Fock solution with a proper impact parameter and an appropriate collective momentum. In the approaching phase, the nuclei might move collectively by keeping the initial static Hartree-Fock solution. Namely, the relative motion between two nuclei is well decoupled from the intrinsic degrees of freedom. When two colliding nuclei are going to interact, the initial boosting energy may be transferred to the intrinsic energy and there appears dissipation in the relative collective motion. As will be discussed in the next section, from a viewpoint of developing the many-body theory in the finite system, especially of understanding the dissipation mechanism of the damped collective motion, it is very important to understand a dynamical process of the heavy-ion deep inelastic collision, like (1) To what extent the relative motion can be treated as the collective motion. (2) How can one divide the total system into a weakly coupled collective and intrinsic subsystems, and how many collective degrees of freedom are necessary in describing the heavy-ion dissipative motion. Since the microscopic structure of the relative motion before and after the collision seems to be different, a macroscopic treatment of using the transport equation for the macroscopic relative coordinate alone might miss some essential dynamics. (3) Can one understand the dissipative diabatic dynamics[7] as a source of the dissipation of the relative motion? Since the single-particle level crossing point approximately corresponds to an instability point of the collective sub-manifold, and there appears other kind of motion bifurcated from the instability point, and this point will develop into a center of chaotic sea, one may have a hope to relate the occurrence mechanism of dissipation with the quantum mechanical single-particle level dynamics. (4) What kinds of essential dynamics are missing in the mean-field treatment. Apparently, the two-body or higher order density matrices are not included in the mean-field treatment. However, the higher order effects only contribute to produce some broadening effects around the mean-field dynamics, it might not be essential to include the higher order density matrices in discussing the zero-th order physics. Namely, one has to make clear what are genuine effects of the quantum mechanics coming from the higher order density matrices. (5) How the intrinsic system develops into such a statistical object that is described as a heat bath. (6) How the classical regime described by the time-dependent mean-field changes into the quantum regime described by e.g., the quantum molecular dynamics, as the initial energy increases, and so forth.

Namely, there still remains many interesting problems on an interplay between the collective and non-collective degrees of freedom from the micro-level dynamics, and an interplay between the classical and quantum level dynamics.

MICROSCOPIC DERIVATION OF TRANSPORT PHENOMENA

The second main issue discussed in the various fields of contemporary science is how to derive the *macro*-level transport equation describing the macroscopic irreversible motion from the *micro*-level reversible dynamics. An evolution of the early universe, that of the chemical reaction, many active processes in biological system, the measurement theory and the quantum correspondent of the classically chaotic system are typical examples, which are described by the Langevin-type transport equations with a stochastic force. As was discussed in the previous section, the heavy-ion deep inelastic collision may also provide us with a good benchmark field on this type of investigation.

In so-far-as one wants to describe the macroscopic phenomena alone, it is not necessary to ask what is the origin of the stochastic force and whether it comes from the thermal fluctuation or from the quantum fluctuation. When one intends to understand a dynamics of evolution of matter which connects the macro-level dynamics with the micro-level dynamics, one has to start with how to divide the total system into the weakly coupled relevant (collective or macro η, η^*) and irrelevant (intrinsic or micro ξ, ξ^*) systems. In a case of the Hamilton system, a division may be performed by applying the SCC method. As discussed in the previous section, however, there still remains a lot of studies how one may divide the collective degrees of freedom from the rest in a case of the heavy-ion collision.

When the system is divided into the collective and intrinsic systems, one may start with the Liouville equation for a total distribution function as

$$\dot{\rho}(\eta, \eta^*, \xi, \xi^*) = -i\mathcal{L}\rho(\eta, \eta^*, \xi, \xi^*), \tag{10}$$

which is changed into a master equation for the separable- and correlated-partial distribution functions defined by

$$\rho_s(t) \equiv \rho_\eta(t)\rho_\xi(t), \quad \rho_c(t) \equiv \rho(\eta, \eta^*, \xi, \xi^*) - \rho_s(t),$$
$$\rho_\eta(t) \equiv \mathrm{Tr}_\xi \rho(\eta, \eta^*, \xi, \xi^*), \quad \rho_\xi(t) \equiv \mathrm{Tr}_\eta \rho(\eta, \eta^*, \xi, \xi^*),$$
$$\mathrm{Tr}_\xi \equiv \int d\xi d\xi^*, \mathrm{Tr}_\eta \equiv \int d\eta d\eta^*, \tag{11}$$

as

$$\dot{\rho}_s(t) = -iP(t)\mathcal{L}\rho_s(t) - iP(t)\mathcal{L}g(t, t_I)\rho_c(t_I)$$
$$- \int_{t_I}^{t} dt' P(t)\mathcal{L}g(t, t')\{1 - P(t')\}\mathcal{L}\rho_s(t'), \tag{12}$$

where

$$P(t) \equiv \rho_\eta(t)\mathrm{Tr}_\eta + \rho_\xi(t)\mathrm{Tr}_\xi - \rho_\eta(t)\rho_\xi(t)\mathrm{Tr}_\eta \mathrm{Tr}_\xi, \tag{13}$$

$$g(t, t') \equiv T \exp\{-i \int_{t'}^{t} [1 - P(\tau)]\mathcal{L}d\tau\}. \tag{14}$$

Here $P(t)$ denotes a time-dependent projection operator and T the Dyson time-ordering operator. The coupled master equation for the partial distribution functions $\rho_\eta(t)$ and $\rho_\xi(t)$ is obtained from Eq. (12) by taking the integration Tr_ξ and Tr_η, respectively, which gives a firm theoretical base for investigating the microscopic dynamics of the damped collective motion [6].

Before discussing the damping phenomena in a finite system, one should be aware of a recent very interesting subject which intends to make clear a relation between the dynamics and the thermodynamics[8]. More specifically, there are many works on what relation are there between a chaotic dynamics and a realization of the statistical state in the finite system, and between the dynamics of individual trajectories and that of the averaged trajectory. Namely , there are many studies on what relations are there between the KS-entropy related to the Lyapunov exponent and the BG entropy, and how one can characterize an evolution process of the chaotic system[9]. In this respect, a necessity of using a non-extensive entropy is also interesting in connecting the microscopic dynamics and the statistical mechanics, because the non-extensive entropy[10] might characterize the non-statistical evolution process more properly than the BG entropy.

By using the numerical simulation, a microscopic dynamics of the damping mechanism of the collective motion coupled with the *strong chaotic* intrinsic system with a finite degrees of freedom has been discussed elsewhere[4]. In this study, a special attention has been paid on how the microscopic mechanism changes depending on a number of degrees of freedom in the intrinsic system. Since an intrinsic system with an infinite number of degrees of freedom may be safely replaced by the heat bath, and since the collective damped motion coupled with the heat bath is well described by the linear response theory satisfying the fluctuation-dissipation theorem, it is very interesting what happens when a number of degrees of freedom in the intrinsic system is finite, and how its dynamics changes as the number increases. Here it should be also mentioned that the damping phenomenon in the finite system is also characterized by the non-extensive entropy [4]. This might suggest us that the damping mechanism in the finite system is a non-statistical process, where the usual fluctuation-dissipation theorem is not applicable.

In a realistic case like the nuclear system, the intrinsic system might have a very intricate phase space structure coming from the shell structure, which is usually considered to prevent a fully developed chaos. Consequently, it is a very interesting subject to explore a damping mechanism of the collective motion coupled with the *mixed* intrinsic system, whose phase space has many regular islands surrounded by the chaotic sea. Here, it should be noted that there still remains many works in the macro-level classical dynamics. A variety of effects originating from different types of noise, and its competition with the friction force are not well known even in the phenomenological Langevin-type transport equation, e.g., in its application to the active processes in the biological system.

Other interesting subjects to be explored are how to extend the present discussions to the fully quantum mechanical system. Even in the case of damped motion coupled with the statistical state (, i.e., the infinite number of degrees of freedom system) there still remain many interesting questions on various roles of the fluctuations, like the *Anderson* noise. A difference between classical and quantum dissipation by using the quantized Brownian motion is being discussed[11][12] to explore various effects coming from fluctuation due to noise, from friction, from competition between noise and friction, decoherence, localization, etc. In this context, the giant resonance on top of the "heated nuclei" might be a good example in understanding these interesting effects.

From our point of view in understanding the quantum dissipation in terms of the microscopic dynamics, one may start with a fully quantum mechanical coupled master equation which is derived in the same way as Eq. (12). This equation is based on the time-dependent projection operator method, which is a generalization of the Zwanzig's time-independent projection operator method[13]. The latter is only valid for a case with the collective motion coupled with the time-independent statistical state like the GOE based on the random matrix theory.

In exploring the microscopic dynamics of the damped quantum motion, there are two main subjects. The first is how the individual quantum states undergo a change depending on the strength of the non-linear interaction and on the excitation energy, and how the finite quantum many-body system reaches to its stationary statistical state, and the second is how the irreversible collective dissipative motion is realized as a result of the reversible underlying Hamilton dynamics.

With regards to the first subject, we have developed the quantum version of the SCC method, which has been used to understand the exceedingly rich structure of the quantum space of states, and to explore to what extent the classical concept like the bifurcation or the nonlinear resonance can survive in the quantum system[5]. Our work in studying the quantum space of states has been limited to a case with rather weak nonlinear interaction. Recently, there are many interesting works on the quantum excited states at the relatively large nonlinear interaction region, which is called *chaos assisted tunneling*[14] observed in the cold atoms. A pair of states located around two stable fixed points (nonlinear resonances or islands of stability in the classical phase space) is constructed by a *dynamical tunneling*[15]. In a regular regime, a pair of eigen-energies of these two states is almost degenerate so as to contribute to a Poissonian distribution, whereas in a mixed or chaotic regime, they split depending on the width of the chaotic sea in between two stable fixed points so as to be favorable for the Wignerian distribution in the nearest neighbor level statistics. It is then very interesting question whether or not one may characterize a structure change in the quantum space of states in terms of the *chaos assisted tunneling*, when its corresponding classical system undergoes a change from regular to chaos.

Even-though we clarified a usefulness of the concepts of *quantum non-linear resonance* and *quantum bifurcation* in understanding the structure of the quantum space of states, and have just mentioned an applicability of the chaos assisted tunneling, it is not satisfactory because these concepts are not genuine quantum mechanical ones but have a classical mechanical remnants. In this respect, there still remains many studies in understanding an inexhaustible rich structure of the quantum space of states.

Since the above discussed dynamical tunneling is a quantum mechanical effect, and since it is not included in the usual mean-field theory, the dissipative dynamics discussed in the time-dependent mean-field theory might be insufficient.

The second subject may be treated by considering two coupled systems; one of which is a collective system characterized by a slow motion, and the other is an intrinsic system composed of more than two degrees of freedom and is able to undergo a change from regular to chaos. According to our experience[4] in exploring the macro-level transport phenomena in terms of the micro-level dynamics within the classical theory, and according to our study how the structure of quantum space of states changes depending on the strength of the nonlinear interaction, we may expect many interesting dynamics in clarifying the quantum irreversible motion in terms of the quantum micro-level reversible dynamics, which may suggest us more fruitful information than that obtained in the quantum Brownian motion.

CONCLUSION

In this talk, we have discussed various open problems, which are suggested from the development of the SCC method and lead us to develop a new dynamics of evolution of matter. These investigations request us to develop not only a new way of numerical simulations to fully exploit the state-of-art computational facility, but also a new theoretical framework which allows us to analytically understand the complex results obtained from the numerical simulation. Since these subjects are developed in various fields of natural science, it is strongly desired to develop the dynamics of evolution of matter under really interdisciplinary environment.

REFERENCES

1. F. Sakata, T. Marumori, Y. Hashimoto and Shiwei Yan, Suppl. Prog. Theor. Phys. **141**(2001) 1.
2. T. Marumori, T. Maskawa, A. Kuriyama and F. Sakata, Prog. Theor. Phys. **64**(1980) 1294.
3. Lu Guo, F. Sakata and En-Guang Zhao, in this proceedings.
4. Shiwei Yan, Fumihiko Sakata and Yizhong Zhuo, in this proceedings.
5. Y. Hashimoto, H. Tsukuma and F. Sakata, in this proceedings.
6. F. Sakata, M. Matsuo, T. Marumori and Y. Zhuo, Ann. Phys. **194** (1989) 30
7. W. Noerenberg, Phys. Lett. **B104** (1981) 107
8. M. Bianucci, R. Mannella, B.J. West and P. Grigolini, Phys. Rev. **E51** (1995) 3002
9. V. Latore and M. Baranger, Phys. Rev. Lett. **82** (1999) 520
10. C. Tsallis, J. Stat. Phys. **52** (1988) 479
11. D. Cohen Phys. Rev. Lett. **78** (1997) 2878
12. E. Lutz and H.A. Weidenmueller, Physica **A 267** (1999) 354
13. R. Zwanzig, Lectures in Theoretical Physics **3** (Interscience, New York, 1961)
14. D.A. Steck, W.H. Oskay and M.G. Raizen, Science **293** (2001) 274
15. M.J. Davis and E.J. Heller, J. Chem. Phys. **75** (1986) 246

Application of time-dependent density-matrix theory

M. Tohyama* and A. S. Umar†

*Kyorin University School of Medicine, Mitaka, Tokyo 181-8611, Japan
†Department of Physics and Astronomy, Vanderbilt University, Nashville, Tennessee 37235, USA

Abstract. Isovector dipole resonances in the oxygen isotopes $^{18\sim24}$O and fusion reactions of ^{16}O + ^{16}O are studied using an extended version of the time-dependent Hartree-Fock theory known as the time-dependent density-matrix theory (TDDM). A newly developed TDDM program which includes spin-orbit force in a mean-field potential is used. Low-lying dipole strength in the oxygen isotopes is compared with experimental data and also with a shell-model calculation. It is found that the observed isotope dependence of low-lying dipole strength is reproduced in TDDM when the strength of the residual interaction is appropriately chosen. For the fusion reactions, it is found that the dissipation of translational energy due to two-body collisions is small when spin-orbit force is included in the mean-field potential.

INTRODUCTION

The time-dependent density-matrix theory (TDDM) is an extended version of the time-dependent Hartree-Fock theory (TDHF) and is formulated in order to determine the time evolution of one-body and two-body density matrices ρ and ρ_2 in a self-consistent manner. TDDM, therefore, includes the effects of both a mean-field potential and nucleon-nucleon correlations. TDDM has been applied to giant resonances[1, 2] and low-energy heavy-ion collisions[3]. However, the TDDM calculations were not highly quantitative since spin-orbit force was neglected in the calculation of a mean-field potential. The main purpose of this report is to present more quantitative TDDM results than the previous ones. Two recent improvements make it possible to perform better TDDM calculations: one is the treatment of ground-state correlations. As will be described below, an adiabatic treatment of the residual interaction enables us to obtain a correlated ground state which is a stationary solution of TDDM[4]. The other improvement is the extension of the TDDM program to include spin-orbit force, which has been done based on the TDHF code with spin-orbit force [5]. This report is organized as follows: in sect.2 we briefly describe the equations of motion in TDDM and explain how the strength function for a giant resonance is calculated. In sect.3 we present the results for isovector dipole ($E1$) resonances in the oxygen isotopes $^{18\sim24}$O and compare them with recent experimental data. The head-on collisions of ^{16}O + ^{16}O are also calculated in TDHF and TDDM to study the effects of spin-orbit force and nucleon-nucleon collisions on fusion reactions. The results are presented in sect.4 and sect.5 is devoted to a summary.

CP597, *Nonequilibrium and Nonlinear Dynamics in Nuclear and Other Finite Systems*,
edited by Z. Li et al.
© 2001 American Institute of Physics 0-7354-0041-5/01/$18.00

TDDM AND METHOD FOR CALCULATING THE STRENGTH FUNCTION

The equations of motion for ρ and C_2, the correlated part of ρ_2, can be derived by truncating the well-known BBGKY hierarchy for reduced density matrices[6]. To solve the equations of motion for ρ and C_2, we expand ρ and C_2 using a finite number of single-particle states ψ_α which satisfy a TDHF-like equation,

$$\rho(11',t) = \sum_{\alpha\alpha'} n_{\alpha\alpha'}(t)\psi_\alpha(1,t)\psi_{\alpha'}^*(1',t), \tag{1}$$

$$\begin{aligned} C_2(121'2',t) &= \rho_2 - A(\rho\rho) \\ &= \sum_{\alpha\beta\alpha'\beta'} C_{\alpha\beta\alpha'\beta'}(t) \\ &\times \psi_\alpha(1,t)\psi_\beta(2,t)\psi_{\alpha'}^*(1',t)\psi_{\beta'}^*(2',t), \end{aligned} \tag{2}$$

where the numbers denote space, spin and isospin coordinates and $A(\rho\rho)$ is an antisymmetrized product of the one-body density matrices. Thus, the equations of motion of TDDM consist of the following three coupled equations[3]

$$i\hbar\frac{\partial}{\partial t}\psi_\alpha(1,t) = h(1,t)\psi_\alpha(1,t), \tag{3}$$

$$i\hbar\dot{n}_{\alpha\alpha'} = \sum_{\beta\gamma\delta}[\langle\alpha\beta|v|\gamma\delta\rangle C_{\gamma\delta\alpha'\beta} - C_{\alpha\beta\gamma\delta}\langle\gamma\delta|v|\alpha'\beta\rangle], \tag{4}$$

$$i\hbar\dot{C}_{\alpha\beta\alpha'\beta'} = B_{\alpha\beta\alpha'\beta'} + P_{\alpha\beta\alpha'\beta'} + H_{\alpha\beta\alpha'\beta'}, \tag{5}$$

where h is the mean-field hamiltonian and v the residual interaction. The terms on the right-hand side of Eq.(5) contain all two-body correlations including those induced by the Pauli exclusion principle. They are explicitly shown in Ref.[3]. The TDDM equations of motion satisfy conservation laws of the total number of particles and the total momentum and energy. The small amplitude limit of TDDM was investigated[7] and it was found that if only the one particle - one hole and one hole - one particle elements of $n_{\alpha\alpha'}$ and the two particle - two hole and two hole - two particle elements of $C_{\alpha\beta\alpha'\beta'}$ are taken, the small amplitude limit of TDDM will be equivalent to the conventional second RPA (SRPA)[8]. Thus, TDDM is a more general framework than the conventional SRPA.

The $E1$ strength function is calculated in the following three steps:

1) A static Hartree-Fock (HF) calculation is performed to obtain the initial ground state. The Skyrme III with spin-orbit force is used as the effective interaction. Unoccupied single-particle states up to the $2s$ and $1d$ orbits are also calculated for both protons and neutrons to solve the TDDM equations for $n_{\alpha\alpha'}$ and $C_{\alpha\beta\alpha'\beta'}$. The single-particle

wavefunctions are confined to a cylinder with length 20fm and radius 10fm. (Axial symmetry is imposed to calculate the single-particle wavefunctions[5].) The mesh size used is 0.5fm. The neutron $1d_{5/2}$ state is assumed to be partially occupied to obtain the HF ground states for ^{18}O and ^{20}O.

2) To obtain a correlated ground state, we evolve the HF ground state using the TDDM equations and the following time-dependent residual interaction of the δ-function form

$$v(t) = v_0(1 - e^{-t/\tau})\delta^3(\vec{r} - \vec{r}'). \tag{6}$$

The TDDM program is written for use with more complicated residual interactions such as the Skyrme force. To save computation time, however, we use the simple residual interaction in this report. The time constant τ should be sufficiently large to obtain a nearly stationary solution of the TDDM equations[4]. We choose τ to be 150fm/c. The strength of the residual interaction is determined to approximately reproduce the observed occupation probability of the proton $1d_{5/2}$ state in ^{16}O. The obtained value of v_0 is -230MeVfm^3, which may be smaller than the value of approximately -300MeVfm^3 found in the literature for pairing-gap calculations. The time step size used to solve the TDDM equations is 0.75fm/c. The calculated occupation probabilities of the neutron $2s_{1/2}$ and $1d_{3/2}$ states are $2 \sim 3\%$ in the oxygen isotopes considered here.

3) The $E1$ mode is excited by boosting the single-particle wavefunctions at $t = 5\tau$ with the dipole velocity field:

$$\psi_\alpha(5\tau) \longrightarrow e^{ikD(z)}\psi_\alpha, \tag{7}$$

where

$$D(z) = \begin{cases} \frac{N}{A}z & \text{for protons} \\ -\frac{Z}{A}z & \text{for netrons.} \end{cases} \tag{8}$$

Here, Z and N are the numbers of protons and neutrons, respectively, and $A = Z + N$. When the boosting parameter k is sufficiently small, the strength function defined by

$$S(E) = \sum_n |\langle \Phi_n | \hat{D} | \Phi_0 \rangle|^2 \delta(E - E_n) \tag{9}$$

is obtained from the Fourier transformation of the time-dependent dipole moment $D(t)$ as

$$S(E) = \frac{1}{\pi k \hbar} \int_0^\infty D(t) \sin \frac{Et}{\hbar} dt, \tag{10}$$

where

$$D(t) = \int D(z)\rho(\vec{r}, t)d^3\vec{r}. \tag{11}$$

In Eq.(9) $|\Phi_0\rangle$ is the total ground-state wavefunction and $|\Phi_n\rangle$ the wavefunction for an excited state with excitation energy E_n. It is very time consuming to solve the TDDM equations (especially Eq.(5)) for a long period. Thus, we stop TDDM calculations at $t = 1200$fm/c. The upper limit of the time integration in Eq.(10) is limited to 450fm/c. To reduce fluctuations in $S(E)$, the dipole moment is multiplied by a damping factor $e^{-\Gamma t/2\hbar}$ with $\Gamma = 1$MeV before the time integration. Since the integration time is limited, the strength function in a very low energy region ($E < 2\pi\hbar/450 \approx 3$MeV) is not well determined. The energy-weighted sum rule (EWSR) is expressed as

$$
\begin{aligned}
\int S(E)EdE &= \frac{1}{2}\langle\Phi_0|[\hat{D},[H,\hat{D}]]|\Phi_0\rangle \\
&= \frac{\hbar^2}{2m}\frac{ZN}{A} + \frac{t_1+t_2}{4}(\int\rho_p(\vec{r})\rho_n(\vec{r})d^3\vec{r} \\
&\quad + \sum_{\alpha\alpha'\in p,\beta\beta'\in n}\int\psi_\alpha^*(\vec{r})\psi_\beta^*(\vec{r}) \\
&\quad \times \psi_{\alpha'}(\vec{r})\psi_{\beta'}(\vec{r})d^3\vec{r}C_{\alpha'\beta'\alpha\beta}),
\end{aligned} \tag{12}
$$

where m is the nucleon mass, and t_1 and t_2 are the parameters for the momentum-dependent parts of the Skyrme interaction. We assume that the hamiltonian H consists of a two-body interaction of the Skyrme type. The first term on the right-hand side of the above equation corresponds to the classical Thomas-Reiche-Kuhn (TRK) sum rule and the second term, the enhancement term, is due to the momentum dependence of the hamiltonian. The contribution of the momentum-dependent part is about 28% of the total EWSR value in the oxygen isotopes considered here and the term proportional to $C_{\alpha\beta\alpha'\beta'}$ describing the effects of ground-state correlations is quite small (less than 1% of the total sum rule value).

DIPOLE RESONANCES IN $^{18\sim24}$O

In Figs.1~4, the $E1$ strength functions of $^{18\sim24}$O calculated in TDDM (solid line) are compared with those in TDHF (dotted line). The TDHF calculations presented here are equivalent to the RPA calculations without any truncation of unoccupied single-particle states because the TDHF equation for the boosted single-particle wavefunctions ψ_α is solved in coordinate space. The boundary condition for the continuum states, however, is not properly taken into account in our calculations because all the single-particle wave functions are confined to the cylinder. Therefore, the calculated strength functions slightly depend on the cylinder size. The difference between the TDDM and TDHF calculations is due to the effects of two-body correlations, which may have two aspects: one is to induce the ground-state correlations which increase the occupation probabilities of weakly bound neutron orbits. The increase in low-lying strength ($E < 15$MeV) seen in the TDDM results for all the isotopes is due to partial occupation of the neutron $2s_{1/2}$ and $1d_{3/2}$ states. The comparison between the TDDM result for ^{24}O and those for other isotopes suggests that the occupation of the neutron $2s_{1/2}$ state is responsible for the increase in the $E1$ strength in the very low energy region (around $E = 5$MeV). This is because the appearance of a prominent peak at 6MeV in ^{24}O is related to the nearly full occupation of the neutron $2s_{1/2}$ state. The other aspect of the two-body correlations is to

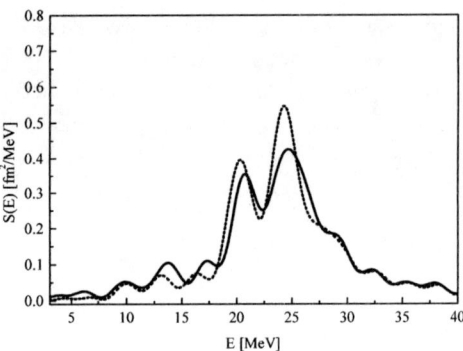

FIGURE 1. Strength functions of isovector dipole resonances in ^{18}O calculated in TDDM (solid line) and in TDHF (dotted line).

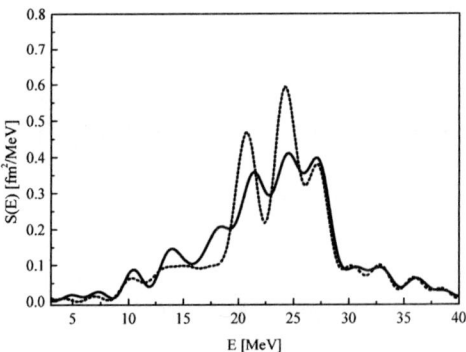

FIGURE 2. Strength functions of isovector dipole resonances in ^{20}O calculated in TDDM (solid line) and in TDHF (dotted line).

increase the width of the giant dipole resonance (GDR) which is located around 25MeV. There are background two particle - two hole states which consist of $1\hbar\omega$ excitations of protons and $0\hbar\omega$ excitations of neutrons. The coupling of the GDR to these states leads to the spreading of the GDR strength. The fractions of the EWSR values depleted over the energy range between 0MeV to 40MeV are about 85% in TDDM and about 90% in TDHF. The EWSR values depleted below 15MeV are shown in Fig.5 for $^{18\sim24}$O. The ordinate indicates the ratio of the EWSR value to the TRK value. The results in TDDM with $v_0 = -230$MeVfm^3 (rhombus) are smaller than the experimental data[9] except for ^{22}O. In TDDM, the EWSR values in the low-energy region crucially depend on the strength of the residual interaction. Since the strength of the residual interaction

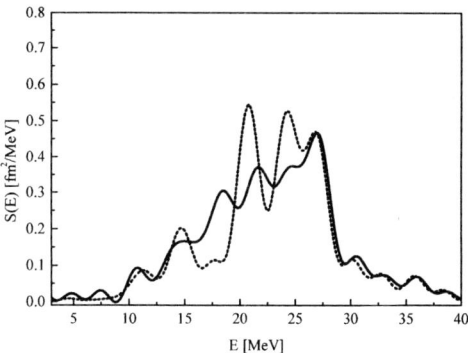

FIGURE 3. Strength functions of isovector dipole resonances in ^{22}O calculated in TDDM (solid line) and in TDHF (dotted line).

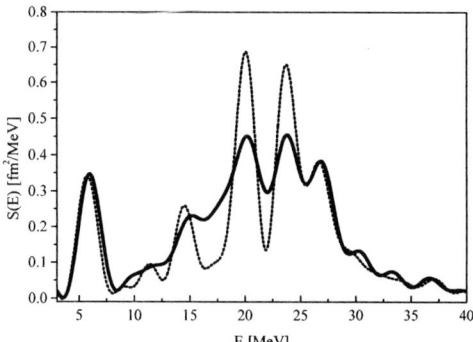

FIGURE 4. Strength functions of isovector dipole resonances in ^{24}O calculated in TDDM (solid line) and in TDHF (dotted line).

is not well-known in unstable nuclei, we performed TDDM calculations using slightly stronger residual interaction with $v_0 = -330\text{MeVfm}^3$ to increase $E1$ strength in the low-energy region. The value of $v_0 = -330\text{MeVfm}^3$ was used in our previous studies [1, 10] and found to give reasonable spreading widths for giant resonances in stable nuclei[10]. The EWSR values depleted below 15MeV are shown by circles in Fig.5. The magnitude of the low-lying strength is increased and now becomes close to the experimental data except for ^{22}O. The fact that the low-lying $E1$ strength in ^{20}O is most increased with the use of the stronger residual interaction may be explained by the large fragmentation of the GDR strength.

The TDDM results for ^{24}O shown in Fig.5 significantly differ from the shell-model

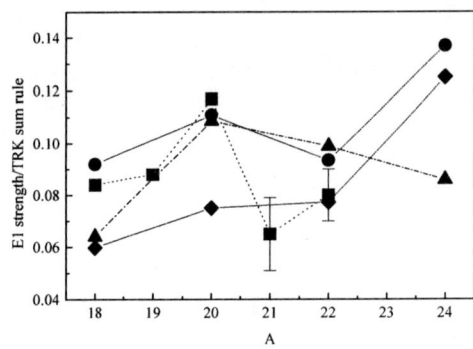

FIGURE 5. Isotope dependence of the low-lying $E1$ strength. Square, triangle, rhombus, and circle indicate the experiment[9], the shell-model calculation[11], and the TDDM results with $v_0 = -230\text{MeVfm}^3$ and -330MeVfm^3, respectively.

calculation[11]. The difference originates in the fact that the shell-model calculation gives much smaller spreading width to the GDR in ^{24}O than the TDDM calculations do. Since a large amount of the $E1$ strength is concentrated in the GDR, the low-lying $E1$ strength calculated in the shell model becomes small in ^{24}O.

HEAD-ON COLLISIONS OF $^{16}\text{O} + {}^{16}\text{O}$

It is well-known that in TDHF, colliding nuclei do not fuse in a low L region when incident energy is higher than a certain threshold value[12]. Experimental search for non-fusion events in nearly central collisions has been done but no evidence for the low L non-fusion window has been found. This suggests that TDHF does not give sufficient dissipation of translational energy. Two different studies have shown that improvement of conventional TDHF calculations dramatically increases the threshold energy in $^{16}\text{O} + {}^{16}\text{O}$: one is work done by Umar et al.[13]. They included spin-orbit force in the mean-field potential which had been neglected in the conventional TDHF calculations and found that the center-of-mass threshold energy E_{th} is increased from 31MeV to 68MeV for the Skyrme II force. The other is a calculation done by Tohyama[14], in which the effects of nucleon-nucleon collisions were taken into account using TDDM, and it was found that E_{th} increases from 27MeV to 72MeV for the BKN force. So far, no extended TDHF calculations which include both spin-orbit fore and the effects of nucleon-nucleon collisions have been reported. So, we performed such calculations using TDDM. The results are summarized in table 1. In these calculations, the Skyrme II force was used as the effective interaction, the strength of the residual interaction was chosen to be $v_0 = -350\text{MeVfm}^3$ and the HF ground state was used as the initial ground state for simplicity. As seen from table 1, an increase in E_{th} due to the effects of nucleon-nucleon collisions is rather small (from 69MeV to 80MeV) when spin-orbit force is included in

TABLE 1. Threshold energy E_{th} in the center-of-mass frame for the head-on collisions of ^{16}O + ^{16}O. Fusion occurs below E_{th}.

Method	$E_{th}[MeV]$
TDHF without $\vec{\ell} \cdot \vec{s}$	30
TDDM without $\vec{\ell} \cdot \vec{s}$	66
TDHF with $\vec{\ell} \cdot \vec{s}$	69
TDDM with $\vec{\ell} \cdot \vec{s}$	80

the mean-field potential.

SUMMARY

We applied TDDM to two nuclear phenomena: one is the isovector dipole resonances in the oxygen isotopes $^{18\sim24}O$ and the other is the fusion reactions of ^{16}O + ^{16}O. The calculations of the dipole resonances are much more improved and quantitative than the previous ones[1] in two aspects: the inclusion of spin-orbit force in the mean-field potential and the use of the correlated ground state. The results were compared with recent experimental data and also with a shell-model calculation. It was found that if the strength of the residual interaction is appropriately chosen, TDDM approximately reproduces the observed isotope dependence of the low-lying $E1$ strength. In the case of the fusion reactions of ^{16}O + ^{16}O, it was found that an increase in the threshold energy E_{th} due to nucleon-nucleon collisions is rather small when spin-orbit force is included in the mean-field potential.

REFERENCES

1. M. Tohyama, Phys. Lett. **B323** (1994) 257.
2. M. Tohyama, Nucl. Phys. **A657** (1999) 343; M. Tohyama, Phys. Lett. **B484** (2000) 231.
3. M. Gong and M. Tohyama, Z. Phys. **A335** (1990) 153.
4. M. Tohyama, Prog. Theor. Phys. **94** (1995) 147.
5. A. S. Umar et al., Phys. Rev. **C40** (1989) 706.
6. S. J. Wang and W. Cassing, Ann. Phys. **159** (1985) 328.
7. M. Tohyama and M. Gong, Z. Phys. **A332** (1989) 269.
8. J. Sawicki, Phys. Rev. **126** (1962) 2231; J. Da Providencia, Nucl. Phys. **61** (1965) 87; C. Yannouleas, Phys. Rev. **C35** (1987) 1159; S. Drożdż et al., Phys. Rep. **197** (1990) 1.
9. T. Aumann et al., Nucl. Phys. **A687** (2001) 103c.
10. M. Tohyama, Prog. Theor. Phys. **99** (1998) 109.
11. H. Sagawa and T. Suzuki, Phys. Rev. **C59** (1999) 3116.
12. P. Bonche et al., Phys. Rev. **C17** (1978) 1700.
13. A. S. Umar et al., Phys. Rev. Lett. **56** (1986) 2793.
14. M. Tohyama, Phys. Rev. **C36** (1987) 187.

PART V

DAMPING PHENOMENA
IN NUCLEAR SYSTEMS

Correlation and Fluctuation Measures for Damped Collective Motion

M. Matsuo

Graduate School of Science and Technology, Niigata University, Niigata 950-2181, Niigata, Japan

Abstract. Strength functions for damped collective motion sometimes exhibit fluctuation that is called fine structure. Using two examples, i.e. the damped collective rotation of deformed compound nuclei and the giant resonances, we discuss statistical measures that characterize the fine structures, and demonstrate how such measures tell us important information on the microscopic mechanism of the damping phenomena.

INTRODUCTION

Collective motion in nuclei often exhibits damping properties as is typically known in the case of the giant resonances [1]. One of the most basic measure of the damped collective motion is the total damping width, which is the FWHM or corresponding quantity of the strength function or the cross section. However, the damping mechanism of the nuclear collective motion in many cases does not arise from a single origin, but it rather involves multiple or multistep processes. Therefore, if one can characterize the strength functions in terms of various measures besides the total width, such information will help us to understand mechanism of the damping. Here we note that the strength functions sometimes exhibit fluctuation or fine structure. An experimentally known example is the high precision electron scattering measurement of the isoscalar giant quadrupole resonance (GQR) in ^{208}Pb [2].

It has been established recently that the collective rotation of deformed nuclei becomes damped motion [3, 4, 5]. The damped rotation occurs for the rapidly rotating deformed compound states which have large angular momentum and internal excitation energy (the relative energy above the yrast line) higher than $U \sim 1$ MeV, whereas the conventional mode of rotation, i.e. the rotation characterized by spectra with rotational band structure, is seen in the region near the yrast line. For the damped rotation, the E2 transition strength from a compound state is spread over many final states with different gamma-ray energies. In this case, the FWHM of the E2 strength distribution is called the rotational damping width Γ_{rot}, and one can describe the E2 transitions in terms of the strength function. It is possible to describe miscroscopically the rotating compound states and the damped rotational transitions using a cranked shell model, which we have developed recently [6]. The model agrees quite well with the observed properties of the quasicontinuum gamma-rays spectra from rotating compound states [6, 7]. The model study furthre predicts a new feature that the E2 transition strength function exhibit significant fine structure, which has not been recognized before.

CP597, *Nonequilibrium and Nonlinear Dynamics in Nuclear and Other Finite Systems,*
edited by Z. Li et al.
© 2001 American Institute of Physics 0-7354-0041-5/01/$18.00

We discuss in this paper the fine structures associated with the two kinds of damped collective motion mentioned above. In either case, one should have proper methods to probe the fine structures. In the case of the damped rotation, we show that correlation among two or three gamma rays emitted from the rotating compound states provides us a useful probe. For the giant resonances, we introduce a new statistical measure called the local scaling dimension.

DAMPED ROTATIONAL MOTION

Compound formation and rotational damping

The damped rotational motion is associated with formation of compound states at high internal excitation energy. Let us first specify the characteristics of the compound states from a microscopic view point.

The rotational band structures observed near the yrast line in rapidly rotating deformed nuclei are described well in terms of the cranked mean-field model. In this model, the single-particle orbits of the cranked mean-field potential are defined in the rotating frame as a function of the rotational frequency ω_{rot}, which is approximately proportional to the angular momentum I. An internal state of the rotating nucleus is specified by a many-particle many-hole (np-nh) configuration or many-quasiparticle configuration defined for these single-particle orbits. Thus, to each rotational band, a np-nh configurations can be assigned. Smooth change of the internal structure as a function of the rotational frequency ω_{rot} is taken into account by the cranking term. We denote the cranked shell model configuration $|\mu(I)\rangle$. As the internal excitation energy increases, the density of np-nh configurations increases exponentially. In such situation, the residual nuclear effective interaction is no more negligible, since it causes mixing among the np-nh configurations. Consequently, the energy eigenstates denoted $|\alpha(I)\rangle$ becomes an admixture of many $|\mu(I)\rangle$ states. If the configuration mixing is complex enough, the energy eigenstates can be regarded as the Bohr's compound states.

We can describe the formation of compound states in terms of "μ-state strength function"

$$S_\mu(E) = \sum_\alpha |\langle \alpha(I) | \mu(I) \rangle|^2 \delta(E - (E_\alpha - \bar{E}_\mu)),$$

which represents how a μ-state component is spread over the energy eigenstates α's. Examples, calculated by means with the cranked shell model, are shown in Fig.1(a). The FWHM of the μ-state strength function may be called the "compound damping width" since its reciprocal $\tau = \hbar/\Gamma_\mu$ represents a equilibration time of a μ-state resolving to compound states. Note here that the strength function of individual μ-state is not a smooth function as shown in Fig.1(a), but rather fluctuate strongly from state to state. In order to extract the compound damping width Γ_μ, a technique is neccesary. The technique is to use the auto-correlation function. Namely, we calculate the auto-correlation function $C_\mu(e) = \int S_\mu(E+e)S_\mu(E)dE$ of individual μ-state strength function, and then calculate the average auto-correlation function from a set of many μ states in an energy range. The compound damping width Γ_μ is defined by a half of the FWHM

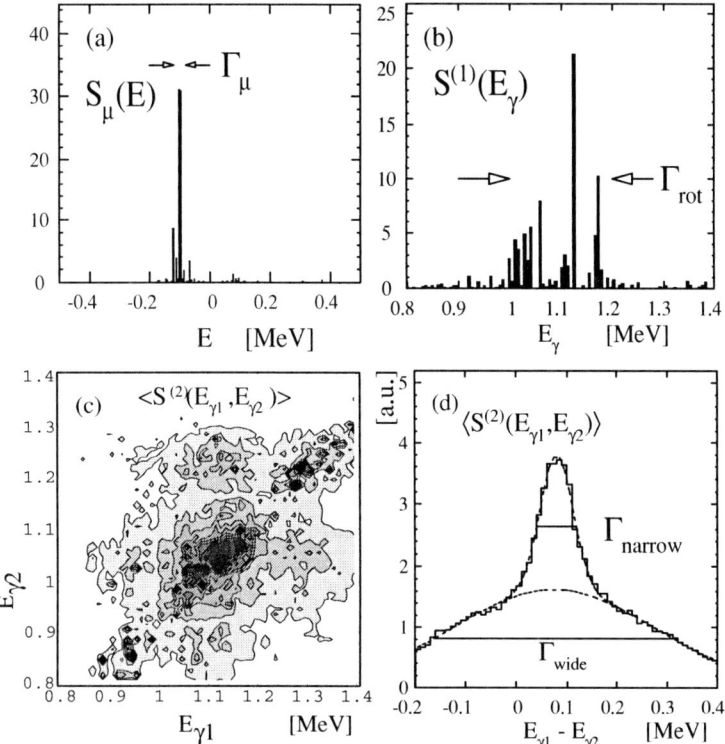

FIGURE 1. Calculated examples of (a) the μ-state strength function $S_\mu(E)$, (b) the "one-step" E2 strength function $S^{(1)}(E_\gamma)$, and (c,d) the "two-step" E2 strength function $S^{(2)}(E_{\gamma 1}, E_{\gamma 2})$ [8]. The two-step strength function plotted is obtained with averaging over 50 initial levels. (d) Projection of the two-step strength function on the $E_{\gamma 1} - E_{\gamma 2}$ axis.

of the average auto-correlation function. The extracted value of Γ_μ is $\Gamma_\mu \sim 40$ keV for the excitation energy $U = 1 - 2$ MeV, and it is approximately independent of spin. Note however that the μ-state strength functions are not experimentally accessible since there is no way to excite pure np-nh states.

$\gamma - \gamma$ correlation as a probe of compound damping width

As the compound states are formed, damping of rotational motion sets in. Fig.1(b) shows the E2 transition strength $S_{\alpha I, \beta I-2}$ for gamma decays from a compound state α (at spin I) to compound states $\beta's$ at spin $I - 2$. Corresponding E2 strength function is given by

$$S^{(1)}(E_\gamma) = \sum_\beta S_{\alpha I, \beta I - 2} \delta(E_\gamma - (E_{\alpha I} - E_{\beta I - 2})),$$

which we call the "one-step" E2 strength function. The E2 strength is strongly fragmented. The fragmentation implies the damping of rotational motion. The damping width Γ_{rot} of the rotational motion is corresponds to the FWHM of the E2 strength function. However, it should be noted again that the strength distribution is not smooth, but there is significant fluctuation or fine structure.

In the experiment, one observe only the strength function averaged over initial compound states α in a certain energy and spin range. This washes out the fine structures. Furthermore, since the gamma-rays from states in wide spin range are summed up in the experiment, it is difficult to extract the rotational damping width Γ_{rot} just by looking the averaged E2 strength function $\langle S^{(1)}(E_\gamma)\rangle$.

One can overcome this difficulty. It is indeed possible to probe the fine structure by look at "correlation" between two gamma-rays emmitted sequentially from I to $I - 2$, and $I - 2$ to $I - 4$. Let us show this with use of the microscopic cranked shell model calculation. Two consecutive transitions are described by the "two step" E2 strength function

$$S^{(2)}(E_{\gamma 1}, E_{\gamma 2}) = \sum_{\beta\beta'} S_{\alpha I, \beta I-2} S_{\beta I-2, \beta' I-4} \delta(E_{\gamma 1} - (E_{\alpha I} - E_{\beta I-2})) \delta(E_{\gamma 2} - (E_{\beta I-2} - E_{\beta' I-4})),$$

for the consecutive transitions $I \rightarrow I - 2 \rightarrow I - 4$. Calculated example is shown in Fig.1(c). It is seen clearly that the distribution is not a simple convolution of two Lorentzian (which may become an uncorrelated broad distribution in the $E_{\gamma 1} \times E_{\gamma 2}$ plane), but rather there is a strongly correlated component which is concentrated along the diagonal line $E_{\gamma 1} = E_{\gamma 2} + 4/J_{eff}$ (J_{eff} being the effective moment of inertia). The correlated component has narrow width Γ_{narrow} with respect to the difference $E_{\gamma 1} - E_{\gamma 2}$ of the two transitions, as seen in Fig.1(d). The narrow correlated part survives even after averaging over many initial states, and is clearly separated from the uncorrelated part.

The origin of the two-component structure in the $E_{\gamma 1} - E_{\gamma 2}$ correlation is as follows. If the compound damping width is smaller than the rotational damping width $\Gamma_\mu < \Gamma_{rot}$, the np-nh basis bands (μ-states) contained in a compound state $|\alpha(I)\rangle$ play a role of doorway states for the E2 decay from this state to $I - 2$ states, as illustrated in Fig.2. Since the compound damping width Γ_μ of these doorway states is smaller than Γ_{rot}, the E2 strength function exhibits the fine structures that reflect the doorway states. The doorway states are further carried over to the next transition $I - 2 \rightarrow I - 4$. Since the doorway states, i.e. the np-nh basis bands, have the rotational correlation $E_{\gamma 1} = E_{\gamma 2} + 4/J_{eff}$, the fine structures originating from the doorway states is not very different between the transitions $I \rightarrow I - 2$ and those $I - 2 \rightarrow I - 4$. Thus there remains strong correlation, which is diffused only by

$$\Gamma_{narrow} \approx 2\Gamma_\mu.$$

We have indeed confirmed that the relation holds generally at least in the case of the microscopic cranked shell model [8]. We can also prove that the width Γ_{wide} of the wide component of the "two step" strength function, on the other hand, is related to the rotational damping width as $\Gamma_{wide} \approx 2\Gamma_{rot}$.

The above relation thus allows us to measure the compound damping width Γ_μ if the narrow component can be identified in the experimental $\gamma - \gamma$ correlation spectra. This provides us with a novel tool to probe the compound damping width in rotating nuclei.

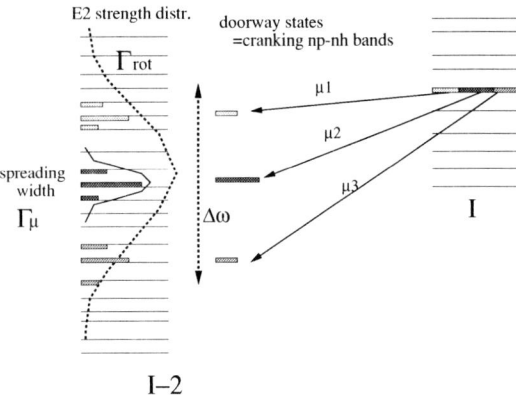

FIGURE 2. The mechanism of rotational damping in the case of $\Gamma_\mu \ll \Gamma_{rot}$.

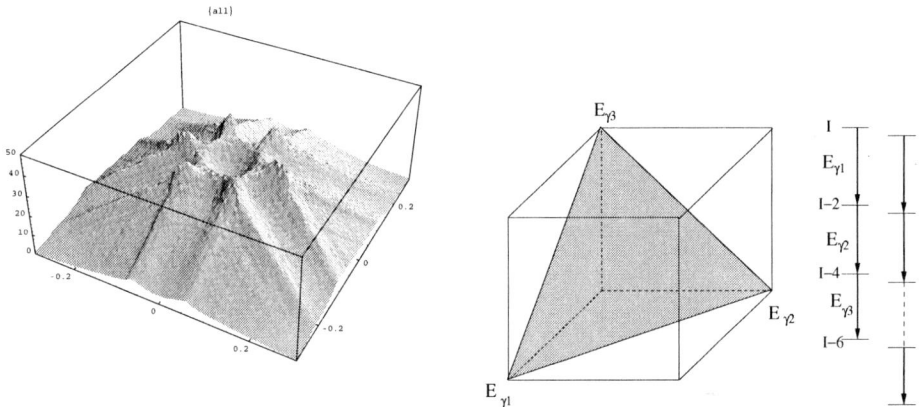

FIGURE 3. Calculated $\gamma-\gamma-\gamma$ correlation spectra (left) projected on the transversal plane (right) of the triple-gamma spectra represented as a cube.

$\gamma-\gamma-\gamma$ **correlation**

Experimental efforts to locate directly the narrow component are now under progress [9, 10]. One of the promising ways is to use the three-gamma spectra, which is available these days with high statistics. The "transversal plane" (see Fig.3) in the three-dimensional cubic spectra recording three gamma-rays is a useful way to see correlation among three gamma's [11]. Microscopically calculated $\gamma-\gamma-\gamma$ spectra projected on the transversal plane exhibit a characteristic pattern looking like a "volcano", as shown in Fig.3. The "ridges" and the "crater rim" of the volcano arise from the narrow component. These prominent structures could be identified in the experiment and be used to quantify the compound damping width Γ_μ.

SCALING ANALYSIS OF FLUCTUATING STRENGTH FUNCTION

The damping of the giant resonance is often described by the doorway coupling picture. Here the collective 1p1h state couples at the first step to doorway states, which are 2p2h states or specific configurations of 2p2h states [1]. The total damping width is mostly governed by this doorway coupling. However, the damping processes go further since the doorway states couple in the next step to more complicated configurations such as 3p3h states. The fine structures, if observed and properly measured, will provide us with information on the second step.

We have proposed a method to characterize the fine structures in the strength function by means of a new statistical measure, called "local scaling dimension"[12]. This method is designed to extract characteristic energy scales involved in fluctuating strength functions, and is based on the following idea. Let us assume that the strength function is partitioned into energy bins with a bin size e. One can then evaluate fluctuation of the binned strength distribution, and ask how the fluctuation changes as a function of the bin size e, i.e. how is the scaling property of strength fluctuation. The idea is borrowed from the scaling analysis of multifractal phenomena [13, 14]. For the fractal cases, a power law scaling is satisfied for the binned strength fluctuations. On the other hand, we expect that the strength fluctuation in the nuclear damping processes has some characteristic energy scales such as the damping widths. Thus we do not assume the global power law scaling, but rather ask how the fluctuation scales locally around a given energy scale e. This idea leads to a new statistical measure, the local scaling dimension $D_m(e)$ [12], which we define as a function of the energy scale (the bin size) e by

$$D_m(e) \equiv \frac{1}{m-1} \frac{\partial \log \chi_m(e)}{\partial \log e}$$

in terms of the partition function $\chi_m(e)$, which is the m-th moment

$$\chi_m(e) \equiv \sum_i p_i^m$$

of the binned strength $\{p_i\}$, where p_i is the strengths in i-th energy bin (and normalized as $\sum_i p_i = 1$). The local scaling dimension $D_m(e)$ exhibits characteristic behaviors as a function of the bin size e for typical strength functions. For the strength fluctuation in the GOE random matrix model (i.e., the uncorrelated Porter-Thomas fluctuation together with the spectral rigidity), $D_m(e) \sim 0$ for $e \sim d$ where d is the average spacing. As e increases, $D_m(e)$ increases monotonically and reach $D_m(e) \sim 1$ for $e \sim 100d$. For the Lorentz-type smooth strength function with FWHM of Γ, the local scaling dimension behaves $D_m(e) \sim 1$ for $e << \Gamma$ and decreases sharply around $e \sim \Gamma$, and $D_m(e) \sim 0$ for $e > \Gamma$. For strength functions with fine structures, $D_m(e)$ exhibits multifold behavior, as shown in Fig.4. Here we adopt a schematic random matrix model which mimics the doorway damping processes. The model consists of a collective state, doorway states, and background states and interaction matrix elements given by Gaussian random variables. The doorway states have their own spreading width because of the coupling to the background states. The fine structure emerges when the spreading width of the

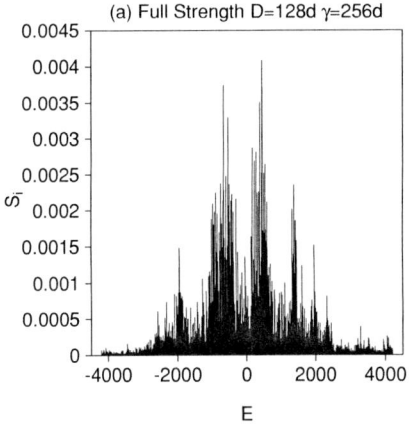
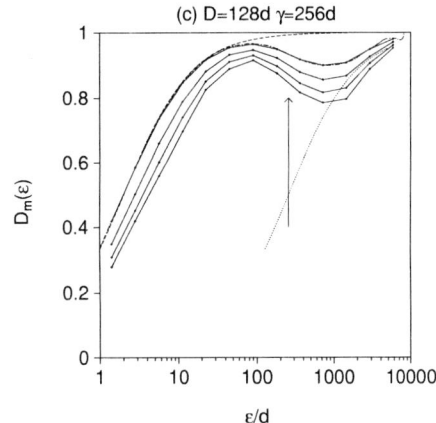

FIGURE 4. The strength function (left) and its local scaling dimension $D_m(e)$ (right) in the schematic doorway damping model [12]. The unit is arbitrary. The arrow points to the value of spreading width ($\gamma = 128d$) of the doorway states. The dashed curve shows the GOE limit.

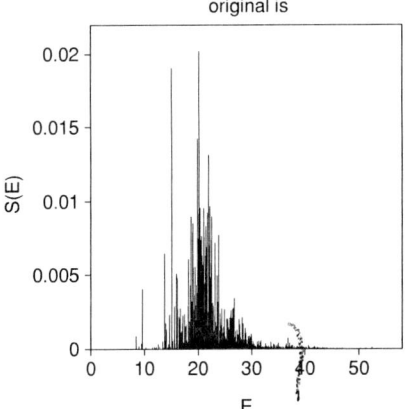
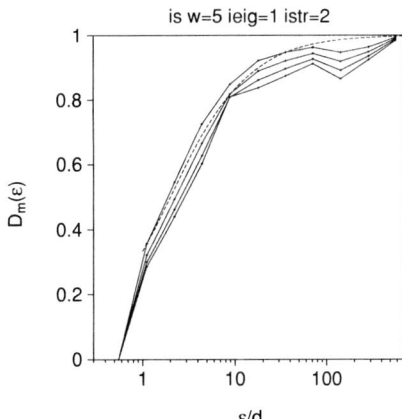

FIGURE 5. The strength function and its local scaling dimension $D_m(e)$ of the isoscalar quadrupole strength function in ^{40}Ca, calculated in the 1p1h+2p2h shell model. The energy unit is MeV.

doorway states γ is small (Fig.4). The local scaling dimension $D_m(e)$ for this strength function is shown in the right panel of Fig.4. For small e, $D_m(e)$ follows the GOE behavior, but starts to deviate at larger value of e. This deviation is due to the fine structure. In this model, the spreading width γ of the doorway states is known, and we found the energy scale e^* where $D_m(e)$ start to deviate from the GOE is related to γ as $e^* \sim \gamma/5$. The energy scale e^{**} where $D_m(e)$ decreases most steeply is turned out to satisfy another relation $e^{**} \sim \gamma$.

The method is applied to a more realistic model of the giant resonance. We use a spherical shell model consisting of the Woods-Saxon potential and the residual two-body interaction with the Landau-Migdal force[15]. All the 1p1h and 2p2h configurations with $I^\pi = 2^+$ up to 50 MeV are included as basis states and the shell model Hamiltonian is diagonalized. The strength function of the isoscalar quadrupole operator, shown in Fig.5(left), is treated with the scaling analysis discussed above. The local scaling dimension, shown in Fig.5(right), clearly points to the presence of the fine structure. Namely $D_m(e)$ follows the GOE curve upto the scale $e \sim 50d$ (d being the average level spacing), but starts to deviate at larger e. The energy scale with largest deviation $e \sim 100d$ correspond to 1.4 MeV, which is the energy scale of the fine structure in the microscopically calculated strength function. A similar scaling analysis that however uses different statistical measure has been proposed [16]. It is interesting to compare with this approach.

ACKNOWLEDGMENTS

This presentation is based on the collaborations with T.Døssing, B. Herskind, E. Vigezzi, S. Leoni, A. Bracco, and R.A. Broglia for the damped rotatinal motion, and with H. Aiba, S. Nishizaki, and T. Suzuki for the scaling analysis.

REFERENCES

1. G.F. Bertsch, P.F. Bortignon, and R.A. Broglia, Rev. Mod. Phys. **55**, 287 (1983).
2. G. Kilgus *et al.*, Z. Phys. **A326**, 41 (1987).
3. B.Lauritzen, T.Døssing and R.A.Broglia, Nucl. Phys. **A457**, 61 (1986).
4. B. Herskind, A. Bracco, R.A. Broglia, T. Døssing, A. Ikeda, S.Leoni, J. Lisle, M. Matsuo, and E. Vigezzi, Phys. Rev. Lett. **68**, 3008 (1992).
5. T. Døssing, B. Herskind, S. Leoni, A. Bracco, R.A. Broglia, M. Matsuo, and E. Vigezzi, Phys. Rep. **268**, 1 (1996).
6. M. Matsuo, T. Døssing, E. Vigezzi, R.A. Broglia, and K. Yoshida, Nucl. Phys. **A617** , 1 (1997).
7. A. Bracco et al., Phys. Rev. Lett. **76**, 4484 (1996).
8. M. Matsuo, T. Døssing, B. Herskind, S. Leoni, E. Vigezzi, R.A. Broglia, Phys. Lett. **B465**, 1 (1999).
9. T. Døssing, B. Herskind, M. Matsuo, S. Leoni, A. Bracco, E. Vigezzi, and R. A. Broglia, Nucl. Phys. **A682** , 439c (2001).
10. T. Døssing, B. Herskind, A. May, M. Matsuo, S. Leoni, A. Bracco, E. Vigezzi, and R. A. Broglia, preprint nucl-th/0104034.
11. B.Herskind, et al., Phys. Lett. **B276**, 4 (1992); Nucl. Phys. **A557**, 191c (1993).
12. H. Aiba and M. Matsuo, Phys. Rev. **C60**, 034307 (1999).
13. J. Feder, *Fractals* (Plenum Press, 1988).
14. H. Aiba and T. Suzuki, Phys. Lett. **A201**, 319 (1995).
15. S. Nishizaki, J. Wambach, Phys. Lett. **B349**, 7 (1995).
16. D. Lacroix et al. Phys. Lett. **B429**, 15 (2000); D. Lacroix and Ph. Chomaz, Phys. Rev. **C60**, 064307 (1999).

Coulomb Excitation of Double Giant Dipole Resonances

J. Z. Gu and H. A. Weidenmüller

Max-Planck-Institut für Kernphysik, Postfach 103980, D-69029 Heidelberg, Germany

Abstract. We implement the Brink–Axel hypothesis for the excitation of the double giant dipole resonance (DGDR): The background states which couple to the one–phonon giant dipole resonance are themselves capable of dipole absorption. These states (and the ones which couple to the two–phonon resonance) are described in terms of the Gaussian Orthogonal Ensemble of random matrices. We use second–order time–dependent perturbation theory and calculate analytically the ensemble–averaged cross section for excitation of the DGDR. Numerical calculations illuminate the mechanism and are specifically performed for the reaction ^{208}Pb + ^{208}Pb. We show that the contribution of the background states to the excitation of the DGDR is significant. At a projectile energy of 640 MeV/nucleon we find that the width of the DGDR, the energy–integrated cross section and the ratio of this quantity over the energy–integrated cross section for the single giant dipole resonance, all agree with experiment within experimental errors.

INTRODUCTION

The double giant dipole resonance (DGDR) has been observed in several nuclei: ^{136}Xe [1], ^{197}Au [2] and ^{208}Pb [3]. Compared to the predictions of the harmonic picture [4, 5], the measured cross sections for the DGDR excitation are found to be enhanced by factors ranging from 1.3 to 2 [1, 2, 3]. The definition of cross sections in the harmonic picture was first defined in [4] then modified in [5]. We shall give our definition for the cross sections later. At the same time the experimentally determined widths of the DGDR are close to $\sqrt{2}$ times the width of the GDR. These discrepancies between simple–minded theoretical predictions and experimental results have attracted much theoretical attention. Several mechanisms have been studied. We mention anharmonicities of the collective Hamiltonian and nonlinearities of the external field. Following several earlier papers, Carlson *et al.* [6] have recently discussed the discrepancy using the Brink–Axel hypothesis. This hypothesis states that a giant resonance is built on top of *every* excited nuclear state [7]. These authors considered the contribution to the cross section of DGDR excitation due to the background states which couple dynamically to the one–phonon state. It was found that this contribution is sizable.

This result is interesting and calls for further study, especially since the approach of Carlson *et al.* [6] uses approximations which are plausible but not based upon an expansion in terms of a small parameter. In the present paper, we apply essentially the same physical picture as Carlson *et al.* but use a formulation which allows us to derive the DGDR excitation cross section within perfectly controlled approximations. The resulting formula is subsequently evaluated numerically. Our approach makes it possible to clearly identify and calculate the modification of the DGDR absorption process due

CP597, *Nonequilibrium and Nonlinear Dynamics in Nuclear and Other Finite Systems,*
edited by Z. Li et al.
© 2001 American Institute of Physics 0-7354-0041-5/01/$18.00

to the Brink–Axel hypothesis.

HAMILTONIAN OF THE PROJECTILE

We consider the relativistic Coulomb excitation of the DGDR of the projectile in a collision with a target.

For the Hamiltonian of the projectile, we use the physical picture shown in Fig. 1. The one–phonon state $|10\rangle$ is the giant dipole mode of the ground state $|00\rangle$, and the two–phonon state $|20\rangle$ is the giant dipole mode of the one–phonon state. The one–phonon state $|10\rangle$ is dynamically coupled by the nuclear Hamiltonian to more complex particle–hole configurations $|0k\rangle$ with $k = 1,\ldots,K$ and $K \gg 1$. This coupling causes the giant dipole resonance to acquire a spreading width $\Gamma_{10}^{\downarrow}$. The two–phonon state $|20\rangle$ is likewise coupled to such configurations. In order to accommodate the Brink–Axel hypothesis, we group these configurations into two classes. States in the first class are denoted by $|1k\rangle$ with $k = 1,\ldots,K$. Each such state represents the dipole mode of the lower state $|0k\rangle$. States in the second class are denoted by $|0\alpha\rangle$ with $\alpha = 1,\ldots,M$ and $M \gg 1$. These latter states are not coupled by the dipole operator to lower–lying configurations. The states $|20\rangle$, $|1k\rangle$ and $|0\alpha\rangle$ are all dynamically coupled to each other.

According to this picture, the Hamiltonian matrix H has the form

$$
H = \begin{pmatrix}
E_0 & \phi & \phi & \phi & \phi & \phi \\
\phi & E_1 & V_{1k} & \phi & \phi & \phi \\
\phi & V_{l1} & H_{lk}^{(0)} & \phi & \phi & \phi \\
\phi & \phi & \phi & E_2 & V_{2k'} & V_{2\alpha} \\
\phi & \phi & \phi & V_{l'2} & H_{l'k'}^{(1)} & W_{l'\alpha} \\
\phi & \phi & \phi & V_{\beta 2} & W_{\beta k'} & \mathcal{H}_{\beta\alpha}^{(0)}
\end{pmatrix}. \tag{1}
$$

The matrix H consists of three diagonal blocks. The first block has dimension one and contains the energy E_0 of the ground state $|00\rangle$. The second block has dimension $1 + K$. It contains the unperturbed energy E_1 of the one–phonon state $|10\rangle$, the elements $H_{kl}^{(0)}$ with $k,l = 1,\ldots,K$ of the Hamiltonian matrix $H^{(0)}$ governing the states $|0k\rangle$, and the real coupling matrix elements V_{1k} connecting the one–phonon state with the states $|0k\rangle$. The third block has dimension $1 + K + M$. It contains the unperturbed energy E_2 of the two–phonon state $|20\rangle$, the Hamiltonian matrices $H_{kl}^{(1)}$ with $k,l = 1,\ldots,K$ and $\mathcal{H}_{\alpha\beta}^{(0)}$ with $\alpha,\beta = 1,\ldots,M$ governing the states $|1k\rangle$ and $|0\alpha\rangle$, respectively, and the real coupling matrix elements V_{2k}, $V_{2\alpha}$ and $W_{k\alpha}$ connecting the three sets of states.

We turn to the statistical assumptions. We describe the K states $|0k\rangle$ with $K \to \infty$ in terms of the Gaussian Orthogonal Ensemble of random matrices (GOE) [8, 9, 10], so that $H^{(0)}$ represents this ensemble. For excitation energies of the giant dipole resonance (which lie typically between 8 and 15 MeV), this description seems eminently reasonable, except perhaps for the lightest nuclei. The spreading width $\Gamma_{10}^{\downarrow} = 2\pi v^2/d$ of the giant dipole resonance is expressed in terms of the mean level spacing d of the states $|0k\rangle$ and of the mean square coupling matrix element $\overline{V_{1k}^2} = v^2$. The matrix $\mathcal{H}^{(0)}$ is also

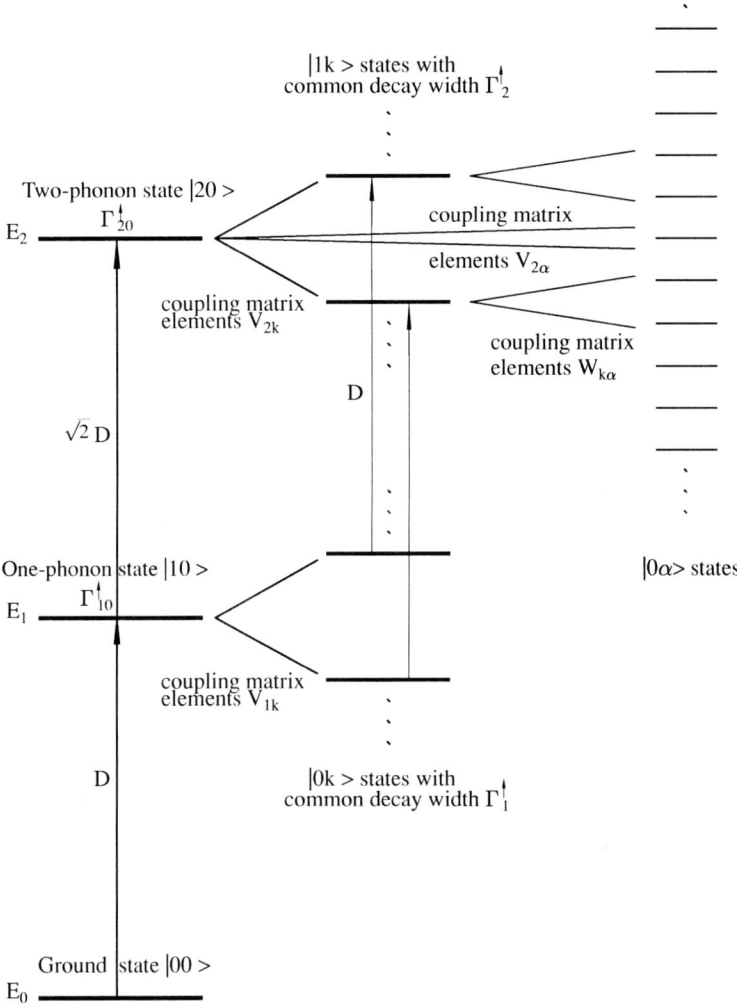

FIGURE 1. Schematic illustration of the DGDR excitation and the couplings between the phonon states and many-particle many-hole configurations. The level spacings of the configurations are exaggerated in the figure.

assumed to represent a GOE, with M taken to infinity. We assume that $\mathcal{H}^{(0)}$ and $H^{(0)}$ are uncorrelated.

The states in the second and third block can decay by particle emission. We take account of this fact by introducing the decay widths $\Gamma_{10}^{\uparrow}$ and $\Gamma_{20}^{\uparrow}$ of the one–phonon and the two–phonon states, respectively, and the decay widths $\Gamma_{1}^{\uparrow}$ and $\Gamma_{2}^{\uparrow}$ of the states $|0k\rangle$ and $|1k\rangle$, respectively. Within the statistical model, it is obviously adequate to assume that $\Gamma_1^\uparrow$ and $\Gamma_2^\uparrow$ are independent of the running index $k = 1,\ldots,K$. The width matrix Σ accordingly has the form

$$\Sigma = -(i/2)\begin{pmatrix} 0 & 0 & 0 & 0 & 0 & 0 \\ 0 & \Gamma_{10}^\uparrow & 0 & 0 & 0 & 0 \\ 0 & 0 & \delta_{lk}\Gamma_1^\uparrow & 0 & 0 & 0 \\ 0 & 0 & 0 & \Gamma_{20}^\uparrow & 0 & 0 \\ 0 & 0 & 0 & 0 & \delta_{l'k'}\Gamma_2^\uparrow & 0 \\ 0 & 0 & 0 & 0 & 0 & 0 \end{pmatrix}. \tag{2}$$

The effective Hamiltonian H_{eff} is then given by

$$H_{\text{eff}} = H + \Sigma. \tag{3}$$

All four decay widths can be calculated with the help of the optical model and the exciton model.

INTENSITIES AND CROSS SECTIONS FOR THE DGDR EXCITATION

We suppress the intrinsic structure of the target and replace it by a point source of the electromagnetic field. To describe the DGDR excitation of the projectile, we use the standard approach to Coulomb excitation [11]: The relative motion of projectile and target is described classically, and the intrinsic excitation of the projectile is treated quantum–mechanically.

We use second -order time-dependent perturbation theory and calculate the intensity for the DGDR excitation. The ensemble averaged total intensity is given by [12]

$$\overline{I_2(E)} = 2\pi i \left(\frac{D}{\hbar}\right)^4 \int_{-\infty}^{+\infty} dt_1 \int_{-\infty}^{0} d\tau_2 \int_{-\infty}^{+\infty} dt_1' \int_{-\infty}^{0} d\tau_2'$$

$$h(t_1)h(t_1+\tau_2)h^*(t_1')h^*(t_1'+\tau_2')\exp[i(E-E_0)(t_1-t_1')/\hbar]$$

$$\times \Big(f_0 \Theta(\tau_2 - \tau_2')\exp[-i(E_1 - E_0 + \tfrac{i}{2}\Gamma_{10})\tau_2'/\hbar]\exp[i(E_1 - E_0 + \tfrac{i}{2}\Gamma_{10} - i\Delta\rho)\tau_2/\hbar]$$

$$\times (f_1^+(E) - f_2^-(E))$$

$$+ f_0\Theta(\tau_2 - \tau_2')\exp[-i(E''-E_0+i\Delta\tau)\tau_2'/\hbar]\exp[i(E''-E_0+i\Delta\sigma)\tau_2/\hbar]$$

$$\times (f_2^+(E) - f_1^+(E))$$

$$- f_0\Theta(\tau_2' - \tau_2)\exp[-i(E''-E_0-i\Delta\sigma)\tau_2'/\hbar]\exp[i(E''-E_0-i\Delta\tau)\tau_2/\hbar]$$

$$\times (f_2^-(E) - f_1^-(E))$$

$$- f_0\Theta(\tau_2' - \tau_2)\exp[-i(E_1-E_0-\tfrac{i}{2}\Gamma_{10}+i\Delta\rho)\tau_2'/\hbar]\exp[i(E_1-E_0-\tfrac{i}{2}\Gamma_{10})\tau_2/\hbar]$$

$$\times (f_1^-(E) - f_2^+(E))$$

$$+ (\sqrt{2} - 1 + i\Delta\sigma_0 f_2^-(E))(-\overline{G_2^-(E'')})$$

$$+ (1 + i\Delta\tau_0 \overline{G_2^-(E'')}) f_1^-(E)) \exp[-i(E_1 - E_0 + \frac{i}{2}\Gamma_{10})\tau_2'/\hbar]$$

$$\times \exp[i(E'' - E_0 - i\Delta\tau)\tau_2/\hbar] + (1 - \sqrt{2} + i\Delta\sigma_0 f_2^+(E))(-\overline{G_2^+(E'')}$$

$$+ (1 - i\Delta\tau_0 \overline{G_2^+(E'')}) f_1^+(E)) \exp[-i(E'' - E_0 + \Delta\tau)\tau_2'/\hbar]$$

$$\times \exp[i(E_1 - E_0 - \frac{i}{2}\Gamma_{10})\tau_2/\hbar] - [(1 - \sqrt{2} + i\Delta\sigma_0 f_2^+(E))(1 - i\Delta\tau_0 \overline{G_2^+(E'')}) f_1^+(E)$$

$$+ (\sqrt{2} - 1 + i\Delta\sigma_0 f_2^-(E))(1 + i\Delta\tau_0 \overline{G_2^-(E'')}) f_1^-(E))$$

$$+ \Delta\rho_0 \frac{1}{\Gamma_{10} - \Delta\rho} (f_2^-(E) - f_2^+(E)) + (\sqrt{2} - 1)(\overline{G_2^-(E'')} - \overline{G_2^+(E'')})]$$

$$\times \exp[-i(E_1 - E_0 + \frac{i}{2}\Gamma_{10})\tau_2'/\hbar] \exp[i(E_1 - E_0 - \frac{i}{2}\Gamma_{10})\tau_2/\hbar] \Big), \qquad (4)$$

with

$$\overline{G_2^\pm(E)} = \frac{1}{E - E_1 \pm \frac{i}{2}\Gamma_{20}}, \quad f_1^\pm(E) = \frac{1}{E - E_2 \mp \frac{i}{2}\Gamma_{10} \pm i\Delta\tau},$$

$$f_2^\pm(E) = \frac{1}{E - E_2 \pm \frac{i}{2}\Gamma_{10} \pm i\Delta\sigma}, \quad f_0 = 1 - \frac{\Delta\rho_0}{\Gamma_{10} - \Delta\rho}. \qquad (5)$$

Here $\Gamma_{10} = \Gamma_{10}^\uparrow + \Gamma_{10}^\downarrow$ and $\Gamma_{20} = \Gamma_{20}^\uparrow + \Gamma_{20}^\downarrow + \Gamma_{10}^\downarrow$. $\Gamma_{20}^\downarrow$ is the spreading width of the two-phonon state due to the coupling to the states $|0\alpha\rangle$. Eq. (4) constitutes the main result of the theoretical part of this paper. Under perfectly controlled approximations, we have derived an expression for the intensity which embodies the Brink–Axel hypothesis and which describes the formation of the DGDR as a transport process.

The intraband intensity $\overline{I_2^{\text{intra}}(E)}$ which would result if only the ground state and the one–phonon state could absorb dipole radiation. In other words, we suppress dipole absorption by the states labelled $|0k\rangle$, although we do keep the dynamical coupling of all the states as described by the matrix H_{eff} defined in Eq. (3). We do so in order to distinguish the dipole excitation taken into account by the usual approach to the problem, from the transport process described by Eq. (4). The ensemble average is

$$\overline{I_2^{\text{intra}}(E)} = 2\pi \times 2 \times \left(\frac{D}{\hbar}\right)^4 \int_{-\infty}^{+\infty} dt_1 \int_{-\infty}^{0} d\tau_2 \int_{-\infty}^{+\infty} dt_1' \int_{-\infty}^{0} d\tau_2' \, h(t_1) h(t_1 + \tau_2)$$

$$\times h^*(t_1') h^*(t_1' + \tau_2') \exp[i(E - E_0)(t_1 - t_1')/\hbar] \frac{\Gamma_{20}}{(E - E_2)^2 + \frac{1}{4}\Gamma_{20}^2}$$

$$\times \exp\{i(E_1 - E_0 - i/2\Gamma_{10})\tau_2/\hbar\} \exp\{-i(E_1 - E_0 + i/2\Gamma_{10})\tau_2'/\hbar\}. \qquad (6)$$

The factor 2 arises because the dipole transition $|10\rangle \to |20\rangle$ is twice as strong as the transition $|00\rangle \to |10\rangle$. When Γ_{10} and Γ_{20} go to zero, $\overline{I_2^{\text{intra}}(E)}$ is denoted by $I_2^{\text{har}}(E)$, which defines our harmonic limit.

The cross section for the DGDR excitation is defined by

$$\sigma_2(E) = 2\pi \int_{b_{\min}}^{\infty} b \, db \, \overline{I_2(E)}. \qquad (7)$$

281

Here b_{min} is the minimal impact parameter which is introduced in order to account for the strong absorption that occurs as soon as the colliding nuclei come within reach of their nuclear forces. We also define

$$\sigma_2(E)^{intra} = 2\pi \int_{b_{min}}^{\infty} b\, db\, \overline{I_2^{intra}(E)} \tag{8}$$

and

$$\sigma_2^{har}(E) = 2\pi \int_{b_{min}}^{\infty} b\, db\, I_2^{har}(E). \tag{9}$$

NUMERICAL RESULTS

For the calculation of the decay widths $\Gamma_{10}^{\uparrow}, \Gamma_1^{\uparrow}, \Gamma_{20}^{\uparrow}$ and $\Gamma_2^{\uparrow}$, we have used the computer code of global optical-model potential developed by E. Sheldon and V. C. Rogers [13] and the exciton model [14].

In the long–wavelength approximation [15] where the impact parameter b is large compared to the nuclear radius r, the time–dependent function $h(t)$ is given by

$$h(t) = \frac{\gamma}{b^2} \frac{1+\gamma\tau}{(1+\tau^2)^{3/2}} - i\, \frac{\gamma v \omega}{b} \frac{1}{(1+\tau^2)^{1/2}}. \tag{10}$$

Here $\tau = \gamma v t/b$ with $\gamma = 1/\sqrt{(1-v^2/c^2)}$, v is the relative velocity, and $\omega = (E_f - E_i)/\hbar$.

We consider the DGDR excitation of a ^{208}Pb projectile, incident on a ^{208}Pb target. We find for the decay widths the values $\Gamma_{10}^{\uparrow} = 0.11$ MeV, $\Gamma_{20}^{\uparrow} = 0.026$ MeV, $\Gamma_1^{\uparrow} = 0.30$ MeV and $\Gamma_2^{\uparrow} = 0.16$ MeV. We set $E_1 - E_0$ and $E_2 - E_0$ equal to their experimental values 13.5 MeV and 27.0 MeV, respectively. The parameter ω appearing in the parametrization (10) of $h(t)$ was accordingly chosen as $\hbar\omega = 13.5$ MeV.

Fig. 2 shows the dependence of the cross sections on projectile energy. σ_2, σ_2^{intra}, σ_2^{har} and σ_2^{extra} are the total, intraband, harmonic limit and extraband energy–integrated cross sections, respectively. σ_2^{extra} is the difference of the total and intraband cross sections. All of the cross sections increase monotonically with E_p/A. The total cross section largely stems from intraband excitation. The extraband transition contributes not more than 10% compared with the harmonic limit. For $E_p/A < 420$ MeV, σ_2^{intra} is larger than σ_2^{har}. This indicates the damping enhancement [12] which occurs at lower projectile energies where the extraband contribution is negligible. For very low projectile energy, the ratio of σ_2^{intra} to σ_2^{har} can reach several hundred percent. This results from the damping enhancement.

We turn to a comparison of our results with experimental data [3]. In keeping with experimental results [16], we use 122% of the sum rule for all dipole matrix elements. Fig. 3(a) shows the differential cross section for the DGDR excitation as a function of excitation energy. The cross section has been normalized to its maximum value. The width of the DGDR is 6.1 MeV. Integrating the differential cross section over excitation energy from neutron threshold at 7.5 MeV up to 40.0 MeV (this corresponds to the experimental situation), we find the value 410 mb. This value is somewhat larger than the experimental one, 380 ± 40 mb [3]. Both our DGDR width and energy–integrated cross section are consistent, however, with the experimental values within experimental

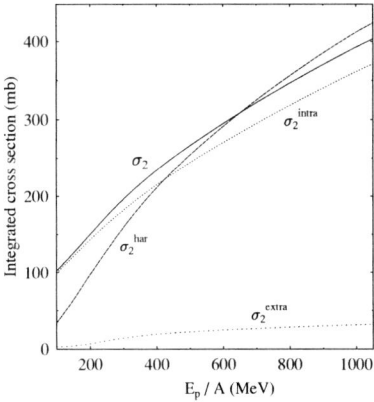

FIGURE 2. Energy–integrated cross sections versus E_p/A. Parameter values are $\Gamma_{10}^{\uparrow} = 0.11$, $\Gamma_{20}^{\uparrow} = 0.026$, $\Gamma_{10}^{\downarrow} = 4.0$, $\Gamma_{1}^{\uparrow} = 0.30$, $\Gamma_{2}^{\uparrow} = 0.16$, $\Gamma_{20}^{\downarrow} = 0.5\Gamma_{10}^{\downarrow}$ and $\Gamma_{2}^{\downarrow} = 0.5$ (MeV).

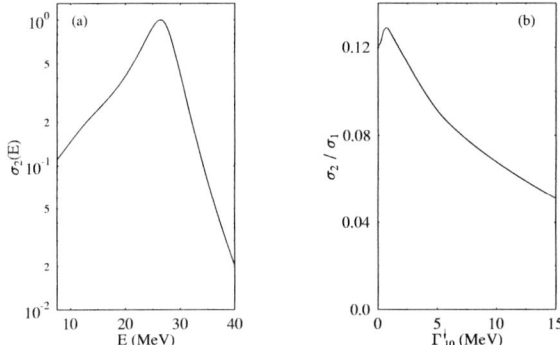

FIGURE 3. (a)Differential cross section for the DGDR excitation versus excitation energy with $\Gamma_{10}^{\downarrow}=4.0$, (b) ratio of energy–integrated cross sections for the DGDR and GDR excitation versus spreading width $\Gamma_{10}^{\downarrow}$. Parameter values are $b_{min} = 15.5$ fm, $E_p/A = 640$, $\Gamma_{10}^{\uparrow} = 0.11$, $\Gamma_{20}^{\uparrow} = 0.026$, $\Gamma_{1}^{\uparrow} = 0.30$, $\Gamma_{2}^{\uparrow} = 0.16$, $\Gamma_{20}^{\downarrow} = 0.5\Gamma_{10}^{\downarrow}$ and $\Gamma_{2}^{\downarrow} = 0.5$ (MeV).

errors. In the present case, the enhancement factor R_1 is 1.03. In Fig. 3(b), we plot the ratio of the energy–integrated cross sections for the DGDR and the GDR as a function of $\Gamma_{10}^{\downarrow}$. The ratio first increases and then decreases as this parameter increases. At the experimental value $\Gamma_{10}^{\downarrow} = 4.0$ MeV, the ratio is 0.0988, which again agrees with experiment (0.11 $\pm$ 0.013 [3]) within the experimental error.

283

SUMMARY

We have studied the DGDR excitation using the Brink–Axel mechanism. The background states which couple to the one–phonon and two–phonon states are described in terms of the Gaussian Orthogonal Ensemble of random matrices. We use second–order time–dependent perturbation theory and calculate analytically the ensemble–averaged cross section for the DGDR excitation. This quantity is a function of the various decay widths and spreading widths of the one–phonon, two–phonon, and background states. The decay widths have been calculated from the optical model and the exciton model. The spreading widths are taken from experiment or used as parameters. We have shown that for realistic values of these parameters, the contribution to the cross section from the Brink–Axel mechanism is significant. This is especially true for the reaction ^{208}Pb + ^{208}Pb at projectile energy 640 MeV/nucleon. For sensible values of the various spreading widths, we find that the width of the DGDR, the value of the energy–integrated cross section, and the ratio of this quantity over the energy–integrated cross section for single GDR excitation, all agree with experiment within experimental errors. We take this as a strong indication that the present approach accounts quantitatively for the DGDR excitation. It would not be difficult to use our Eq. (4) for the analysis of other data sets. Clearly, the formalism can be extended to triple–phonon excitation.

ACKNOWLEDGMENTS

We are grateful to A. Bulgac, B. Carlson, L. Canto, H. Emling, M. Hussein, C. Lewenkopf, and J. Li for discussions, and to H. Emling for very helpful suggestions.

REFERENCES

1. R. Schmidt et al., Phys. Rev. Lett. **70** (1993) 1767.
2. T. Aumann et al., Phys. Rev. C **47** (1993) 1728.
3. K. Boretzky et al., Phys. Lett. B **384** (1996) 30.
4. G. Baur and C. A. Bertulani, Phys. Lett. B **174** (1986) 23.
5. H. Emling, Prog. Part. Nucl. Phys. **33** (1994) 729.
6. B. V. Carlson et al., Ann. Phys. (N. Y.) **276** (1999) 111.
7. D. Brink, D. Phil. thesis, Oxford University (unpublished), 1955; P. Axel, Phys. Rev. **126** (1962) 671.
8. T. Guhr, A. Müller-Groeling and H. A. Weidenmüller, Phys. Rep. **299** (1998) 189.
9. J. J. M. Verbaarschot, H. A. Weidenmüller and M. R. Zirnbauer, Phys. Rep. **129** (1985) 367.
10. J. Z. Gu and H. A. Weidenmüller, Nucl. Phys. A **660** (1999) 197.
11. K. Alder and A. Winther, Coulomb Excitation, Academic Press (New York) 1965.
12. J. Z. Gu and H. A. Weidenmüller, Nucl. Phys. A **690** (2001) 382.
13. E. Sheldon and V. C. Rogers, Computer Phys. Commun. **6** (1973) 99.
14. G. Mantzouranis, H. A. Weidenmüller and D. Agassi, Z. Phys. A **276** (1976) 145.
15. C. A. Bertulani and V. Zelevinsky, Nucl. Phys. A **568** (1994) 931.
16. D. Schelhaas et al., Nucl. Phys. A **489** (1988) 189.

Phase transitions above and along the yrast line in ^{154}Dy

W.C. Ma [1]

Department of Physics, Mississippi State University, Mississippi State, MS 39762, USA

Abstract. Spectra of the E2 quasicontinuum γ rays feeding different spin regions of the ^{154}Dy yrast line have been extracted. These are compared with corresponding theoretical spectra obtained by numerical simulations based on temperature-dependent Hartree-Fock theory, with thermal shape fluctuations. In this manner, different regions of the spin-energy plane can be examined. The results support the predictions of a smeared-out phase transition at high spin above the yrast line. Along the yrast line band terminations were identified in the spin range $I = 36 - 48\hbar$, and found to compete with collective rotational bands up to the highest observed spins. The data can be understood within the framework of configuration-dependent cranked Nilsson-Strutinsky calculations without pairing.

INTRODUCTION

Mean-field theory suggests [1, 2, 3] that nuclei can undergo phase transitions. An especially interesting example is the transitional nucleus ^{154}Dy, which is predicted [4] to change from a collective to an oblate aligned-particle structure with increasing spin and temperature, as sketched in Fig. 1. In a mesoscopic system, such as a nucleus, thermal shape fluctuations are expected to smear the phase transition and a question is whether remnant signatures of the phase transition persist at high excitation energies. A sudden change from collective to aligned-particle configurations at spin 34$\hbar$ has indeed been confirmed along the zero-temperature yrast line in ^{154}Dy [5]. However, in excited states, where thermal fluctuations are dominant even at moderate temperature, it remains a challenge to obtain unambiguous signatures for a phase transition. Recently we carried out an investigation [6] and obtained clear signatures for phase transitions in both experimental and theoretical spectra.

Calculations [7] based on results of mean-field theory and incorporating fluctuations have suggested that the spectrum of quasicontinuum (QC) collective electric quadrupole (E2) transitions can provide such a signature. Specifically, an E2 spectrum consisting of a unique two-peak feature was predicted [7] for ^{154}Dy, when the gamma cascade straddles the two regions on either side of the phase-transition boundary. In contrast, only one broad QC peak is detected in the vast majority of nuclei [8]. This two-peak feature in the QC E2 spectrum has been, in fact, previously observed [9]. The calculations of ref. [7] suggest that, while fluctuations indeed smear out the phase transition, they also play a critical role in providing an observable signature via the E2 spectrum. That is because

---

[1] E-mail: mawc@ra.msstate.edu.

CP597, *Nonequilibrium and Nonlinear Dynamics in Nuclear and Other Finite Systems,*
edited by Z. Li et al.

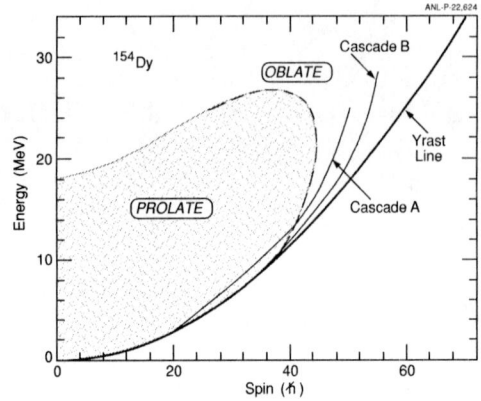

FIGURE 1. Theoretical yrast line and regions of prolate (dotted) and oblate phases in ^{154}Dy. The phase boundary (dashed line) corresponds to the $\gamma = -60°$ line in finite-temperature Hartree-Fock-Bogoliubov calculations without fluctuations [4]. Sketches of two cascade paths (A,B) are shown, which connect the experimental entry and exit points for cascades feeding into two selected regions of the yrast line, I=16-22$\hbar$ and I=34-36$\hbar$.

no collective E2 spectrum would normally be expected in the oblate phase; it arises only because thermal fluctuations cause admixtures of collective (prolate and triaxial) shapes. This has been amply demonstrated in ref. [7], where the calculated E2 spectra reproduce the different features observed [9, 10] in ^{152,154,156}Dy. A stringent test of the calculations consists of proving that the different E2 peaks indeed arise from the two phase regions, as was done in our recent study [6].

INTEGRAL ANALYSIS OF THE QC GAMMA RAYS

The hot excited states in nuclei remain a little explored frontier. This is partly because there has been no clear-cut way to specify the spin-excitation energy being studied. We have developed a new, but simple, way to provide this specification. By demanding that cascades feed the yrast line at a particular spin, the precursor QC spectrum then arises from states of higher spins. This can be simply accomplished by setting coincidence gates on specific yrast transitions. For example, cascade A in Fig. 1, which feeds into a low-spin region of the yrast line, can be selected; in this case we predict two E2 peaks, one from each phase region. On the other hand, cascade B, selected by demanding entry into the yrast line at high spin, picks out cascades which traverse only the nominally oblate region; only the upper E2 peak is then expected. Furthermore, the average initial point of the cascade is also deduced from the resulting QC spectrum. It is this knowledge of *both the initial and final points* that allows the decay pathways to be constrained. A second critical requirement of our method is to directly compare, under *identical gating conditions*, the experimental and theoretical spectra, the latter obtained as described in ref. [7].

The experiment was performed at the Lawrence Berkeley 88" Cyclotron. Excited states in ^{154}Dy were populated via the ^{36}S(^{122}Sn, 4n) reaction with a 165-MeV beam. The target consisted of a stack of three 350 μg/cm^2 self-supporting foils. The γ rays were detected with the early-implementation phase of the Gammasphere spectrometer [11], which consisted of 36 Compton-suppressed Ge detectors at that time. A total of 1.3 x 10^9 events was collected, with a requirement of $\geq$ 3 suppressed Ge detectors in prompt coincidence.

Spectra were obtained for ^{154}Dy at three detector angles (forward, 90°, backward), coincident with selected pairs of yrast transitions. This was achieved by first constructing $\gamma - \gamma$ matrices after coincidence gates were set on low-spin yrast transitions from $I^\pi = 2^+$ to 8^+. One-dimensional spectra were then obtained from the γγ matrices by selecting energies on the x-axis corresponding to yrast transitions of different spins. Several spectra originating from states with similar spins were summed to increase statistics.

The spectra were corrected for neutron interaction, coincidence summing, detector response (unfolding) and photopeak efficiency. Discrete peaks (originating from the yrast region) were removed to obtain the QC portion of the γ-ray spectra (from excited states above the yrast region) and statistical γ rays were subtracted. The dipole and quadrupole components were then separated, based on the angular distribution coefficients. Each spectrum was normalized so that the ground-state transition has unit intensity; the integral then gives the multiplicity. From the multiplicity and average energy of each component, the total spins and energies removed by all γ rays entering each selected spin region could be deduced, with the assumption that 0.5, 1.0, and 2.0$\hbar$ angular momenta were removed by statistical, dipole, and quadrupole transitions, respectively. The average entry point for the total ^{154}Dy channel was thus found to be 56.1 $\hbar$ and 31 MeV. Table I summarizes the multiplicities for the E2 QC components feeding different spin regions. More details for data analysis can be found in ref. [6].

The spectra feeding the different regions of the yrast line are given in Fig. 2(a), where they are compared to equivalent theoretical spectra. A distinct and unusual feature is the occurrence of two broad peaks. The spectrum of E2 transitions from excited states of a

TABLE 1. Theoretical and experimental quasicontinuum E2 multiplicities for different gates. The numbers in the first row denote the selected initial spins of the gating transitions. The uncertainty in the experimental multiplicity is 10%, obtained by adding in quadrature the uncertainties in each step of the analysis.

Gates	2-10	12-14	16-24	26-32	34-38	40-46
Exp.	10.7	10.2	9.0	7.7	6.9	6.4
Theor.	9.8	9.8	9.2	7.3	5.9	4.6

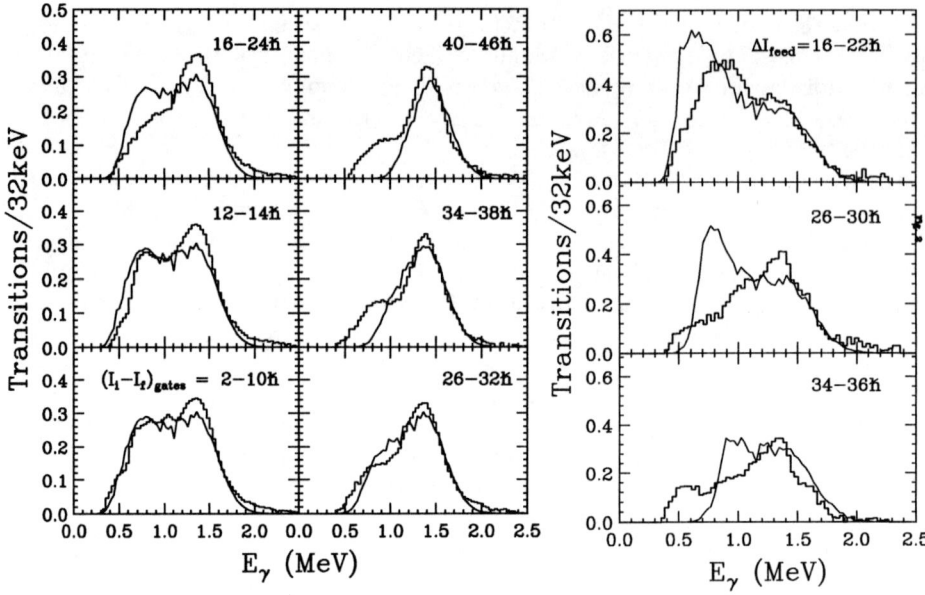

FIGURE 2. (a) Left: Integral spectra of the quasicontinuum E2 transitions feeding into and above the indicated spin intervals. Histograms represent experimental results, solid lines the theoretical calculations. The labels $(I_i - I_f)_{gates}$ denote the spin interval of the transitions used for coincidence gates. (b) Right: Differential E2 spectra feeding into the yrast line *only* in the indicated spin region ΔI_{feed}. Histograms and solid lines correspond to experiment and theory. The approximate decay pathways corresponding to the top and bottom spectra are shown as Cascades A and B, respectively, in Fig. 1.

nucleus of approximately fixed deformation normally consists of a single broad peak [8], since the transition energy, $E_\gamma = (2I - 1)\hbar^2/\mathcal{J}$, grows monotonically with spin I. ($\mathcal{J}$ is the moment of inertia.) Hence, the two-peak spectra provide a clear signal of a deviation from constant deformation.

To quantitatively understand the origin of the two peaks, we have performed theoretical calculations to simulate the QC decay of this nucleus by using a Monte Carlo method [7]. The collective E2 strengths, which carry information about the structure and phase transition, are microscopically calculated as a function of spin I and excitation energy E from finite-temperature Hartree-Fock theory, with the standard set of Pairing-Plus-Quadrupole force parameters and configuration space [12]. In order to reproduce the features of a finite system, we have to go beyond mean field theory by including fluctuations in the shape degrees of freedom [4].

The main characteristic of the calculation is that the number of adjustable parameters is kept to a minimum (see ref. [6] for more detail). The γ cascade is followed until it reaches a specified energy, U_{cut}, above the yrast line; the reduced level density at lower energy will lead to discrete transitions. The procedure produces spectra that are equivalent to the experimental ones, where the discrete lines are removed. The cut-off is chosen to mimic the top of the pair gap; U_{cut} has a value of 1.5 MeV at I = $0\hbar$ and linearly decreases to a value of 0.4 MeV for I = $30\hbar$, remaining constant thereafter.

Fig. 2(a) shows the experimental and theoretical QC E2 spectra from coincidence gates from different spin regions. The spectra are normalized such that the area under each spectrum gives the total multiplicity of the associated γ rays. As can be seen in Table I, when the gating spin increases, the QC E2 multiplicity decreases. In the low spin region, an almost constant E2 multiplicity was obtained since most feeding cascades have already entered the yrast line at higher spins. In the high-spin region the QC E2 multiplicity decreases since the cascades are shorter as they are "forced" into the yrast line at higher spin. Two peaks, centered around 0.8 and 1.4 MeV, are prominent in the spectra from low-spin gates – as found in a previous study [9]. Since the lowest gate, $2 - 10\hbar$, (see Fig. 2(a)) collects all the cascades, this spectrum is similar to that in Fig. 2(a) of ref. [9], which was obtained under similar conditions. (Small differences may be attributed to differences in the beam energy.)

The two-peak feature indicates a redistribution of γ-ray energies along the deexcitation pathways. In the later decay stages, when the cascades approach the yrast line at medium and low spins, the moment of inertia increases and transition energies shift downwards, causing the clustering around 800 keV and the dip around 1.1 MeV. In the spectra gated at higher spins, contributions of cascades traversing the low-spin prolate region of the I-E plane are eliminated, thus reducing the strength of the low-energy component. This demonstrates that the low-energy component originates mainly from regions of lower spins. The high-energy peak begins to dominate over the low-energy one for cascades feeding regions as low as $16 - 24\hbar$ and, in the theoretical spectra, becomes almost the only component for feeding above $32\hbar$. This clearly identifies the region $I > 32\hbar$ as the one with the smaller average moment of inertia.

DIFFERENTIAL ANALYSIS OF THE QC GAMMA RAYS

The spectra in Fig. 2(a) might be termed integral spectra. Each spectrum, which is obtained from coincidence gates on yrast transitions of specific spin, selects all QC cascades that feed into the yrast line above that spin. As the spin threshold increases, the QC pathways become more narrowly defined in spin, thus better specifying the I-E region being studied. However, as the threshold spin is lowered, the selectivity of such integral spectra decreases since all contributions from higher spins persist. It is possible to emphasize the particular QC cascades from a lower-spin region, by using differential spectra that feed *only* a narrow spin interval. For example, the QC spectrum feeding into the yrast states with only $I = 16 - 22\hbar$ can be extracted by subtracting the spectra gated on the transitions depopulating the 16^+ and 24^+ levels (after correct normalization) – see Fig. 2(b). (The result is reliable only when the γ intensities and the differences are large.) The low-energy E2 component is now significantly stronger than in the integral $I = 16 - 24\hbar$ spectrum (Fig. 2(a)). Again this emphasizes that this component must arise from the low-spin region. In contrast, in the differential spectrum feeding the spin interval 26-30$\hbar$, this low-energy component is markedly smaller, indicating that the γ cascade crosses the phase-transition boundary around spin 34-36$\hbar$. Although the trend

289

of a low-energy component that diminishes with yrast entry spin is reproduced by theory, discrepancies in details indicate that the theoretical phase boundary is at higher spin than found in the experiment, so that the γ cascade crosses it at $I \sim 40\hbar$ – also see Fig. 1.

PHASE TRANSITION ALONG THE YRAST LINE

A sudden onset of collectivity occurs at neutron number N=90 in the Dy isotopes. Transitional nuclei with N=88 straddle the well deformed region (N$\geq$90, e.g., ^{156}Dy) and the single-particle domain (N$\leq$86, e.g., ^{152}Dy), and exhibit properties of both classes of nuclei. In ^{154}Dy the structure changes rapidly with spin I along yrast line [5]. Normal prolate deformation ($\beta_2 \sim 0.2$) is seen at low spins. A pronounced reduction in the transition quadrupole moment from $\sim$200 to $\sim$1 W. u. occurs at spin 34 $\hbar$, as the prolate rotational band terminates and the oblate aligned particle configurations become yrast, signifying a phase transition along the yrast line. In order to study the band terminating structures [13, 14] at higher spins, a new experiment was performed at the 88" Cyclotron at the Lawrence Berkeley National Laboratory [15]. Excited states in ^{154}Dy were populated via the ^{36}S(^{122}Sn, 4n) reaction at 165 MeV. The target consisted of a stack of three 330 μg/cm^2 isotopically enriched, self-supporting foils. Decay γ rays were detected with the Gammasphere spectrometer, which consisted of 103 Compton-suppressed Ge detectors. A total of 1.5 x 10^9 events was collected with five or more suppressed Ge detectors required to fire in prompt coincidence. After a detailed off-line data analysis, most bands established earlier [5] have been extended to higher spins. Among them band $(+,0)_1$ was extended from 32 to 46 and band $(-,1)_1$ from 33 to 51. Several new bands have been observed for the first time. A partial level scheme from this study is shown in Fig. 3(a).

Band terminating structures were observed in both positive and negative parity configurations. One of the prominent level sequences of positive parity is band $(+,0)_3$ which collects a large fraction of the flux in the spin region above 36$\hbar$. Remarkably, the γ-ray intensity drops quickly at the highest spins: while the 1126 keV, 44$^+\rightarrow$42$^+$ transition carries 9.5% of the intensity of the 2$^+\rightarrow$0$^+$ ground state transition, the 1311- and 1410 keV γ rays located just above it drop to 3.2% and 0.4%, respectively. This is due to the large fragmentation of the decay pathways that is noticeable along this sequence. Nevertheless, the band has now been observed up to its termination point.

Theoretical calculations within the configuration-dependent cranked Nilsson-Strutinsky (CNS) approach [13, 16] have been used to predict many of the band structures discovered in the present work. The calculated configurations are labeled as $[p_1p_2(p_3p_4), (n_0)n_1n_2]$, where p_1 is the number of proton holes of $(g_{7/2}d_{5/2})$ character, p_2 the number of $h_{11/2}$ protons, n_1 is the number of $h_{11/2}$ neutron holes and n_2 the number of $i_{13/2}$ neutrons. Furthermore, the labels in parenthesis, given only when they are different from zero, denote the number of $(h_{9/2}f_{7/2})$ (p_3) and $i_{13/2}$ (p_4) protons, and $N = 4$ neutron holes (n_0), respectively. The number of $(d_{3/2}s_{1/2})$ protons and $(h_{9/2}f_{7/2})$ neutrons is not given explicitly, but can be determined from the total number of nucleons. Some calculated configurations, as well as the experimental ones, were presented in Fig. 3(b). Excellent agreement between the calculations and the experimental results

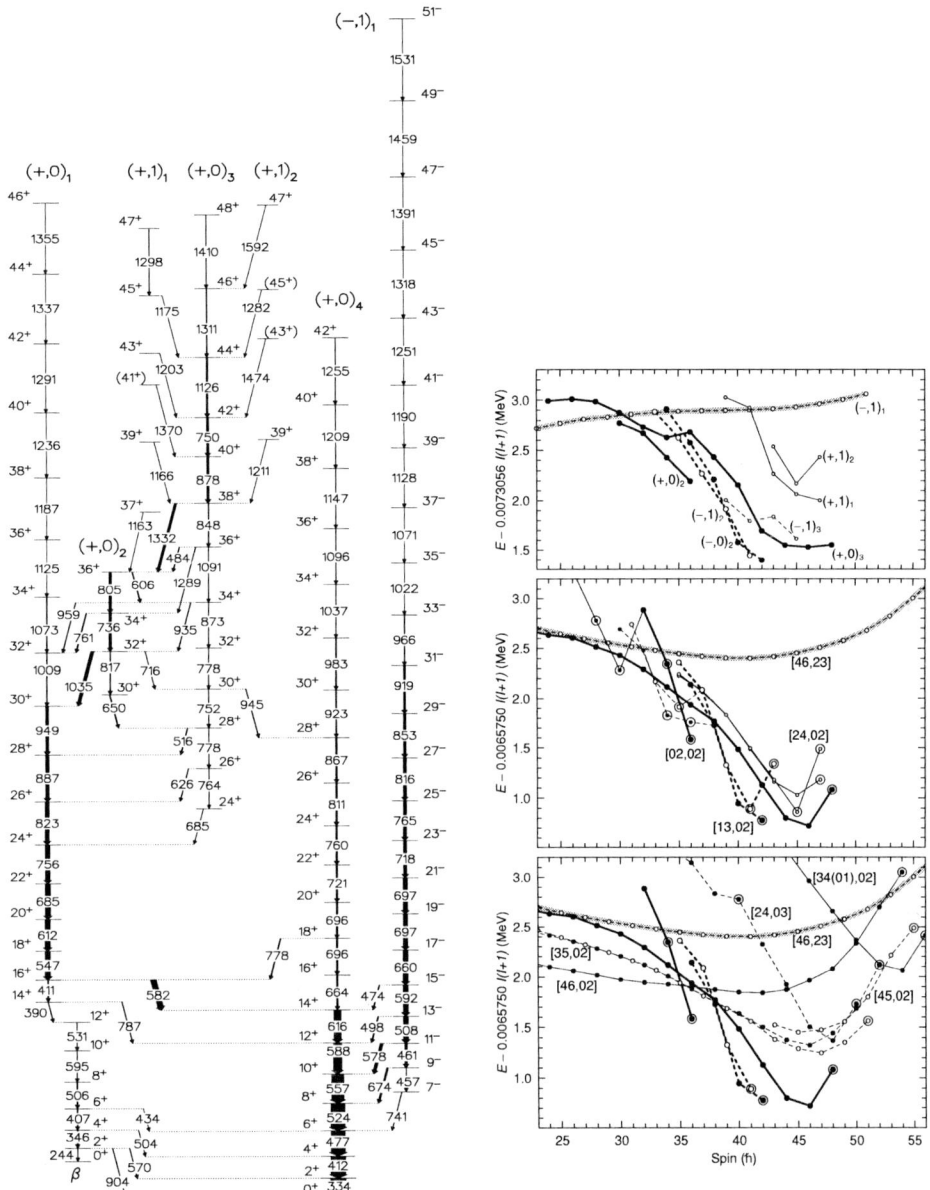

FIGURE 3. (a) Left: A partial level scheme of ^{154}Dy from recent studies [15]. The γ-ray intensities are indicated by the arrow widths. (b) Right: Observed terminating bands of ^{154}Dy in the upper panel with their calculated counterparts in the middle panel. Note that the [24,02] label refers to the three bands terminating at 47^+ and 48^+. In the lower panel, some configurations predicted to terminate in the $I = 50 - 55$ range are added and the collective $(-,1)_1$ band and its theoretical counterpart are drawn in this figure (thick shaded line) to facilitate the comparison between the two types of configurations [15].

291

was achieved for most configurations.

CONCLUSIONS

In conclusion, nuclear phase transitions above and along the yrast line of ^{154}Dy were studied. *both* experimental and theoretical quasicontinuum spectra exhibit two broad peaks, with pronounced changes in relative strengths when going from low- to high-spin gates. The varying strengths provide an incisive probe of the change in nuclear structure in the two zones of the I-E plane. The low-energy E2 component originates from the lower spin region, with prolate deformation. The high-energy component is largely from the nominally oblate region of the phase diagram, where strong fluctuations give rise to E2 transitions. Despite the fluctuations present in a finite system, signatures of a phase transition are seen to clearly persist in both the experimental and theoretical spectra.

Along the yrast line, high spin states up to $51\hbar$ have been established. Some of the bands were found to maintain their smooth, collective rotational behavior up to the highest spins, while others were shown to terminate in the $I = 36 - 48$ spin range. Configuration-dependent cranked Nilsson-Strutinsky calculations are able to account for most of the observations. With an increased experimental sensitivity, very rich structures of interacting bands have been revealed in the unpaired, very high-spin ($I \geq 50$) regime.

ACKNOWLEDGMENTS

The author would like to thank all collaborators in the research projects discussed. The work at Mississippi State University was supported by US Department of Energy under grant number DE-FG02-95ER40939.

REFERENCES

1. J.L. Egido *et al.*, Phys. Lett. B **178**, 139 (1986).
2. A.L. Goodman, Phys. Rev. C **39**, 2478 (1989).
3. Y.Alhassid, B. Bush, and S. Levit, Phys. Rev. Lett. **61**, 1926 (1988); Y.Alhassid and B. Bush, Nucl. Phys. **A509**, 461 (1990).
4. V. Martin and J.L. Egido, Phys. Rev. C **51**, 3084 (1995).
5. W.C. Ma *et al.*., Phys. Rev. Lett. **61**, 46 (1988).
6. W.C. Ma *et al.*., Phys. Rev. Lett. **84**, 5967 (2000).
7. V. Martin *et al.*, Phys. Rev. C **51**, 3096 (1995).
8. R. M. Diamond and F. S. Stephens, Ann. Rev. Nucl. Part. Sci. **30**, 85 (1980).
9. R. Holzmann *et al.*, Phys. Rev. Lett. **62**, 520 (1989).
10. R. Holzmann *et al.*, Phys. Lett. B **195**, 321 (1987).
11. I.-Y. Lee, Nucl. Phys. **A520**, 641c (1990).
12. K. Kumar and M. Baranger, Nucl. Phys. **A110**, 529 (1968).
13. A.V. Afanasjev, D.B. Fossan, G.J. Lane and I. Ragnarsson, Phys. Rep. **322** 1 (1999).
14. T. Bengtsson and I. Ragnarsson, Phys. Scr. **T 5** 165 (1983).
15. W.C. Ma *et al.*, to be published.
16. T. Bengtsson and I. Ragnarsson, Nucl. Phys. **A436** 14 (1985).

Study of various resonances within the phonon damping model

Nguyen Dinh Dang

RI-beam factory project office, RIKEN,
2-1 Hirosawa, Wako, 351-0198 Saitama - Japan
and Institute for Nuclear Science and Technique,
Vietnam Atomic Energy Commission - Hanoi, Vietnam
E-mail: dang@rikaxp.riken.go.jp

Abstract. The main successes of the phonon damping model (PDM) are presented in the description of the damping of: 1) The giant dipole resonance (GDR) in highly excited nuclei, 2) The double giant dipole resonance (DGDR) and multiple phonon resonances, 3) The Gamow-Teller resonance (GTR), 4) The damping of pygmy dipole resonance (PDR) in neutron-rich nuclei. The analyses of the results of numerical calculations within the PDM will be discussed in comparison with the experimental systematics on i) the width and the shape of the GDR at finite temperature and angular momentum for tin isotopes, ii) the electromagnetic cross sections for ^{136}Xe and ^{208}Pb on a lead target at relativistic energies, iii) the strength function of GTR, and iv) the PDR in oxygen isotopes.

INTRODUCTION

Although giant resonance is a subject of more than 50 years old, it continues to amaze us with new surprises. Those are the studies of GDR in hot nuclei, the observation of the multiple-phonon giant resonance, the extraction of the high-lying tail of Gamow-Teller resonance (GTR), which recovers the missing part of the Ikeda sum rule, and the measurements of pygmy dipole resonance (PDR) in neutron-rich nuclei. In this talk I will show you how the features of these resonances, especially the PDR and GDR, can be well described within a simple approach called the phonon damping model (PDM), which has been proposed in 1998 ago by Akito Arima and myself[1].

QUASIPARTICLE REPRESENTATION OF THE PDM

The quasiparticle representation of the PDM Hamiltonian[1,2] is obtained by adding the superfluid pairing interaction and expressing the particle (p) and hole (h) creation and destruction operators, $a_s^\dagger$ and a_s ($s = p, h$), in terms of the quasiparticle operators, $\alpha_s^\dagger$ and α_s, using the Bogolyubov's canonical transformation. The equation for the propagation of the GDR phonon, which is damped due to coupling to quasiparticle field, is given below making use of the double-time Green's function method. The final equation of the Green function for the one-phonon propagation

CP597, *Nonequilibrium and Nonlinear Dynamics in Nuclear and Other Finite Systems,*
edited by Z. Li et al.
© 2001 American Institute of Physics 0-7354-0041-5/01/$18.00

has the form [3]

$$G_{\lambda i}(E) = \frac{1}{2\pi} \frac{1}{E - \omega_{\lambda i} - P_{\lambda i}(E)} \ , \tag{1}$$

where the explicit form of the polarization operator $P_{\lambda i}(E)$ is

$$P_{\lambda i}(E) = \frac{1}{\hat{\lambda}^2} \sum_{jj'} [f_{jj'}^{(\lambda)}]^2 \Big[\frac{(u_{jj'}^{(+)})^2(1 - n_j - n_{j'})(\epsilon_j + \epsilon_{j'})}{E^2 - (\epsilon_j + \epsilon_{j'})^2}$$

$$- \frac{(v_{jj'}^{(-)})^2(n_j - n_{j'})(\epsilon_j - \epsilon_{j'})}{E^2 - (\epsilon_j - \epsilon_{j'})^2} \Big]. \tag{2}$$

The presence of polarization operator (2) due to ph – phonon coupling in the PDM Hamiltonian, and the analytic property of the double-time Green's function allows the damping to be calculated in an explicit and microscopic way. Namely, the phonon damping $\gamma_{\lambda i}(\omega)$ (ω real) is obtained as the imaginary part of the analytic continuation of the polarization operator $P_{\lambda i}(E)$ into the complex energy plane $E = \omega \pm i\varepsilon$:

$$\gamma_{\lambda i}(\omega) = \frac{\pi}{2\hat{\lambda}^2} \sum_{jj'} [f_{jj'}^{(\lambda)}]^2 \Big\{ (u_{jj'}^{(+)})^2(1 - n_j - n_{j'})[\delta(E - \epsilon_j - \epsilon_{j'}) - \delta(E + \epsilon_j + \epsilon_{j'})] -$$

$$(v_{jj'}^{(-)})^2(n_j - n_{j'})[\delta(E - \epsilon_j + \epsilon_{j'}) - \delta(E + \epsilon_j - \epsilon_{j'})] \Big\}. \tag{3}$$

The energy $\bar{\omega}$ of giant resonance (damped collective phonon) is found as the pole of the Green's function (1):

$$\bar{\omega} - \omega_{\lambda i} - P_{\lambda i}(\bar{\omega}) = 0 \ . \tag{4}$$

The width Γ_λ of giant resonance is calculated as twice of the damping $\gamma_\lambda(\omega)$ at $\omega = \bar{\omega}$, i.e.

$$\Gamma_\lambda = 2\gamma_\lambda(\bar{\omega}), \tag{5}$$

where $\lambda = 1$ corresponds to the GDR. The line shape of the GDR is described by the strength function $S_{\mathrm{GDR}}(\omega)$, which is derived from the spectral intensity in the standard way using the analytic continuation of the Green function (1) and by expanding the polarization operator (2) around $\bar{\omega}$:

$$S_{\mathrm{GDR}}(\omega) = \frac{1}{\pi} \frac{\gamma_{\mathrm{GDR}}(\omega)}{(\omega - \bar{\omega})^2 + \gamma_{\mathrm{GDR}}^2(\omega)} \ . \tag{6}$$

The photoabsorption cross section $\sigma(E_\gamma)$ is calculated from the strength function $S_{\mathrm{GDR}}(E_\gamma)$ as

$$\sigma(E_\gamma) = c_1 S_{\mathrm{GDR}}(E_\gamma)E_\gamma \ , \tag{7}$$

where $E_\gamma \equiv \omega$ is used to denote the energy of γ-emission. The normalization factor c_1 is defined so that the total integrated photoabsorption cross section $\sigma = \int \sigma(E_\gamma) dE_\gamma$ satisfies the GDR sum rule $\mathrm{SR}_{\mathrm{GDR}}$, hence

$$c_1 = \mathrm{SR}_{\mathrm{GDR}} \Big/ \int_0^{E_{\max}} S_{\mathrm{GDR}}(E_\gamma) E_\gamma dE_\gamma \ . \tag{8}$$

In heavy nuclei with $A \geq 40$, the GDR exhausts the Thomas-Reich-Kuhn (TRK) sum rule $\mathrm{SR}_{\mathrm{GDR}} = \mathrm{TRK} \equiv 60\ NZ/A$ (MeV·mb) at the upper integration limit $E_{\max} \simeq 30$ MeV, and exceeds TRK ($\mathrm{SR}_{\mathrm{GDR}} > \mathrm{TRK}$) at $E_{\max} > 30$ MeV due to the contribution of exchange forces. In some light nuclei, such as ^{16}O, the observed photoabsorption cross section exhausts only around 60% of TRK up to $E_{\max} \simeq 30$ MeV[4]. The EM cross section σ_{EM} is calculated from the corresponding photoabsorption cross section $\sigma(E_\gamma)$ and the photon spectral function $N(E_\gamma)$ according to[5,6]. In the numerical results discussed below the GTR and PDR have been calculated including pairing. The results for the hot GDR and DGDR are presented neglecting pairing.

GDR IN HIGHLY EXCITED NUCLEI
AND MULTIPLE PHONON GDR

The PDM has been proved to be quite successful in the description of the width and the shape of the GDR as a function of temperature T[1,2,7] and angular momentum J[8]. An example is shown in Fig. 1. The PDM has resolved the long-standing

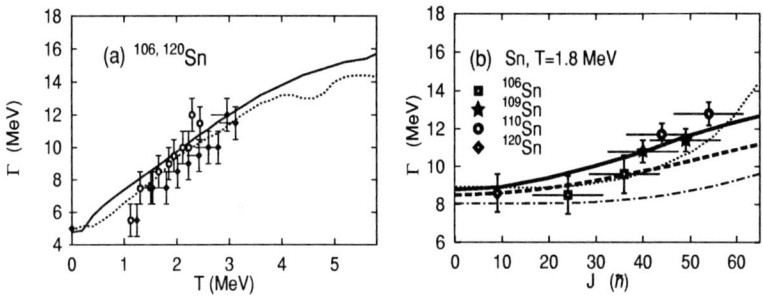

Figure 1: GDR width as a function of temperature T (a) and angular momentum J (b) in tin isotopes. In (a) the solid line is the width for ^{106}Sn, while the one for ^{120}Sn is given by the dotted line. In (b) the thick solid and dashed lines are results obtained within PDM for ^{106}Sn and ^{120}Sn, respectively, while the corresponding results within the thermal fluctuation model are given by dotted and dash-dotted lines, respectively. The data points in (a) and (b) are taken from[9] and[10], respectively.

problem with the electromagnetic (EM) cross sections of the DGDR in ^{136}Xe and

^{208}Pb, in which the prediction by the non-interacting phonon picture underestimated significantly the observed DGDR cross sections by the LAND collaboration. The prediction using the strength functions obtained within PDM [11] is given in Fig. 2 in comparison with the latest results of data analyses by LAND collaboration [12]. The agreement between the PDM prediction and the data is remarkable. The PDM also predicted the DGDR at non-zero temperature [13], which has been confirmed experimentally in a very recent work by Viesti et al [14].

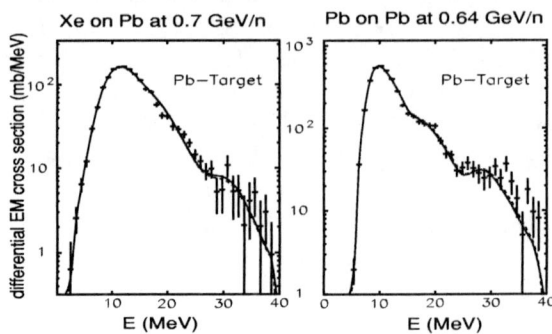

Figure 2: EM cross sections of GDR and DGDR for ^{136}Xe and ^{208}Pb. The solid lines are theoretical predictions, in which the DGDR strength functions within PDM are used. The data points are results of the LAND collaboration [12]. The dashed lines show the best fit using χ^2. The theoretical results have been folded with the detector response by K. Boretzky [12].

GAMOW-TELLER RESONANCE

Shown in Fig. 3 is the prediction of [15] within the PDM-1 [1] (solid line), and PDM-2 [2] (dashed line) for the GTR in ^{90}Nb in comparison with the result obtained within a microscopic theory which explicitly includes coupling to $2p2h$ configurations in terms of two-phonon configurations (dotted line) [16], and the experimental data (data points with errorbars) [17]. Again, the agreement between theory and experiment is quite reasonable.

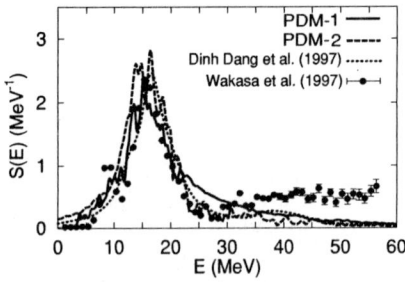

Figure 3: Strength functions of the GTR in 90Nb. See text for the notation.

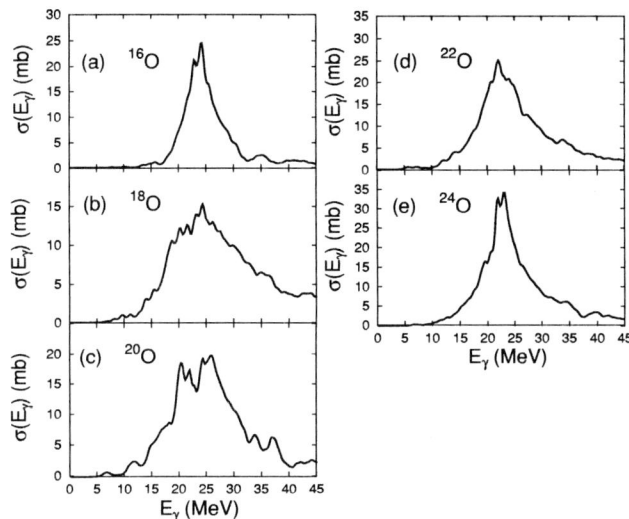

Figure 4: Photoabsorption cross sections for oxygen isotopes obtained within PDM.

Shown in Fig. 4 are the photoabsorption cross sections $\sigma(E_\gamma)$, which have been obtained within PDM for oxygen isotopes. The shapes of the photoabsorption cross section calculated for stable isotopes ^{16,18}O and ^{40,48}Ca are found in overall reasonable agreement with available experimental data as shown in the left panels of Fig. 5 ((a) - (d)). This agreement is better than those given by several other models shown in the right panels of Fig. 5 ((e) - (h)), namely the large-scale shell model (LSSM) [18] (thin solid and thin dashed lines in (e) and (f) with $T_< =$ 1 and $T_> = 2$ denoting two isospin components of GDR in ^{18}O), the surface coupling model (SCM) [19] (dotted lines in (e) and (g)), the second RPA (SRPA) [20] (thick dashed line in (g)), and a microscopic model including $1p1h\otimes$phonon plus continuum (phPC) [21] (dash-dotted lines in (g) and (h)). It is seen from Fig. 4 that the GDR becomes broader for isotopes with $N > Z$. Its width is particularly large for isotopes between the double closed shell ones, such as ^{18,20}O. The increase of GDR spreading enhances both of its low- and high-energy tails. In the region $E_\gamma \leq 15$ MeV, some weak structure of PDR is visible for ^{18,20,22}O. In the rest of isotopes under study, except for an extension of the GDR tail toward lower-energy, there is no visible structure of PDR.

The fractions of the EWS of strength exhausted by the low-energy tail of GDR are shown in Figs. 6. The trend obtained within PDM for oxygen isotopes reproduces the one observed in the recent experiments at GSI [22], which shows a clear deviation from the prediction by the cluster sum rule (CSR). The agreement

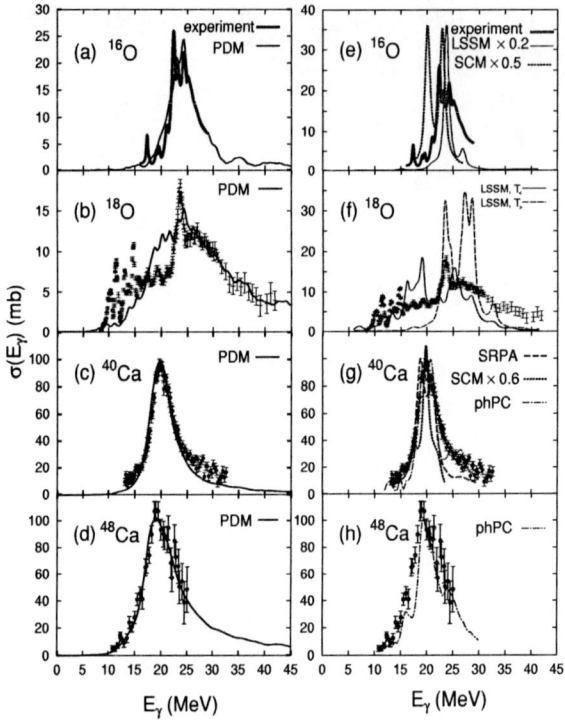

Figure 5: Photoabsorption cross sections for ^{16,18}O and ^{40,48}Ca obtained within PDM (left panels) and within other models mentioned in the text (right panels) in comparison with experimental data for ^{16}O (thick solid line in (a) and (e)), ^{18}O ((b) and (f)), ^{40}Ca ((c) and (g)), and ^{48}Ca ((d) and (h)).

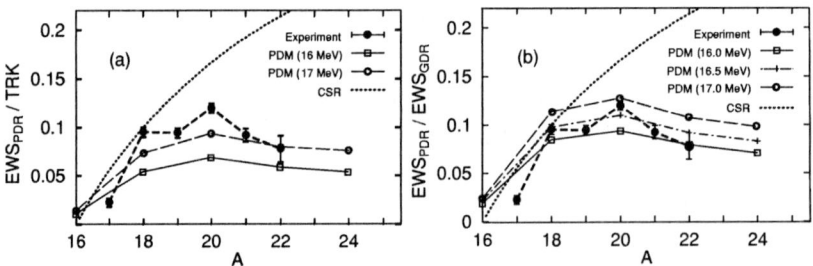

Figure 6: EWS of PDR strength up to excitation energy E_{max} for oxygen isotopes. Results obtained within PDM with $E_{max} = 16, 16.5$, and 17 MeV are displayed as open boxes connected with solid line, crosses connected with dash-dotted line, and open circles connected with thin dashed line, respectively. In (a) the PDM results are shown in units of TRK, while in (b) they are in units of the total GDR strength integrated up to 30 MeV. Experimental data (in units of TRK), obtained with $E_{max} = 15$ MeV, are shown by full circles connected with thick dashed line. The dotted line is the prediction by the cluster sum rule (CSR) (in units of TRK).

between the PDM prediction and the experimental data for the photoabsorption cross sections as well as for the EWS of PDR strength suggests that the mechanism of the damping of PDR is dictated by the coupling between the GDR phonon and noncollective *ph* excitations rather than by the oscillation of a collective neutron excess against the core. Strong pairing correlations also prevent the weakly bound neutrons to be decoupled from the rest of the system [23]. Only when the GDR is very collective so that it can be well separated from the neutron excess, the picture of PDR damping becomes closer to the prediction by the CM. The photon

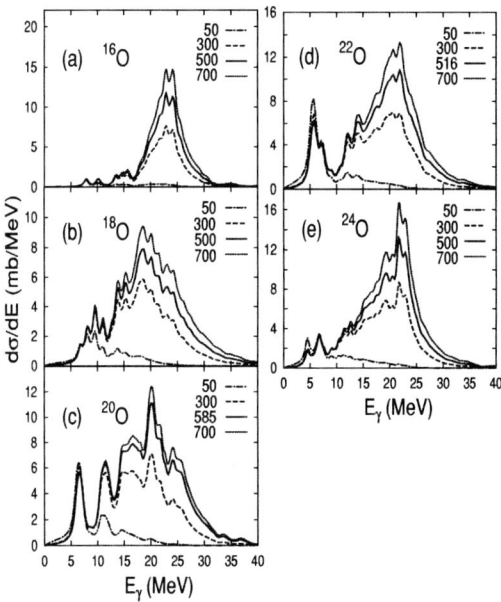

Figure 7: Electromagnetic cross sections of E1-excitations within PDM for oxygen isotopes on ^{208}Pb target. Different lines display results obtained at different projectile energies, whose values (in MeV/nucleon) are indicated in the panels.

spectral function $N(E_\gamma)$ in the EM differential cross section $d\sigma_{EM}/dE_\gamma$ contains an exponentially decreasing factor $e^{-m(b)}$ with increasing E_γ. This behavior enhances the low-energy part of the E1 strength in the EM differential cross section. These EM differential cross sections obtained within PDM are shown in Fig. 7 for oxygen isotopes. The PDR shows up in the EM cross sections of all neutron-rich isotopes, especially for ^{20,22}O, where a well isolated peak located at around 7 MeV is clearly seen. The PDR becomes depleted when the neutron number approaches a magic number (where the GDR is most collective) as can be seen in calcium isotopes when N increases from 20 to 28. Therefore, the decrease of PDR strength in ^{24}O can be understood as the depletion on the way toward the next magic number $N = 28$. Because of the behavior of $N(E_\gamma)$, the decrease of the projectile energy from

700 MeV/nucleon to 300 MeV/nucleon strongly reduces the EM cross section in the GDR region (above 10 MeV), but leaves the PDR almost intact, as can be seen in Fig. 7. A further decrease of projectile energy to 50 MeV/nucleon suppresses completely the GDR peak and even enhances the PDR peak in some very neutron-rich nuclei [20,22,24]O (dash-dotted lines in Fig. 7). This observation shows that, using high-intensity RI beams at low energies of around 50 – 60 MeV/nucleon, one may be able to observe a rather clean and sharp PDR peak in [20,22]O in the region around 7 MeV.

CONCLUSIONS

The PDM is a simple yet microscopic model, which can describe rather well various resonances and has resolve several long standing problems including the width and shape of the hot GDR, the electromagnetic cross section of the DGDR, the quenching of the GTR. It also predicts the PDR in neutron-rich nuclei and the DGDR in hot nuclei.

REFERENCES

1. N. Dinh Dang and A. Arima, Phys. Rev. Lett. **80** (1998) 4145, Nucl. Phys. **A 636** (1998) 427.
2. N. Dinh Dang. K. Tanabe, and A. Arima, Phys. Rev. **C 58** (1998) 3374, Nucl. Phys. **A 645** (1999) 536.
3. N. Dinh Dang, V. Kim Au, T. Suzuki, and A. Arima, Phys. Rev. **C 63** (2001) 44302.
4. J.G. Woodworth et al., Phys. Rev. **C 19** (1979) 1667.
5. W. J. Llope and P. Braun-Muzinger, Phys. Rev. **C 41**, (1990) 2644.
6. I.A. Pshenichnov *et al.*, Phys. Rev. **C 60** (1999) 044901.
7. N. Dinh Dang, K. Eisenman, J. Seitz, and M. Thoennessen, Phys. Rev. **C 61** (2000) 027302
8. N. Dinh Dang, Nucl. Phys. **A 687** (2001) 253c.
9. T. Baumann et al., Nucl. Phys. **A 635** (1998) 428, **A649** (1999) 173c.
10. A. Bracco et al., Phys. Rev. Lett. **74** (1995) 3748, M. Matiuzzi et al., Nucl. Phys. **A 612** (1997) 262.
11. N. Dinh Dang, V. Kim Au, and A. Arima, Phys. Rev. Lett. **85** (2000) 1827.
12. K. Boretzky, private communication (September, 2000).
13. N. Dinh Dang, K. Tanabe, and A. Arima, Phys. Rev. **C 59** (1999) 3128.
14. G. Viesti et al., Phys. Rev. **C 63** (2001) 034611.
15. N. Dinh Dang. T. Suzuki. and A. Arima, Phys. Rev. **C 64** (2001) 27303.
16. N. Dinh Dang et al., Nucl. Phys. **A 621** (1997) 719.
17. T. Wakasa et al., Phys. Rev. **C 55** (1997) 2909.
18. H. Sagawa and T. Suzuki, Phys. Rev. **C 59** (1999) 3116.
19. P.F. Bortignon and R.A. Broglia, Nucl. Phys. **A 371** (1981) 405.
20. S. Nishizaki and J. Wambach, Phys. Lett. **B 349** (1995) 7.
21. S. Kamerdzhiev, J. Speth, and G. Tertychny, Nucl. Phys. **A 624** (1997) 328.
22. T. Aumann et al., GSI Scientific Report 1999 (2000) 27.
23. S. Mizutori et al., Phys. Rev. **C 61** (2000) 044326.

Invariant Manifolds and Collective Motion in Many-Body Systems

T. Papenbrock*,† and T. H. Seligman*,**

*Centro Internacional de Ciencias, Cuernavaca, Mexico
†Physics Division, Oak Ridge National Laboratory, Oak Ridge, TN 37831, USA
**Centro de Ciencias Físicas, University of México (UNAM), Cuernavaca, Mexico

Abstract. Collective modes of interacting many-body systems can be related to motion on classically invariant manifolds. We introduce suitable coordinate systems. These coordinates are Cartesian in position and momentum space. They are collective since several components vanish for motion on the invariant manifold. We make a connection to Zickendraht's collective coordinates and also obtain shear modes. The importance of collective configurations depends on the stability of the manifold. We present an example of quantum collective motion on the manifold.

INTRODUCTION

Interacting many-body systems such as atomic nuclei display regular and collective motion as well as complex and chaotic behavior. The interplay of chaotic motion and collective behavior is of particular interest and has been studied for many years. Several authors have investigated models which exhibit chaotic single-particle dynamics in slowly oscillating mean-field potentials [1, 2, 3, 4, 5, 6, 7]. While these models may yield damping and equilibration of the collective mode, they neglect the residual two-body interaction. For the special case of attractive, billiard-like two-body interactions the many-body aspects of collective motion can be studied within the framework of classical dynamics [8].

Here we present an alternative and more general approach to the problem. It is based on the observation that any rotationally invariant system of identical interacting particles possesses low dimensional invariant manifolds in phase space [9]. On such manifolds, the classical motion displays largely collective behavior and decouples from more complex single-particle behavior. The importance of a given invariant manifold depends crucially on its stability properties. If the manifold under consideration is sufficiently stable in transverse directions, the quantum system may exhibit wave function scarring [10, 11, 12] or display a strong revival for wave-packets localized to the vicinity of the manifold [13]. These findings may be directly associated with the slow decay of collective motion despite of the coupling between collective and single-particle motion.

Suitably adapted coordinates for motion on the invariant manifold separate single-particle motion from the collective motion on the manifold [14]. For some types of collective motion, the adapted coordinates can be related to collective coordinates introduced by Zickendraht [16] about thirty years ago. However, the natural coordinates for invariant manifolds are capable of more complicated collective motion such as shearing

CP597, *Nonequilibrium and Nonlinear Dynamics in Nuclear and Other Finite Systems,*
edited by Z. Li et al.

modes [15].

This contribution is divided as follows. First, we define invariant manifolds in interacting many-body systems and introduce suitable coordinates. Second, we make a connection with the Zickendraht coordinates. Third, we present an example and show that quantum collective motion decays slowly close to invariant manifolds which are not too unstable. Finally we give a summary.

INVARIANT MANIFOLDS AND ADAPTED COORDINATES

Consider rotationally invariant systems of N identical particles in d spatial dimensions ($d = 2$ or $d = 3$). The Hamiltonian is invariant under both, the action of the rotation group $O(d)$ and the group of permutations S_N. One may now take a finite subgroup $G \subset O(d)$ with elements g and properly chosen permutations $P(g)$ such that

$$gP^{-1}(g)(\vec{p},\vec{q}) = (\vec{p},\vec{q}), \quad \forall g \in G \qquad \vec{p} \equiv (p_1,\ldots,p_{Nd}), \vec{q} \equiv (q_1,\ldots,q_{Nd}) \quad (1)$$

for points $(\vec{p},\vec{q})$ on some invariant submanifold in phase space. On such a manifold, the action of the rotations $g \subset G$ can be canceled by permutations. These permutations clearly form a subgroup isomorphic to G.

Fig. 1 shows a configuration of four particles in two spatial dimensions that corresponds to a point on an invariant manifold. The operations of elements from the discrete symmetry group $G = C_{2v}$ can be undone by suitable permutations. This leads to a collective motion with two degrees of freedom which will be shown to be vibrations.

FIGURE 1. Collective configuration on invariant manifold. Positions are indicated by filled circles and momenta by arrows.

Fig. 2 shows two spatial configurations of eight (2a) and six (2b) particles, respectively which display a D_{4h} symmetry. If initial momenta display the same symmetry the motion on the invariant manifold will have two degrees of freedom. For eight particles the radii of the two circles will oscillate synchronously, and the two circles will vibrate against each other. For the six particles we will have a vibration of the radius of the circle and of the two particles along the vertical axis. We may choose initial momenta to reduce the symmetry group to C_{4h} which will allow rotations around the vertical axis and thus add an additional degree of freedom. For eight particles we could alternatively choose initial conditions that are limited to a D_4 symmetry. Besides the vibrations discussed above this allows for a shearing motion of the two circles thus yielding again three degrees of freedom. Further reductions of symmetry will yield different invariant manifolds with varying degrees of freedom. We will see this exemplified by explicit construction of coordinates.

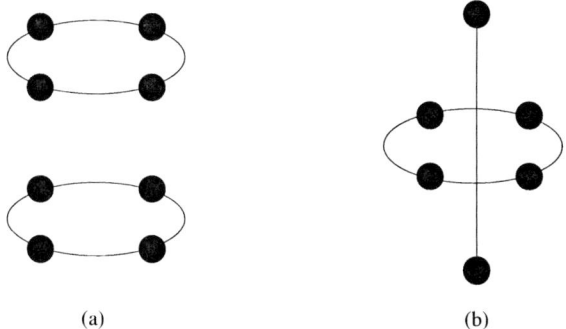

(a)   (b)

FIGURE 2. Configurations of eight (a) or six (b) particles in three dimensions that correspond to invariant manifolds. Positions are indicated by filled circles.

We may use the definition (1) directly for the construction of coordinate systems where non collective coordinates vanish for motion on the invariant manifold. For this purpose we consider the many-body system in Cartesian coordinates in momentum and position space. In what follows we will consider orthogonal transformations in configuration space only; momenta will be subject to the same transformation.

In a Cartesian coordinate system each element $g \in G$ and each permutation $P(g)$ can be represented by an orthogonal matrix $\mathbf{M}_g$ and $\mathbf{P}_g$ of dimension Nd. It is clear that the products $\mathbf{M}_g \mathbf{P}_g^{\mathrm{T}}$ form a matrix group $\mathcal{H}$ that acts onto position and momentum space, respectively. The construction of the coordinate system is now straightforward. Every vector $\vec{p}$ and $\vec{q}$ may be expanded in basis vectors of the irreducible representations (IRs) of $\mathcal{H}$ by means of projectors [17].

$$\Pi_v = \sum_{g \in G} \chi_g^{(v)} \mathbf{M}_g \mathbf{P}_g^{\mathrm{T}}. \qquad (2)$$

Here $\chi_g^{(v)}$ denotes the character of g in the v'th IR. The projection onto the identical IR defines the invariant manifold. Note that the identical representation is one-dimensional while the invariant manifolds of interest typically have higher dimensionality. We can find independent vectors on the manifold by projecting from different vectors, but in practice the construction of the independent vectors seems to be unproblematic as we shall see in the example.

ZICKENDRAHT'S COORDINATES AND INVARIANT MANIFOLDS

About thirty years ago Zickendraht [16] introduced a set of collective coordinates to describe nuclear vibrations and rotations, as well as their coupling with single particle motion. We shall compare these coordinates correspond to the ones we introduced in

303

the previous sections. On one hand this will allow to identify certain vibrational modes of a many-body system with invariant manifolds. On the other hand we shall see that our procedure proposes collective movements that are not of the type described easily in Zickendraht's coordinates.

Following Zickendraht [16] we write the coordinates $\vec{r}_i$ of the ith particle in the center of mass system as

$$\vec{r}_i = s_{i1}\,\vec{y}_1 + s_{i2}\,\vec{y}_2 + s_{i3}\,\vec{y}_3, \quad i = 1,\ldots,N \tag{3}$$

where the $\vec{y}_i$ span the inertia ellipsoid and s_{ik} are called single-particle coordinates. The newly introduced coordinates $\vec{y}_i$ and s_{ij} are not independent. The constraints are $\vec{y}_i \cdot \vec{y}_j = y_i y_j \delta_{ij}$, $i,j = 1,2,3$; $\sum_{i=0}^{N} s_{ij} = 0$, $j = 1,2,3$, and $\sum_{i=0}^{N} s_{ij} s_{ik} = \delta_{jk}$, $j,k = 1,2,3$. The first six equations ensure the orthogonality and normalization of the principal axis of the inertia ellipsoid whereas the next three equations fix the origin at the center of mass system. The last six equations are orthogonality relations of the single-particle coordinates. In the center of mass system, one may therefore characterize the N-body system by its inertia ellipsoid (e.g. three Euler angles of the principle axis and three moments of inertia) and $3N - 9$ single particle coordinates. The moments of inertia I_i are related to the coordinates y_i by

$$I_1 = m(y_2^2 + y_3^2), \qquad I_2 = m(y_1^2 + y_3^2), \qquad I_3 = m(y_1^2 + y_2^2), \tag{4}$$

where m denotes the mass of the particles.

It is interesting to determine those configurations, where the motion of the many-body system may be described in terms of the collective coordinates y_i only. While such motion would be restricted to *some* invariant manifold in phase space it would not obviously be one of those defined by eq. (1). We may however determine invariant manifolds (1) such that the motion on the manifold changes only the inertia ellipsoid of the system and hence may be described entirely by Zickendraht's collective coordinates y_i. Two necessary conditions define this situation. First, the number of coordinates on such invariant manifolds may not exceed six. Second, every motion on such invariant manifolds has to change the inertia ellipsoid of the many-body system.

For simplicity let us start with the a system of four particles in two spatial dimensions and the invariant manifold displayed in Fig. 1, i.e.

$$\vec{r}_1 = \begin{bmatrix} x \\ y \end{bmatrix}, \quad \vec{r}_2 = \begin{bmatrix} x \\ -y \end{bmatrix}, \quad \vec{r}_3 = \begin{bmatrix} -x \\ -y \end{bmatrix}, \quad \vec{r}_4 = \begin{bmatrix} -x \\ y \end{bmatrix},$$

and the momenta are chosen by replacing $x \to p_x, y \to p_y$. Computation of the moments of inertia yield the collective Zickendraht coordinates $y_1 = 2x$, $y_2 = 2y$. On the invariant manifold the remaining coordinates are given by $s_{11} = s_{12} = s_{21} = -s_{22} = -s_{31} = -s_{32} = -s_{41} = s_{42} = 1/2$. This shows that every motion on the invariant manifold changes the moments of inertia only and therefore decouples from the single-particle motion.

Consider next the example of an eight-body system in three dimensions. Let

$$\vec{r}_1 = \begin{bmatrix} x \\ y \\ z \end{bmatrix}, \quad \vec{r}_2 = \begin{bmatrix} -y \\ x \\ z \end{bmatrix}, \quad \vec{r}_3 = \begin{bmatrix} -x \\ -y \\ z \end{bmatrix}, \quad \vec{r}_4 = \begin{bmatrix} y \\ -x \\ z \end{bmatrix}, \quad \vec{r}_{4+i} = \vec{r}_i(z \leftrightarrow -z) \tag{5}$$

denote a configuration restricted to the invariant manifold displayed in Fig. 2 (a) with C_{4h} symmetry. (The momenta are chosen by replacing $x \to p_x, y \to p_y, z \to p_z$ in eq.(5).) The moments of inertia are $I_1 = I_2 = 4m(x^2 + y^2) + 8mz^2, I_3 = 8m(x^2 + y^2)$ and yield collective coordinates (4) $y_1^2 = y_2^2 = 4(x^2 + y^2), y_3^2 = 8z^2$. Since the inertia ellipsoid is symmetric we have freedom in the choice of two of its principle axis. Using

$$
\vec{y}_1 = 2 \begin{bmatrix} x \\ y \\ 0 \end{bmatrix}, \quad \vec{y}_2 = 2 \begin{bmatrix} -y \\ x \\ 0 \end{bmatrix}, \quad \vec{y}_3 = \sqrt{8} \begin{bmatrix} 0 \\ 0 \\ z \end{bmatrix},
$$

one obtains constant single-particles coordinates $s_{11} = -s_{31} = s_{51} = -s_{71} = s_{22} = -s_{42} = s_{62} = -s_{82} = 1/2, s_{13} = s_{23} = s_{33} = s_{43} = -s_{53} = -s_{63} = -s_{73} = -s_{83} = 1/\sqrt{8}$ for the motion on the invariant manifold. Thus, on the invariant manifold the single-particle motion decouples from the collective one. Similar results hold for the six particle configuration displayed in Fig. 2.

It is also instructive to consider a more complex situation. The configuration

$$
\vec{r}_1 = \begin{bmatrix} x \\ y \\ z \end{bmatrix}, \quad \vec{r}_2 = \begin{bmatrix} -y \\ x \\ z \end{bmatrix}, \quad \vec{r}_3 = \begin{bmatrix} -x \\ -y \\ z \end{bmatrix}, \quad \vec{r}_4 = \begin{bmatrix} y \\ -x \\ z \end{bmatrix},
$$

$$
\vec{r}_5 = \begin{bmatrix} x \\ -y \\ -z \end{bmatrix}, \quad \vec{r}_6 = \begin{bmatrix} y \\ x \\ -z \end{bmatrix}, \quad \vec{r}_7 = \begin{bmatrix} -x \\ y \\ -z \end{bmatrix}, \quad \vec{r}_8 = \begin{bmatrix} -y \\ -x \\ -z \end{bmatrix},
$$

displays D_4 symmetry and differs from configuration (5) by a shearing motion. Like in the previous example, the moments of inertia are given by $I_1 = I_2 = 4m(x^2 + y^2) + 8mz^2, I_3 = 8m(x^2 + y^2)$ and the ellipsoid of inertia is symmetric. However, no choice of the principal axis allows to fulfill eqs. (3) with *constant* single-particle coordinates s_{ij}. Therefore, single-particle degrees of freedom depend on collective degrees of freedom and a decoupling does not exist using Zickendraht's coordinate system. However, decoupling is achieved if we use the projector (2) and find coordinates on the invariant manifold and perpendicular to it. The collective motion does not correspond to vibrations or rotations of the inertia ellipsoid only. These findings are interesting e.g. in relation with with the magnetic dipole mode in nuclei [15] since this type of collective behavior is associated with a shearing motion.

A SIMPLE ILLUSTRATION

Let us consider a system of four particles in two dimensions with quartic one- and two-body interactions. The invariant manifold of Fig. 1 is defined as those points which are invariant under $\mathcal{H} = \{E, \sigma_x P_{(12)(34)}, \sigma_y P_{(14)(23)}, C_2 P_{(13)(24)}\}$. Here E denotes the identity, P a permutation of particles as indicated, σ a reflection at the axis indicated, and C_2 a rotation about π. Thus, $\mathcal{H} = C_{2v}$ with four IRs labeled by $\nu = A_1, B_1, A_2, B_2$ [17]. Let $\vec{q} = (x_1, x_2, x_3, x_4, y_1, y_2, y_3, y_4)$ denote a coordinate vector in position space

$(x_i, y_i$ denote the coordinates of the i'th particle). We have

$$
\begin{aligned}
E\vec{q} &= (x_1, x_2, x_3, x_4, y_1, y_2, y_3, y_4), \\
\sigma_x P_{(12)(34)}\vec{q} &= (x_2, x_1, x_4, x_3, -y_2, -y_1, -y_4, -y_3) \\
C_2 P_{(13)(24)}\vec{q} &= (-x_3, -x_4, -x_1, -x_2, -y_3, -y_4, -y_1, -y_2) \\
\sigma_y P_{(14)(23)}\vec{q} &= (-x_4, -x_3, -x_2, -x_1, y_4, y_3, y_2, y_1).
\end{aligned}
$$

From the character table of C_{2v} [17] and the projectors (2) one obtains the following basis vectors corresponding to the IR labeled by

$$
\begin{array}{lll}
A_1 : & e_1' = (1,1,-1,-1,0,0,0,0)/2, & e_2' = (0,0,0,0,1,-1,-1,1)/2, \\
B_1 : & e_3' = (1,1,1,1,0,0,0,0)/2, & e_4' = (0,0,0,0,1,-1,1,-1)/2, \\
A_2 : & e_5' = (1,-1,-1,1,0,0,0,0)/2, & e_6' = (0,0,0,0,1,1,-1,-1)/2, \\
B_2 : & e_7' = (1,-1,1,-1,0,0,0,0)/2, & e_8' = (0,0,0,0,1,1,1,1)/2.
\end{array}
$$

The vectors associated with the identical IR A_1 span the two-dimensional invariant manifold and the vectors associated with the IRs B_1, A_2, B_2 span the transverse directions.

Having specified the invariant manifold and appropriate coordinates we now consider the interacting four-body system with the Hamiltonian

$$
H = \sum_{i=1}^{4} \left((p_{x_i}^2 + p_{y_i}^2)/2 + 16(x_i^2 + y_i^2)^2 \right) - \sum_{i<j} \left[(x_i - x_j)^2 + (y_i - y_j)^2 \right]^2. \tag{6}
$$

The stability of the invariant manifold defined above has been studied by computing the full phase space monodromy matrix of several periodic orbits that are inside the invariant manifold [13]. It was found that several orbits are linearly stable in transverse directions or possess rather small transverse stability exponents, while the motion inside the manifold is strongly chaotic.

Let us investigate the decay of collective motion in the corresponding quantum system. To this purpose we consider the time evolution of a Gaussian wave packet $\Psi(\mathbf{r}, t)$ that initially is localized on the invariant manifold. The autocorrelation function $C(t) = \langle \Psi(t=0) | \Psi(t) \rangle$ is computed in semiclassical approximation. Within the manifold we used Heller's cellular dynamics [18] which takes into account the nonlinearity of the classical motion. In the transverse direction the time–propagation was done using linearized dynamics only. This approximation neglects any recurrences from the transverse directions and implies a permanent flux of probability out of the manifold and its vicinity. On the time scales considered here, the linearization is justified since the classical return probability to the manifold of transversely escaping trajectories is negligible. It is also important to note that the loss of probability inside the manifold is not severe since the transverse stability exponents are not too large.

We launch wave packets along periodic or aperiodic orbits lying within the invariant manifold and consider their revival as measured by the autocorrelation function. To achieve shorter recurrence times the initial packet was symmetrized with respect to the reflection symmetry of the system within the invariant manifold.

We propagate such wave packets for the invariant manifold. For not too unstable periodic orbits we expect a fairly strong revival after one period, known as the linear revival [18]. As an example, we show in Fig. 3 the real part of the autocorrelation function, calculated for a wave packet on a periodic orbit. We indeed find strong linear revival. However, at larger times we find randomly scattered strong revivals. The revival corresponding to twice the period is not dominant. This implies that a significant fraction of the original amplitude remains near the invariant manifold, and that this fact is not related to the periodic orbit we started on. Revivals calculated for packets started on aperiodic orbits show similar features except for the obvious absence of the linear revival.

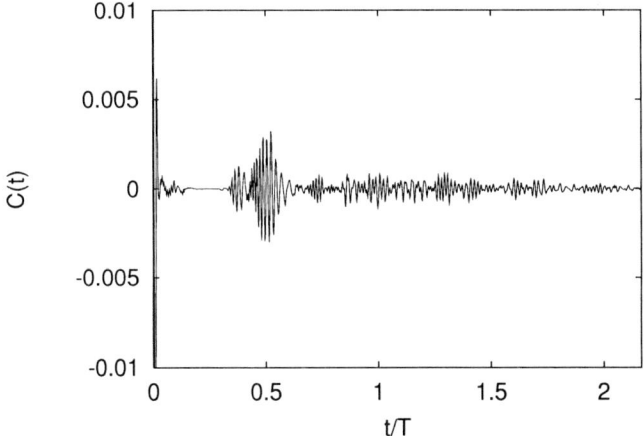

FIGURE 3. Autocorrelation function $C(t)$ of a symmetrized wave packet launched on a periodic orbit with period T inside the weakly unstable manifold. In addition to the linear revival around $t = \frac{T}{2}$, a strong nonlinear revival is seen for larger times.

Thus, wave packets may have unusually long life times on certain invariant manifolds characterized by small classical transverse instabilities. This is a quantum and not a classical phenomenon and constitutes an extension of the concept of a scar [19].

SUMMARY

Low-dimensional invariant manifolds are part of the phase space for interacting many-body systems with permutation and rotation symmetry. We presented coordinates that are naturally adapted to an invariant manifold. There are several configurations of few-body systems, where the motion on the invariant manifold corresponds to a vibration or rotation and may be described in terms of Zickendraht's coordinates, but differs when the collective motion goes beyond that. These results are independent of the details of the Hamiltonian of the N-body system, and are entirely determined by rotational and permutational symmetry.

The importance of a given invariant manifold depends on its stability for motion that is close to the manifold. There are few-body systems with invariant manifolds that have vanishing [10, 12] or small [13] instability exponents in transversal directions.

Examples include collinear Helium [10], a chaotic, self-bound three-body system [12] and a four-body systems with anharmonic interactions [13]. In these systems one finds an enhancement in wave function amplitude close to the corresponding invariant manifolds, or large revival probabilities for wave packets that are launched on such manifolds. For low level densities the strong revival can result in individual, strongly scarred states which should be identified with low-lying collective states. For high level densities the strength of these modes will be divided among many individual states thus giving rise to intermediate structure.

We thank T. Guhr for useful discussions and acknowledge financial support by CONACyT project 25192E and DGAPA (UNAM) project IN112200. TP acknowledges support as a Wigner Fellow and staff member at Oak Ridge National Laboratory, managed by UT-Battelle, LLC for the U.S. Department of Energy under contract DE-AC05-00OR22725.

REFERENCES

1. T. Guhr, H. A. Weidenmüller, Ann. Phys. **193** (1989) 472
2. J. Blocki, F. Brut, T. Srokowski, and W. J. Swiatecki, Nucl. Phys. A **545** (1992) 551c
3. W. D. Heiss, R. G. Nazmitdinov, and S. Radu, Phys. Rev. Lett. **72** (1994) 2351
4. S. Drozdz, S. Nishizaki, and J. Wambach, Phys. Rev. Lett. **72** (1994) 2839
5. W. Bauer, D. McGrew, V. Zelevinski, and P. Schuck, Phys. Rev. Lett. **72** (1994) 3771
6. V. R. Manfredi, L. Salasnich, Int. J. Mod. Phys. E **4** (1995) 625
7. V. Zelevinski, Ann. Rev. Nucl. Part. Sci. **46** (1996) 237
8. T. Papenbrock, Phys. Rev. C **61** (2000) 034602
9. T. Papenbrock and T. H. Seligman, Phys. Lett. A **218** (1996) 229
10. D. Wintgen, K. Richter, and G. Tanner, Chaos **2** (1992) 19
11. T. Prosen, Phys. Lett. A **233** (1997) 332
12. T. Papenbrock and T. Prosen, Phys. Rev. Lett. **84** (2000) 262
13. T. Papenbrock, T. H. Seligman, and H. A. Weidenmüller, Phys. Rev. Lett. **80** (1998) 3057
14. T. Papenbrock and T. H. Seligman, preprint nucl-th/0003049, submitted to J. Phys. A.
15. T. Guhr, H. Diesener, A. Richter, C. W. de Jager, H. de Vries, and P. K. A. de Witts-Huberts, Z. Phys. A **336** (1990) 159
16. W. Zickendraht, J. Math. Phys. **12** (1970) 1663
17. M. Hammermesh, *Group theory and its application to physical problems*, Dover Publications, N.Y., 1989
18. E.J. Heller, J. Chem. Phys. **94** (1991) 2723; M.A. Sepulveda, E.J. Heller, J. Chem. Phys. **101** (1994) 8004
19. E.J. Heller, Phys. Rev. Lett. **53** (1984) 1515

PART VI

ORDER TO CHAOS
PHASE TRANSITION
AND COMPLEXITY

Exceptional Points: Global and Local Aspects

W.D Heiss

Department of Physics, University of the Witwatersrand, PO Wits 2050, South Africa

Abstract. The intricate connection between the distribution of exceptional points and particular fluctuation properties of level spacings is discussed. The distribution of the exceptional points of the problem $H_0 + \lambda H_1$ is given for the situation of hard chaos. Predictions of local properties of exceptional points have recently been confirmed experimentally. Chiral properties associated with exceptional points are pointed out.

INTRODUCTION

We address the questions: which are the common properties of the *quantum mechanical Hamilton operators* that give rise to spectral properties that are ascribed to quantum chaos. If a matrix representation of a Hamiltonian which originates from a classically chaotic analogous case is given, which is the mathematical mechanism that yields the special features of the spectrum within the particular range of the parameter where classical chaos is discerned. While cases with a classical analogy have no intrinsic statistical property, a further puzzling question is: why is it that a statistical approach without physical input (GOE) reproduces such good agreement of the statistical properties of the spectrum.

We believe that the common root to the answer of these questions lies in what is called the exceptional points[1] of an operator. Most physical problems in quantum mechanics can be formulated by the Hamiltonian $H_0 + \lambda H_1$ where the parameter λ can play the role of a perturbation parameter, or it may serve to effect a phase transition, or it may under variation steer the system from an ordered into a chaotic regime. The exceptional points of the full operator are the points λ for which two eigenvalues coalesce. Here we exclude genuine degeneracies of the self-adjoint problem, in other words, the eigenvalues coincide for no real λ. The exceptional points occur in the complex λ-plane. Note that the operator is not self-adjoint for complex λ-values.

The definition of exceptional points is general and applies in particular to operators in an infinite dimensional space, also when the spectrum of the operator has a continuum part. In the present work we restrict ourselves to finite dimensional matrices H_0 and H_1 as in this case the role of the exceptional points and the associated Riemann sheet structure is thoroughly understood[2, 3]. We do not believe that restriction to matrices has a major impact on our conclusions since virtually all the practical work even in connection with quantum chaos is done in a finite dimensional matrix space.

The physical significance of the exceptional points is due to their relation with avoided level crossing for real λ-values. The spectrum $E_k(\lambda)$, $k = 1, \ldots, N$ has branch point singularities at the exceptional points, in fact, any two of the N levels are connected via a

CP597, *Nonequilibrium and Nonlinear Dynamics in Nuclear and Other Finite Systems,*
edited by Z. Li et al.

square root branch point. If this happens near to the real λ-axis, a level repulsion will occur for the two levels for real λ-values. Globally, all the exceptional point singularities determine the shape of the whole spectrum. There is a nice analogy to the more widely known connection between the pole singularities of the scattering function and the shape of the cross section: in a similar way as the positions of the poles including their statistical properties determine the measurable cross section, the exceptional points determine the shape of the spectrum and in particular the occurrences of avoided level crossings. The distribution of the exceptional points will therefore determine the fluctuation properties of level spacings.

The positions of the exceptional points are fixed in the complex λ-plane and are determined solely by H_0 and H_1. For large matrices it is prohibitive to determine the positions of the exceptional points. However, it is possible to determine the distribution reasonably well from the knowledge of the two operators. A high density of exceptional points is a sufficient prerequisite for the occurrence of quantum chaos if they are randomly distributed according to a specific distribution function.

In the following section we recapitulate the basics about exceptional points. Section three presents matrix models to exemplify the distribution of exceptional points and level spacing fluctuations. Section four presents a recent experiment where the local topological structure has been shown to be a physical reality. Section five presents a summary and outlook.

EXCEPTIONAL POINTS AND UNPERTURBED LINES

Avoided level crossing is always associated with exceptional points[2, 4, 5] if it occurs for the levels $E_k(\lambda)$ of the Hamiltonian $H_0 + \lambda H_1$. The exceptional points are square root branch point singularities in the complex λ-plane. We give an elementary example for illustration and briefly list the essential aspects with regard to exceptional points.

Consider a two dimensional matrix problem where H_0 is diagonal with eigenvalues ε_1 and ε_2, while H_1 is represented in the form

$$H_1 = U \cdot D \cdot U^{-1}. \tag{1}$$

Here, the diagonal matrix D contains the eigenvalues ω_1 and ω_2 of the matrix H_1 and U is the rotation

$$U = \begin{pmatrix} \cos\phi & -\sin\phi \\ \sin\phi & \cos\phi \end{pmatrix}. \tag{2}$$

The eigenvalues of the problem $H_0 + \lambda H_1$ are

$$E_{1,2}(\lambda) = \frac{\varepsilon_1 + \varepsilon_2 + \lambda(\omega_1 + \omega_2)}{2} \pm R \tag{3}$$

where

$$R = \left\{ \left(\frac{\varepsilon_1 - \varepsilon_2}{2}\right)^2 + \left(\frac{\lambda(\omega_1 - \omega_2)}{2}\right)^2 + \frac{1}{2}\lambda(\varepsilon_1 - \varepsilon_2)(\omega_1 - \omega_2)\cos 2\phi \right\}^{1/2}. \tag{4}$$

Clearly, when $\phi = 0$ the spectrum is given by the two lines

$$E_k^0(\lambda) = \varepsilon_k + \lambda \omega_k \qquad k = 1, 2 \tag{5}$$

which intersect at the point of degeneracy $\lambda = -(\varepsilon_1 - \varepsilon_2)/(\omega_1 - \omega_2)$. When the coupling between the two levels is turned on by switching on ϕ the degeneracy is lifted and avoided level crossing occurs. Now the two levels coalesce in the complex λ-plane where R vanishes which happens at the complex conjugate points

$$\lambda_c = -\frac{\varepsilon_1 - \varepsilon_2}{\omega_1 - \omega_2} \exp(\pm 2i\phi). \tag{6}$$

At these points, the two levels $E_k(\lambda)$ are connected by a square root branch point, in fact the two levels are the values of one analytic function on two different Riemann sheets.

These considerations carry over to an N-dimensional problem[2]. The diagonal matrix H_0 contains the elements ε_k and D the elements ω_k, $k = 1, \ldots, N$; the matrix U is now an N-dimensional rotation which can be parameterized by $N(N-1)/2$ angles. (In the quoted paper a parameterization was chosen so that U is unity when all angles are zero.) The exceptional points are determined by the simultaneous solution of the equations

$$\begin{aligned} \det(E - H_0 - \lambda H_1) &= 0 \\ \frac{d}{dE} \det(E - H_0 - \lambda H_1) &= 0. \end{aligned} \tag{7}$$

There are generically $N(N-1)$ solutions which occur in complex conjugate pairs in the λ-plane. At those points the N levels $E_k(\lambda)$ are connected in pairs by square root branch points when they are analytically continued into the complex λ-plane. Since the positions of the singularities determine the shape of the spectrum, and in particular the fluctuation properties, a closer analysis is indicated. As is exemplified in the next section, the crucial condition for the occurrence of level statistics ascribed to quantum chaos is a high density of exceptional points in the complex plane within a small window of real λ-values.

To get an idea about the density of exceptional points we introduce the concept of *unperturbed lines*. Clearly, when U is the unit matrix (all angles are zero), the spectrum of $H_0 + \lambda H_1 = H_0 + \lambda D$ is given by the lines $\varepsilon_k + \lambda \omega_k$ with $k = 1, \ldots, N$. The $N(N-1)/2$ intersection points of the N lines depend on the relative order of the numbers ε_k and ω_k. If both sequences are in ascending order, all intersections occur at negative λ-values; conversely, if one sequence is ascending and the other descending all intersections occur at positive λ-values. In general, the order which is appropriate for the actual problem, is expected to lie between the two extremes. To find out the appropriate order we are guided by the asymptotic behavior of the levels $E_k(\lambda)$ of the full problem. For large values of λ the leading terms are given by

$$E_k(\lambda) = \lambda \omega_k + \alpha_k + \ldots \tag{8}$$

where the dots stand for first and higher order terms in $1/\lambda$. Neglecting these terms, Eq.8 yields just the unperturbed lines with the appropriate association of slopes ω_k and intercepts α_k. From perturbation theory we find the latter to be the diagonal elements of the 'backwards' rotated H_0, *viz.*

$$\alpha_k = (U^{-1} \cdot H_0 \cdot U)_{k,k}. \tag{9}$$

DISTRIBUTION FUNCTION

We begin with the distribution function of the intersection points of the unperturbed lines as this can be derived rigorously. Assuming a uniform random distribution for the ω_k and the ε_k (the eigenvalues of H_1 and H_0, respectively), the intersection points $\lambda_{i,k} = -(\varepsilon_i - \varepsilon_k)/(\omega_i - \omega_k)$ are distributed according to [6]

$$P(x) = \frac{\text{const}}{x^2}. \tag{10}$$

The fact that $P(x)$ cannot be normalized is due the fact that we assumed a uniform distribution of the ω_k over the whole range of the real numbers; for a finite sample there will therefore be a deviation of $P(x)$ around $x = 0$ from the form given in Eq.(10) as no $|\omega_k|$ occurs beyond a finite large number and hence an arbitrary small value of $1/\omega$ is unlikely to occur.

The intersection points are the points of degeneracies of $H_0 + \lambda H_1$ as long as U is equal to the identity matrix. We now gradually switch on the interaction between the levels by switching on the angles randomly of the random orthogonal matrix U [6] but keeping initially the interval from which the angles are chosen very small. In this way the degeneracies become level repulsions as the exceptional points start moving out into the complex λ-plane, a complex conjugate pair from each degeneracy. We conjecture that their distribution function remains unchanged when they move out into the plane. For small values of the mixing angles in U they will of course be concentrated around the real axis. The distribution function is now a function of two variables, say the real and imaginary part of a complex number. We parameterize this number by $r \exp(i\alpha)$. For a fixed angle α the exceptional points are distributed according to

$$P(r) = \frac{\text{const}}{r^2}. \tag{11}$$

This has been confirmed numerically in numerous cases. If the mixing angles of U are very small, there is an obvious dependence on α as the exceptional points cluster around the real axis; yet for fixed α the distribution law is always given by Eq.(11). Moreover, for the nearest neighbor distribution of the energy levels it turns out that a proper Wigner distribution is obtained only when the exceptional points have fanned out into the plane completely so that their distribution becomes independent of the angle α. In Fig.1 examples are illustrated how the exceptional points fan out into the plane when the mixing angles of U are turned on.

The law of Eq.(11) is verified numerically even if the exceptional points originate out of a degeneracy of high multiplicity. In Fig.2 the emergence of exceptional points out of a point of degeneracy is illustrated. Here, Eq.(11) is obtained once they fill the plane in an isotropic way, that is when the distribution has become independent of α.

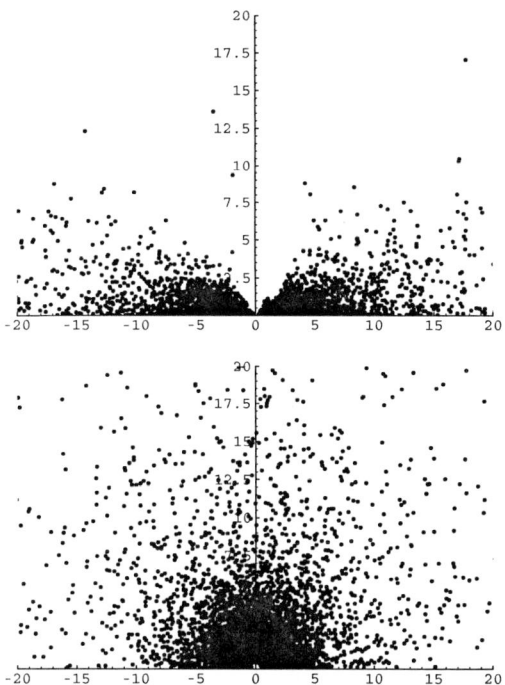

FIGURE 1. Exceptional points in the upper complex λ-plane. The top illustrates the α-dependence associated with U near to the identity matrix; the bottom corresponds to an arbitrary random orthogonal matrix for U.

To summarize: hard chaos, that is a Wigner distribution for the NND of the energy levels, is associated with a distribution of exceptional points in the complex plane that is independent of the angle α of the complex number $r\exp(i\alpha)$ and depends only on the distance r according to Eq.(11). If an α-dependence prevails, the NND is not Wigneresque. In the limiting case (U identity) of the example used above the NND is of course a Poisson distribution.

We only mention that in special non-generic cases the exceptional points are arranged in a geometrically ordered pattern like in the integrable Lipkin model [7]. However, a small perturbation leads to the generic situation as discussed above [5].

LOCAL PROPERTIES OF EXCEPTIONAL POINTS

The topological structure of the square root branch point associated with an exceptional point has been shown to be a physical reality in a recent experiment [8]. As a particular consequence it has been established experimentally that the phases of the wave functions that take part in the coalescence of the two energy levels show a phase behavior that is

315

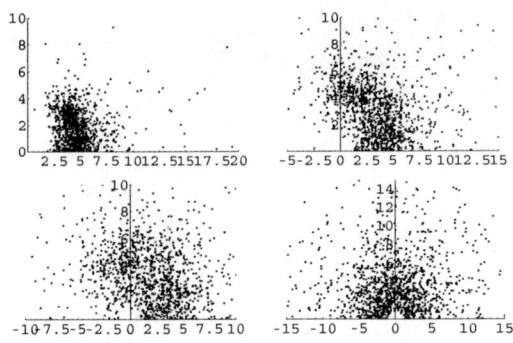

FIGURE 2. Exceptional points in the upper complex plane associated with the lifting of a multiple degeneracy. The size of the angles in U is increasing when going from the top left to the bottom right figure.

distinctly different from that of a usual degeneracy, i.e. from that at a diabolic point [9]. The experiment [8] yielded three major results which have been predicted in [10]:

1. If a loop is performed in the λ-plane around the EP, the eigenenergies E_1 and E_2 are interchanged.

2. The wave functions $|\psi_1\rangle$ and $|\psi_2\rangle$ are interchanged by the loop and, in addition, one of them changes sign. In other words, a loop in the λ-plane transforms the pair $\{\psi_1, \psi_2\}$ into $\{-\psi_2, \psi_1\}$. Therefore the two possible directions of looping yield different phase behavior. In fact, encircling the EP a second time in the same direction, we obtain $\{-\psi_1, -\psi_2\}$ while the next loop yields $\{\psi_2, -\psi_1\}$ and only the fourth loop restores the original pair. It follows that by going in the opposite direction, one finds after the first loop what is obtained after three loops in the former direction.

3. The eigenvalues E_1, E_2 have been studied as functions of λ for two paths that were not closed. One path was just above, the other one just below λ_c. The results were different. Let us call resonance energy the real part of an eigenvalue and resonance width its imaginary part. On one of the paths, the widths cross while the resonance energies avoid each other. On the other path, the resonance energies cross while the

widths avoid each other.

We conclude the discussion with a further interesting local property of exceptional points. The wave function at the exceptional point has always a specific chiral behavior [11]. An experiment at the TU-Darmstadt is in preparation to directly detect this feature. In similar context this has been discussed for acoustic waves in a medium [12] and indirectly observed in optics [13]. The latter has been explained in terms of exceptional points in [14]. The essential finding of [11] is that the wave function at the exceptional point is *always*, i.e. irrespective of a particular physical situation, of the form

$$|\psi_{EP}\rangle = |\psi_1\rangle \pm i|\psi_2\rangle. \tag{12}$$

where the plus or minus sign refers to a specific exceptional point. The $|\psi_i\rangle$ are the two wave functions that coalesce at the exceptional point. No other superposition is possible. For a time dependent problem this signals chiral behavior. If the two wave functions relate to different parities or to different linear polarizations, the superposition is obviously chiral; in the latter case it is a circularly polarized wave of specific orientation.

SUMMARY AND OUTLOOK

Exceptional points are a fascinating subject of theoretical physics. As they are the only singularities of the spectrum for a matrix problem, they 'make' the spectrum, so to speak. They are directly associated with level repulsion. As a consequence, their statistical properties relate directly to that of the spectrum itself. Integrable systems give rise to a geometrical ordering while chaotic systems relate to disordered arrangements, yet with a specific distribution function.

The local behavior is fascinating on its own. The topological structure of a square root branch point is a physical reality with all its consequences for the wave functions. In addition, with its intrinsic chiral behavior it may even hold some promise to shed light on the ubiquitous left-right asymmetry of our macroscopic world. On a speculative note: the kinematic relation $E = \pm\sqrt{\vec{p}^2 + m^2}$ bears all properties of an exceptional point at $|\vec{p}| = \pm im$. It was this relationship that led Dirac to his famous equation including spin and the properties of chirality.

ACKNOWLEDGMENTS

The author acknowldges pertinent discussions with the Darmstadt group and with HL Harney (MPI-Heidelberg).

REFERENCES

1. T. Kato, *Perturbation theory of linear operators* Springer, Berlin 1966
2. W.D. Heiss and A.L. Sannino, J. Phys. **A23** 1167 (1990)
3. W.D. Heiss and W.-H. Steeb, J. Math. Phys. **32** 3003 (1991)

4. P.E. Shanley, Ann. Phys. (N.Y.) **186** 292 (1989); C. Bender and T.T. Wu, Phys. Rev. **D7** 1620 (1973).
5. W.D. Heiss and A.L. Sannino, Phys. Rev. **A43** 4159 (1991)
6. W.D. Heiss Z. Phys. A-Atomic Nuclei **349** 9 (1994)
7. A.J. Glick, H.J. Lipkin and N. Meshkov, *Nucl. Phys.* **62** 199 (1965)
8. C. Dembowski, H.-D. Gräf, H.L. Harney, A. Heine, W.D. Heiss, H. Rehfeld and A. Richter, Phys.Rev.Lett. **86**, 787 (2001)
9. M.V. Berry, *Quantum Chaos*, ed. by G. Casati (London: Plenum) 1985; Proc. R. Soc. A **239**, 45 (1983); M.V. Berry and M. Wilkinson, Proc. R. Soc. A **392**, 15 (1984)
10. W.D. Heiss, Eur.Phys.J. D7, 1 (1999); W.D. Heiss, Phys. Rev. E **61**, 929 (2000)
11. W.D. Heiss and H.L. Harney, Eu.Phys.J. **D**, submitted
12. A.L. Shuvalov and N.H. Scott, Acta Mech. **140**, 1 (2000)
13. S Pancharatnam, Proc.Ind.Acad.Sci. **XLII**, 86 (1955)
14. M.V. Berry, Current Science **67**, 220 (1994)

Spontaneous coherence and non-equilibrium correlation phase transitions in microscopic and mesoscopic systems

Sergey Yu. Kun

Department of Theoretical Physics, RSPhysSE, IAS, The Australian National University, Canberra ACT 0200, Australia

Abstract. We calculate the time-dependence of the micro-channel coherence in complex quantum collisions. It is argued that this micro-channel coherence undergoes correlation phase transitions: initially the micro-channel coherence is absent and it switches on spontaneously by abrupt jumps at precisely defined moments of time. As time proceeds further the micro-channel correlations decay exponentially. A possibility for experimental test of the predicted effect is proposed.

INTRODUCTION

In the modern theory of quantum chaotic scattering and complex quantum collisions it is commonly assumed (see, e.g. Refs. [1-4]) that the decoherence time, i.e. the time it takes to lose all the initial phase correlations, is the shortest time scale of the problem. This implies the absence of correlation between the partial width amplitudes that carry different total spin and exit micro-channel quantum numbers for collisions proceeding through the formation of highly excited intermediate system (IS). Some examples of complex quantum collisions and chaotic scattering are: compound nucleus reactions [5-7], heavy-ion collisions [8], electron-ion collisions [9], atomic and molecular collisions [10], the photo-dissociation of polyatomic molecules [11], collisions between atomic clusters [12,13] and electron propagation through nanostructures in the presence of electron-electron interaction, e.g., through many-electron quantum dots [1,14,15,16].

The idea of a rapid phase randomization has been originally developed and successfully applied in the context of the random-matrix theory (RMT) of highly excited strongly interacting systems [1]. Yet the domain of its applicability for many-body systems has been remained an open question. Therefore experimental and theoretical [17] identifications of the deviations from the predictions of RMT represents a challenging area in microscopic and mesoscopic many-body physics.

Experimentally, distinct deviations from the RMT have been revealed in strongly dissipative heavy-ion collisions [18] and fusion [19]. These deviations manifests themselves in strong micro-channel correlations (MC) leading to the non-self-averaging of excitation function oscillations in these complex microscopic collisions. An experimental indication for a similar phenomenon is noticeable in the cross section for a fusion of C_{60} atomic clusters [13], i.e. in nanoscale regime. The manifestation of quantum coherence effects for hot C_{60} atomic clusters has been unambiguously demonstrated in recent

CP597, *Nonequilibrium and Nonlinear Dynamics in Nuclear and Other Finite Systems*, edited by Z. Li et al.

experiments [20].

In this contribution we present a derivation of the MC in complex quantum collisions by developing an approach proposed in Refs. [21,22]. This approach suggests that the spin off-diagonal S-matrix correlations and MC in highly excited many-body systems occur spontaneously. We concentrate specifically on the time-dependence of the MC. We argue that the MC undergo phase transitions, i.e., the MC switch on spontaneously by abrupt jumps at precisely defined moments of time which are determined by the value of spin decoherence width β. The finite β-values are intimately related to a phenomenon of the localization, $\beta \ll \Gamma_{spr}$, in orthogonal subspaces of the Hilbert space [21], where Γ_{spr} is the spreading width [1,7]. Physically, the inverse decoherence width, $\hbar/\beta$, sets up a new time scale for quantum many-body systems. This new scale is analogous to the inverse Thouless energy for a single-particle motion in disordered systems. For times shorter than this scale the RMT of quantum many-body systems ceases to apply [1]. While RMT develops "a new statistical mechanics" (Dyson) of fully equilibrated finite systems, our previous work [21,22,23] and this contribution aim at the description of critical and non-equilibrium phenomena in finite highly-excited many-body systems.

An identification of the MC correlation phase transitions in this contribution reinforces the physical picture [23,24,25] of the extreme sensitivity and quantum chaos in finite highly-excited many-body systems and its qualitative analogy with the sensitivity of the direction of the spontaneous magnetization vector to the direction of an infinitesimally small external magnetic field.

THEORETICAL FORMALISM

We treat the collision in terms of the formation and decay of a highly excited IS assuming rotational invariance of the total Hamiltonian of the system. The normalized S-matrix elements are taken [22] in the pole form $\bar{S}^J_{\bar{a}\bar{b}}(E) = (\Gamma D/2\pi)^{1/2} \sum_\mu \bar{\gamma}^{J\bar{a}}_\mu \bar{\gamma}^{J\bar{b}}_\mu / (E - E^J_\mu + i\Gamma/2)$. Here, E is the total energy, J is the total spin, Γ is the total width of the resonance levels, $D \ll \Gamma$ is the average level spacing and E^J_μ are the resonance energies of the highly excited IS. The indices $\bar{a}(\bar{b})$ specify intrinsic micro-states of the reaction partners in the entrance (exit) channels. The normalized partial width amplitudes, $\bar{\gamma}^{J\bar{a}(\bar{b})}_\mu$, are real random Gaussian variables with a mean value of zero. Both $\bar{\gamma}$'s and $\bar{S}^J_{\bar{a}\bar{b}}(E)$ are taken to be independent (or smoothly dependent) of the orbital momenta and channel spins due to the continuum-continuum correlation [21,22]. The above pole expansion for the S-matrix elements is not unitary. However under the condition $N \gg 1$, where N is the number of open channels, the restrictions due to unitarity relax as comparing with the case of few open channels. Moreover, in what follows, we shall consider subsets of $N_{\bar{a}(\bar{b})} \gg 1$ micro-channels but with $N_{\bar{a}(\bar{b})} \ll N$. We can show that the micro-channel correlation phase transitions can also be obtained using a unitary S-matrix [6] (to be reported elsewhere).

Our purpose is to calculate first the time (t) dependence of the spin off-diagonal, $J \neq J'$, MC

$$\Pi^{JJ'}(t) = \overline{\mathcal{P}^J(t)\mathcal{P}^{J'}(t)^*}. \tag{1}$$

Here, the overbar stands for ensemble averaging, $\mathcal{P}^J(t) = (1/N_{\bar{a}}N_{\bar{b}})\sum_{\bar{a}\bar{b}}\mathcal{P}^J_{\bar{a}\bar{b}}(t)$, and $\mathcal{P}^J_{\bar{a}\bar{b}}(t) = \lim_{I\to\infty}(2\pi\hbar I)^{-1/2}\int_{-I/2}^{I/2}dE\exp(-iEt/\hbar)\bar{S}^J_{\bar{a}\bar{b}}(E)$.

We first calculate the MC (1) in terms of the partial MC, $\Pi^{JJ'}_{\bar{a}\bar{b},\bar{a}'\bar{b}'}(t) = \overline{\mathcal{P}^J_{\bar{a}\bar{b}}(t)\mathcal{P}^{J'}_{\bar{a}'\bar{b}'}(t)^*}$. This means that, in the r.h.s. of Eq. (1), we perform, first, the ensemble averaging leading to establishing of the spin off-diagonal MC, and, then, perform the micro-channel averaging: $\Pi^{JJ'}(t) = (1/N_{\bar{a}}N_{\bar{b}})^2\sum_{\bar{a}\bar{b}\bar{a}'\bar{b}'}\Pi^{JJ'}_{\bar{a}\bar{b},\bar{a}'\bar{b}'}(t)$.

In order to calculate $\Pi^{JJ'}_{\bar{a}\bar{b},\bar{a}'\bar{b}'}(t)$ one needs an explicit expression for the correlator $r^{JJ'}_{\mu\nu}(\bar{a}\bar{b},\bar{a}'\bar{b}') = \overline{\bar{\gamma}^{J\bar{a}}_\mu\bar{\gamma}^{J\bar{b}}_\nu\bar{\gamma}^{J'\bar{a}'}_\mu\bar{\gamma}^{J'\bar{b}'}_\nu}$, which is unknown yet. However this correlator does not vanish due to the non-vanishing [21,22] of the spin off-diagonal MC:

$$\overline{\bar{\gamma}^{J\bar{a}}_\mu\bar{\gamma}^{J'\bar{b}}_\nu} = \rho^{JJ'}_{\mu\nu}(\bar{a}\bar{b}), \tag{2}$$

where $\rho^{JJ'}_{\mu\nu}(\bar{a}\bar{b})$ have random signs and

$$(\rho^{JJ'}_{\mu\nu}(\bar{a}\bar{b}))^2 = (D/2\pi)\mathrm{Re}\{1/[\beta|J-J'|-i(E^J_\mu-E^{J'}_\nu-\hbar\omega(J-J'))]\}. \tag{3}$$

Here $\beta \gg D$ is the spin decoherence width and ω is the angular velocity of the coherent rotation of the IS [22].

Let us consider first the MC for $J > I > J'$ or $J > I > J'$, where I is the average total spin of the IS. Clearly, the non-vanishing of $r^{JJ'}_{\mu\nu}(\bar{a}\bar{b},\bar{a}'\bar{b}')$ itself does not yet guarantee the non-vanishing of $\Pi^{JJ'}_{\bar{a}\bar{b},\bar{a}'\bar{b}'}(t)$. In order to have non-zero off-diagonal MC one requires a certain degree of correlation between the signs of $\rho^{JJ'}_{\mu\nu}(\bar{a}\bar{b})$ and $\rho^{JJ'}_{\mu\nu}(\bar{a}'\bar{b}')$. Such a correlation is expected to arise from the spontaneous origin of the correlations between $\bar{\gamma}^J_\mu$ and $\bar{\gamma}^{J'}_\nu$ [21,22]. Therefore it is reasonable, as a working hypothesis, to assume that the quantities $\rho^{JJ'}_{\mu\nu}(\bar{a}\bar{b})$ are micro-channel independent: $\rho^{JJ'}_{\mu\nu}(\bar{a},\bar{b}) = \rho^{JJ'}_{\mu\nu}$. This leads to the undistinguishability between the MC (1) and the partial MC: $\Pi^{JJ'}_{\bar{a}\bar{b},\bar{a}'\bar{b}'}(t) = \Pi^{JJ'}(t)$. Thus writting $r^{JJ'}_{\mu\nu}(\bar{a}\bar{b},\bar{a}'\bar{b}') = r^{JJ'}_{\mu\nu} = (\rho^{JJ'}_{\mu\nu})^2$ we obtain

$$\Pi^{JJ'}(t) = \theta(t)(\Gamma/\hbar)\exp[-(\Gamma+\beta|J-J'|+i\hbar\omega(J-J'))t/\hbar], \tag{4}$$

where $\theta(t)$ is the step function.

It should be stressed that the MC does not occur within a single, I, total spin value [21,22]:

$$\overline{\bar{\gamma}^{J\bar{a}}_\lambda\bar{\gamma}^{J\bar{b}}_{\lambda'}} = \delta_{\lambda\lambda'}\delta_{\bar{a}\bar{b}}. \tag{5}$$

The deviation from the conventional RMT can only be originated from the *spin off-diagonal*, $J \neq J'$, MC (2). In other words, the requirement that more than one spin value I contributes to the collision is a precondition for the MC to occur. If the only one single spin value I contributes to the collision, our approach reproduces the results of RMT regarding the absence of MC,

$$\Pi^{II}(t) = 0. \tag{6}$$

321

for arbitrary t. Moreover, we shall see below (Eq. (9)) that Eq. (6) leads to

$$\Pi^{JI}(t) = 0 \qquad (7)$$

for arbitrary t and J, implying that at least three different total spin values (I and $I \pm 1$) are required for the MC to occur.

Because of the constraints (6) and (7), Eq. (4) has the meaning of a conditional equality: it holds *provided* it does not contradict to Eqs. (6) and (7). Otherwise, Eq. (4) is not valid. Therefore, in calculating $\overline{\mathcal{P}_{\bar{a}\bar{b}}^{J}(t)\mathcal{P}_{\bar{a}'\bar{b}'}^{J'}(t)^*}$ we have to consider it as the correlation moment of a joint *conditional* probability distribution of $\mathcal{P}_{\bar{a}\bar{b}}^{J}(t)$ and $\mathcal{P}_{\bar{a}'\bar{b}'}^{J'}(t)$ provided that the correlation between $\mathcal{P}_{\bar{a}\bar{b}}^{\tilde{J}}(t)$ and $\mathcal{P}_{\bar{a}'\bar{b}'}^{I}(t)$ is absent for an arbitrarily $\tilde{J}$ and $(\bar{a}\bar{b}) \neq (\bar{a}'\bar{b}')$.

Clearly Eq. (4) and Eqs. (6) and (7) are not consistent on the short time interval, $t \ll \hbar/\beta|J - I|, \hbar/\beta|J' - I|$. Indeed, taking into account that $\overline{|\mathcal{P}_{\bar{a}\bar{b}}^{J}(t)|^2} = \theta(t)(\Gamma/\hbar)\exp(-\Gamma t/\hbar)$, Eq. (4) yields $\exp(-i\omega t J)\mathcal{P}_{\bar{a}\bar{b}}^{J}(t) \simeq \exp(-i\omega t J')\mathcal{P}_{\bar{a}'\bar{b}'}^{J'}(t)$. However, since $\overline{\mathcal{P}_{\bar{a}\bar{b}}^{J}(t)\mathcal{P}_{\bar{a}\bar{b}}^{J'}(t)^*}$ is given by the r.h.s. of Eq. (4) for arbitrary J, J' and t [22], we also have $\exp(-i\omega t J)\mathcal{P}_{\bar{a}\bar{b}}^{J}(t) \simeq \exp(-i\omega t I)\mathcal{P}_{\bar{a}\bar{b}}^{I}(t)$ and $\exp(-i\omega t J')\mathcal{P}_{\bar{a}'\bar{b}'}^{J'}(t) \simeq \exp(-i\omega t I)\mathcal{P}_{\bar{a}'\bar{b}'}^{I}(t)$ for $t \ll \hbar/\beta|J - I|, \hbar/\beta|J' - I|$. This yields $\mathcal{P}_{\bar{a}\bar{b}}^{I}(t) \simeq \mathcal{P}_{\bar{a}'\bar{b}'}^{I}(t)$ and $\exp(-i\omega t J)\mathcal{P}_{\bar{a}\bar{b}}^{J}(t) \simeq \exp(-i\omega t I)\mathcal{P}_{\bar{a}'\bar{b}'}^{I}(t)$. Thus, for $t \ll \hbar/\beta$, Eq. (4) and Eqs. (6) and (7) can not be satisfied simultaneously, invalidating Eq. (4) for short times. It is also clear that the constraints (6) and (7) results in the vanishing of the MC at the initial moment of time, $\Pi^{JJ'}(t = 0) = 0$, for arbitrary J, J'.

In order to find the earliest moment of time at which Eq. (4) becomes valid we use the relation $\bar{\gamma}_{\mu}^{J\bar{a}}\bar{\gamma}_{\mu}^{I\bar{b}} = \sum_{\lambda}\bar{\gamma}_{\lambda}^{J\bar{a}}\bar{\gamma}_{\lambda}^{I\bar{b}}(\rho_{\mu\lambda}^{JI})^2 + R_{\mu,\bar{a}\bar{b}}^{JI}$ (see Eqs. (8.1-8.3) in Ref. [21]). This accordingly yields $\bar{S}_{\bar{a}\bar{b}}^{J}(E) = \Delta S_{\bar{a}\bar{b}}^{J}(E) + \delta S_{\bar{a}\bar{b}}^{J}(E)$, where $\Delta S_{\bar{a}\bar{b}}^{J}(E)$ and $\delta S_{\bar{a}\bar{b}}^{J}(E)$ are uncorrelated Gaussian stochastic processes. We also have $\mathcal{P}_{\bar{a}\bar{b}}^{J}(t) = \Delta \mathcal{P}_{\bar{a}\bar{b}}^{J}(t) + \delta \mathcal{P}_{\bar{a}\bar{b}}^{J}(t)$, and when $D \ll \Gamma, \beta$ we change the μ-summation to E_{μ}-integration to obtain $\Delta \mathcal{P}_{\bar{a}\bar{b}}^{J}(t) = \exp[i\omega t(J - I) - \beta t|J - I|t/\hbar]\mathcal{P}_{\bar{a}\bar{b}}^{I}(t)$. This means that, because of the constraints (6) and (7), the MC can only originate from the correlation between $\delta \mathcal{P}_{\bar{a}\bar{b}}^{J}(t)$ and $\delta \mathcal{P}_{\bar{a}'\bar{b}'}^{J'}(t)$. Therefore, for $J - I = I - J'$, the intensity of this correlation must not exceed $\overline{|\delta \mathcal{P}_{\bar{a}\bar{b}}^{J}(t)|^2} = (\Gamma/\hbar)\exp(-\Gamma t/\hbar)[1 - \exp(-2\beta t|J - I|/\hbar)]$. This requires $|\Pi^{JJ'}(t)| \leq \overline{|\delta \mathcal{P}_{\bar{a}\bar{b}}^{J}(t)|^2}$. One finds that the above inequality holds for $t \geq \tau_J$, where $\tau_J = \hbar \ln 2/(2\beta|J - I|)$. For $t < \tau_J$, Eq. (4) and Eqs. (6) and (7) can not be satisfied simultaneously invalidating Eq. (4) with $J - I = I - J'$ for $t < \tau_J$.

Recall that Eq. (4) has been obtained by performing, first, ensemble averaging, and, second, the micro-channel averaging. Clearly this order of the two averaging procedures is inconsistent with the constraints (6) and (7) losing its physical meaning for $t < \tau_J$. This leads us to calculate $\Pi^{JJ'}(t < \tau_J)$ by changing the order of the two averaging procedures and performing, first, the micro-channel averaging, and, second, ensemble averaging. Since $\overline{\Delta \mathcal{P}_{\bar{a}\bar{b}}^{J}(t)}^{\bar{a}\bar{b}} = 0$, we have $\mathcal{P}^{J}(t) = \overline{\mathcal{P}_{\bar{a}\bar{b}}^{J}(t)}^{\bar{a}\bar{b}} = \overline{\delta \mathcal{P}_{\bar{a}\bar{b}}^{J}(t)}^{\bar{a}\bar{b}}$, where $\overline{(...)}^{\bar{a}\bar{b}} = $

$(1/N_{\bar a}N_{\bar b})\sum_{\bar a\bar b}(...)$. This yields $\Pi^{JJ'}(t)=\overline{\delta\mathcal{P}^J_{\bar a\bar b}(t)\delta\mathcal{P}^{J'}_{\bar a'\bar b'}(t)^*}$ guaranteeing that $|\Pi^{JJ'}(t)|\le$
$\overline{|\delta\mathcal{P}^J_{\bar a\bar b}(t)|^2}$ for $J-I=I-J'$. We use the decomposition $\delta\mathcal{P}^{J(J')}_{\bar a\bar b}(t)=\delta\tilde{\mathcal{P}}^{J(J')}_{\bar a\bar b}(t)+\mathcal{P}^{J(J')}(t)$,
where, by definition, $\overline{\delta\tilde{\mathcal{P}}^{J(J')}_{\bar a\bar b}(t)}^{\bar a\bar b}=0$. We obtain

$$\overline{\mathcal{P}^J(t)\mathcal{P}^{J'}(t)^*}=\overline{\delta\mathcal{P}^J_{\bar a\bar b}(t)\delta\mathcal{P}^{J'}_{\bar a\bar b}(t)^*}-\overline{\delta\tilde{\mathcal{P}}^J_{\bar a\bar b}(t)\delta\tilde{\mathcal{P}}^{J'}_{\bar a\bar b}(t)^*}.\qquad(8)$$

Using the decomposition $\delta S^J_{\bar a\bar b}(E)=\bar S^J_{\bar a\bar b}(E)-\Delta S^J_{\bar a\bar b}(E)$ and Eqs. (3,5) we find that $\overline{\delta S^J_{\bar a\bar b}(E+\varepsilon)\delta S^{J'}_{\bar a\bar b}(E)^*}=0$ for arbitrary ε. Therefore, the first term in the r.h.s. of Eq. (8) vanishes. Let us change $\delta\tilde{\mathcal{P}}^{J(J')}_{\bar a\bar b}(t)\to\exp(i\alpha_{J(J')})\delta\tilde{\mathcal{P}}^{J(J')}_{\bar a\bar b}(t)$, where $\alpha_J\ne\alpha_{J'}$ are arbitrary real $(\bar a\bar b)$-independent phases. This transformation changes the r.h.s. of Eq. (8) if $\overline{\delta\tilde{\mathcal{P}}^J_{\bar a\bar b}(t)\delta\tilde{\mathcal{P}}^{J'}_{\bar a\bar b}(t)^*}$ is finite. However, since $\overline{\delta\tilde{\mathcal{P}}^{J(J')}_{\bar a\bar b}(t)}^{\bar a\bar b}=0$, $\mathcal{P}^{J(J')}(t)$ must be $\alpha_{J(J')}$-independent. Therefore the r.h.s. of Eq. (8) must vanish yielding $\overline{\mathcal{P}^J(t)\mathcal{P}^{J'}(t)^*}=0$, and, thus, $\Pi^{JJ'}(t<\tau_J)=0$ with $J-I=I-J'$.

Let us consider the obvious inequality $|\overline{\mathcal{P}^J(t)\mathcal{P}^{J'}(t)^*}|\le(\overline{|\mathcal{P}^J(t)|^2}\ \overline{|\mathcal{P}^{J'}(t)|^2})^{1/2}$, i.e.

$$|\Pi^{JJ'}(t)|\le[\Pi^{JJ}(t)\Pi^{J'J'}(t)]^{1/2},\qquad(9)$$

which demonstrates that Eq. (6) leads to Eq. (7). The inequality (9) is valid independent of the order of the micro-channel averaging and ensemble averaging. This inequality shows that the spin off-diagonal MC, $\Pi^{JJ'}(t)\ne 0$, leads necessarily to the spin diagonal MC, $\Pi^{JJ}(t)\ne 0$ and $\Pi^{J'J'}(t)\ne 0$. However the explicit form of $r^{J=J'}_{\mu\mu'}$ is unknown. Therefore we can not calculate $\Pi^{JJ}(t)$ in a straightforward way similar to the calculation of $\Pi^{J\ne J'}(t)$. Yet, $\Pi^{JJ}(t)$ can be calculated without the knowledge of the $r^{J=J'}_{\mu\mu'}$.

Consider the relation (9) for $J-I=I-J'$, so that $\Pi^{JJ}(t)=\Pi^{J'J'}(t)$. We have seen that the non-vanishing of the micro-channel averaging, i.e. the non-vanishing of the spin diagonal MC, can be obtained only after the spin off-diagonal MC are switched on by means of ensemble averaging. This means that the spin diagonal MC originates exclusively from the spin off-diagonal MC implying that the strength of the former should not exceed the intensity of the latter one. Therefore we consider the relation (9) as an equality and obtain $\Pi^{JJ}(t)=\theta(t-\tau_J)(\Gamma/\hbar)\exp[-(\Gamma+2\beta|J-I|)t/\hbar]$. Consequently, using Eq. (4) we have for arbitrary $J>I>J'$ and $J<I<J'$

$$\Pi^{JJ'}(t)=\theta(t-\tau^{max}_{JJ'})(\Gamma/\hbar)\exp[-(\Gamma+\beta(|J-I|+|J'-I|)+i\hbar\omega(J-J'))t/\hbar],\qquad(10)$$

where $\tau^{max}_{JJ'}=\max(\tau_J,\tau_{J'})$.

Let us consider such $\tilde J,J,J'$-values that $\tilde J<I<J,J'$ or $\tilde J>I>J,J'$ with $|\tilde J-I|\ge|J-I|,|J'-I|$. From Eq. (10) we have $\exp[i\omega t(J-\tilde J)]\Pi^{\tilde J J'}(t)=[\Pi^{\tilde J\tilde J}(t)\Pi^{JJ}(t)]^{1/2}$ and $\exp[i\omega t(J'-\tilde J)]\Pi^{\tilde J J'}(t)=[\Pi^{\tilde J\tilde J}(t)\Pi^{J'J'}(t)]^{1/2}$. We can show that these two relations yield $\exp[i\omega t(J'-J)]\Pi^{JJ'}(t)=[\Pi^{JJ}(t)\Pi^{J'J'}(t)]^{1/2}$ demonstrating that Eq. (10) holds for arbitrary J,J'.

To interpret the abrupt switching on of the MC at $t = \tau_J$, we represent $\bar{S}^J_{\bar{a}\bar{b}}(E)$ in the form $\bar{S}^J_{\bar{a}\bar{b}}(E) = \tilde{S}^J_{\bar{a}\bar{b}}(E) + S^J_{\bar{a}\bar{b}}(E)$, where $\tilde{S}^J_{\bar{a}\bar{b}}(E) = \int_0^{\tau_J} dt \exp(iEt/\hbar)\mathcal{P}^J_{\bar{a}\bar{b}}(t)$ and $S^J_{\bar{a}\bar{b}}(E) = \int_{\tau_J}^{\infty} dt \exp(iEt/\hbar)\mathcal{P}^J_{\bar{a}\bar{b}}(t)$. We observe that the MC originates from $S^J_{\bar{a}\bar{b}}(E)$ while $\tilde{S}^J_{\bar{a}\bar{b}}(E)$ do not contribute to the MC. Physically, the absence of the MC for $t < \tau_J$ indicates that the associated characteristic energy scale, $\hbar/\tau_J = 2\beta|J - I|/\ln 2$, is not resolved on the time interval shorter than τ_J.

Our result regarding a coherence between wave functions of either different micro-states and even different masses and charges of the collision fragments may seem to contradict one of the basic principle of quantum mechanics: quantum coherence-interference is allowed only for identical systems. This objection can be removed by considering the two asymptotically separated highly excited collision fragments not as independent systems but rather as correlated *entangled* [26,27] components of the same single system. This entanglement occurs in the collision zone before the decay of the highly excited many-body IS. The entanglement of the collision fragments implies that quantum interference-coherence between different micro-channel configurations is analogous to the interference-coherence between different resonance configurations of the highly excited IS.

It follows from Eq. (10) that the MC can be neglected for $t \gg \tau_J \sim \hbar^2/\beta$, where we have taken into account that J, I are in $\hbar$ units. Therefore, the criterion of applicability of RMT is $\tau_J \ll \hbar/\Gamma$. It should be noted that absence of the MC for $t < \tau_J$ does not imply the overall applicability of the RMT on this short time interval. This is because of the strong micro-channel diagonal, $(\bar{a}\bar{b}) = (\bar{a}'\bar{b}')$, but spin off-diagonal, $J \neq J'$, correlation for $t < \tau_J, \tau_{J'}$ [21,22].

The MC (10) has been obtained in two steps by changing the order of ensemble and micro-channel averaging at $t = \tau^{max}_{JJ'}$. This order has been changed to satisfy the constraints (6) and (7) in the calculation of $\Pi^{JJ'}(t)$ by taking $r^{JJ'}_{\mu\nu} = (\rho^{JJ'}_{\mu\nu})^2$ for $J < I < J'$ and $J > I > J'$. Let us ask what is the expression for $r^{JJ'}_{\mu\nu}$ which leads to the MC (10) and is consistent with the constrains (6) and (7) on the entire time interval, so that there is no need to change the order of ensemble and micro-channel averaging. One can find that the answer is:

$$r^{JJ'}_{\mu\nu} = (D/2\pi)\exp[-\beta(|J - I| + |J' - I|)\tau^{max}_{JJ'}/\hbar]$$

$$\mathrm{Re}\{\exp[i(E^J_\mu - E^{J'}_\nu - \hbar\omega(J - J'))\tau^{max}_{JJ'}/\hbar]/ \quad\quad (11)$$

$$[\beta(|J - I| + |J' - I|) - i(E^J_\mu - E^{J'}_\nu - \hbar\omega(J - J'))]\}.$$

Having expressed $\Pi^{JJ'}(t)$ by (10), we can calculate the MC energy cross correlation:

$$\overline{S^J(E + \varepsilon)S^{J'}(E)^*} = \int_{\tau^{max}_{JJ'}}^{\infty} dt \exp(i\varepsilon t/\hbar)\Pi^{JJ'}(t) = i(\Gamma/\lambda^{JJ'})\exp(i\tau^{max}_{JJ'}\lambda^{JJ'}/\hbar). \quad (12)$$

Here, $\lambda^{JJ'} = \varepsilon - \hbar\omega(J - J') + i\beta(|J - I| + |J' - I|) + i\Gamma$, and $S^J(E) = \overline{\bar{S}^J_{\bar{a}\bar{b}}(E)}^{\bar{a}\bar{b}} = \overline{S^J_{\bar{a}\bar{b}}(E)}^{\bar{a}\bar{b}}$, where the micro-channel averaging is performed after ensemble averaging in Eq. (12).

324

For a numerical simulation of the spontaneous coherence and non-equilibrium phase transitions in complex quantum collisions one can use the following expression

$$S^J(E) = (\Gamma D/2\pi)^{1/2} \sum_v \bar{\gamma}_v^{(1)} \bar{\gamma}_v^{(2)} \exp(i\tau_J Q_v^J/\hbar)/Q_v^J, \tag{13}$$

where $Q_v^J = E - E_v^J - \hbar\omega(J - I) + i\beta|J - I| + i\Gamma/2$, $\bar{\gamma}_v^{(1)}$ and $\bar{\gamma}_v^{(2)}$ are real random Gaussian variables with zero mean value and the following non-correlation properties: $\overline{\bar{\gamma}_v^{(1)}\bar{\gamma}_{v'}^{(1)}} = \overline{\bar{\gamma}_v^{(2)}\bar{\gamma}_{v'}^{(2)}} = \delta_{vv'}$ and $\overline{\bar{\gamma}_v^{(1)}\bar{\gamma}_{v'}^{(2)}} = 0$. Indeed, one can check that $\lim_{I\to\infty}(1/2\pi\hbar I)\overline{\mathcal{L}^J(t)\mathcal{L}^{J'}(t)^*} = \Pi^{JJ'}(t)$ (10), where $\mathcal{L}^{J(J')}(t) = \int_{-I/2}^{I/2} dE \exp(-iEt/\hbar)$ $S^{J(J')}(E)$ and $S^{J(J')}(E)$ are given by Eq. (13).

In molecular and mesoscopic physics, e.g., for photo-disintegration of highly-excited polyatomic molecules and atomic clusters, the micro-channel correlation phase transitions can be searched for by employing the experimental methods of femtochemistry, i.e. short femtosecond pump and probe laser pulses [28]. Alternatively, one can use long laser pulses with well defined photon energy and study the detailed energy dependence of the cross section. Indeed, Eqs. (12) and (13) demonstrate that the non-equilibrium correlation phase transitions distinctly affect the energy dependence of the cross sections, which are summed over a very large number of exit micro-channels. In particular, one should concentrate on the study of the Fourier components of the cross sections as well as on the analysis of the cross section energy autocorrelation and angle cross correlation functions. This method, which is based on the studies of the detailed high resolution energy dependence of the cross sections, is especially relevant in the search for the micro-channel correlation phase transitions in complex nuclear collisions, such as strongly dissipative heavy ion collisions [18] and heavy ion fusion [19]. In these processes, one deals with the characteristic time scales $\simeq 10^{-20} - 10^{-21}$ sec and the direct measurement of time evolution is not accessible.

CONCLUSION

We have studied the time dependence of the micro-channel coherence in complex quantum collisions. Our analysis indicates that the spontaneous micro-channel coherence undergoes a non-equilibrium correlation phase transitions. The possibility for an experimental test of the effect is suggested.

ACKNOWLEDGMENTS

I am grateful to Hans Weidenmüller for useful discussions and suggestions. I would like to thank members of the Max-Planck-Institut für Kernphysik, Heidelberg for a warm hospitality during my visit on May-July 2001, when this work was completed. This visit was supported by the Max Planck Society Fellowship.

REFERENCES

1. T. Guhr, A. Müller-Groeling, and H.A. Weidenmüller, Phys. Rep. **299**, 189 (1998), and references therein.
2. Y.V. Fyodorov and H.-J. Sommers, J. Math. Phys. **38**, 1918 (1997), and references therein.
3. U. Smilansky, in *Chaos and Quantum Physics*, Proc. of the Les Houches Summer School, Session LII, edited by M.J. Giannoni, A. Voros, and J. Zinn-Justin, (North-Holland, Amsterdam, 1991), p. 372, and references therein.
4. R. Blümel and U. Smilansky, Phys. Rev. Lett. **60**, 477 (1988); Phys. Rev. Lett. **64**, 241 (1990).
5. T. Ericson, Ann. Phys. (N.Y.) **23**, 390 (1963).
6. J.J.M. Verbaarschot, H.A. Weidenmüller, and M.R. Zirnbauer, Phys. Rep. **129**, 367 (1985), and references therein.
7. O. Bohigas and H.A. Weidenmüller, Annu. Rev. Nucl. Part. Sci. **38**, 421 (1988).
8. W. Nörenberg and H.A. Weidenmüller, *Introduction to the Theory of Heavy-Ion Collisions*, Lecture Notes in Physics, vol. 51 (Springer, Berlin, 1980).
9. V.V. Flambaum, A.A. Gribakina, and G.F. Gribakin, Phys. Rev. A**54**, 2066 (1996).
10. R. Blümel and W.P. Reinhardt, *Chaos in Atomic Physics*, (Cambridge University Press, 1997).
11. Y.V. Fyodorov and Y. Alhassid, Phys. Rev. A**58**, R3375 (1998).
12. R. Schmidt and H.O. Lutz, Comments At. Mol. Phys. **31**, 461 (1995).
13. F. Rohmund and E.E.B. Campbell, Phys. Rev. Lett. **76**, 3289 (1995).
14. B.L. Altshuler, Y. Gefen, A. Kamenev, and L. Levitov, Phys. Rev. Lett. **78**, 2803 (1997).
15. C. Mejia-Monasterio, J. Richter, T. Rupp, and H.A. Weidenmüller, Phys. Rev. Lett. **81**, 5189 (1998).
16. J.P. Bird, J. Phys.: Condens. Matter **11**, R413 (1999), and references therein.
17. T. Papenbrock, T.H. Seligman, and H.A. Weidenmüller, Phys. Rev. Lett. **80**, 3057 (1998); T. Papenbrock and T. Prosen, Phys. Rev. Lett. **84**, 262 (2000); T. Papenbrock, Phys. Rev. C **61**, 034602 (2000).
18. A. De Rosa *et al.*, Phys. Lett. B**160**, 239 (1985); G. Pappalardo, Nucl. Phys. A**488**, 395c (1988); T. Suomijärvi *et al.*, Phys. Rev. C**36**, 181 (1987); A. De Rosa *et al.*, Phys. Rev. C**37**, 1042 (1988); *ibid.* C**40**, 627 (1989); *ibid.* C**44**, 747 (1991); Wang Qi *et al.*, Chin. Phys. Lett. **10**, 656 (1993); Wang Qi *et al.*, Chin. J. Nucl. Phys. **15**, 113 (1993); F. Rizzo *et al.*, Z. Phys. A**349**, 169 (1994); Wang Qi *et al.*, High Energy Phys. and Nucl. Phys. **18**, 25 (1994); Lu Jun *et al.*, Chin. Phys. Lett. **12**, 661 (1995); M. Papa *et al.*, Z. Phys. A**353**, 205 (1995); Wang Qi *et al.*, High Energy Phys. and Nucl. Phys. **20**, 289 (1996); Lu Jun *et al.*, Chin. J. Nucl. Phys. **18**, 91 (1996); Wang Qi *et al.*, Phys. Lett. B**388**, 462 (1996); S.Yu. Kun *et al.*, Z. Phys. A**359**, 263 (1997); I. Berceanu *et al.*, Phys. Rev. C**57**, 2359 (1998); Wang Qi, Nucl. Phys. Review **15**, 74 (1998); Tian Wendong *et al.*, High Energy Phys. and Nucl. Phys. **23**, 334 (1999); I. Berceanu *et al.*, Pramana J. Phys. **53**, 419 (1999); M. Papa *et al.*, Phys. Rev. C **61**, 044614 (2000).
19. R.A. Racca, P.A. Deyoung, J.J. Kolata, and R.J. Thornburg, Phys. Lett. B**129**, 294 (1983).
20. M. Arndt, O. Nairz, J. Vos-Andreae, C. Keller, G. van der Zouw, and A. Zeilinger, Nature **401**, 680 (1999).
21. S.Yu. Kun, Z. Phys. A**357**, 255 (1997).
22. S.Yu. Kun, Z. Phys. A**357**, 271 (1997).
23. S.Yu. Kun, Phys. Rev. Lett. **84**, 423 (2000).
24. Wang Qi *et al.*, High Energy Phys. and Nucl. Phys. **24**, 1060 (2000).
25. Wang Qi *et al.*, "Experimental test of slow phase randomization and quantum chaos in finite highly excited many-body systems", in these Proceedings.
26. E. Schrödinger, Naturwissenschaften **23**, 807,823,844 (1935).
27. P. Knight, Nature **395**, 12 (1998); S. Dürr, T. Nonn, and G. Rempe, Nature **395**, 33 (1998).
28. A.H. Zewail, *Femtochemistry - Ultrafast Dynamics of the Chemical Bond*, vol. I and II, World Scientific, New Jersey, Singapore, 1994.

Phase Transition from Order to Chaos in Nuclei [1]

Xizhen Wu,[*] Zhuxia Li[*], Yingxun Zhang[*], Renfa Feng[*], Yizhong Zhuo[*] and Jianzhong Gu[†]

[*] China Institute of Atomic Energy, P. O. Box 275(18), Beijing 102413, P. R. China
[†] Center for Nonlinear Studies, Hong Kong Baptist University, Hong Kong, China

Abstract. Some aspects of quantum chaos in a finite system have been studied based on the analysis of statistical behaviors of quantum spectra in nuclei.The dependence of the order to chaos transition on nuclear deformation and nuclear rotating has been described. The influence of pairing effect on the statistical properties of spectra is also discussed. Some important experiment phenomena in nuclear physics have been understood from the point of view of interplay between order and chaos.

INTRODUCTION

Nowadays, the fruitful achievements on investigation of the classical chaos have opened our eyes [1], so that nobody doubts that in the quantum mechanics there is also a very important and unknown research field to be explore. A nucleus is to be considered as a typical quantum system which consists of several or several hundred of particles. In order to explore this puzzle in nuclei , one paid a great attention to study the statistical properties of quantum spectra. The behavior of nucleonic motion is dominated by the quantum mechanics and their energies are discrete. The energy levels (eigenvalues) in a nucleus form a set of spectra. For each of energy level, one can find a corresponding wave function (eigenvector) which describes its existence probability. From the study of eigenvalues and eigenstates in nuclei we have obtained a lot of information about irregular behavior which is related to chaos in nuclei. These quantum irregularity in nuclei can be divided into three kinds: the time evolving feature in non-stationary wave function; the statistical property of level spacing distribution and the distribution behavior of eigenstates. In this paper we devote ourselves to the statistical properties of neighbouring level spacing distribution in nuclei. We know that the property of level spacing distribution depends on the coupling between the levels and the coupling is closely related to the symmetry as well as structure of nuclei. If the system under consideration has high symmetry and the energy levels are fully random, i.e. there is no coupling between levels, then the neighboring level spacing distribution will obey a Poisson distribution

$$P(s) = \exp(-s), \tag{1}$$

[1] Supported by National Natural Science Foundation of China under Grant No. 19975073 and the Nuclear Industry and Science Foundation of China as well as Major State Basic Reseach Development Program in China under contract No. G20000774

CP597, *Nonequilibrium and Nonlinear Dynamics in Nuclear and Other Finite Systems,*
edited by Z. Li et al.

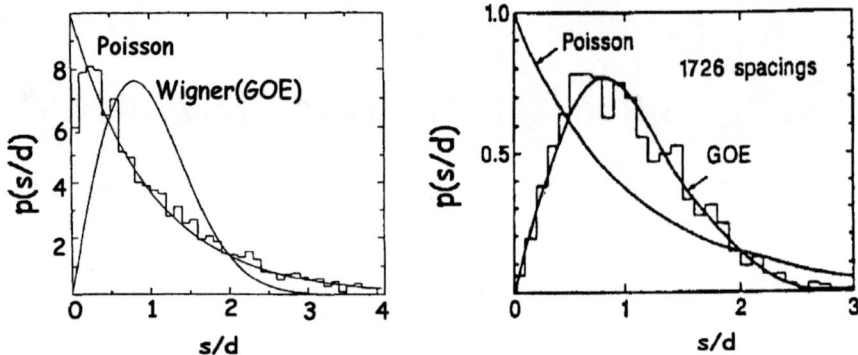

FIGURE 1. The nearest neighbor level spacing distribution for nuclei. (a) at the ground state and along the Yrast line; (b) at the excitation energy around the neutron separation energy. The histograms in (a) and (b) are for experiment data.

where, s is for the relative level spacing, i.e. the neighboring level spacing divided by the average level spacing. If the symmetry of system is broken, the coupling between levels will appear (i.e. the repulsion between levels exists), then the level spacing distribution will become a Wigner distribution

$$P(s) = \frac{\pi}{2} s \exp(-\frac{\pi s^2}{4}), \tag{2}$$

From the Random Matrix Theory [2, 3], one can deduce that the level spacing distribution in Gaussion orthognal ensemble(GOE) obeys a Wigner distribution. Therefore, one can consider that the fact that the level spacing distribution obeys a Wigner distribution is one of the characteristics of chaotic motion in quantum systems. Based on this point of view, people have studied the level spacing distribution for a quite long time and want to explore the physical contents of order to chaos transition in nuclei.

The statistical properties of spectra in spherical nuclei have been theoretically and experimentlly investigated. Since there are definite energy and angular momentum as wells as good parity at the groud states(along the Yrast line) of nuclei, the spectra in the ground states have a strong degeneracy and quite a simple structure, it is expected for these systems to be at regular states [4]. An experimentalist, Garrett [5], collected 2522 level spacing data and ploted the level spacing distribution (Fig.1(a)), The level spacings come from the rare-earth nuclear region between mass number 155 to 185 and proton number 62 to 77. This distribution, as we will see, is characterized by what is called a Poisson distribution. However, with increasing external energy input, nuclei are excited and the structure of nuclear levels become complicated. The repulsion between levels appears and it makes the degeneracy of levles decrease, therefore the level spacing distribution approaches to a Wigner distribution. Bohigas and his colleagues [6] collected 1726 level spacing data and gave a Wigner distribution of level spacing from experiment data, as shown Fig. 1(b). From the above analysis the transition from order to chaos with increasing an excitation energy in spherical nuclei has been observed and

328

recognized widely.

TRANSITION FROM ORDER TO CHAOS AND NUCLEAR FISSION

It is very interesting to explore whethere the phase transition from order to chaos exits or not for a deformed nucleus, especially for a large deformed nucleus explored by the fission [7]. Recently, our group [8] carried out the study about this kind of phase transition for the heavy nuclei, for instance, ^{252}Cf. The results showed a beautiful picture about the transition from order to chaos. In order to describe this phenomenon clearly, we employed the two center shell model (TCSM) ,the Hamiltonian of which is given by

$$h_0 = -\frac{\hbar^2 \nabla^2}{2m_0} + V(\rho, z) + V_{LS}(p, s) + V_{L^2}(l). \tag{3}$$

Where, V_{LS} and V_{L^2} are the spin - orbit term and L^2 term, respectively. The potential of TCSM is axially symmetric with respect to z-axis and is taken to be

$$V(\rho, z) = \begin{cases} \frac{1}{2}m_0\omega_{z_1}^2 z'^2 + \frac{1}{2}m_0\omega_{\rho_1}^2 \rho^2, & z < z_1 \\ \frac{f_0}{2}m_0\omega_{z_1}^2 z'^2 \left(1 + c_1 z' + d_1 z'^2\right) + \frac{1}{2}m_0\omega_{\rho_1}^2 \left(1 + g_1 z'^2\right)\rho^2, & z_1 < z < 0 \\ \frac{f_0}{2}m_0\omega_{z_2}^2 z'^2 \left(1 + c_2 z' + d_2 z'^2\right) + \frac{1}{2}m_0\omega_{\rho_2}^2 \left(1 + g_2 z'^2\right)\rho^2, & 0 < z < z_2 \\ \frac{1}{2}m_0\omega_{z_2}^2 z'^2 + \frac{1}{2}m_0\omega_{\rho_2}^2 \rho^2, & z > z_2 \end{cases} \tag{4}$$

with the abbreviation

$$z' = \begin{cases} z - z_1, & z < 0, \\ z - z_2, & z > 0. \end{cases}$$

All parameters appearing in above formula are related to five shape parameters, by which the nuclear shape can be described very well. They are the separation of the two centers $\Delta z = z_2 - z_1$, the neck parameter ε (ε=0 corresponds to no-neck shape and ε=1 to well necked-in shape), the mass asymmetry $X_i = \frac{(A_1 - A_2)}{(A_1 + A_2)}$ with A_1 and A_2 the mass numbers of the fragments (X_i ranges from 0 to 1) and finally the ellipsoidal deformations of the fragments, β_1 and β_2 .

Based on this model, we can calculate the liquid drop energy and shell correction, so that we can obtain the fission potential, as shown in Fig.2, for the important deformations. We also give the schematic diagram of the corresponding nuclear shapes in this figure. In Fig. 2, we can find three saddle points and three potential wells, among them the third saddle point and third well result from the mass-asymmetry fission. Since one can consider that the fissioning system undergoes fission along the fission path shown in this Figure, the investigation on the transition from order to chaos in this system has an extremely important significance. Here,we show the nearest neighbor level spacing distributions in Fig.3 for the different deformations along the fission path, Where (a)

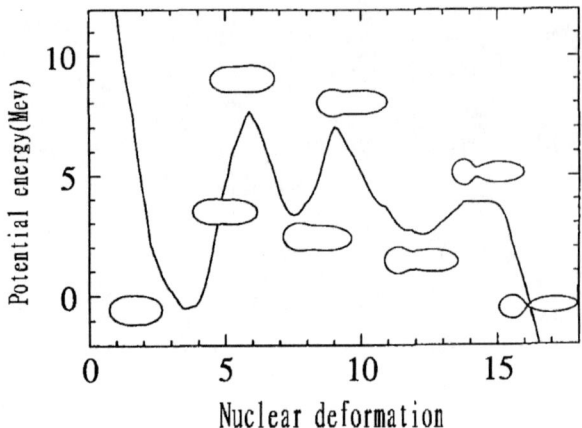

FIGURE 2. The fission potential of ^{252}Cf. Some important nuclear configurations are denoted.

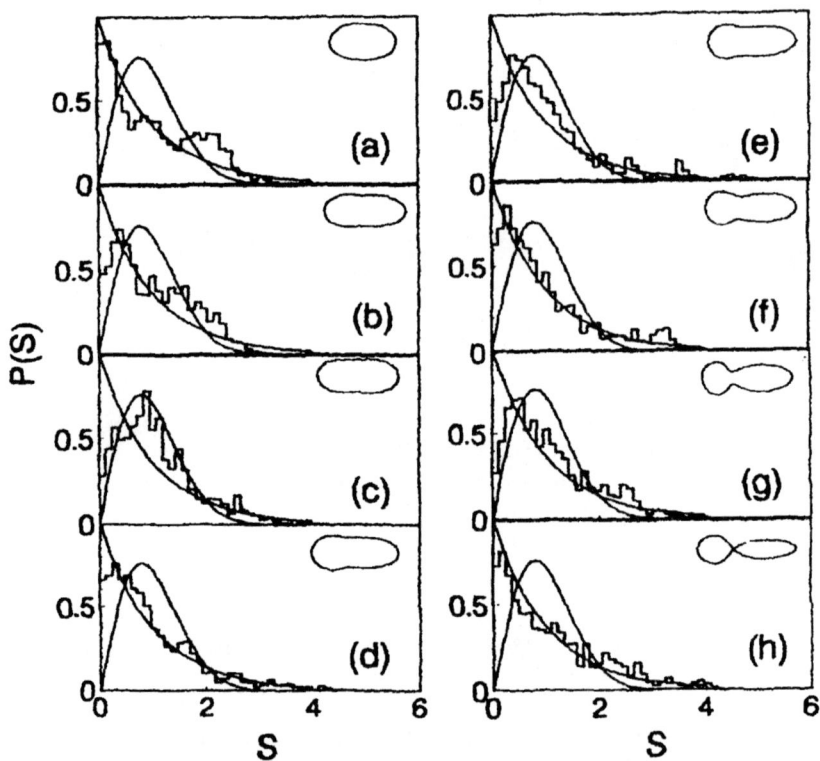

FIGURE 3. The level spacing distributions of spectra for the deformed space explored by the fission process of ^{252}Cf. The histograms are our numerical results. The solid lines denote Wigner distribution and Poisson distribution. The corresponding nuclear configurations are denoted in (a)-(h).

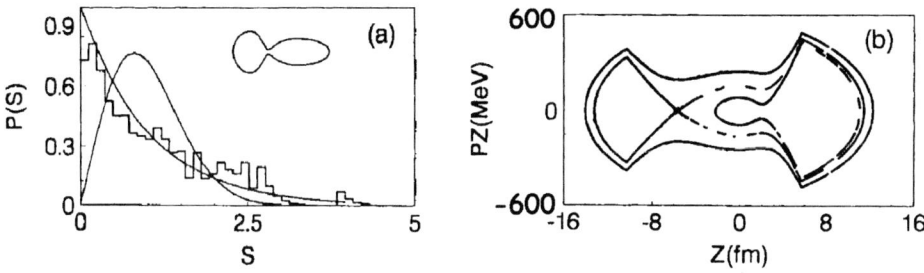

FIGURE 4. (a) The nearest neighbor level spacing distribution at the scission configuration of ^{252}Cf; (b)Poincarè section map at the scission configuration of ^{252}Cf.

is for the ground state, (b) for the configuration between the ground state and the first saddle point, (c) for the first saddle point, (d) for the second minimum, (e) for the second saddle point, (f) for the third minimum, (g)for the third saddle point, and (h) for the scission point. From Fig. 3 we can see that with the system from the ground state approaching to the scission point, the nucleons inside nuclei undergo the motion from order (at the ground state) to chaos (at the first saddle point), then back to order (at the second minimum), then to chaos (at the second saddle point), then back to order again (at the third minimum) and further approaching to chaos (at the third saddle point), finally back to order (at the scission point). These results provide us with a new point for understanding the fission process in addtion to the statics of fission. At the same time, we also explore that the configuration of scission point is at a regular state. In order to make sure of this point, we illustrate the statistical properties of quantum spectra at the scission point and the corresponding classical-like Poincarè section map in Fig. 4(a), 4(b). Fig.4 clearly shows the Poisson-like distribution of the level spacings (a) and the close orbits in Poincarè map (b) at the scission point. Based on the above general investigation on statistical properties of spectra for the deformed space explored by the fission process, we try to understand phenomena appearing in the fission and heavy ion collisions. For example, the hyperdeformed $^{144-146}Ba$ [9] that appeared at the spontaneous fission of ^{252}Cf has been observed. In order to explain this phenomena theoretically, we can examine the statistical properties of spectra and Poincarè section map at the scission point of ^{252}Cf. The observed regular feature in the scission configuration means that the byperdeformed states can be populated at this configuration.

To give further proof of above theoretical explaination, we calculated the potential energy surface of fission for ^{252}Cf, in which one shallow minimum just corresponds to the scission configuration (i. e. the ratio of two axes of fragment Ba being about 3:1 and the distance between two fragments Ba and Mo being 18.2 fm) of ^{252}Cf. Therefore, we have given a possible understanding for the observed hyperdeformed fragment Ba appearing in spontaneous fission of ^{252}Cf from the point of view of the interplay between order and chaos.

For synthesis of the superheavy elements, special attention has been paid to the mechanism of di-nuclear fusion reaction [10]. In order to obtain the cross section of fusion reaction agreeing with experiment, one has to suppose that the nuclear matter is transfered at a fixed neck size between the di-nuclear system. This requires that the mass parameter along the neck coordinate has to be very large. However, According to the point of view of the ergodic state at the scission point, there seems to be a problem in obtaining a large mass parameter along the neck coordinate. Since within the cranking model, the main contribution to the mass parameter comes from diagonal terms and in terms of the linear response theory the diagonal components vanish when the system is at the ergodic state (according to the random matrix model, diagonal components do not appear), the mass parameter along the neck coordinate will become very small and the fast growth of the neck will appear. In this way, the calculated cross section for fussion is much larger than that of the experiment value. Based on our conclussion that well necked-in configuration at the scission point is not ergotic and stable against chaos, the diagonal component of mass parameter along the neck coordinate should not vanish [11] and the mass parameter should be large. This makes us obtain the cross section of fusion which is in agreement with the experiment value. Since the cross section of fusion is closely related to the cross section of synthesis of superheavy elements, the investigation on the transition from order to chaos is of great significance for the synthesis of superheavy elements and the formation of hyperdeformation.

TRANSITION FROM ORDER TO CHAOS AND NUCLEAR ROTATION

Since in the compound systems formed by heavy ion collisions, there exists a considerable angular momentum, the investigation on the transition from order to chaos in rotating systems is one of the most interesting subjects. As is well known, the totating will destroy the time-reversal symmetry of nuclear systems. For the systems without time-reversal symmetry (Gaussian unitory ensemble (GUE)), the condition that the system reaches to chaotic state is to satisfy the GUE distribution of level spacings, as shown in formula.

$$P(s) = \frac{32}{\pi^2} s^2 \exp(-s^2 4/\pi) \tag{5}$$

In order to study the statistical properties of spectra in the rotating system, the rotating two center shell Model (RTCSM) is employed, in which the Hamiltonian of the system is given by

$$h = h_0(\Delta z, X_i, \varepsilon, \beta_1, \beta_2) - \Omega j_x, \tag{6}$$

where j_x means the x-component of the single particle angular momentum, Ω is the cranked frequency measured in units of $\hbar\Omega_0 = 41$ Mev/$A^{\frac{1}{3}}$ (A is the mass number of the nucleus), the term of Ωj_x is known as Coriolis term. h_0 is the singel-particle Hamiltonian in the laboratory frame and is axially symmetric with respect to the z axis, as described in section 2. We have studied the level spacing distribution at different deformations as a function of rotating frequencies of nuclei [12, 13]. It has been shown that for the nucleus at the ground state even if there is a relative large rotating velocity, the level spacing distribution is still a Poisson-like, i.e. the motion of nucleons inside nuclei is in the order state. The reason for this is because the shell effect is strong enough at the ground state. For deformed nuclei, the shell effect, in general speaking, becomes weak. When there is no rotating or very small rotating in these systems the regular motion will be still kept ,

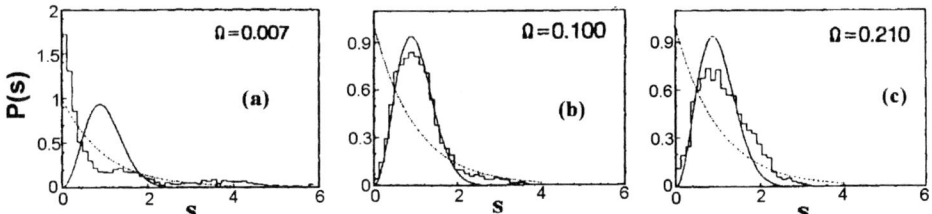

FIGURE 5. The proton level spacing distribution of ^{238}U at the rotating system. (a) low rotating frequency, for example, $0.007\ \Omega_0$; (b) rotating frequency $0.100\ \Omega_0$; (c) high rotating frequency, for example, $0.210\ \Omega_0$.

as shown in Fig. 5(a). When increasing the rotating frequency in these deformed nuclei, i.e. increasing the Coriolis force, the coupling between single-particle levels appears and the level spacing distribution will approach to GUE, as shown in Fig 5(b). In this case the motion of nucleons will demanstrate the chaotic character. If we continue to increase the rotating frequency for the deformed nuclei, the motion of nucleons will be dominated by rotating and the paralleled rearrangement of spin of nucleons will appear. This will make the level spacing distribution deviate from the distribution of GUE, as shown in Fig. 5(c). From above analysis, we can see that the transition from order to chaos appears indeed with increasing the rotating frequency. This kind of study is of great significance for the high-spin physics.

PAIRING INTERACTION AND TRANSITION FROM ORDER TO CHAOS

We know that the experimental spectra contain the energy levels from collective states, for example, rotation and vibration states, in addition to the single-particle states. Thus, in order to make a comparison between the analysis of experiment spectra and theoretical calculations, we need to develop a model which contains the correlation among single-particle motions. As the first step, we examine the effect of the pairing correlation by studying the statistical properties of the quasiparticle spectra. To calculate the quasiparticle levels, the BCS approach is employed. In the simplest case one may assume a constant matix element for the pairing interaction -G. Then the Hamiltonian is given by

$$\hat{H} = \sum_k \varepsilon_k^0 \hat{a}_k^+ \hat{a}_k - G \sum_{k,k'>0} \hat{a}_k^+ \hat{a}_{-k}^+ \hat{a}_{-k'} \hat{a}_{k'}. \tag{7}$$

Based on the BCS approximation, the solution of above Hamiltonian is

$$|BCS> = \prod_{k>0}^{\infty} (u_k + v_k \hat{a}_k^+ \hat{a}_{-k}^+)|0>. \tag{8}$$

In this state each pair of single-particle levels (k, -k) is occupied with a probability $|v_k|^2$ and is unoccupied with probability $|u_k|^2$. The parameter u_k and v_k are determined through the variational principle. The pairing gap can be obtained iteratively by the so-called gap-equation

$$\Delta = G/2 \sum_{k>0} \left(\Delta / \sqrt{(\varepsilon_k - \varepsilon_f)^2 + \Delta^2} \right), \tag{9}$$

where, ε_k are single-particle energy levels, which are calculated by TCSM. The value of the Fermi energy ε_f is determined from the conservation of the nucleon number. Finally, the quasi-particle energy is given by ε_k

$$E_k = \sqrt{(\varepsilon_k - \varepsilon_f)^2 + \Delta_k^2}. \tag{10}$$

Based on this model , We calculate the energy levels for the heavy nuclei of ^{252}Cf, in which there is the partially filled shell and the pairing force is expected to be important. From the statistical analysis of the quasi-particle spectra, we have observed that at the ground state configuration the level spacing distribution of quasiparticles show a perfect Poisson distribution and nucleons demonstrate order motion even if there is pairing interaction. This implied that the pairing force can not change the statistical properties of levels when the system is stabilized by the strong shell effect. We now turn to a large deformed configuration of ^{252}Cf [4], especially to an asymmetric saddle point configuration , which is of crucial importance for the study of the fission and hyperdeformation . At this deformation the pairing correction has been known to be of minor importance compared to the shell correction from the point of view of the contribution to the potential energy. However, from the point of view of the interplay between order and chaos, pairing effects may play an important role. So that we study the statistical properties of single-particle and quasiparticle spectra of ^{252}Cf at the asymmetric saddle point both with and without the pairing force. Fig.6 shows the results. In Fig. 6(a) a Wigner distribution of the single-particle level spacings (there is no pairing force) is shown, whereas in fig. 6(b) a Poisson-like distribution of quasiparticle level spacings (there is a pairing force) is illustrated. This figure clearly shows that the pairing effect makes the quasiparticle level spacing distribution diviate from the original Wigner distribution of single-particle spectra, which characterizes the chaotic motion of nucleons. Therefore, the pairing force enchances the stability of the system aganist chaos. With respect to this point, we may understand this as follows: Because of the transfromation from a real single-particle to a quasiparticle in the BCS model, the residual interaction or complexity in the orignal single-particel Hamiltonian is taken into account and has been partially canceled. Therefore, the coupling between the quasiparticle levels is weaker than that between the single-particle levels.

SUMMARY

In summary, the invistigation on the transition from order to chaos not only opens up a new research field for the classical physics, but also brings a great challenge to the quantum physics, especially to the nuclear physics and solid state physics. From this preliminary study, We can conclude that there exists a rich field on the transition from

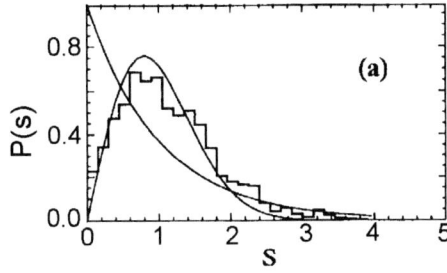 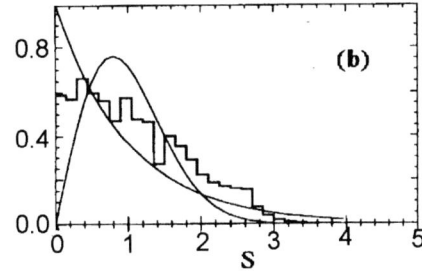

FIGURE 6. (a) The level spacing distribution of single-particle spectra of ^{252}Cf. (b) The level spacing distribution of quasi-particle spectra of ^{252}Cf.

order to chaos in the nuclear systems. We expect that the study scope of the order to chaos phase transition will expand greatly with the further development of study on the nuclear physics.

REFERENCES

1. Wu Xizhen, Sakata F.,Zhuo Yizhong,Li Zhuxia and Dang N. D., Dynamic realization of statistical state in finite systems, Phys. ReV. C 53, 1233-1244 (1996)
2. Weidenmüller H. A., statistical theory of nuclear reactions and the Gaussian Othogonal Ensamble, Annals of Physics 158 , 120 141 (1984).
3. Hag R.U., Pandey A., and Bohigas O., Fluctuation properties of nuclear energy levels: Do theory and experiment agree? , Phys. Rev. Lett, 48, 1086-1089 (1982).
4. Wu Xizhen ,Gu Jianzhong and Iwamoto Ahira, Statistical properties of quasiparticle spectra In deformed nuclei, Phys. Rev. C 59 , 215-220 (1999).
5. Garrett J.D., Robinson J. Q.,. Foglia A. J and Jin H. Q. , Nuclear level repulsion and Order vs chaos , Phys. Lett.B 392, 24-29 (1997).
6. Bohigas O.,Hag R.U., and Pandy A, Fluctuation properties of nuclear energy levels and widths comparison of theory with experiment , in Nuclear Data for Science and Technology, ed. by Bockhoff K. H., Dordrecht: Reidel (1983), P. 809-813.
7. Heiss W. D., Nazmitdinov R. G.and Radu, Chaos in axially symmetric potentials with Octupole deformation, Phys. Rev. Lett. , 72, 2351-2354 (1994).
8. Wu Xizhen , Gu Jianzhong , Zhuo Yizhong, Li Zhuxia, Chen Yongshou , and Greiner W., Possible understanding of hyperdeformed $^{144-146}Ba$ nuclei appearing in the spontaneous fission of ^{252}Cf , Phys. Rev. Lett. 79, 4542-4545 (1997).
9. Ter-Akopian G. M., Hamilton J.H., Oganessian Yu . Ts. et al., New spontaneous fission Mode for ^{252}Cf : Indication of hyperdeformed $^{144,145,146}Ba$ at scission, Phys. Rev. Lett. 77,32-35 (1996).
10. Adamian G. G., Antonenko N. V., Ivanova S. P. and et al., Problems in description of fusion of heavy nuclei in the two-center shell model approach, Nucl. Phys. A 646, 29-52 (1999).
11. Hofmann H., A quantal transport theory for nuclear collective motion: the metrits of a locally harmonic approximation method, Phys. Rep., 284, 139-380 (1997).
12. Gu Jianzhong, Wu Xizhen, Zhuo Yizhong, Quantum chaotic motion of a single particle in heavy nuclei, Nucl. Phys. A 625, 621-632 (1997).
13. Gu Jianzhong, Wu Xizhen, Zhuo Yizhong , The single-particle spectrum and its spacing and curvature distributions in rotating heavy nuclei , Nucl. Phys. A 611, 315-331 (1996).

Correspondence Between Propagating Characters of Coherent States and Energy Spectral Statistics in Chaotic Systems [1]

Junqing Li[*,†], Fang Liu[**], Yongzhong Xing[*,†], Wei Zuo[*,†] and Wenfei Li[†]

[*]Center of Theoretical Nuclear Physics, National Laboratory of Heavy Ion Accelerator, 730000, Lanzhou
[†]Institute of Modern Physics, The Chinese Academy of Sciences, Lanzhou 730000, P.R.China
[**]Institute of Modern Physics, North-West University, Xian 710069, P.R.China

Abstract. Taking coherent states as initial quantum states, the spatio-temporal evolution of the wave packet under the action of perturbed Hamiltonian showed one-to-one correspondence to the behavior of classical phase points. The character of the coherent state propagation in classically soft chaotic system is found to be associated with the statistical property of the spectrum contained in the coherent state wave packet. The spreading width of wave packet increases with time exponentially if the statistics of the spectrum contained in the wave packet show GOE distribution due to avoided level crossings.

The quantum mechanical behavior of classically chaotic systems, usually referred to as quantum chaos, is of interest, since the classical limit is still poorly understood for soft chaos[1,2]. It is well known that quuantum chaos does not manifest itself at any special state or energy level, but show distinct statistical character of its energy spectrum or scars of its wave function. On the other hand, classical chaos is characterized by the exponential separation of initially neighboring trajectories in phase space. This peculiar exponential instability can be described by using coherent states corresponding to a definite dynamical symmetry as an initial state. The coherent state will evolve with time and show different behaviors under the action of the system with or without the same dynamical symmetry, and show one to one correspondence to corresponding classical trajectories[3]. Our recent numerical studies indicated that the spreading width of the coherent state wave packet increased with time linearly or exponentially was associated with the statistical properties of the energy spectrum contained in the coherent wave packet. Here the spatio-temporal evolution of axially symmetric harmonic oscillator coherent states under the action of perturbed Hamiltonian with octupole deformation is studied for a classically soft chaotic systems. The Hamiltonian under investigation is

$$H = H_0 + \lambda V' \quad , \quad H_0 = -\frac{\hbar}{2m}\nabla^2 + \frac{1}{2}m\omega^2(x^2 + \frac{z^2}{b^2}),$$

---

[1] The work is supported by Chinese National Science Foundation 19847002, and one hundred person subject of Chinese academy of sciences.

CP597, *Nonequilibrium and Nonlinear Dynamics in Nuclear and Other Finite Systems,*
edited by Z. Li et al.

$$V'(x,z) = \frac{1}{2}m\omega^2 \frac{2z^3 - 3zx^2}{\sqrt{x^2 + z^2}}. \tag{1}$$

Where H_0 is the Hamiltonian of a two-dimensional axially symmetric harmonic oscillator, $V'(x,z)$ couples the two degrees of freedom, and generates octupole deformation with the deformation strength λ. Thus the spatial reversal symmetry of the system is destroyed. For any given b, there exists a critical value λ_c, for $\lambda > \lambda_c$, the potential no longer binds a particle. When b is equal to 0.5, λ_c is equal to 1.64, and hereafter we always take $b = 0.5$. Varying λ from zero to λ_c, the behavior of the classical particle changes from regular to completely chaotic type(see ref.[4]). H_0 represents an integrable system of which the analytical eigensolution $| n_{1x}n_{2z} \rangle$ of the stationary schrodinger equation

$$H_0 | n_{1x}n_{2z} \rangle = \varepsilon_{1x2z} | n_{1x}n_{2z} \rangle. \tag{2}$$

can be readily found. By taking the octupole deformation into account, we have to solve the eigenequation of H

$$H | \psi_m(x,z) \rangle = E_m \psi_m(x,z) \rangle,$$
$$| \psi_m(x,z) \rangle = \sum_n C_{mn} | n_{1x}n_{2z} \rangle. \qquad n = (n_{1x}, n_{2z}) \tag{3}$$

numerically. If the system is initially in a coherent state with the dynamical symmetry of the axially symmetric harmonic oscillator, its specific expression is taken as

$$| \alpha_0 \rangle = exp[-\frac{1}{2}(| \alpha_x |^2 + | \alpha_z |^2)] \sum_{n_{1x}=0}^{\infty} \sum_{n_{2z}=0}^{\infty} \frac{(\alpha_x)^{n_{1x}}(\alpha_z)^{n_{2z}}}{\sqrt{n_{1x}! \, n_{2z}!}} | n_{1x}n_{2z} \rangle, \tag{4}$$

where $\alpha_x = x_0 + ip_{x0}$, $\alpha_z = z_0 + ip_{z0}$ are determined by the initial position of the wave packet in four-dimensional phase space. $| \alpha_0 \rangle$ is characterized with almost precisely determined expectation values of the whole set of basic dynamical variables at the initial instant. When the initial coherent state evolves under the action of H which violates the dynamical symmetry of H_0 and hence the dynamical symmetry of the initial coherent state, the wave packet at time t becomes

$$| \alpha(\lambda,t) \rangle = \sum_{n_{1x},n_{2z}} exp[-\frac{i}{\hbar}H(\lambda)t] | n_{1x}n_{2z} \rangle \langle n_{1x}n_{2z} | \alpha_0 \rangle,$$
$$exp[-\frac{i}{\hbar}H(\lambda)t] = \sum_m | \psi_m(\lambda,x,z) \rangle exp[-\frac{i}{\hbar}E_m(\lambda)t] \langle \psi_m(\lambda,x,z) |. \tag{5}$$

$| \psi_m(\lambda,x,z) \rangle$ and $E_m(\lambda)$ in equ.(5) are obtained by solving the eigenequation of H numerically in a suitably chosen truncated space such that relative errors in relevant results do not exceed 0.13%.

Since $H(\lambda)$ is an autonomous system, the wave packet $| \alpha(\lambda,t) \rangle$ has a time independent value for the energy expectation value and spreading width, which read

$$\langle H(\lambda) \rangle = Tr[H(\lambda) | \alpha(\lambda,t) \rangle \langle \alpha(\lambda,t) |] = Tr[H(\lambda) | \alpha_0 \rangle \langle \alpha_0 |] \tag{6}$$

$$\langle(\Delta H(\lambda))^2\rangle = Tr[(\Delta H(\lambda))^2 \mid \alpha(\lambda,t)\rangle\langle\alpha(\lambda,t) \mid] = Tr[(\Delta H(\lambda))^2 \mid \alpha_0\rangle\langle\alpha_0 \mid] \qquad (7)$$

If the deformation strength λ is infinitesimally small, the spreading widths of the wave Packet(SWWP) of the whole set of basic dynamical variables $(\Delta x)^2$, $(\Delta p_x)^2$, $(\Delta z)^2$, $(\Delta p_z)^2$ will all keep their initial values.

Hence the corresponding expectation values for this set of basic dynamical values just describe the "quantum trajectory" inside the energy region. The spatio-temporal evolution behaves regularly as in the corresponding classical case. This is shown in Fig.1:Left, where λ/λ_c=0.1, the initial coherent state $\mid \alpha_1\rangle$ starting at $\{x,z;p_x,p_z\}$ = { 0.0,2.67; 2.0,0.0 }, and the corresponding classical motions are all regular[4]. Due to the j-dependent frequency, however, there appears dispersion effect in SWWP. They show slow correlated amplitude modulated features. Only after a rather long time interval, they return almost to their initial values simultaneously. The corresponding values in z direction are all similar with those in x direction, so they are not shown. According to equs.(6) and (7), during the whole evolution process, the coherent state should be always inside of a definite energy region, which only depends on the initial conditions. Fig.2:Left showed the distribution function of nearest neighboring level distance in this region. Since there are too little levels contained in the wave packet, especially for regular case(usually there are only 20 to 30 leveles contained in the wave packet), the statistics is poor, but one may still find the Poission-like figure.

If λ is efficiently large, the spreading widths will all increase exponentially during the initial stage. This is shown in Fig.1:Right, where λ/λ_c=0.95, with the initial state $\mid \alpha_1\rangle$, and the corresponding classical motions are all chaotic. The Poincare sections are dominated by chaotic sea, there is hardly any structure. But $\mid \alpha(\lambda,t)\rangle$ must be always situated inside the prescribed energy region in which the spectral statistics is of GOE type, (see Fig.2:Right, where the level repulsion is clearly seen.) The spreading widths fluctuate about their respective asymptotic values. These asymptotic values themselves

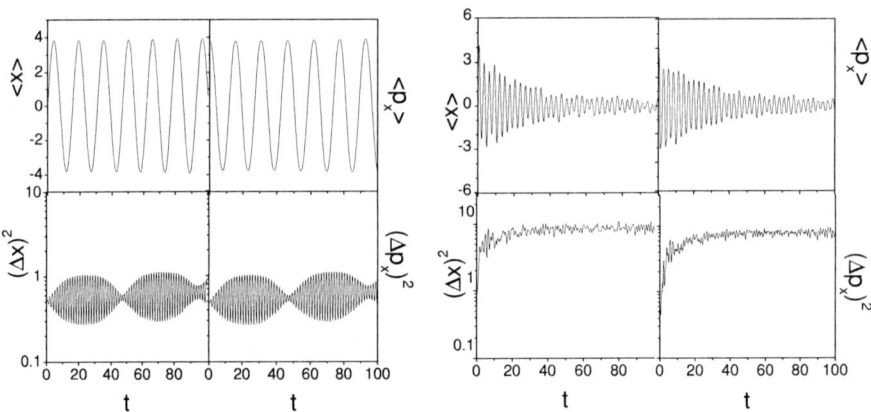

FIGURE 1. Left:The expectation values $< x >,< p_x >$ and spreading width $< \Delta x >^2,< \Delta p_x >^2$ of dynamical variables during the spatio-temporal evolution of the wave packet under the condition $\lambda = 0.1\lambda_c$ and with the initial state α_1. Right: The same as Fig.1 left, but under the condition $\lambda = 0.95\lambda_c$,and with the initial state α_1.

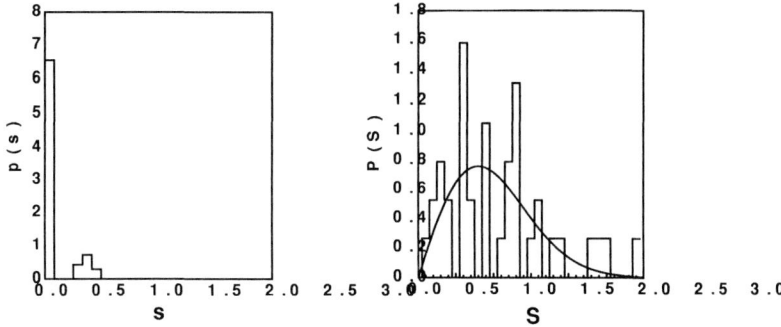

FIGURE 2. Left: The distribution function of the nearest neighboring level distance of spectrum contained in the wave packet of Fig.1:Left. Right:The distribution function of the nearest neighboring level distance of spectrum contained in the wave packet of Fig.1:Right.

must be much greater than the initial values. Consequently, the expectation values for the set of basic dynamical variables will thus tend to definite asymptotic values with small rapid fluctuations. These properties just imply the utterly irregular spreading of the wave packet $|\alpha(\lambda,t)\rangle$ inside the prescribed energy region, and is almost independent of the complex parameters α_x, α_z of the initial coherent state $|\alpha_0\rangle$ provided that they are all situated in the same prescribed energy region. These properties altogether can be regarded as quantum behaviors corresponding to classical ergodicity and thus can be reasonably used to characterize the deterministic quantum chaotic motions. For $\lambda/\lambda_c=0.5$, classically both regular and chaotic behaviors may appear in Poincare sections at different phase space(see Fig.1 and Fig.3 of ref.[4]). In the corresponding quantum case, initial coherent states residing in different phase space may exhibit different quantum behaviors. Fig.3:Left shows the expectation values $\langle x\rangle_{\lambda,t}$, $\langle p_x\rangle_{\lambda,t}$ and the corresponding SWWP $\langle(\Delta x)^2\rangle_{\lambda,t}$, $\langle(\Delta p_x)^2\rangle_{\lambda,t}$ for the initial coherent state $|\alpha_1\rangle$ and a classical trajectory starts at the point behaves regular. Fig.3:Right shows the same plots for the initial coherent state $|\alpha_2\rangle$ which is $\{x,p_x;z,p_z\}=\{$ 0.0,1.335;2.0,2.31 $\}$ and a classical trajectory starts at the point behaves chaotic. The plots in Fig.3:Left are very similar with those

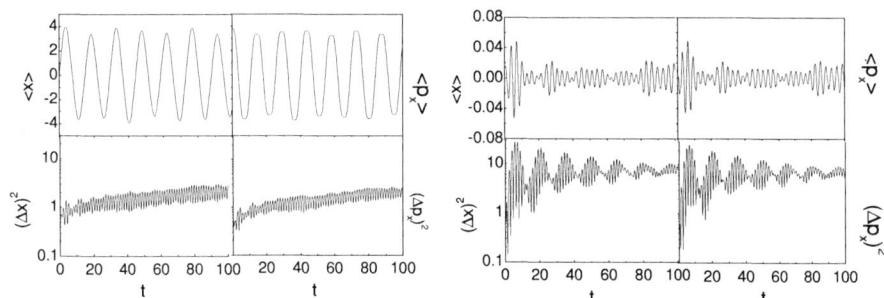

FIGURE 3. Left:The same as Fig.1 left, but under the condition $\lambda=0.5\lambda_c$,and with the initial state α_1. The same as Fig.1 left, but under the condition $\lambda=0.5\lambda_c$,and with the initial state α_2.

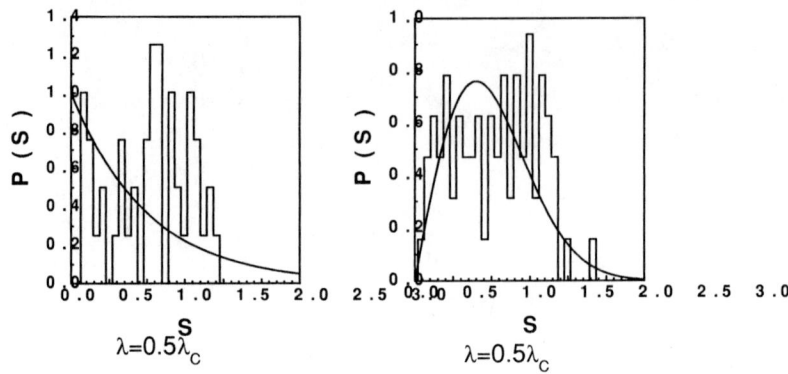

FIGURE 4. Left: The distribution function of the nearest neighboring level distance of spectrum contained in the wave packet of Fig.3:Left. Right: The distribution function of the nearest neighboring level distance of spectrum contained in the wave packet of Fig.3:Left

in Fig.1:Left. The SWWP in Fig.3:Left show severe dispersion, but never increase with time exponentially. While the SWWP in Fig.3:Right increase with time exponentially. The spectral statistics are shown in Fig.4:Left and Fig.4:Right which contained in coherent state wave packets in Fig.3:Left and Fig.3:Right, respectively. They are not very much close to standard Poission or GOE distributions due to fewer levels contained in the wave packets which give rise to rather poor statistics, but still one may find level repulsion for chaotic case.

To sum up, with the algebraic property of dynamical symmetry of quantum integrable system H_0, we are able to use the corresponding minimum uncertainty states one-to-one correspondent to classical phase points as initial states. Then the spatio-temporal evolution of the wave packet under the action of perturbed Hamiltonian can be studied causally as in classical mechanics[5]. And furthermore the propagating property of the coherent state is corresponding to the statistical property of the energy spectrum containd in the coherent state, and due to a set of avoided level crossings of the spectrum[3], where the neighboring states will strongly mix each other, so that there will subsequently appear nonlinear resonances between different pairs of neighboring levels which are responsible both to the exponential instability of the coherent state disperssion and the level repulsion in spectrum.

REFERENCES

1. O.Bohigas, S.Tomsovic, and D. Ullmo, Phys.Rep.**223**(1993)43.
2. W.D.Heiss,R.G.Nazmitdinov and S.Radu,Phys.Rev.Lett.**72**(1994)2351.
3. Liu Fang, Li JunQing, Luo YiXiao, Xu GongOu, and Zuo Wei, Commun. Theor.Phys. (Beijing, China)**35**(2001)531.
4. Jun-qing Li, J.Phys. G24(1998)1021 and Phys.Rev.Lett.79(1997)2387.
5. Xu Gong-ou, Yang Ya-tian and Xing Yong-zhong, Phys. Rev. **A61**, 042104(2000).

The world according to Rényi: thermodynamics of fractal systems

Petr Jizba* and Toshihico Arimitsu*

*Institute of Theoretical Physics, University of Tsukuba, Ibaraki, 305-8571, Japan

Abstract. We discuss a basic thermodynamic properties of systems with multifractal structure. This is possible by extending the notion of Gibbs–Shannon's entropy into more general framework - Rényi's information entropy. We show a connection of Rényi's parameter q with the multifractal singularity spectrum $f(\alpha)$ and clarify a relationship with the Tsallis–Havrda–Charvat entropy. Finally, we generalize Hagedorn's statistical theory and apply it to high–energy particle collisions.

INTRODUCTION

One of the fundamental observations of information theory is that the most general functional form for the mean transmitted information (i.e., information measure) is that of Rényi. Although Rényi's information measure offers perhaps the most general and conceptually cleanest setting for the entropy, it has not found so far as much applicability as its Shannon's counterpart. To clarify the position of Rényi's entropy in physics, we resort to systems with a multifractal structure. Such systems are very important and highly diverse, including phase transitions, turbulent flow of fluids, irregularities in heartbeat, population dynamics, chemical reactions, plasma physics, and most recently the motion of groups and clusters of stars. We shall argue that for the aforementioned the Rényi parameter is connected via a Legendre transformation with the multifractal singularity spectrum. To put some flesh on bones we generalize Hagedorn's statistical theory and subsequently apply to a differential cross section in high–energy scattering experiments. More thorough investigation will be published elsewhere.

RÉNYI'S ENTROPY

Motivation

From information theory follows that the most general information entropy is that of Rényi [1]. In discrete cases where the probability distribution $\mathcal{P} = \{p_n\}$ the Rényi entropy is defined as

$$I_q(\mathcal{P}) = \frac{1}{(1-q)} \log_2 \left(\sum_{k=1}^{n} p_k^q \right).$$

CP597, *Nonequilibrium and Nonlinear Dynamics in Nuclear and Other Finite Systems*, edited by Z. Li et al.

On the other hand, in continuous probability cases the entropy must be properly renormalized - with an arbitrary precision of measurement comes infinity of information. If $f(x)$ is an arbitrary positive density function, say in the interval $[a,b]$ one may define the integrated probability

$$p_{nk} = \int_{k/n}^{(k+1)/n} f(x)dx,$$

then [1]

$$I_q(f) \equiv \lim_{n \to \infty} \left(I_q(\mathcal{P}_{nk}) - \log_2 n \right) = \frac{1}{(1-q)} \log_2 \left(\int_a^b f^q(x)dx \right). \tag{1}$$

Eq.(1) might be generalized to any Lebesgue or Hausdorff measurable sets [2].

In the former context a natural question arises; how comes that there are other information entropies apart from Shannon's one. To understand this we should go to information theory. The latter asserts that the amount of information received by learning that an event of probability p took place (in bit units) is $I(p) = -\log_2(p)$. In general, if the possible outcomes of an experiment are $\mathcal{A}_1, \mathcal{A}_2, \ldots, \mathcal{A}_n$ with corresponding probabilities $p_1, p_2, \ldots, p_n$, and $\mathcal{A}_k$ conveys I_k bits, then the mean conveyed information is

$$I(\mathcal{P}) = \sum_{k=1}^n p_k I_k.$$

However, the linear averaging is only a specific case of a more general mean! It has been recognized be A.Kolmogorov [3] and M.Nagumo [4] that the most general mean compatible with postulates of probability theory gives the entropy

$$I_g(\mathcal{P}) = g^{-1} \left(\sum_k^n p_k \, g(I_k) \right),$$

where g is an arbitrary invertible function.

Applying the postulate of additivity of independent information one obtains only two possible classes of g [1], namely $g(x) = cx + d$ which implies $I(\mathcal{P}) = -\sum_{k=1}^n p_k \log_2(p_k)$ (i.e., Shannon's information measure) and $g(x) = c2^{(1-q)x} + d$ which implies directly Rényi's information measure $I_q(\mathcal{P})$.

Among the basic properties of Rényi's entropy we may mention; positivity ($I_q \geq 0$), for $q \leq 1$ Rényi's entropy is concave but for $q > 1$ is not pure convex nor pure concave and, in addition, when we continue q to 1, Rényi's entropy equals Shannon's one, i.e., $\lim_{q \to 1} I_q = I$.

In physics Rényi's entropy has been sporadically, albeit successfully applied in various non–equilibrium dynamical systems, e.g., fully developed turbulence (q directly relates with Reynolds number), percolating clusters (q directly describes p_c), etc. It also provides a consistent mathematical setting for Tsallis–Havrda–Charvat (THC) entropy and, as we shall see, it is a correct entropy for (multi)fractal systems.

Connection with Tsallis–Havrda–Charvat entropy

THC entropy introduced originally be J.H.Havrda and F.Charvat [5] and later applied to physical problems by C.Tsallis [6] is currently fruitfully used in many statistical systems; 3–dimensional fully developed hydrodynamic turbulence, 2–dimensional turbulence in pure electron plasma, Hamiltonian systems with long–range interactions, granular systems, systems with strange non–chaotic attractors, peculiar velocities in galactic clusters, etc. Its form reads

$$S_q = \frac{1}{(1-q)} \left[\sum_{k=1}^{n} (p_k)^q - 1 \right], \quad q > 0.$$

Among important properties of THC entropy we can mention positivity ($S_q \geq 0$), concavity in $\mathcal{P}$, gibbsian limit for $q \to 1$ (i.e., $\lim_{q \to 1} S_q = I$) and peculiar non–extensive behaviour

$$S_q(A+B) = S_q(A) + S_q(B) + (1-q)S_q(A)S_q(B),$$

for two independent events A and B.

Rényi's entropy vs.THC entropy

To find a connection between Rényi and THC entropies we utilize the expansion of $\log_2(1+x)$. Then we may write

$$I_q = \frac{1}{(1-q)} \log_2 \left[(1-q)S_q + 1 \right] = \frac{1}{k}S_q - \frac{1}{2k}(1-q)S_q^2 + O\left[(1-q)^2 S_q^3 \right], \quad (2)$$

with the scale factor $k = \ln 2$. So $I_q \approx S_q$, provided

$$\frac{1}{|1-q|} \left[\sum_{l}^{n} (p_l)^q - 1 \right]^2 \ll 1. \tag{3}$$

Condition (3) is fulfilled in numerous ways. For instance, when $q \approx 1$, $q \gg 1$, for systems with large deviations or for rare events systems [7].

At this stage some comments are in order. First of all we see from (2) that THC entropy and Rényi's entropy are monotonic functions of each other and, as a result, both are extremized by the same $\mathcal{P}$. However, while Rényi's entropy is additive, THC entropy is not, so it appears that the additivity property is not important for entropies required for extremizing purposes. Thus from thermodynamic point of view both entropies give the same predictions!

Secondly, as we show in [7], Rényi's entropy provides a consistent renormalization prescription for a continuous THC entropy, we find that

$$S_q(f) \equiv \lim_{n \to \infty} \left(\frac{S_q(\mathcal{P}_{nk})}{n^{(1-q)}} - \frac{S_q(1/n)}{n^{(1-q)}} \right) = \frac{1}{(1-q)} \int_a^b dx\, f(x) \left(f^{q-1}(x) - 1 \right).$$

FRACTAL AND MULTIFRACTAL SYSTEMS

Brief introduction into (multi)fractal sets

Fractals

Let us begin to illustrate the basic features of fractals sets on a simple example - triadic Koch curve (TKC). The latter is defined iteratively in the following way: in 0th iteration ($n = 0$) we start with a straight line - *initiator* - with length $r_0 = a$. In the following step ($n = 1$) we raise an equilateral triangle over the middle third of initiator. The result is *generator*. Its four straight line segments ($N_1 = 4$) have length $r_1 = a/3$ and total length $L[a/3] = 4a/3$. The construction of the Koch curve proceeds by replacing each segment of initiator with generator, i.e., for $n = 2$, $r_2 = (1/3)^2 a$, $L[(1/3)^2 a] = (4/3)^2 a$ and $N_2 = 16$, etc. So when $n = k$ we have $r_k = (1/3)^k a$, $L[(1/3)^k a] = (4/3)^k a$ and $N_k = 4^k$. Note that the length L diverges as $k \to \infty$!

Can we define somehow a finite length for the triadic Koch curve? The answer is yes, provided we extend the notion of euclidean dimension. To see this let us note that for ordinary smooth curves the approximative length is $L[r] \sim N(r)r$, and as r goes to zero $L[r]$ approaches the finite limit - length; $L = \lim_{r \to 0} N(r)r$.

Generalization to any D–dimensional volumes is then natural: $V = \lim_{r \to 0} N(r)r^D$. However, in order to get V finite, the following scaling must apply

$$N(r) \sim \frac{c}{r^D} \Leftrightarrow \log N(r) \sim c + D \log \frac{1}{r},$$

and so

$$\lim_{r \to 0} \frac{\log N(r)}{\log \frac{1}{r}} = D. \tag{4}$$

As the LHS of (4) is well defined for wider class of sets than just usual metric spaces one may accept it as a generalized definition of dimension. The latter is usually called the *Hausdorff–Besicovitch* or *fractal* dimension. It should be stressed that D in (4) is not necessarily integer - price which is paid for the finiteness of the volume.

Thus, for instance, in the case of TKC the fractal dimension is

$$D = \lim_{r \to 0} \frac{\log N(r)}{\log \frac{1}{r}} = \lim_{n \to \infty} \frac{\log 4^n}{-\log \frac{a}{3^n}} = \frac{\log 4}{\log 3} = 1.26\ldots.$$

One may often write (e.g., for strictly self–similar fractals), after n iterations $N = N_G^n$ (N_G is the number of pieces of the generator) $r = a r_G^n$ (r_G is the length of the segments of the generator). In such cases the fractal dimension follows from a simple analysis:

$$\lim_{r \to 0} N(r)r^D = \lim_{n \to \infty} \left(N_G r_G^D \right)^n = const. \Rightarrow D = \frac{\log N_G}{\log \frac{1}{r_G}}. \tag{5}$$

Relation (5) allows to recover fairly simply some standard results; e.g., the well known triadic Cantor dust ($r_G = 1/3$, $N_G = 2$) has $D = \log 2 / \log 3$.

Multifractals

Multifractals are related to the study of a distribution of physical or other quantities on a generic support (be it or not fractal) and thus provide a move from the geometry of sets as such to geometric properties of distributions. An intuitive picture about an inner structure of multifractals is obtained by introducing the $f(\alpha)$ spectrum [8]. To elucidate the latter let us suppose that some support (usually a subset of a metric space) is covered by probability of a certain phenomenon. If we pave the support with boxes of size l and denote the integrated probability in the ith box as p_i, we may define the scaling exponent α_i by $p_i(l) \sim l^{\alpha_i}$ and the corresponding density as $\rho_i = \frac{p_i}{l} \propto l^{\alpha_i-1}$. The factor α_i is called *Lipshitz–Hölder* exponent. Counting boxes $N(\alpha)$ where p_i has $\alpha_i \in (\alpha, \alpha + d\alpha)$, then the singularity spectrum $f(\alpha)$ is defined as $N(\alpha) \sim l^{-f(\alpha)}$. Accordingly, we may view a multifractal as the ensemble of intertwined (mono)fractals each with its own fractal dimension $f(\alpha_i)$. It is thus suggestive to define the "partition function"

$$Z(q) = \sum_i p_i^q = \int d\alpha' \rho(\alpha') l^{-f(\alpha')} l^{q\alpha'}. \tag{6}$$

In the small l limit the partition function (6) scales as $Z(q) \sim l^\tau$, where

$$\tau(q) = \min_\alpha (q\alpha - f(\alpha)), \ f'(\alpha(q)) = q. \tag{7}$$

Eq.(7) represents defining relations for the Legendre transformation.

Rényi's entropy - entropy of self–similar systems

Let us now turn to the question whether there is any connection of Rényi's entropy with (multi)fractal systems. At present it seems to us that there are at lest two such connections.

a) Formal connection - generalized dimensions

Generalized dimensions are defined as:

$$D_q = \lim_{l \to 0} \left(\frac{1}{(q-1)} \frac{\log Z_q}{\log l} \right) = -\lim_{l \to 0} I_q(l).$$

For example, D_0 is the usual fractal dimension - dimension of the support, D_1 is known as information dimension and D_2 is correlation dimension. D_0, D_1 and D_2 are usually sufficient to describe simple fractals (e.g., strictly self similar ones). However, in general all D_q are necessary to pinpoint fractals uniquely. This is typical e.g., for strange attractors [10]! The situation is somehow analogous to statistical physics when the whole tower of correlation function equations (BBGKY hierarchy) is needed to get the full information on density matrix.

b) Direct physical connection

We will show now that from the maximal entropy (MaxEnt) point of view, extremizing the Gibbs–Shannon entropy on fractals is equivalent to extremizing directly Rényi's entropy without invoking the underlying fractal structure explicitly.

Let us have a multifractal with a measure $p(x)$. Shannon's entropy for the corresponding process is $I = -\sum p_k \log_2 p_k$. The Billingsley theorem then states [9] that there is an intimate connection between Shannon's entropy and the Hausdorff dimension of the measure theoretic support $\mathcal{M}$ of $p(x)$ (i.e., the infimum of the dimensions of all sets on which $p(x)$ lives). Namely,

$$d_h(\mathcal{M}) = -\lim_{N\to\infty} \frac{1}{\log_2 N} \sum p_k \log_2 p_k \sim \frac{1}{\log_2 \varepsilon} \sum p_k(\varepsilon) \log_2 p_k(\varepsilon),$$

with the cutoff scale ε. In this connection it is useful to introduce a one–parametric family of normalized measures $\mu(q)$ (escort or zooming distributions)

$$\mu_i(q,l) = \frac{[p_i(l)]^q}{\sum_j [p_j(l)]^q}.$$

It is important to notice that the parameter q provides a microscope for exploring different regions of the singular measure. Indeed, for $q > 1$, $\mu(q)$ amplifies the more singular regions of p, while for $q < 1$, $\mu(q)$ accentuates the less singular ones. So one may zoom into any required regions of fractality. The corresponding "zooming" entropy is $\tilde{I}(q) = -\sum_k \mu_k \log_2 \mu_k$, and the Hausdorff dimension of the measure theoretic support of $\mu(q)$ then reads

$$f(q) = -\lim_{N\to\infty} \frac{1}{\log_2 N} \sum_k^N \mu_k \log_2 \mu_k \sim \frac{1}{\log_2(\varepsilon)} \sum_k \mu_k(\varepsilon) \log_2 \mu_k(\varepsilon). \tag{8}$$

In addition, the *average value* of the singularity exponent $\alpha_i = \log_2(p_i)/\log_2(\varepsilon)$ with respect to $\mu(q)$ is

$$\alpha(q) = \frac{\sum_k \mu_k(\varepsilon) \log_2 p_k(\varepsilon)}{\log_2(\varepsilon)}. \tag{9}$$

Eqs.(8) and (9) establish a relationship between a Hausdorff dimension $f(q)$ and an *average* singularity exponent $\alpha(q)$ via functional dependence on the parameter q. Note that $f = q\alpha - \tau$, $\alpha = d\tau/dq$ is precisely the Legendre transformation. Thus Shannon's entropy on a multifractal with a given $f(\alpha)$

$$-\sum_k p_k \log_2 p_k \bigg|_{f(\alpha)} = -\sum_k \mu_k \log_2 \mu_k \bigg|_{\alpha(q)}$$

$$= -q\alpha(q)\log_2(\varepsilon) + (1-q) I_q(\mathcal{P}). \tag{10}$$

So as long as we fix the "fractality" condition, Shannon's entropy I turns out to be (up to an additive constant) Rényi's entropy. Namely, from thermodynamic point of view $I|_{f(\alpha)}$ and I_q are completely interchangeable .

GENERALIZED HAGEDORN'S STATISTICAL THEORY

Hagedorn's statistical theory is applicable whenever the density of quantum states grows exponentially with temperature, i.e., when

$$v(E) \propto \exp[\beta_H E]. \tag{11}$$

Assumption (11) applies for example to (quantized) string theory [11], to cosmic string theory [12], or to high–energy particle collisions [13].

Except a proliferation of the states near $T_H = 1/\beta_H$, the state space acquires an approximately multifractal structure which is exact at critical point [13, 14]. This suggests that Rényi's rather than Gibbs entropy could better grasp vital features near T_H [1].

Differential cross section in high–energy scattering experiments

Recent experiments suggest that Hagedorn's theory is not adequate for generic high–nergy scattering experiments. For instance, in e^+e^- processes Hagedorn's description is satisfactory provided CMS energies are small ($E < 10 Gev$) but it fails at large energies. Hagedorn's approach at $E > 10 GeV$ predicts an exponential decay of differential cross sections while experiments observe a power–law behavior [15].

It addition, the latest applications of THC entropy in the context of high energy collisions [16], high energy cosmic ray physics [17] or a Focker–Planck equation treatment of charmed quarks in quark–gluon plasma [18] fit far better with experimental data than predictions based on the Gibbs–Hagedorn approach.

Hagedorn's theory basically predicts that the differential cross section in high–energy collisions should be

$$\frac{1}{\sigma}\frac{d\sigma}{dp_T} = c p_T \int_0^\infty dx e^{-\beta\sqrt{x^2+\mu^2}} , \quad \mu = \sqrt{p_T^2 + m^2} ,$$

which for large p_T asymptotically behaves as

$$\frac{1}{\sigma}\frac{d\sigma}{dp_T} \sim p_T^{3/2} e^{-\beta p_T} .$$

Yet, for ep high–energy collisions the results are best fitted by a power law [15];

$$\frac{1}{\sigma}\frac{d\sigma}{dp_T} \sim (1 + const\, p_T)^{-\gamma}, \quad \gamma = 5.8 \pm 0.5. \tag{12}$$

Whereas ordinary thermodynamics (and hence Hagedorn's theory itself) is derived by extremizing Gibbs–Shannon entropy we extremize now I_q instead. Following Hagedorn [13] we arrive at the probability density of transverse momenta $\rho(p_T)$ [7];

$$\rho(p_T) \propto u \int_0^\infty dx \left(1 + (q-1)\beta\sqrt{x^2 + u^2 + m_\beta^2}\right)^{-\frac{q}{q-1}} ,$$

---

[1] Actually Rényi's statistics also modifies T_H. Usually $T_R \leq T_H$

where $u = \beta p_T$ and $m_\beta = \beta m_0$. Using the fact that $\rho(p_T) \propto \sigma^{-1} d\sigma/dp_T$ we have

$$\frac{1}{\sigma}\frac{d\sigma}{dp_T} \sim c\sqrt{2(q-1)}\, B\left(\frac{1}{2}, \frac{q}{q-1} - \frac{1}{2}\right) u^{3/2}(1 + (q-1)u)^{-\frac{q}{q-1}+\frac{1}{2}}. \qquad (13)$$

Formula (13) agrees well with the experimental fit (12), provided one suitably determines parameters c, q, T. This in turn may determine the multifractal structure of the state space.

Finally note that (13) has its maximum at $p_T = T\left(\frac{3}{3-q}\right)$, on the other hand, the experimentally measured cross sections, e.g., in relativistic heavy ion–collisions have maximum at roughly the same value of $p_T \approx 180 MeV$, so $T_R \approx (1 - q/3)p_T \approx 105 MeV < T_H = 158 MeV$.

OUTLOOK AND OPEN QUESTIONS

Because of a build–in predisposition to account for self–similar systems Rényi's entropy naturally aspires to be an effective tool to describe phase transitions. It is thus a challenging task to find a closer connection with such typical tools of critical–phenomena physics as are conformal and renormalization groups.

In addition, witnessing an encouraging agreement of THC non–extensive statistics predictions with the cosmic ray experiments [17] or heavy–ion collisions [19] we may naturally ask whether there is a physical grounding for Rényi's entropy in similar (non–equilibrium) systems. Work along those lines is currently in progress.

REFERENCES

1. A. Rényi, *Selected Papers of Alfred Rényi, Vol.2* (Akadémia Kiado, Budapest, 1976).
2. M.F. Barnsley, *Fractals everywhere* (Academic Press, Boston, 1993).
3. A. Kolmogorov, *Atti della R. Accademia Nazionale dei Lincei* **12** (1930) 388.
4. M. Nagumo, *Japanese Jour. Math.* **7** (1930) 71.
5. J.H. Havrda and F. Charvat, *Kybernatica* **3** (1967) 30.
6. C. Tsallis, *J. Stat. Phys.* **52** (1988) 479; C. Tsallis, *Braz. J. Phys.* **29** (1999) 1.
7. P. Jizba and T. Arimitsu, in preparation.
8. T.C. Halsey, M.H. Jensen, L.P. Kadanoff, I. Procaccia and B.I. Schraiman, *Phys. Rev.* A**33** (1989) 1327.
9. P. Billingsley, *Ergodic Theory and Information* (Wiley, New York, 1965).
10. H.G.E. Hentschel and I. Procaccia, *Physica* **8**D (1983) 435.
11. R.D. Carlitz, *Phys. Rev.* D**5** (1972) 3231.
12. D. Mitchell and N. Turok, *Nucl. Phys.* B**294** (1987) 1138; M. Sakellariadou and A. Vilenkin, *Phys. Rev.* D**37** (1988) 885.
13. R. Hagedorn, *Nuovo Cim. Suppl.* **3** (1965) 147.
14. L.P. Kadanoff, *Physics* **2** (1966) 263.
15. DELPHI collab., Z. *Phys* **73**C (1997) 229; ZEUS collab., *Eur. Phys. J.* C**11** (1999) 251.
16. C. Beck, *Physica* A **286** (2000) 164.
17. D. Wilk and Z. Wlodarczyk, *Nucl. Phys.* B **75**A (1999) 191.
18. D.B. Walton and J. Rafelski, *Phys. Rev. Lett.* **84** (2000) 31.
19. W.M. Alberico, A. Lavagno and P. Quarati, *Eur. Phys. J.* C**12** (2000) 499; NA44 collab, *Phys. Lett.* B**467** (1999) 21.

Experimental test of slow phase randomization and quantum chaos in finite highly excited many-body systems

Qi Wang*, Sergey Yu. Kun†, Wendong Tian*, Songlin Li*, Yuchuan Dong*,
Zhichang Li**, Xiuqin Lu**, Kui Zhao**, Changbo Fu**, Jiancheng Liu**,
Hua Jiang** and Guiqing Hu**

*Institute of Modern Physics, Chinese Academy of Sciences, 730000, Lanzhou, People's Republic
of China
†Department of Theoretical Physics, RSPhysSE, IAS, The Australian National University,
Canberra ACT 0200, Australia
**China Institute of Atomic Energy, 102413, Beijing, People's Republic of China

Abstract. Two independent measurements of cross sections for the ^{19}F+^{93}Nb strongly dissipative heavy-ion collisions have been performed at incident energies from 102 to 108 MeV in steps of 250 keV. In the two measurements we used different, independently prepared, ^{93}Nb target foils with the thickness $\simeq 70$ $\mu g/cm^2$. All other experimental conditions were identical in both experiments. The data indicate non-reproducibility of the non-self-averaging oscillating yields in the two measurements. This supports recent theoretical predictions of spontaneous coherence, slow phase randomization and extreme sensitivity in highly-excited complex quantum systems.

INTRODUCTION AND MOTIVATION FOR THE EXPERIMENT

For highly excited interacting many-body systems the independent particle picture has very little validity when the mean spacing between the many-body energy levels is much smaller than the spacing of the single-particle levels [1,2]. For high excitations, the interaction leads to a quick decay of the single-particle as well as of collective modes [3] which are not eigenstates of the total Hamiltonian of the system. This decay results in a formation of highly complicated many-body configurations. Each of these many-body states is characterized by uniform occupation of all accessible parts of phase space and sharing of energy between many particles of the system. The characteristic time for the formation of such ergodic, independent of the initial conditions, many-body states is given by the inverse spreading width, $\tau_{erg} = \hbar/\Gamma_{spr}$ [3].

Consider a highly excited many-body system, whose spectrum obeys Wigner-Dyson statistics, for the time interval $t \gg \tau_{erg}$. Does ergodicity of all *individual* many-body eigenstates necessarily imply that their superposition is incoherent random superposition? Can superposition of spatially extended ergodic modes of a highly excited many-body system produce localized or non-equilibrium non-ergodic patterns? In the absence of a theory for phase randomization in an isolated (disconnected from a heat bath) systems one may conventionally rely on the hypothesis emerging from the foundations and modern developments of the random matrix theory (RMT) of highly excited many-body

CP597, *Nonequilibrium and Nonlinear Dynamics in Nuclear and Other Finite Systems*,
edited by Z. Li et al.

systems. This hypothesis implies that the energy relaxation, i.e. the mere formation of ergodic individual many-body configurations, is a sufficient condition for a phase randomization between these ergodic eigenstates [3]. If true, this conjecture should validate universal applicability of the RMT for the energy interval $\Delta E \leq \Gamma_{spr}$ and for the time interval $t \geq \tau_{erg}$, accordingly.

Consider the decay of a highly excited many-body system with strongly overlapping resonances, $\Gamma \gg D$, where $\hbar/\Gamma \gg \tau_{erg}$ is the average life-time and D is the mean level spacing of the system. This regime, $D \ll \Gamma \ll \Gamma_{spr}$, is known as a regime of Ericson fluctuations [4,5] for the decay of equilibrated nuclear, atomic and molecular systems and in coherent electron transport through nanostructures [3]. Suppose that RMT universally applies for $t \gg \tau_{erg}$. This implies absence of correlations between transition amplitudes (partial width amplitudes), known as Bethe's random signs hypothesis, for the decay of different ergodic states to either the same or different quantum micro-channels [3]. Consider, e.g., a strongly dissipative heavy-ion collision (DHIC) characterized by a high intrinsic excitation energy (≥ 15 MeV) of the intermediate system (IS). Since for nuclear systems $\Gamma_{spr} \simeq 5$ MeV and, for DHIC, $\Gamma \simeq 100$ keV [6] we deal with the decay of a superposition of ergodic strongly overlapping ($\Gamma \gg D \leq 10^{-8}$ MeV) many-body configurations. Then the RMT hypothesis, $\tau_{dec} \leq \tau_{erg}$, τ_{dec} being the phase randomization (decoherence) time between ergodic states, implies that the cross sections for the DHIC, summed over a very large number of partial cross sections, corresponding to different micro-states of the reaction fragments, should show a smooth energy dependence with the characteristic energy variation $\geq \Gamma_{spr} \simeq 5$ MeV. In contrast, experimental studies [7-9] present overwhelming evidence for the persistence of rapid ($\simeq 100$ keV) energy oscillations in the cross sections for DHIC. This manifests itself in the correlations between different transition amplitudes indicating that the phase randomization between individual ergodic configurations should be a much slower process than energy relaxation ($\tau_{dec} \gg \tau_{erg}$) in sharp contrast with the RMT hypothesis.

In attempting to interpret the non-self-averaging of excitation function oscillations in DHIC one faces a non-straightforward task of realization of Wigner's dream [10], namely to modify RMT by taking into account level-level and channel-channel correlations between the transition amplitudes. Such a possible modification has been presented in Refs. [11,12] in terms of spontaneous coherence and slow phase randomization in highly excited many-body systems. While RMT develops "a new statistical mechanics" (Dyson) of, by purpose, fully equilibrated finite systems, the work [14,15] discusses critical and non-equilibrium phenomena in finite highly excited many-body systems.

It has been found [11-13] that a precondition for micro-channel correlations (MC) in complex quantum collisions is $\tau_{dec} \gg \tau_{erg}$. Physically, τ_{dec} sets up a new time scale for quantum many-body systems. For times $t < \tau_{dec}$, the RMT ceased to apply even though $t \simeq \hbar/\Gamma \gg \tau_{erg}$. Since the physical picture [11-13] for the MC is in a sharp contrast with the RMT and the theory of quantum chaotic scattering [3] it is highly desirable to have an additional independent test of the approach [11-13]. Such a possibility does indeed exist. It has been argued [14] that the spontaneous origin of MC should result in the cross sections for DHIC being sensitive to an infinitesimally small perturbation.

The discussion of Ref. [14] has not taken into account nonuniform spatial distribution of electro-magnetic field, defects etc. within the target. How does the consideration [14] apply in the presence of such non-homogeneous "target-environmental" perturbations?

Consider a simple case of spinless reaction partners in the entrance and exit channels. A generalization for the case of the reaction partners having intrinsic spins is straightforward. Consider an experiment in which we do not determine which particular nucleus of the target participates in the single collision act. For a realistic situation of a finite angular resolution, $\Delta\theta \gg \lambda/R$, where λ is a wave length of the reaction products and R is a characteristic distance between nearest target nuclei, interference between collision amplitudes corresponding to different target nuclei is destroyed. Then the measured cross section, per a single target nucleus and for a fixed single exit micro-channel $\bar{b}$ (microscopic states of the reaction products), is given by $\sigma_{\bar{b}}(E,\theta) = (1/\mathcal{N})\sum_{j=1}^{\mathcal{N}} \sigma_{\bar{b}}^{(j)}(E,\theta)$, where $\sigma_{\bar{b}}^{(j)}(E,\theta) = |f_{\bar{b}}^{(j)}(E,\theta)|^2$. Here (j) labels individual target nuclei whose number is $\mathcal{N} \gg 1$ and $f_{\bar{b}}^{(j)}(E,\theta)$ is the amplitude of a collision involving (j) target nucleus. The difference between $f_{\bar{b}}^{(j)}(E,\theta)$ with different (j) originates from a nonuniform distribution of "target-environmental" perturbations. This introduces different local perturbations, $V_j \neq V_i$, in the purely nuclear Hamiltonian H of highly excited nuclear molecules created in the collision of the incident ion with different $(j \neq i)$ target nuclei. We evaluate the strength of the "target-environmental" perturbations to be of the order of the atomic electron effects [14] in DHIC. We employ the perturbation theory [14] and use the decomposition $f_{\bar{b}}^{(j)}(E,\theta) = f_{\bar{b}}(E,\theta) + \delta f_{\bar{b}}^{(j)}(E,\theta)$, where $f_{\bar{b}}(E,\theta)$ is the collision amplitude in the absence of the "target-environmental" perturbations. We also drop the incoherent sum $(1/\mathcal{N})\sum_{j=1}^{\mathcal{N}} |\delta f_{\bar{b}}^{(j)}(E,\theta)|^2$. This sum is about fourteen orders of magnitude smaller than $\sigma_{\bar{b}}(E,\theta)$ and it also does not contribute to the micro-channel correlation which is responsible for both non-self-averaging and sensitivity of the cross sections. We obtain $\sigma_{\bar{b}}(E,\theta) = |F_{\bar{b}}(E,\theta)|^2 - |(1/\mathcal{N})\sum_{j=1}^{\mathcal{N}} \delta f_{\bar{b}}^{(j)}(E,\theta)|^2 \to |F_{\bar{b}}(E,\theta)|^2$, where $|(1/\mathcal{N})\sum_{j=1}^{\mathcal{N}} \delta f_{\bar{b}}^{(j)}(E,\theta)|^2 \leq 10^{-14}|F_{\bar{b}}(E,\theta)|^2$, and $F_{\bar{b}}(E,\theta)$ is the collision amplitude corresponding to the Hamiltonian $(H+v)$ with $v = (1/\mathcal{N})\sum_{j=1}^{\mathcal{N}} V_j$.

It is reasonable to assume that a distribution of the local "target-environmental" perturbations V_j is random throughout the target. This means that $\delta f_{\bar{b}}^{(j)}(E,\theta)$ with different (j) have random phases. In this case we have

$$|F_{\bar{b}}(E,\theta) - f_{\bar{b}}(E,\theta)| \sim (1/\mathcal{N})^{1/2}|\delta f_{\bar{b}}^{(j)}(E,\theta)| \sim (1/\mathcal{N})^{1/2}10^{-7}|f_{\bar{b}}(E,\theta)|,$$

where we used the estimate $|\delta f_{\bar{b}}^{(j)}(E,\theta)| \sim 10^{-7}|f_{\bar{b}}(E,\theta)|$ from Ref. [14].

Suppose we perform two independent measurements with two different targets. The "target-environmental" perturbations, V_j in the first target and $\tilde{V}_j$ in the second one, are different. The cross sections are given by the different amplitudes, $F_{\bar{b}}(E,\theta)$ and $\tilde{F}_{\bar{b}}(E,\theta)$, corresponding to different Hamiltonians, $(H+v)$ and $(H+\tilde{v})$, accordingly. Let $\mathcal{N} \sim 10^{18}$, as it was the case in our experiment. Then we have $|F_{\bar{b}}(E,\theta) - \tilde{F}_{\bar{b}}(E,\theta)| \sim 10^{-16}|F_{\bar{b}}(E,\theta)|$. Therefore one does not expect any detectable difference for the cross sections measured with two different targets. Indeed, such a detectable difference does not occur if one considers $\sigma_{\bar{b}}(E,\theta)$ for a single fixed $\bar{b}$ independently from the cross sections for the decay to other $\bar{b}' \neq \bar{b}$ micro-channels. However, as suggested in Ref.

351

[14], the situation may change drastically for the cross sections summed over very large number of exit micro-channels. This is the case for DHIC where the collision products have high excitation energies and the measured cross section, $\sigma(E,\theta) = \sum_{\bar{b}} \sigma_{\bar{b}}(E,\theta)$, is the sum over very large number of micro-channels, $N_{\bar{b}} \gg 1$.

The above consideration explicitly demonstrates that (i) the presence of non-homogeneous "target-environmental" perturbations should not affect the atomic-electron effects discussed in Ref. [14], and (ii) the physical origin of the non-reproducibility of the cross sections for different targets is completely analogous to the origin of the atomic-electron effects in DHIC discussed in Ref. [14]. As a result the spontaneous origin of MC suggests up to 100% difference between the non-self-averaging oscillating components of the cross sections for DHIC for two measurements with different targets. The key element in the interpretation of the spontaneous co-herence, non-self-averaging and extreme sensitivity in complex quantum collisions is introduction of the infinitesimally small off-diagonal MC between different *model* transition amplitudes which couple *model* single-particle states (Slater determinants) of the quasi-bound IS and the continuum states [11-14]. It has been argued that the limit of the vanishing of this infinitesimally small correlation properly supplemented by the limit of the infinite dimensionality of the Hilbert space does not destroy correlation between different *physical* transition amplitudes which couple the many-body configurations of the IS and the continuum states. As a result, the highly-excited thermalized ($\hbar/\Gamma \gg \tau_{erg}$) matter displays coexistence of two distinct phases. The decay of the disordered phase is associated with the $\Delta S_{\bar{b}}^{J}$-matrix [11], where J is the total spin of the IS, and, thereby, with the amplitude $\Delta F_{\bar{b}}(E,\theta)$ which is a linear combination of $\Delta S_{\bar{b}}^{J}$ with different J. Since $\Delta F_{\bar{b}}(E,\theta)$ with different $\bar{b} \neq \bar{b}'$ do not correlate, this disordered phase does not contribute to the MC producing the stable reproducible self-averaging energy smooth background in cross sections. The non-self-averaging (MC) and sensitivity originate from decay of the ordered phase corresponding to the micro-channel independent δS^{J}-matrix [13,14] and, thereby, the micro-channel independent $\delta F(E,\theta)$. It is this micro-channel independent $\delta F(E,\theta)$ which is so sensitive and, therefore, non-reproducible due to the spontaneous origin of the MC so that $|\delta F(E,\theta) - \delta \tilde{F}(E,\theta)| \sim |\delta F(E,\theta)| \sim |F_{\bar{b}}(E,\theta)|$, where $\delta F(E,\theta)$ and $\delta \tilde{F}(E,\theta)$ correspond to different targets with different "target-environmental" perturbations. Pictorially, the sensitivity of the δS^{J}-matrix and $\delta F(E,\theta)$ resembles the sensitivity of the direction of the spontaneous magnetization vector, below the Curie point, to the direction of an infinitesimally small external magnetic field.

EXPERIMENTAL METHOD AND THE DATA ANALYSIS

In order to test the sensitivity two independent measurements of excitation functions for the strongly dissipative collision for the same reaction system of ^{19}F+^{93}Nb have been carried out at the China Institute of Atomic Energy (CIAE), Beijing. In these measurements, the ^{19}F^{8+} beam was provided by the HI-13 tandem accelerator. The beam incident energies were varied from 102 to 108 MeV in steps of 250 keV. For both measurements the same accelerator parameters and the same electronic and the acquisition systems were selected. The same three sets of gas-solid ($\Delta E - E$) telescopes, with a charge resolution $Z/\Delta Z \geq 30$, were set at $38°$, $45°$ and $53°$. However, only the data for the $38°$ and $53°$ were obtained with sufficient statistical counts. Only the N and O fragments were chosen so as to avoid possible contamination from carbon. In the two measurements we used different, independently prepared, self-supporting ^{93}Nb target foils with the thickness $\simeq 70\ \mu g/cm^2$. Both the target foils were produced by the sputtering method. The thickness of each of the two foils was determined by the spectrophotometry. It was found that the difference in thickness of the two foils ≤ 5 $\mu g/cm^2$. This difference results in different stopping energy losses in the two different targets. However, this should not affect reproducibility of the cross sections since this difference in stopping energy losses ~ 15 keV is smaller than the energy spread (~ 50 keV) in the beam.

Absolute cross sections were not determined, though great care was taken to ensure no spurious sources of oscillations were introduced into the relative cross sections. To examine a possible effect of the carbon build up in the target, we measured cross sections for N and O fragments from the ^{19}F+^{12}C reaction at $E_{lab}(^{19}F) = 100.25$ MeV. These cross sections were found negligible for the N and O outgoing energies ≥ 45 MeV for $\theta_{lab} = 38°$, and ≥ 40 MeV for $\theta_{lab} = 53°$. We also note that, for $E_{lab}(^{19}F) = 108$ MeV and $\theta_{lab} = 53°$, the production of the N and O fragments with the outgoing energy ≥ 39 MeV in the ^{19}F+^{12}C reaction is kinematically forbidden. Since the energies of the N and O yields in our measurements of the ^{19}F+^{93}Nb DHIC exceed 45 MeV we conclude that the carbon build up does not produce uncontrolled systematic errors and does not affect our data. The stability of the beam direction was controlled as follows: (i) TV monitor screen was used before each energy step to check and correct the position of the beam spot on the target. (ii) Two silicon detectors were placed at $\theta_{lab} = \pm 12°$. (iii) The beam charge was collected using a Faraday cup placed at $\theta = 0°$ and was compared with the counting rates of the silicon detectors. The data were normalized both with respect to the count rates of each of the silicon detectors and the integrated beam current. All the three normalizations produced the relative cross sections, for each individual experiment, which agree within the statistical errors, $1/N^{1/2}$, where N is a count rate. This shows that no uncontrolled spurious effects and systematic errors should occur due to the instability of the beam spot position within each experiment as well as upon the change of the targets. In order to control possible target cracking and development of local defects during exposure of the target 25 repeat points (one repetition for each of all 25 different energies measured) for the first experiment and 5 repeat points (one repetition for 5 different energies) for the second experiment, were taken. Before to repeat each point the TV monitor screen was used to check and correct a position of the beam spot. All the repeated points demonstrated the reproducibility, within the statistical errors, for both individual experiments. We consider this as a solid indication that no damages of the targets, which could bring about uncontrolled spurious effects, occurred in our experiments. All the above procedures indicate that the systematic uncertainties can be neglected and the data errors can be evaluated as statistical only.

The cross sections $\sigma(E)$ for the products O and N in the ^{19}F+^{93}Nb DHIC are presented in Fig.1, where the error bars are statistical only. Although Fig. 1 presents energy integrated yields over the whole dissipative spectra with width $\simeq 30$ MeV (i.e. these

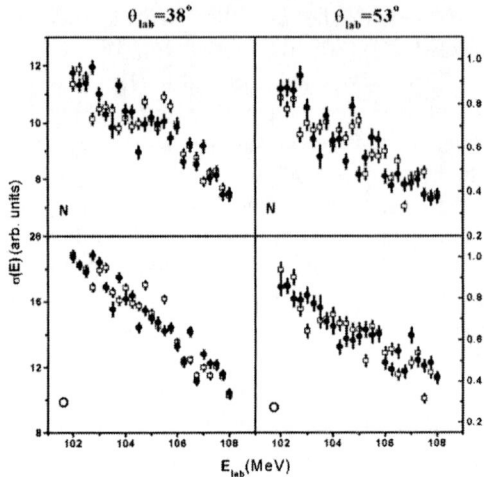

FIGURE 1. Excitation functions for the O and N yields of the $^{19}F+^{93}Nb$ strongly dissipative heavy-ion collisions obtained in the two independent experiments. Full dots correspond to the first experiment and open squares to the second one. The error bars are statistical only.

yields are summed over huge number of different final micro-channels of the highly excited collision products) the characteristic non-self-averaging oscillating structures of the excitation functions in DHIC are clearly identified.

In order to confirm that the oscillations shown in Fig. 1 are true oscillations we calculate the experimental normalized variances of the oscillations, $C(\varepsilon = 0)$. Here $C(\varepsilon) = < \Delta\sigma(E+\varepsilon)\Delta\sigma(E) >$ is a cross section energy autocorrelation function, $\Delta\sigma(E) = (\sigma(E)/ < \sigma(E) > -1)$ is a relative oscillating yield, and $< \sigma(E) >$ is an energy averaged smooth cross section. E.g., for the two independent measurements of the N and O oscillating yields (Fig. 1) at $\theta = 53°$ we obtain $C(\varepsilon = 0) = 0.013 \pm 0.0025$. This is to be compared with the quantities $1/N$, which represent $C(\varepsilon = 0)$ corresponding only to statistical uncertainties due to the finite average counting rate N. For $\theta = 53°$ we have $1/N = 0.004$. It can be seen that the oscillations, reflected in the experimental value of $C(\varepsilon = 0) = 0.013 \pm 0.0025$, are clearly larger than $1/N = 0.004$ expected based on statistics. This demonstrates that the oscillations shown in Fig. 1 are true oscillations and do not result from insufficient statistics.

The cross sections are non-reproducible outside of the range of the statistical errors. This is clearly seen in Fig. 2 from a comparison of the relative non-self-averaging oscillating yields $\Delta\sigma(E)$, for the N products, as an example. Here the energy smooth cross sections $< \sigma(E) >$ were obtained from the best second order polynomial fit of the original data. We observe that, for some energies, the differences between the relative non-self-averaging oscillating yields exceed the amplitude of oscillations in the individual yields. We also find that all the correlation coefficients between the corresponding $\Delta\sigma(E)$'s for the different measurements are insignificant, $\simeq (0.05 - 0.2) \pm 0.1$, where the uncertainty is due to the finite data range only [5]. This indicates that the non-self-averaging non-reproducible components of the cross sections for the two measurements

354

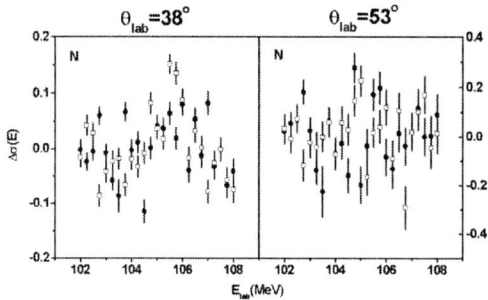

FIGURE 2. Relative non-self-averaging oscillating yields deduced from the excitation functions (Fig. 1) for the N products of the ^{19}F+^{93}Nb strongly dissipative heavy-ion collisions obtained in the two independent experiments. Full dots correspond to the first experiment and open squares to the second one. The error bars are statistical only.

oscillate around each other in almost statistically independent uncorrelated way, supporting the prediction [14].

For a quantitative analysis of the data we construct the cross section energy autocorrelation functions $C(\varepsilon)$ for the two independent measurements of the N and O oscillating yields at $\theta = 53°$ (are not shown, to be reported elsewhere). All the four $C(\varepsilon)$'s demonstrate similar oscillating structures for $\varepsilon \geq 0.5$ MeV and similar Lorentzian-like shapes for $\varepsilon \leq 0.5$ MeV with a half-width $\simeq 55$ keV. We proceed in complete analogy with the detailed analysis [9] of the ^{19}F+^{89}Y system. $C(\varepsilon)$ depends on the values of Γ, β, $\hbar\omega$ and d. Here, $\hbar/\Gamma$ is the average life-time of the IS, $\beta = \hbar/\tau_{dec}$ is the phase randomization (decoherence) width, ω is the angular velocity of the coherent rotation of the IS, and $2d$ is the effective number of partial waves contributing to the peripheral collision (the J-window width) [9]. We find that the theoretical value of $C(\varepsilon = 0)$ depends only on the ratio $\beta d/\Gamma$. From the experimental value of $C(\varepsilon = 0) = 0.013$ we obtain $\beta d/\Gamma = 0.23$. We also find that the theoretical $C(\varepsilon)$ has a half-width $\simeq (\Gamma + \beta d/2)$. Having already extracted the value of $\beta d/\Gamma$ we deduce that $\Gamma = 50$ keV and $\beta d = 11.5$ keV. Taking $2d = 7$ [9] we obtain $\beta = 3.3$ keV. Variation of $2d$ from 10 to 5 with fixed $\beta d = 11.5$ keV does not affect $C(\varepsilon \leq 0.5$ MeV) noticeably. Thus the extracted decoherence width is close to the value of $\beta = 3.5$ keV obtained from analysis of the ^{19}F+^{89}Y system [9]. Finally, from the period of the oscillating structures in $C(\varepsilon)$ for $\varepsilon \geq 0.5 - 1$ MeV we find $\hbar\omega = 1 \pm 0.2$ MeV. For comparison, the semi-classical estimate for an impact parameter close to grazing yields $\hbar\omega \simeq 1.3$ MeV.

CONCLUSIONS

In conclusion, the two statistically significant measurements indicate non-reproducibility of the cross sections for the ^{19}F+^{93}Nb DHIC. This is in contradiction with standard quantum-mechanical calculations which predict reproducibility for the experimental setup employed. However, the non-reproducibility is consistent with the recent theoretical predictions of spontaneous coherence, slow phase randomization and extreme sensitivity in finite highly excited quantum systems. If this non-reproducibility is confirmed in future experiments it will certainly signal that a realization of Wigner's dream [10], a theory for the transition amplitude correlations, will require conceptual revision of modern understanding of microscopic and mesoscopic quantum many-body systems.

ACKNOWLEDGMENTS

This work was supported by the Natural Science Foundation of China (N. 19775057). The authors wish to thank the staff of the HI-13 accelerator at CIAE. The work by S. Kun was supported by the Max Planck Society Fellowship. S. Kun is grateful to members of the Max-Planck-Institute für Kernphysik, Heidelberg for a warm hospitality during his visit on May-July 2001. We thank Prof. Lewis T. Chadderton for useful discussions and suggestions.

REFERENCES

1. E. Wigner, in *Statistical Properties of Nuclei*, edited by J.B. Garg, (Plenum Press, New York-London, 1972), p. 11.
2. P.W. Anderson, *Basic Notions of Condensed Matter Physics*, Frontiers in Physics, vol. 55, (The Benjamin-Cummings, 1984), pp. 71-72.
3. T. Guhr, A. Müller-Groeling, and H.A. Weidenmüller, Phys. Rep. **299**, 189 (1998), and references therein.
4. T. Ericson, Ann. Phys. (N.Y.) **23**, 390 (1963).
5. A. Richter, in *Nuclear Spectroscopy and Reactions*, edited by J. Cerny (Academic, New York, 1974), vol. B, p. 343.
6. H.A. Weidenmüller, Progr. Part. Nucl. Phys. **3**, 49 (1980), and references therein.
7. A. De Rosa *et al.*, Phys. Lett. B**160**, 239 (1985); G. Pappalardo, Nucl. Phys. A**488**, 395c (1988); A. De Rosa *et al.*, Phys. Rev. C**37**, 1042 (1988); *ibid.* C**40**, 627 (1989); *ibid.* C**44**, 747 (1991); Wang Qi *et al.*, Chin. J. Nucl. Phys. **15**, 113 (1993); F. Rizzo *et al.*, Z. Phys. A**349**, 169 (1994); M. Papa *et al.*, Z. Phys. A**353**, 205 (1995); Wang Qi *et al.*, Phys. Lett. B**388**, 462 (1996); I. Berceanu *et al.*, Phys. Rev. C**57**, 2359 (1998); M. Papa *et al.*, Phys. Rev. C**61**, 044614 (2000).
8. T. Suomijärvi *et al.*, Phys. Rev. C**36**, 181 (1987).
9. S.Yu. Kun *et al.*, Z. Phys. A**359**, 263 (1997).
10. E. Wigner, in reference [1], pp. 621,622,635,636.
11. S.Yu. Kun, Z. Phys. A**357**, 255 (1997).
12. S.Yu. Kun, Z. Phys. A**357**, 271 (1997).
13. S.Yu. Kun and A.V. Vagov, Z. Phys. A**359**, 137 (1997).
14. S.Yu. Kun, Phys. Rev. Lett. **84**, 423 (2000).
15. S.Yu. Kun, "Spontaneous coherence and non-equilibrium correlation phase transitions in microscopic and mesoscopic systems", in these Proceedings.

Bifurcation Structure of Eigenstates and Periodic Trajectories in TDHF Phase Space — Interference effects in eigenstates —

Yukio Hashimoto*, Hidehiko Tsukuma† and Fumihiko Sakata**

*Institute of Physics, University of Tsukuba, Ibaraki 305-8571
†Medical Hospital, Hiroshima University, Hiroshima 734-8551
**Department of Mathematical Sciences, Ibaraki University, Mito, Ibaraki 310-8512

Abstract. The resonant periodic trajectories in time-dependent Hartree-Fock phase space are found to be bifurcated from simple vibrational trajectories in a three-level model Hamiltonian. Corresponding to the bifurcated periodic trajectories, there are series of resonant eigenstates with characteristic profile function shapes. In terms of resonant basis system, it is pointed out that the quantum mechanical resonance overlap mechanism determines the stability of the eigenstates.

INTRODUCTION

In investigating the finite quantum many-body system as nuclei, one of the fundamental subjects has been to understand variety of eigenstates in the quantum space of states. The time-dependent Hartree-Fock (TDHF) is one of the time-dependent mean-field frameworks in the nuclear many-body theories and has been made use of in describing the microscopic mechanism of nuclear collective motions which are represented as classical trajectories in the TDHF phase space[1]. Since the TDHF phase space is a classical correspondent of the quantum space of states (i.e., a coherent state representation of the boson-mapped fermion space of states), the classical informations obtained from the TDHF trajectories may provide us with useful informations on the structure change taking place in the quantum space of states. As a candidate of the microscopoic mechanism causing the structure changes of the TDHF trajectories, we have paid special attention to the effects of nonlinear resonances in the TDHF phase space[2, 3].

In order to characterize the nonlinear resonance, we take the advantage of periodic trajectories in phase space. Within a simple three-level model, we have found several sets of periodic trajectories[4]. The periodic trajectories have turned out to come out from special points called bifurcation points where drastic character changes as stable-unstable transitions occur among the periodic trajectories.

In parallel with this classical bifurcation property, we have shown that there are sets of eigenstates with resonance structure which correspond to the bifurcating periodic trajectories in TDHF phase space.

In order to understand how the resonant eigenstates come out, change themselves and finally disappear, we introduced local *resonant basis* in the space of states[3]. On the basis of the local resonant basis, we illustrate "resonance overlap mechanism"

CP597, *Nonequilibrium and Nonlinear Dynamics in Nuclear and Other Finite Systems,*
edited by Z. Li et al.
© 2001 American Institute of Physics 0-7354-0041-5/01/$18.00

determines the structure changes in the resonant eigenstates.

MODEL HAMILTONIAN AND PERIODIC TRAJECTORIES

The model Hamiltonian we have used consists of single-particle part $\hat{H}_0$ satisfying $\varepsilon_0 < \varepsilon_1 < \varepsilon_2$ and two-body interaction part $\hat{H}_V$ [4]:

$$
\begin{aligned}
\hat{H} &= \hat{H}_0 + \hat{H}_V, \\
\hat{H}_0 &= \varepsilon_0 \hat{K}_{00} + \varepsilon_1 \hat{K}_{11} + \varepsilon_2 \hat{K}_{22}, \\
\hat{H}_V &= \frac{V_1}{2}\left(\hat{K}_{10}\hat{K}_{10} + h.c.\right) + \frac{V_2}{2}\left(\hat{K}_{20}\hat{K}_{20} + h.c.\right).
\end{aligned}
\tag{1}
$$

The operators $\hat{K}_{ij}(i,j=0,1,2)$ are defined as $\hat{K}_{ij} = \sum_{m=0}^{N} \hat{C}_{im}^{\dagger}\hat{C}_{jm}$, where $\left\{\hat{C}_{im}^{\dagger}, \hat{C}_{im}\right\}$ are Fermion operators creating (annihilating) particles at the single-particle state with $\varepsilon_i (i = 0, 1, 2)$. Each of the three single-particle levels has N-fold degeneracy. The parameters in the Hamiltonian are set as follows: $N = 30$, $\varepsilon_0 = 0.0$, $\varepsilon_1 = 1.0$, $\varepsilon_2 = 2.0$, $V_1 = V_2 = -0.020$.

In the TDHF phase space, the trajectories are determined by the equations of motion for a set of canonical variables (q_j, p_j),

$$
\dot{q}_j = \frac{\partial \mathcal{H}}{\partial p_j}, \quad \dot{p}_j = -\frac{\partial \mathcal{H}}{\partial q_j}, \quad \mathcal{H} = \langle \phi(f)|\hat{H}|\phi(f)\rangle, \quad (j = 1, 2),
\tag{2}
$$

where the TDHF wave function $|\phi(f)\rangle$ is given as

$$
|\phi(f)\rangle = e^{\hat{F}(t)}|\phi_0\rangle, \quad \hat{F}(t) = \sum_{i=1}^{2}\left(f_i(t)\hat{K}_{i0} - h.c.\right),
\tag{3}
$$

where $|\phi_0\rangle$ is a Hartree-Fock ground state. We have put $\hbar = 1$ and the complex parameters $f_i(t)(i = 1, 2)$ are related to the canonical variables (q_j, p_j) implicitly[4]. The TDHF trajectories are obtained by integrating the equations of motion in Eq.(2) in the TDHF phase space spanned by the variables (q_j, p_j).

Finding out special sets of initial values for the canonical variables (q_j, p_j), we have calculated several series of periodic trajectories which represents bifurcation property from a set of trajectories with simple one-dimensional oscillating character.

In Fig.1, we show the energies and periods of the typical sets of bifurcating periodic trajectories. In this energy-period(E-T) plot, we can see that the two pairs of trajectories, i.e., inner and outer pairs of elliptic and hyperbolic trajectories, start from points on a branch corresponding to the simple periodic trajectories and come to each other as the energies go up. The inner elliptic(IE) type trajectories meet the outer hyperbolic(OH) ones and the inner hyperbolic(IH) trajectories meet the outer elliptic(OE) ones in a critical energy region, which are similar to the *pair annihilations* between two objects. This is the process of structure changes of this system, i.e. generation, transfiguration and disappearance of the periodic motions. The approaching of the two pairs of resonant trajectories is just the realization of the resonance overlap proposed by Chirikov[5].

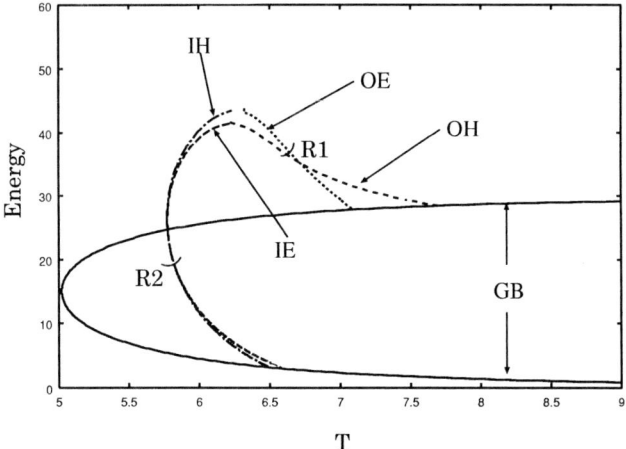

FIGURE 1. Plot of energies and periods for typical sets of periodic trajectories. IE(IH) stands for inner elliptic(hyperbolic) type trajectories and OE(OH) stands for outer elliptic(hyperbolic) type trajectories. GB stands for the simple one-dimensional periodic trajectories.

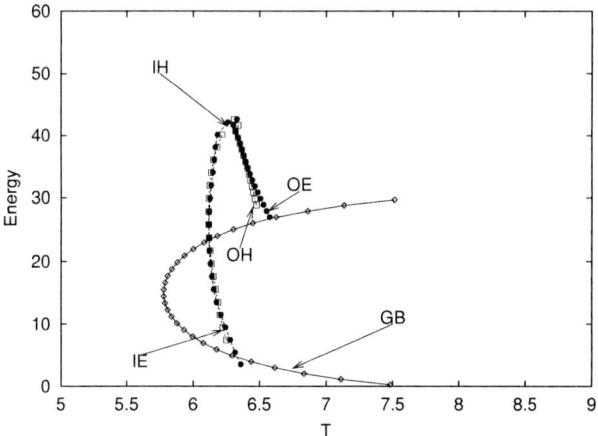

FIGURE 2. Plot of energies and periods for typical sets of eigenstates. IE(IH) stands for inner elliptic(hyperbolic) type states and OE(OH) stands for outer elliptic(hyperbolic) type states. GB stands for ground state band.

OVERLAP OF NONLINEAR RESONANCE

In parallel with the classical case, we have found quantum mechanical bifurcation phenomena, in which series of eigenstates with elliptic and hyperbolic type properties come out from ground state band. The quantum E-T plot for the bifurcated eigenstates are shown in Fig.2. The structure of each of the members in the series of eigenstates changes drastically in the critical energy region where the elliptic type eigenstates meet with the hyperbolic type ones just as in the classical case.

In order to study how the structure change in each of the eigenstates occurs, we introduced local basis states named as $\mu - v$ basis and resonant basis[3]. Both of the $\mu - v$ basis and resonant basis are determined so that approximate(local) quantum numbers are kept and included in the basis states to a good extent.

The local resonant basis states are made from the $\mu - v$ basis states $|\mu, v\rangle$ and span subspace $\mathbf{D}^{\mathcal{N}}$ characterized by a local quantum number $\mathcal{N} = \mu + 2v$,

$$\mathbf{D}^{\mathcal{N}} : \left\{ |\mu, v\rangle; \mathcal{N} = \mu + 2v, 0 \leq \mu + v \leq 30 \right\}, \tag{4}$$

where each of the resonant basis states is labelled as $\{ |\mathcal{N}, a\rangle; a = 0, 1, 2, \cdots \}$. The auxiliary index a is a label of the member in the subspace $\mathbf{D}^{\mathcal{N}}$.

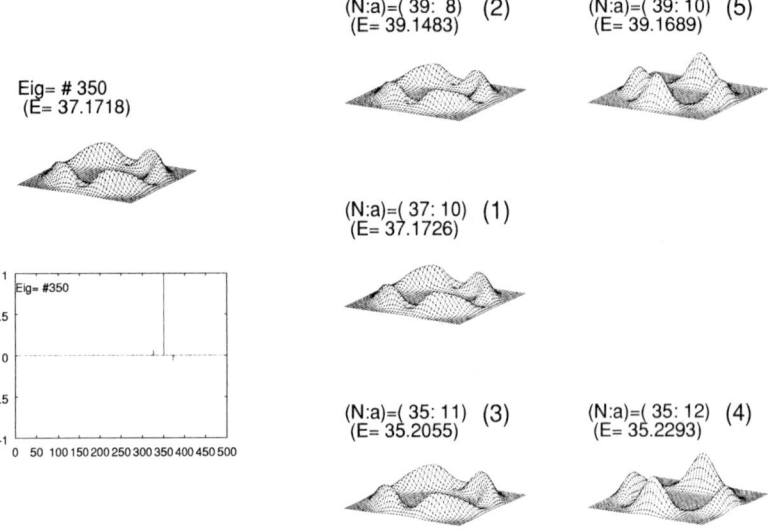

(N:a)=(39: 8) (2)
(E= 39.1483)

(N:a)=(39: 10) (5)
(E= 39.1689)

Eig= # 350
(E= 37.1718)

(N:a)=(37: 10) (1)
(E= 37.1726)

Eig= #350

(N:a)=(35: 11) (3)
(E= 35.2055)

(N:a)=(35: 12) (4)
(E= 35.2293)

FIGURE 3. Components of hyperbolic type eigenstates #350 represented in terms of resonant basis states. Profile functions[4] of the eigenstates and component states with respect to the resonant basis states are displayed.

In terms of the resonant basis states, we have calculated component distributions in each of the eigenstates. As examples, we pick up hyperbolic type eigenstates with labels #350 and #391, whose eigenvalues are 37.172 and 41.134, respectively. In Fig.3 for lower energy eigenstate #350, the strengths are carried by almost one main component. There are a few accompanying components carrying the small part of the total strength. The profile functions[4] of the main components as well as the accompanying components are of the hyperbolic type. They contribute constructively to the hyperbolic type profile functions of the eigenstates. As a result, this eigenstate #350 has well developed four peaks in the profile function, representing clearly that it is the hyperbolic type resonant eigenstate.

On the other hand, for the higher energy eigenstate #391 in Fig.4, we can see that there are two main components together with several accompanying ones. The largest compo-

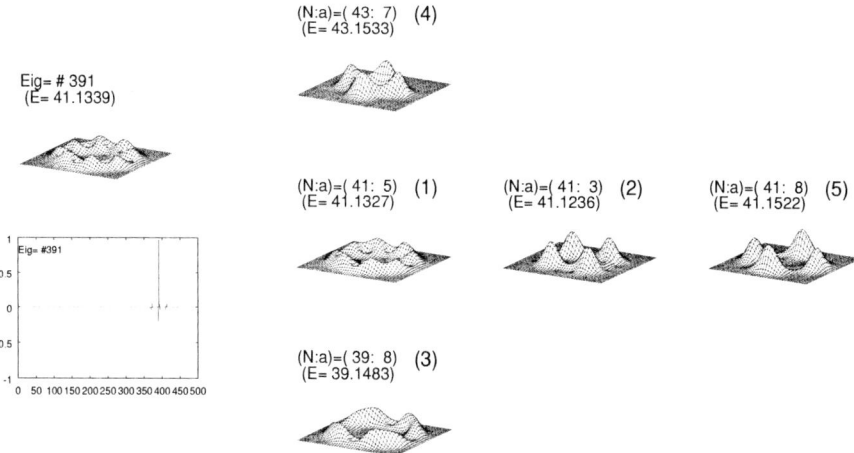

FIGURE 4. Components of hyperbolic type eigenstates #391 which has higher eigen energy than #350(Fig.3) represented in terms of resonant basis states. Profile functions[4] of the eigenstates and component states with respect to the resonant basis states are displayed.

nent displays the typical hyperbolic type profile function. The second largest component represents the *inner elliptic* type state with different phase from the largest component. The two main components are on the same energy plane around 41. The accompanying components are outer and inner hyperbolic type states, and their energies are around 39 and 43, respectively. Thus the hyperbolic type eigenstates #391 is a mixture of two different types of components; mixture of outer and inner type components together with mixture of components with two units higher and lower energies.

As a result, the profile function of the state #391 is complicated and looks unstable when we follow the series of the hyperbolic line toward the critical region of the pair annihilations as in Fig.2. Thus, the resonant basis states make it clear that the hyperbolic type eigenstates change their structure to a large extent in the critical region owing to the mixing of different types of resonant components. This is regarded as the quantum mechanical representation of the resonance overlap proposed in the classical mechanical case[5].

The microscopic analyses of the stability of the eigenstates are now in progress and will be reported elsewhere.

REFERENCES

1. J.W.Negele, *Rev. Mod. Phys.* **54** (1982), 913.
2. F.Sakata, T.Marumori, Y.Hashimoto, Y.Yamamoto, H.Tsukuma and K.Iwasawa, *Nucl. Phys.* **A519**(1990), 93.
3. H.Tsukuma, F.Sakata, T.Marumori, K.Iwasawa, H.Itabashi, Y.Hashimoto and T.Tanaka, *Prog. Theor. Phys.* **91**, 1135(1994).
4. H.Tsukuma, Y.Hashimoto, F.Sakata and K.Iwasawa, *Prog. Theor. Phys.* **100**, 1203(1998).
5. B.V.Chirikov, *Phys. Rep.* **52** (1979), 263.

Nuclear chaotic behavior in particles-rotor model and cranking model

Xian Rong Zhou[1], Lu Guo[1], Jie Meng[2,3], En Guang Zhao[1,3]

[1]Institute of Theoretical Physics, Academia Sinica, P.O. Box 2735, Beijing 100080, P. R. China
[2]Department of Technical Physics, Peking University, Beijing 100871, P. R. China
[3]Center of Theoretical Physics, National Laboratory of Heavy Ion Accelerator, Lanzhou 730000,
P. R. China

Abstract. The chaotic properties for six particles interacting by delta force in a two-j model coupled with a deformed core are studied by replacing the scalar rotation energy of particles-rotor model by a one-body cranking term. The nearest-neighbor distribution of energy levels and spectral rigidity are studied as the function of the spin or cranking frequency, respectively. The results of single-j shell are compared with those in two-j case. The system becomes more regular when single-j ($i_{13/2}$) space is replaced by two-j ($g_{7/2} + d_{5/2}$) shell although the basis size of the configuration space is unchanged. However, the degree of chaoticity of the system changes slightly when configuration space is enlarged by extending single-j ($i_{13/2}$) shell to two-j ($i_{13/2} + g_{9/2}$) shell.

INTRODUCTION

The statistical study of quantum systems is almost as old as quantum mechanics itself. There exists a well established theory to describe fluctuation properties of quantum spectra, namely the random matrix theory (RMT)[1, 2, 3, 4, 5, 6]. Statistical properties of the observed spectra of nuclei are consistent with the predictions of RMT. Exact numerical calculations and analytical investigations support the conclusion that Poissonian and GOE spectral statistics are signs of fully ordered and chaotic motion respectively. For generic system Berry and Tabor showed[7] that the nearest-neighbor distribution (NND) is Poissonian if the system is classically integrable. Although there is no formal proof that NND is of GOE form for chaotic systems, this suggestion [8] is now widely accepted and is supported by several numerical studies. As far as spectral rigidity is concerned, Berry proved[9] that $\triangle_3$ statistics display GOE or Poissonian form depending on whether the classical motion is fully chaotic or fully ordered.

Fluctuating properties of quantum systems fall into one of the four universality classes of Dyson and Mahta depending on the space-time symmetries of the Hamiltonian. In general, various approximation to the Hamiltonian are invoked in different nuclear models. The aim of these approximation is usually to obtain simplified calculations by throwing away certain symmetries. This may in turn have significant effects on the chaotic behavior of the system.

The most appropriate realistic nuclear models are the cranking shell model (CSM) and the particles-rotor model (PRM) for studying the coupling of the single-particle degrees of freedom to the collective motion through Coriolis force . Lots of works have been

CP597, *Nonequilibrium and Nonlinear Dynamics in Nuclear and Other Finite Systems,*
edited by Z. Li et al.
© 2001 American Institute of Physics 0-7354-0041-5/01/$18.00

done to study the difference of CSM and PRM [10, 11, 12, 13]. In the Ref[13] PRM and CSM were compared from the viewpoint of spectral statistics. This comparison, however, was made in the single-j shell space. This paper is to study if there are any changes about the conclusions in Ref.[13] when the configuration space is changed. The first change is that we replace the single-j $(i_{13/2})$ space by a two-j $(g_{7/2}+d_{5/2})$ space and don't change the basis size. The second change is that the configuration space is enlarged by extending single-j $(i_{13/2})$ shell to two-j $(i_{13/2}+g_{9/2})$ shell.

CRANKING AND PARTICLES-ROTOR HAMILTONIANS

An even-even nucleus is visualised as an axially symmetric rotor coupled with a few valence particles outside the core. The spin $\vec{I}$ of the system is the sum of the angular momentum $\vec{R}$ of the core and $\vec{J}$, the sum of the angular momenta of the valence particles. The total Hamiltonian is divided into two parts:

$$H_{PRM} = H_{intr} + H_{coll}. \tag{1}$$

$$H_{intr} = \sum_{j'm'jm} <j'm'|H_{sp}|jm> a^+_{j'm'} a_{jm} \tag{2}$$

$$+\frac{1}{4} \sum_{j'_1m'_1 j'_2m'_2 j_1m_1 j_2m_2} V^{(2)}_{j'_1m'_1 j'_2m'_2 j_1m_1 j_2m_2} a^+_{j'_1m'_1} a^+_{j'_2m'_2} a_{j_2m_2} a_{j_1m_1},$$

where H_{intr} describes microscopically the motion of valence particles, H_{sp} is the single particle Hamiltonian for the potential of axially symmetrical Harmonic oscillator. We take two-body interaction $V^{(2)}$ as delta force $-G\delta(\vec{r_1}-\vec{r_2})$, which can be characterized by the interaction strength G . Here we make the restriction that for single-j case, the particles are put in single-j orbitals (such as $i_{13/2}$) and for two-j case the particles can occupy two-j shell orbitals (such as $i_{13/2}$ and $g_{9/2}$, $g_{7/2}$ and $d_{5/2}$ ect.). In single-j case[14],

$$\sum_{j'm'jm} <j'm'|H_{sp}|jm> a^+_{j'm'} a_{jm} = \sum_m \kappa \frac{3m^2 - j(j+1)}{j(j+1)} a^+_{jm} a_{jm}, \tag{3}$$

where κ is different in different shell. In two-j shell, there are nondiagonal matrix elements between two-j shell in Eq.(3). In Ref.[15], these nondiagonal matrix elements were not considered. It will be seen from this paper that they play an important role in affecting the degree of chaoticity of the system in two-j case.

The collective motion of the core is written as:

$$H_{coll} = \sum_{i=1}^{3} \frac{R_i^2}{2\mathcal{J}_i}$$

$$= \frac{I^2 - I_3^2}{2\mathcal{J}} + \frac{j_1^2 + j_2^2}{2\mathcal{J}} - \frac{I_1 j_1 + I_2 j_2}{\mathcal{J}}$$

$$= H_{rot} + H_{rec} + H_{cor} \qquad (4)$$

where $H_{rot} = \frac{I^2 - I_3^2}{2\mathcal{J}}$ indicates the nuclear collective rotation, $H_{rec} = \frac{j_1^2 + j_2^2}{2\mathcal{J}}$ is the recoil term and $H_{cor} = -\frac{I_1 j_1 + I_2 j_2}{\mathcal{J}}$ is the coriolis term. Until now we have three parameters in the model: quadrupole deformation κ, delta interaction strength G and the moment of inertia $\mathcal{J}$.

The eigenfunctions of particles-rotor Hamiltonian are:

$$\varphi_{IM}^i = \sum_k C_k^i \psi_{IMK}, \qquad (5)$$

where ψ_{IMK}^α is the symmetrized state

$$\psi_{IMK}^\alpha = \sqrt{\frac{2I+1}{16\pi^2(1+\delta_{K0})}} \left\{ \mathcal{D}_{MK}^{I*}(\Omega)\phi_{K\alpha}^{(12\cdots N)} + (-)^{I+K}\mathcal{D}_{M-K}^{I*}(\Omega)\phi_{\bar{K}\alpha}^{(12\cdots N)} \right\}. \qquad (6)$$

$D_{MK}^I(\Omega)$ is usual rotation matrix and Ω describes the orientation of the core with respect to space-fixed axes. $\phi_{K\alpha}^{(12\cdots N)}$ is the N-body anti-symmetric wavefunction for the valence particles and α is for the other quantum numbers.

We invoke cranking approximation as in Ref.[13], which simply involves a consideration of the Routhian or "energy in the rotating frame". The Hamiltonian is

$$H_{cr} = H_{intr} - \omega \sum_{j'm'jm} < j'm'|j_1|jm > a_{j'm'}^+ a_{jm}, \qquad (7)$$

where ω is the cranking frequency and j_1 is the one-component of the single-particle angular momentum.

EVALUATION OF SPECTRAL STATISTICS

Levels of given spins are obtained by diagonizing the Hamiltonian in Eq.(1). To remove the influence of the local fluctuation of level density on level spacings, we map the spectra $\{E_i\}$ onto the spectra $\{X_i\}$ through $X_i = \bar{N}(E_i)$, where X_i are the unfolded levels. The unfolding of the spectra $\{E_i\}$ are carried out according to the following procedures[16]: choosing an interval of levels with $E_i \cdots E_{i+n}$ (e.g. n=7), we can calculate their average spacing: $d_i = \sum_{k=1}^n \frac{E_{i+k} - E_{i+k-1}}{n} = \frac{E_{i+n} - E_i}{n}$, then the unfolded levels X_i can be defined as $X_i = \frac{E_i}{d_i}$. The same procedure are repeated until all the levels are covered. The level spacing of the unfolded levels is defined as $S_i = X_{i+1} - X_i$.

The first spectral statistics we studied is the nearest neighbor distribution $P(S)$. It is obtained by counting the spacings S_i that lie in a certain interval $(S, S+dS)$ and normalizing the resulting distribution. We considered the NND in the interval $S \in (0,2)$.

Spectral statistics show Poissonian and GOE forms for fully ordered and chaotic systems. One interpolation formula between these two distributions was proposed by Brody[2]:

$$P_B(b,S) = (1+b)AS^b \exp(-AS^{1+b}), \tag{8}$$

where

$$A = \left\{ \Gamma(\frac{2+b}{1+b}) \right\}^{1+b} \tag{9}$$

and Γ is the usual gamma function. $q = b = 0$ corresponds to the Poisson distribution, while $q = b = 1$ corresponds to the GOE distribution.

NND emphasizes short-range correlation characteristics. The long-range correlation is measured by the spectral rigidity $\Delta_3(L)$, which is defined as:

$$\Delta_3(\alpha, L) = \frac{1}{L} \underset{A,B}{\text{Min}} \int_\alpha^{\alpha+L} [N(E) - (AE+B)]^2 dE, \tag{10}$$

where A and B are fitting parameters, α and $\alpha + L$ describe the order number of the unfolded levels. To get $\overline{\Delta}_3(L)$, we proceed as described in Ref.[8]. For a given stretch of levels, we calculate $\Delta_3(\alpha, L)$ for the interval $[\alpha, \alpha+L]$, $[\alpha + \frac{L}{2}, \alpha + \frac{3L}{2}]$, $[\alpha + L, \alpha + 2L]$, $[\alpha + \frac{3L}{2}, \alpha + \frac{5L}{2}]$... until the stretch $[a, b]$ has been covered, then we average α to get

$$\overline{\Delta}_3(L) = \frac{1}{N'} \sum_i \Delta_3(\alpha + \frac{i-1}{2}L, L), \tag{11}$$

where N' is the number of sums in the numerator.

RESULTS

In our calculations six particles were put in a single $j = 13/2$ orbitals in single-j case, while in two-j case six particles occupied two-j $(i_{13/2} + g_{9/2}$ or $g_{7/2} + d_{5/2})$ orbitals. The spectral statistics were carried out using states with a given spin in the particles-rotor model. In cranking model, the statistics were analyzed for fixed cranking frequency and signature. Unless stated we used the following parameters in the calculation: $G = 0.45 MeV$, $\mathcal{J} = 24\hbar^2 MeV^{-1}$ and $\kappa = 2.5 MeV$, $2.4 MeV$, $2.2 MeV$ and $2.0 MeV$ for $i_{13/2}$, $g_{9/2}$, $g_{7/2}$ and $d_{5/2}$ orbitals.

In Ref.[13], chaotic behavior of PRM and CSM in single $j = 13/2$ space was studied. Now we don't change the dimension of configuration space and consider two-j $(g_{7/2} + d_{5/2})$ shell instead of single-j $(i_{13/2})$ shell. The total basis size is 1519 for high even angular momenta (signature +1) in PRM. This number is smaller in single-j $(i_{13/2})$ case for $I < 24$, e.g., for I=20 one has 1512 states and it falls to just 93 configurations for I=0. In two-j $(g_{7/2} + d_{5/2})$ case, this number is smaller for $I < 12$, e.g., for $I = 10$ the number is 1513 and 165 for $I = 0$. In the cranking model the basis size is independent of ω and always contains 1519 states in single-j $(i_{13/2})$ and two-j $(g_{7/2} + d_{5/2})$ case. In our spectral statistics, all of the configurations are taken into account.

The calculated NND in single-j $(i_{13/2})$ and two-j $(g_{7/2} + d_{5/2})$ shell were fitted with formula (8). In Fig.1 we show the best-fit Brody parameters as functions of spin and

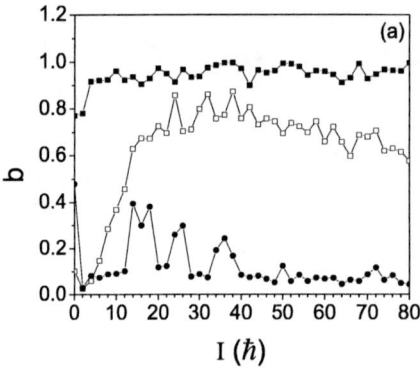 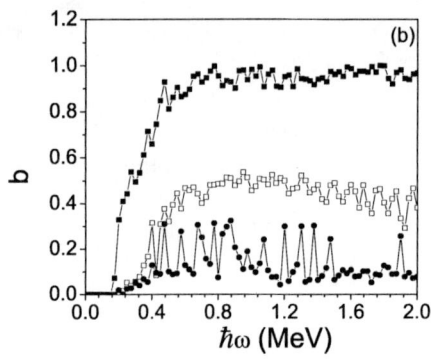

FIGURE 1. The best-fit Brody parameters for the degree of chaoticity with mixed statistics are shown for (a) the particles-rotor model (b) the cranking model as functions of spin and cranking frequency. The results for single-j ($i_{13/2}$) case, two-j ($g_{7/2} + d_{5/2}$) case and two-j ($g_{7/2} + d_{5/2}$) case without considering the off-diagonal matrix elements in Eq.(3) are indicated by solid square, open square, solid circle, respectively.

cranking frequency,respectively. In Fig.1 we notice that the Brody parameters b are larger in single-j case than that in two-j case, which means that the system becomes more regular when the configuration space is changed from single-j ($i_{13/2}$) to two-j ($g_{7/2} + d_{5/2}$) shell. It can be explained as follows: in single-j case, the chaos of the system is caused by the coriolis force, delta interaction and two-body recoil term. However, all these off-diagonal elements which result in chaos only exist between two states whose quantum number j are the same. So, when the space is changed from single-j ($i_{13/2}$) to two-j ($g_{7/2} + d_{5/2}$) shell, although the total number of states is unchanged, the number of off-diagonal elements decreases relative to the single-j case, which decreases the degree of chaoticity of the system. Another factor which causes chaos in two-j case is the off-diagonal matrix elements in Eq.(3). We studied the effects of these matrix elements. The Brody parameter values as functions of the spin and cranking frequency in two-j ($g_{7/2} + d_{5/2}$) case without considering the off-diagonal elements in Eq.(3) are given in Fig.1. In this case, the system becomes more regular because these off-diagonal matrix elements disappear which couple two subspace in two-j case and make the system more chaotic. This phenomena can also be seen from Fig.5, where the single-j ($i_{13/2}$) shell is extended to two-j ($i_{13/2} + g_{9/2}$) shell.

The spectral rigidity in PRM in the case of single-j and two-j case is displayed for different spins in Fig.2. The same conclusion can be drawn. That is the system becomes more regular when the space is changed from single-j ($i_{13/2}$) shell to two-j ($g_{7/2} + d_{5/2}$) shell.

The effect of the moment of inertia is investigated by changing $\mathcal{J}$ and fixing the values of the other parameters. In Ref.[13] this effect was studied in single-j case ($i_{13/2}$) and their conclusion is that the larger the spin the smaller the chaoticity holds true only for smaller values of $\mathcal{J}$ and the opposite trend prevails for large moments of inertia. We

366

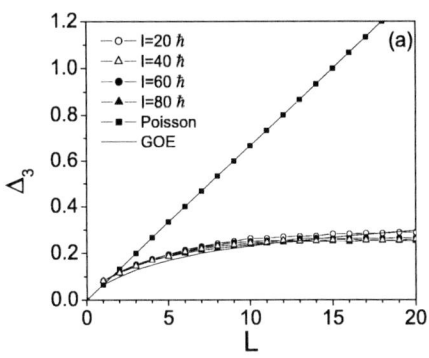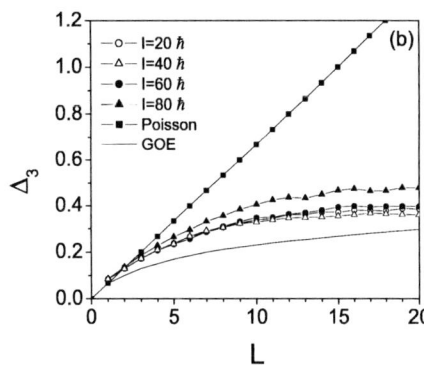

FIGURE 2. The spectral rigidity of the particles-rotor Hamiltonian is presented for different values of spins in (a) single-j $(i_{13/2})$ case and (b) two-j $(g_{7/2}+d_{5/2})$ case.

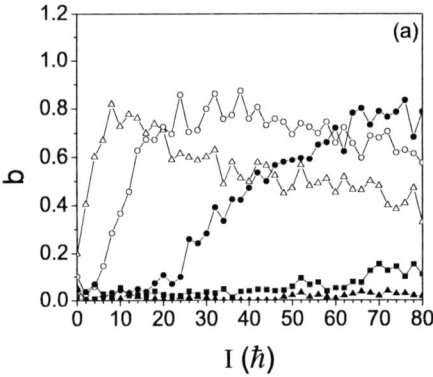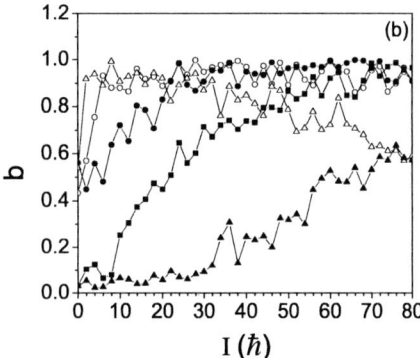

FIGURE 3. The degree of chaoticity in (a) two-j $(g_{7/2}+d_{5/2})$ case and (b) two-j $(i_{13/2}+g_{9/2})$ case is displayed as a function of spin for different values of the moment of inertia $\mathcal{J}$. The results for $\mathcal{J} = 8, 24, 72, 216$ and $648\hbar^2 MeV^{-1}$ are indicated by open triangle, open circle, solid circle, solid square and solid triangle, respectively.

investigate the effect of the moment of inertia in two-j $(g_{7/2}+d_{5/2})$ case. The results are shown in Fig.3(a). The same conclusion can be drawn as the Ref.[13]. Fig.3(b) can also give the same conclusion, where the single-j $(i_{13/2})$ shell is extended to two-j $(i_{13/2}+g_{9/2})$ shell. It is expected that for very large moments of inertia the PRM should give very similar results to the cranking approximation since $\omega = I/\mathcal{J}$[13] . Figure 3 confirms this expectation for the spectral statistics in two-j case.

We next varied the deformation parameter and fixed other parameters in the standard parameters. The results of PRM and CSM in two-j $(g_{7/2}+d_{5/2})$ shell are shown in Fig.4.

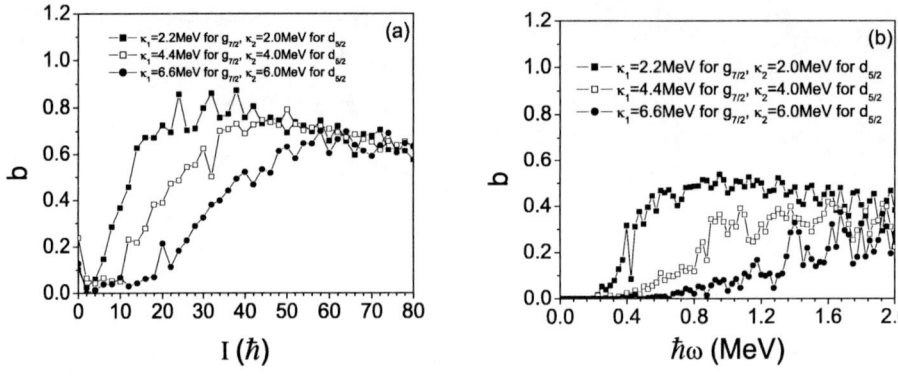

FIGURE 4. The degree of chaoticity is depicted as a function of (a) spin in PRM and and (b) the cranking frequency in CSM for two-j $(g_{7/2} + d_{5/2})$ case.

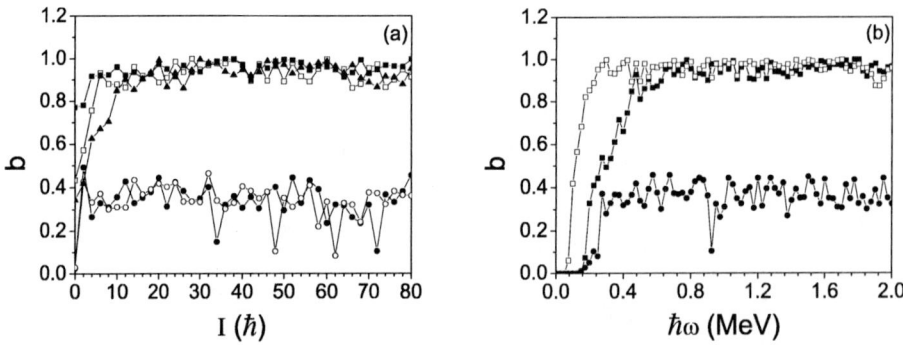

FIGURE 5. The same as Fig.1, but for two-j $(i_{13/2} + g_{9/2})$ case. In Fig.5(a) the results of truncation energy $E = 18.5$ and $19.0 MeV$ are indicated by open square and solid triangle in two-j $(i_{13/2} + g_{9/2})$ case and by open circle and solid circle in two-j $(i_{13/2} + g_{9/2})$ case without considering the off-diagonal matrix elements in Eq.(3). The results correspondind to different truncation energy are nearly the same, which indicates our results are reliable. The results for single-j $(i_{13/2})$ case, two-j $(i_{13/2} + g_{9/2})$ case and two-j $(i_{13/2} + g_{9/2})$ case without considering the off-diagonal matrix elements in Eq.(3) are indicated by solid square, open square and solid circle.

For large κ, an decrease in chaoticity can be observed. In PRM the degree of chaoticity becomes independent of κ quickly, while this convergence is slower in CSM.

Now we want to see what will happen if we enlarge the configuration space by extending single-j $(i_{13/2})$ shell to two-j $(i_{13/2} + g_{9/2})$ shell. In two-j $(i_{13/2} + g_{9/2})$ case, configuration truncation is needed. In our calculation, we took the truncation energy for valence particles as E=18.5MeV and E=19.0MeV and the corresponding basis size is

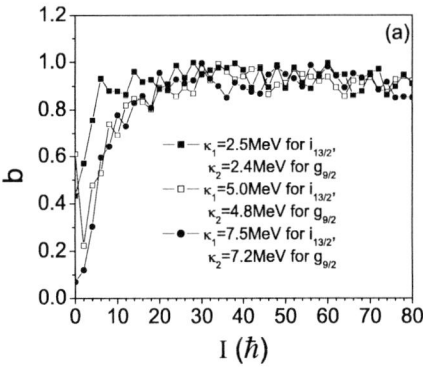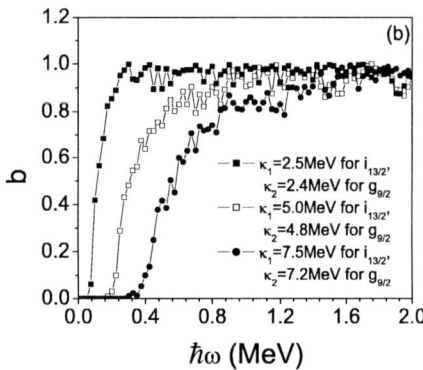

FIGURE 6. The same as Fig.4(b) and 4(c), but for two-j $(i_{13/2} + g_{9/2})$ case.

1667 and 2047, respectively. In the cranking model the basis size is independent of ω and always contains 1519 states in single-j case and 1667 states or 2047 states in two-j case for truncation energy $E = 18.5 MeV$ or $E = 19.0 MeV$.

The fitted Brody parameters as the functions of spin and cranking frequency for single-j $(i_{13/2})$ and two-j $(i_{13/2} + g_{9/2})$ shell are given in Fig.5(a) and 5(b). Fig.5 shows that few changes happen when we enlarge the configuration space. The results of two kinds of truncation are nearly the same, which indicate our truncation is reliable.

Fig.6 gives the Brody parameters in two-j $(i_{13/2} + g_{9/2})$ shell changing with the spins for different deformation κ. In PRM, the results seems independent of deformation, while in CSM, the trend is the same as Fig.4(c) and the "convergence" is slower.

CONCLUSIONS

A systematic study of the chaotic behavior in two realistic models has been carried out. The difference of NND and spectral rigidity between single-j $(i_{13/2})$ case and two-j $(g_{7/2} + d_{5/2})$ case shows that the system is more regular in two-j $(g_{7/2} + d_{5/2})$ model than that in the single-j $(i_{13/2})$ case although the basis size of the configuration space is unchanged. However, when the single-j $(i_{13/2})$ space is enlarged by extending single-j $(i_{13/2})$ shell to two-j $(i_{13/2} + g_{9/2})$ shell, the degree of chaoticity of the system changes slightly. In two-j case, the conclusion[13] also holds true that chaos is more pronounced for normal deformation and low spins than for superdeformations and high spins.

It is known that two-body interaction plays an important role in the appearance of chaotic motion[17, 18]. The present study, however, reveals that the off-diagonal matrix elements of one-body interaction (single-particle Hamiltonian) also play an important role in affecting the degree of chaoticity of the system.

ACKNOWLEDGMENTS

We are grateful to H. D. Heiss and I. Hamamoto for helpful discussions and suggestions. This work is partly supported by National Natural Science Foundation of China and the Major State Basic Research Development Program under contract No. G2000-0774-07.

REFERENCES

1. C. E. Porter, Statistical Theories of Spectra: Fluctuations (Academic Press, New York, 1965).
2. T. A. Brody, J. Flores, J. B. French, P. A. Mello, A. Pandey, and S. S. M. Wong, Rev. Mod. Phys. 53, 385 (1981).
3. R. U. Haq, A. Pandey, and O. Bohigas, Phys. Rev. Lett. 48, 1086(1982).
4. O. Bohigas, R. U. Haq, and A. Pandey, in Nuclear Data for Science and Technology, edited by K. H. Bockhoff (Reider, Dordecht, 1983), p. 809.
5. O. Bohigas, R. U. Haq, and A. Pandey, Phy. Rev. Lett. 54, 1645 (1985).
6. G. E. Mitchell, E. G. Bilpuch, P. M. Endt, and J. F. Shriner, Jr., Phys. Rev. Lett. 61, 1473(1988).
7. M. V. Berry and M. Tabor, Proc. Roy. Soc. London A 356, 375 (1977).
8. O. Bohigas, M. J. Giannoni, and C. Schmit, Phys. Rev. Lett. 52, 1 (1984).
9. M. V. Berry, Proc. Roy. Soc. London A 400, 229(1985).
10. Abraham Klein, Phys. Rev. C62,014316(2001).
11. I. Ray, P. Banerjee and S. Bhattacharya et al., Nucl. Phys. A646, 141(1999).
12. S. Frauendorf, J. Meng, Z. Phys. A356,263(1996).
13. A. T. Kruppa, K. F. pal and N. Rowley, Phys. Rev. C52, 1818(1995).
14. I. Hamamoto, Nucl. Phys. A271,15(1976); R. Bengtsson and H. Hakansson, Nucl. Phys. A357, 61(1981).
15. Lu Guo, Xianrong Zhou, Jie Meng, Enguang Zhao, To be published.
16. Li Junqing, Zhu Jieding and Gu Jinnan, Phys. Rev. B52, 6458(1995).
17. S. Aberg, Phys. Rev. Lett. 64, 3119(1990); Prog. Part. Nucl. Phys. 28, 11(1992).
18. M. Matsuo, T. Dossing, E. Vigezzi, and R. A. Broglia, Phys. Rev. Lett. 70, 2694(1993).

The Effects of Quasiparticle Upon Spectra Statistics

Renrong Zheng, Shunquan Zhu and Nanpu Cheng

Department of Physics, Shanghai Normal University, Shanghai 200234, China

Abstract. Based on the energy levels of the odd-odd nucleus [84]Y at low spins calculated by using two quasi-particles plus rotor model and the distribution of the nearest-neighbor spacing (NNS) and the spectral rigidity (SR) Δ_3 got from random matrix theory (RMT), the effect of quasi-particle treatment upon spectra statistics are studied. We find that the quasi-particle characteristic makes the energy spectra statistics tends to chaotic and the chaos of the spectra mainly comes from the common action of the quasiparticles in the Coriolis and recoil terms of the collective Hamiltonian.

INTRODUCTION

Spectra statistics, as one of quantum signatures of chaos, has renewed attentions recently [1-4]. Most of the works are concentrated on the concrete quantum systems, such as: nucleus, atoms and molecules; ect., rather than some pure toy models as earlier. Within the calculations of the spectral statistics of nuclei, the studies of even-even, even-odd, or odd-even nuclei were more often than that of odd-odd ones to which we will pay attention in the present paper.

In our previous works, the *signature inversion* of odd-odd nuclei in *A*=80, A=130 and *A*=160 mass regions have been investigated by using two quasi-particles plus an axially symmetric rotor model [5-6], and the results of the theoretical calculation coincide with the experimental data well.

We calculate the NNS distribution function $p(s)$, the SR Δ_3 statistics and draw the solid lines of full chaotic Gaussian orthogonal ensemble (GOE) distribution and full regular Poisson distribution limits in figures with expecting to know the chaotic degree of the results by the comparison with the standard GOE and Poisson distributions.

THEORETICAL MODEL FOR ENERGY SPECTRA CALCULATION

Since the two quasi-particles plus rotor model adopted in this paper has been described in detail in Refs. [5], only a concise review is given here.

An odd-odd nucleus can be visualized as an axially symmetric rotor with two

CP597, *Nonequilibrium and Nonlinear Dynamics in Nuclear and Other Finite Systems*,
edited by Z. Li et al.
© 2001 American Institute of Physics 0-7354-0041-5/01/$18.00

valance particles outside the core, the total Hamiltonian reads as

$$H = H_{coll} + H_{sp} + H_{sn} \qquad (1)$$

where H_{coll} is the collective Hamiltonian, H_{sp} and H_{sn} are the quasi-proton and -neutron Hamiltonians. The collective Hamiltonian H_{coll} consists of three terms:

$$H_{coll} = H_{rot} + H_{recoil} + H_{coriolis} \qquad (2)$$

i.e., the pure rotational term H_{rot}, the recoil term H_{recoil}, and the Coriolis term $H_{coriolis}$. The eigenvalues ε_{vi} of H_{sp} and H_{sn} are obtained with axially symmetric potential:

$$V_i = k_0 \beta r^2 Y_{20}, \quad (i = p, n) \qquad (3)$$

pairing effects are considered with the standard BCS theory [5] , single quasi-particle energy e_{vi} reads as:

$$e_{vi} = \sqrt{(\varepsilon_{vi} - \lambda_i)^2 + \Delta_i^2}, \quad (i = p, n) \qquad (4)$$

$$(v_v) = \frac{1}{2}(1 - \frac{\varepsilon_v - \lambda}{e_v}), \quad (i = p, n) \qquad (5)$$

with $(v_v^i)^2$ the occupation probability of v th state, and

$$(v_v^i)^2 + (u_v^i)^2 = 1 \qquad (6)$$

The corresponding Fermi energy λ_i and energy-gap parameters Δ_i are the same as in Ref. [5].

The bases used for diagonalization of the total Hamiltonian (1) take the form as

$$\left| IMK v_p v_n \right\rangle = \sqrt{\frac{2I+1}{16\pi^2}} \sum_{\Omega_p \Omega_n} S_{\Omega_p v_p}^{j_p} S_{\Omega_n v_n}^{j_n} \left[D_{MK}^{I*} \chi_{\pm\Omega_p}^{j_p} \chi_{\pm\Omega_n}^{j_n} + (-1)^{I - j_p - j_n} D_{M-K}^{I*} \chi_{\mp\Omega_p}^{j_p} \chi_{\mp\Omega_n}^{j_n} \right] \quad (7)$$

The matrix element of H_{coll} in the using representation, is the matrix element of H_{coll} between two states of (7) times factors relating to u and v, i.e.,

$$H_{\mu,v}^{KK} = h_{\mu,v}^{KK} (u_\mu^p u_v^p + v_\mu^p v_v^p)(u_\mu^n u_v^n + v_\mu^n v_v^n) \qquad (8)$$

and the eigenvalue equation of the total Hamiltonian is

$$\sum_{K' v_p v_n} \left[H_{\mu,v}^{k,k'} + \left(e_{v_p} + e_{v_n}\right) \delta_{KK'} \delta_{\mu v} \right] t_{K'v}^{(I,j_p,j_n)} = E^{(I,j_p,j_n)} t_{K\mu}^{(I,j_p,j_n)} \qquad (9)$$

By solving Eq. (9) [5], we can obtain the energy eigenvalues, construct an energy sequence $\{E_i\}$ with the same spin I and parity π and study the energy spectra statistics.

THE EFFECTS OF QUASIPARTICLE UPON SPECTRA STATISTICS

Taking [86]Nb and [84]Y as examples, we have explored the statistical properties of the energy levels, and find the chaotic degree of the energy spectra increases with the increasing of spin and reaches a maximum near the signature inversion at $I = 10$;

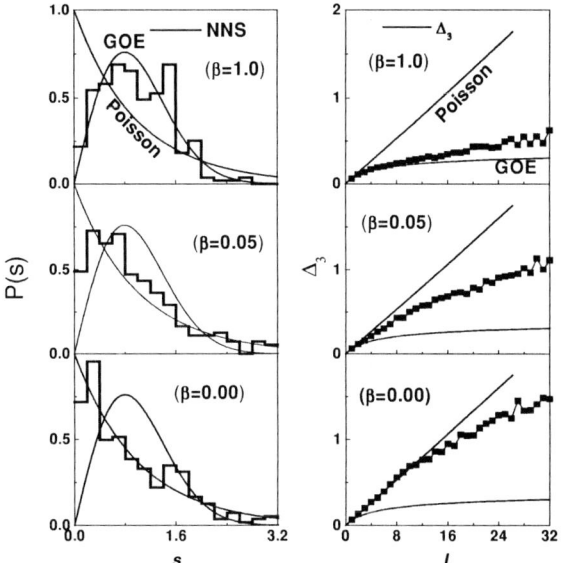

then it decreases gradually for the spins above $I = 10$. The recoil term in the model Hamiltonian makes the energy spectra slightly regular; the Coriolis force, however, brings the spectra to chaotic and plays a major role in the spectral statistics of both odd-odd nuclei. In this paper, we will mainly study the effect of quasiparticle treatment in our theoretical model upon spectra statistics. Taking the NNS and SR of

FIGURE 1. The P(s) distribution of NNS (left column) and SR statistics Δ_3 (right column) for different values of parameter β with histograms (r) and points connected lines as the numeric results.

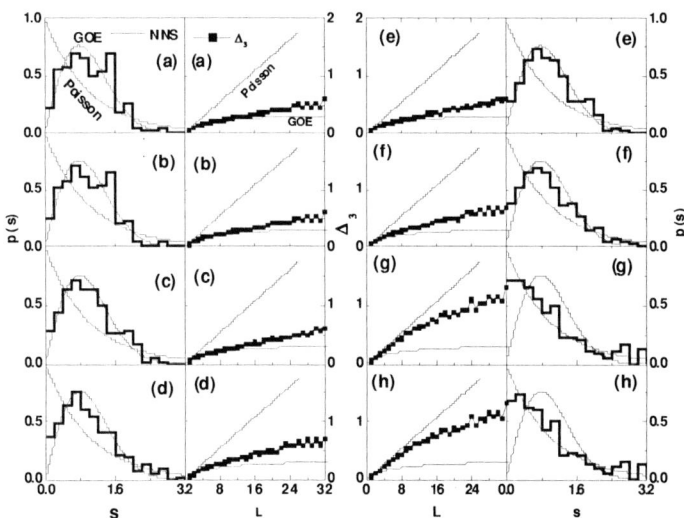

FIGURE 2. The same as Fig.1 but with quasiparticle effect being neglected in some term(s) of formula (2). (see text)

[84]Y at spin I=10 as an example, we see how the spectra statistics changes, when the quasi-particle energy equals real particle energy and the BCS   u, v factors are useless in the matrix elements $H_{coll.}$ Let $\Delta' = \beta\Delta$, $\lambda'_i = \beta\lambda_i$,(i = p, n),   replace Δ, λ by Δ' λ' respectively, and change β from 1 to 0, which means the quasiparticle effect is taken away, gradually, we get figure 1., as we can see in the figure, the spectra statistics change from chaotic (near GOE distributions) to regular (Poisson distributions) both in NNS and SR.

How does the quasiparticle treatment affects the spectra statistics is shown in figure 2, in which we eliminate, one by one, the quasiparticle influence in certain term(s) of (2). In the subfigure (a) of Fig.2, the action of quasiparticle includes in all three terms of (2), however, in all other subfigures of Fig. 2, no one has quasiparticle effect in the whole rotor Hamiltonian (2). The action of quasiparticle in the H_{rot} is taken away in (b) of fig.2; there is no quasiparticle action in the term of H_{recoil} in (c); the action of quasiparticle in the $H_{corioliis}$ is eliminated in (d). The u, v factors in the quasiparticle treatment are given up simultaneously in more than one terms in the Hamiltonian (2) in subfigure (e) to (h) of Fig. 2, no u,v factors in the H_{rot} +H_{recoil} terms for fig.(e);   no u,v factors in the H_{rot} +H_{recoil} terms for fig.(e); no u,v factors in the H_{rot} +$H_{coriolis}$ terms for fig.(f); no u,v factors in the H_{recoil} +$H_{coriolis}$ terms for fig.(g); no u,v factors in the whole H_{coll} of (2) for fig.(h). The fig.2 shows that the most remarkable change takes place in the subfigure (g), the spectra statistics becomes regular Poisson distribution from chaotic GOE distribution in (a) to (f). This means the quasiparticle treatment in our two quasiparticles plus rotor model affects the spectra statistics mainly through the quasiparticle factors in the H_{recoil} and $H_{coriolis}$ two terms of the total rotor Hamiltonian.

CONCLUSIONS

1) The quasiparticle treatment brings spectra statistics into chaotic distribution.
2) The effect of the quasiparticle characteristic is mainly through the common action of the off diagonal matrix elements H_{recoil} and $H_{coriolis}$ in the rotor Hamiltonian

This work is supported by the National Natural Science Foundation of China.

REFERENCES

[1] Giannoni M J , Voros A, and AinnJustin Z   *Chaos and Quantum Physics*, (Amsterdan Press, North-Holland). 1991
[2]Yang X Z and Burgdörfer J *Phys. Rev. Lett.*, **66**, 1991 p982.
[3] Bijker R, . Frank A and Pittel S, *Phys. Rev.*, **C 60**, 1991, p1.
[4] Garrett J D, Robinson J Q and   Foglia A J, *Phys. Lett.*, **B 392**, 1998, p271.
[5] Zheng R R,   Zhu S Q and   Pu Y W,   *Phys. Rev.*, **C 56**, 1997, p175.
[6] Zheng R R,   Zhu S Q et al., *Phys.Rev.*, **C64** , 2001, p014313

Microscopic Theory of Transport Phenomenon in Finite System

Shiwei Yan*, Fumihiko Sakata* and Yizhong Zhuo†

*Department of Mathematical Sciences, Ibaraki University, Mito, Ibaraki 310-8512, Japan
†China Institute of Atomic Energy, P. O. Box 275(18), Beijing, 102413, China

Abstract. We have systematically studied the characteristic features of nonequilibrium transport processes for a microscopic Hamilton system with finite degrees of freedom without introducing any statistical ansatz. It is found that the dissipation of collective motion nonlinearly coupled with the finite intrinsic system is realized through the dephasing, the statistical relaxation and the equilibrium stages. Dominant stage changes depending on the number of intrinsic degrees of freedom. It turns out that the statistical relaxation in the finite system is the anomalous diffusion and the fluctuation effects have finite correlation time.

I. INTRODUCTION

It has been one of the important fields of contemporary science to explore the microscopic origin of the damping phenomenon of dissipative collective motion in the finite many fermions system, which is well described by the phenomenological transport equation. The basic theoretical problems underlying this research are focused on what kinds of microscopic dynamics are there in realizing the statistical state, and how the irreversible macro-level process is generated from the reversible micro-level dynamics[1].

In the conventional approach, as Fokker-Planck or Langevin equations, the ergodic and irreversible property is assumed for the intrinsic system with infinite number of degree of freedom (DOF). Thermodynamics and temperature are introduced by hand through expressing the intrinsic system by a statistical object like the thermal heat bath or the time independent canonical ensemble. In the system where a total number of DOF can be approximately treated as an infinite, there does not arise any serious problem how to introduce the relevant DOF. In the nuclear system where a number of DOF is not large enough, and a time scale of the relevant motion and that of the irrelevant one is typically less than one order of magnitude difference, there arises an important problem how to distinguish the relevant ones from the rest. Intensive studies on this subject have been mainly carried out within the linear response theory (LRT) [2, 3, 4], because a validity of the linear approximation for the macro-level dynamics does not necessarily justify that for the micro-level dynamics. When one intends to derive the phenomenological transport equation from the fundamental level dynamics, moreover, it is not obvious whether the total system is divided into the relevant and irrelevant DOF by leaving a resultant *linear* coupling between them[1, 5, 6, 7, 8].

The SCC method[6] has been developed to dynamically divide the total system into a set of relevant and another set of irrelevant DOF for a given trajectory or for a group

CP597, *Nonequilibrium and Nonlinear Dynamics in Nuclear and Other Finite Systems,*
edited by Z. Li et al.

of trajectories exhibiting the same dynamical property, in a way consistent with the underlying microscopic Hamiltonian. The resultant Hamiltonian has the following form;

$$H = H_{col} + H_{in} + H_{coupl}, \tag{1}$$

where H_{col} depends on the collective (relevant), H_{in} on the intrinsic (irrelevant), and H_{coupl} on both the collective and intrinsic variables. An important point of using the SCC method[6] for dynamically dividing the total system into two subsystems is a form of the resultant coupling between them, where a *linear* coupling is eliminated by the maximal decoupling condition imposed by the method. Based on the SCC method, a general theory of coupled-master equation has been formulated[9] for exploring the microscopic dynamics responsible for the macroscopic transport phenomena, .

In this paper, with the microscopic Hamiltonian (1), we will discuss how to derive the transport equation from the general theory of coupled-master equation and how to realize the dissipation phenomena in the finite system on the basis of the microscopic dynamics, what kinds of necessary conditions there are in realizing the dissipative process, what kinds of dynamical relations there are between the micro-level and phenomenological-level descriptions, without introducing the any statistical anastz.

II. MACROSCOPIC TRANSPORT EQUATION

The transport, dissipative and damping phenomena appearing in the Hamiltonian system may involve a dynamics described by the wave packet rather than that by the single eigenstate. Within the mean-field approximation, these phenomena may be expressed by the collective behavior of the ensemble of trajectories rather than the single trajectory. A difference between the dynamics described by the single-trajectory and by the bundle of trajectories might be related to a long-standing controversy on the effects of one-body and two-body dissipation[10, 11, 12, 13, 14].

In the classical theory of dynamical system, the order-to-chaos transition is usually regarded as the microscopic origin of an appearance of the statistical state in the finite system. Since one may express the heat bath by means of the infinite number of *integrable* systems like the harmonic oscillators whose frequencies have the Debye distribution, it may not be a relevant question whether the chaos plays a decisive role for the dissipation mechanism and for the microscopic generation of the statistical state in a case of the infinite system. In the finite system where the large number limit is not secured, the order-to-chaos is expected to play a decisive role in generating some statistical behavior.

To deal with an ensemble of trajectories, one starts with the Liouville equation for the distribution function $\rho(t)$,

$$\dot{\rho}(t) = -i\mathcal{L}\rho(t), \qquad \mathcal{L}* \equiv i\{H, *\}_{PB}, \qquad \rho(t) \equiv \rho(\eta, \eta^*, \xi, \xi^*, t). \tag{2}$$

and obtains a coupled-master equations[9]

$$\dot{\rho}_\eta(t) = -i\mathcal{L}_\eta^{mean}(t)\rho_\eta(t) - iTr_\xi[\mathcal{L}_\eta + \mathcal{L}_{coupl}]g(t, t_I)\rho_c(t_I)$$

376

$$-\int_{t_I}^{t} d\tau Tr_\xi \mathcal{L}_\Delta(t)g(t,\tau)\mathcal{L}_\Delta(\tau)\rho_\eta(\tau)\rho_\xi(\tau), \tag{3}$$

$$\dot\rho_\xi(t) = -i\mathcal{L}_\xi^{mean}(t)\rho_\xi(t) - iTr_\eta[\mathcal{L}_\xi + \mathcal{L}_{coupl}]g(t,t_I)\rho_c(t_I)$$
$$-\int_{t_I}^{t} d\tau Tr_\eta \mathcal{L}_\Delta(t)g(t,\tau)\mathcal{L}_\Delta(\tau)\rho_\eta(\tau)\rho_\xi(\tau),$$

for the reduced distribution functions $\rho_\eta(t)$ and $\rho_\xi(t)$. The definition of variables in Eq. (3) can be found in Ref.[1]. The coupled-master equation (3) is still equivalent to the original Liouville equation (2). In comparison with the usual time-independent projection operator method of Nakajima-Zwanzig [15, 16] where the irrelevant distribution function ρ_ξ is assumed to be a stationary heat bath, the present coupled-master equation (3) is rich enough to study the microscopic origin of the large-amplitude dissipative motion.

For a general form of coupling interaction, $H_{coupl} = A(\eta)B(\xi)$, recently, it was clarified[1] that a macroscopic transport equation can be obtained from the fully microscopic master equation (3),

$$\dot\rho_\eta(t) = -i[\mathcal{L}_\eta + \mathcal{L}_\eta(t)]\rho_\eta(t) - iTr_\xi[\mathcal{L}_\eta + \mathcal{L}_{coupl}]g(t,t_I)\rho_c(t_I)$$
$$+\zeta\left\{A, (A - <A>_t)\cdot\rho_\eta(t)\right\}_{PB}$$
$$+\zeta\int_0^\infty d\tau c(\tau)\left\{A, \frac{d}{d\tau}(\exp(-i\mathcal{L}_\eta^{mf}\tau)(A - <A>_t))\cdot\rho_\eta(t)\right\}_{PB} \tag{4}$$
$$+\int_0^\infty d\tau\phi(\tau)\left\{A, \left\{\exp(-i\mathcal{L}_\eta^{mf}\tau)A, \rho_\eta(t)\right\}_{PB}\right\}_{PB}.$$

provided the following microscopic conditions are satisfied: (I) The intrinsic time scale is much shorter than the collective time scale. (II) The external effects on the collective system coming from the coupling with intrinsic one are mainly expressed by an averaged effect over the intrinsic distribution function. (III) A weakness of the coupling interaction. That is, the fluctuation effects coming from the coupling interaction is small enough to be treated as a perturbation.

Eq. (5) is a Fokker-Planck type equation. The first term on the right-hand side represents the contribution from the mean-field part, and the second term a contribution from the correlated part of the distribution function at time t_I. The last three terms represent contribution from the dynamical fluctuation effects H_Δ. The friction as well as fluctuation terms are supposed to emerge as a result of those three terms.

III. DYNAMIC REALIZATION OF TRANSPORT PHENOMENON

In this section, we will explore the microscopic dynamics responsible for the macroscopic transport phenomenon with numerical simulation.

A. Microscopic Model

Without any loss of generality, the collective subsystem is represented by a harmonic oscillator with a coordinate q, momentum p, mass M and frequency ω given by

$$H_{col}(q,p) = \frac{p^2}{2M} + \frac{1}{2}M\omega^2 q^2. \tag{5}$$

and the intrinsic subsystem mimicking the environment is described by the β Fermi-Pasta-Ulam (FPU) system (sometime called the β-FPU system because of its quadratic interaction), which was posed in the famous paper [17] and reviewed in [18]:

$$H_{in} = \sum_{i=1}^{N_d} \frac{p_i^2}{2} + \sum_{i=2}^{N_d} W(q_i - q_{i-1}) + W(q_{N_d}), \quad W(q) = \frac{q^4}{4} + \frac{q^2}{2} \tag{6}$$

N_d being the number of intrinsic DOF (i.e., the number of nonlinear oscillators). According to the related literature[2, 18, 19], when the energy per degree of freedom ε is chosen as larger than a certain value (called as the critical value[19]), say $\varepsilon_c \approx 0.1$, the system is strongly chaotic and a "fully developed chaos" is expected. In this case, the appearance of statistical behavior in their chain of oscillators and the equipartition of energy among the modes is realized. In this paper, ε is chosen as 10.

For the coupling interaction, we use the following nonlinear interaction given by

$$H_{coupl} = \lambda \left\{ q^2 - q_0^2 \right\} \left\{ q_1^2 - q_{1,0}^2 \right\}. \tag{7}$$

A physical meaning of introducing new quantities q_0 and $q_{1,0}$ in Eq. (7) was discussed in Ref.[1]. Such a choice of the coupling interaction means that q_1 is considered as a doorway variable, through which the intrinsic subsystem exerts its influence on the collective subsystem[2].

In performing the numerical simulation, the time evolution of the distribution function $\rho(t)$ is evaluated by using the pseudo-particle method as:

$$\rho(t) = \frac{1}{N_p} \sum_{n=1}^{N_p} \prod_{i=1}^{N_d} \delta(q_i - q_{i,n}(t)) \delta(p_i - p_{i,n}(t)) \delta(q - q_n(t)) \delta(p - p_n(t)) \tag{8}$$

where N_p means a total number of pseudo-particles. The distribution function in Eq. (8) defines an ensemble of the system, each member of which is composed of a collective degree of freedom coupled to a single intrinsic trajectory. The collective coordinates $q_n(t)$ and $p_n(t)$, and the intrinsic coordinates $q_{i,n}(t)$ and $p_{i,n}(t)\{i = 1, \cdots, N_d\}$ determine a phase space point of the n-th pseudo-particle at time t, whose time dependence is described by a set of canonical equations of motion given by

$$\dot{q}_i = \frac{\partial H}{\partial p_i}, \quad \dot{p}_i = -\frac{\partial H}{\partial q_i}, \quad \dot{q} = \frac{\partial H}{\partial p}, \quad \dot{p} = -\frac{\partial H}{\partial q} \quad \{i = 1, \cdots, N_d\} \tag{9}$$

We use the fourth order symplectic Runge-Kutta algorithm[20] for integrating the canonical equations of motion, and N_p is chosen to be 10,000.

In our numerical calculation, the used parameters are M=1, ω^2=0.2. In this case, a collective time scale τ_{col} characterized by the harmonic oscillator in Eq. (5) and an intrinsic time scale τ_{in} by the harmonic part of the intrinsic Hamiltonian in Eq.(6) satisfies a relation $\tau_{col} \gg \tau_{in}$. The coupling interaction is switch on after the statistical stationary state has established in the intrinsic system. A switch-on time τ_{sw} [1] is set to be $\tau_{sw} = 100\tau_{col}$. The coupling strength parameter is chosen as $\lambda \sim 0.002$.

B. Energy Dissipation and Fluctuation-Dissipation Relation

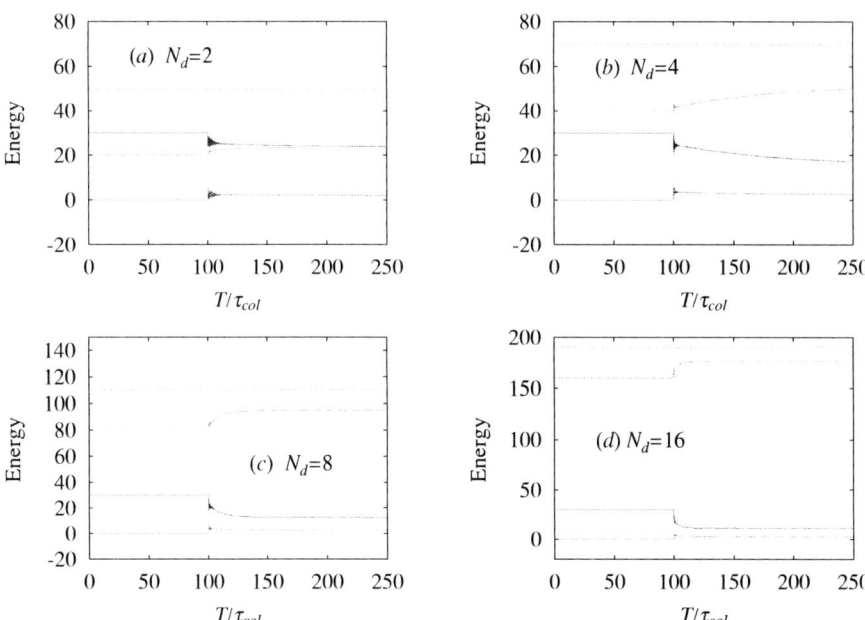

FIGURE 1. Time-dependence of the averaged partial Hamiltonian $\langle H_{col}\rangle$, $\langle H_{in}\rangle$, $\langle H_{coupl}\rangle$ and the total Hamiltonian $\langle H \rangle$. Solid line refers to $\langle H_{col}\rangle$; long dashed line refers to $\langle H_{in}\rangle$; short dashed line refers to $\langle H_{coupl}\rangle$ and dotted line refers to $\langle H \rangle$.

Figures 1 (a-d) show various time-dependent averaged values of the partial Hamiltonian $\langle H_{col}\rangle$, $\langle H_{in}\rangle$ and $\langle H_{coupl}\rangle$, and the total Hamiltonian $\langle H \rangle$ defined through

$$\langle X \rangle = \int X \rho(t) dq dp \prod_{i=1}^{N_d} dq_i dp_i, \qquad (10)$$

for four typical cases with $E_{col} = 30$, λ=0.002, N_d=2, 4, 8 and 16, respectively. In order to clearly show the number dependence of the transport process, in Fig. 2, only $\langle H_{col}\rangle$ is shown for the cases with N_d=2, 4, 8 and 16. It may be easily seen that almost the same result is obtained for the case with N_d=2 as fully discussed in Ref. [1]. That is, the main change occurs in the collective energy as well as the interaction energy. The col-

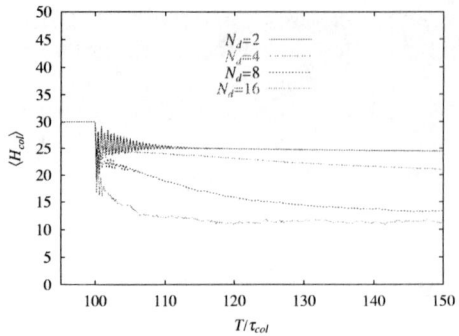

FIGURE 2. The time-dependent averaged values of the partial Hamiltonian $\langle H_{col} \rangle$ for the cases with N_d=2, 4, 8 and 16, respectively. The other parameters are the same as Fig. 1

TABLE 1. An asymptotic averaged energy for each degree of freedom in the intrinsic system and that for the collective system

N_d	2	4	8	16
$\langle H_\xi \rangle$	11.92	12.54	11.851	10.996
$\langle H_\eta \rangle$	24.03	17.15	12.499	11.32

lective energy oscillates around a well defined asymptotic value[1], and the oscillating amplitude decreases due to the dephasing mechanism which is induced by the fluctuation coming from the environment. In this case, the conventional fluctuation-dissipation relation is not satisfied and there is substantial difference between our microscopic simulation and the phenomenological transport equation, even though they produce almost the same macroscopic behaviors. Here, it is worthwhile mentioning that the *decoherence* or *dephasing* process due to the interaction with the environment has also been discussed in the quantum system[21, 22].

One may understand from Fig. 2 that the collective energy transport process is divided into three different phases: (1) *Dephasing regime.* In this regime, the fluctuation originated from the coupling interaction reduces the coherence of collective trajectories and damps the averaged amplitude of collective motion. This regime appears at the first stage, and is the main process for the system with small number of DOF (say, two)[1]. When the number of DOF increases, this regime lasts for a very short time. (2) *Nonequilibrium relaxation regime.* In this regime, the energy of collective motion is irreversibly transferred to the "environment". (3) *Saturation regime,* where the total system reaches to an equilibrium state. In this case, the total energy is equally distributed over each degree of freedom in the collective as well as the intrinsic systems.

Table 1 shows an asymptotic averaged energy for each degree of freedom in the intrinsic system and that for the collective system for N_d=2, 4, 8, 16 at $t \approx 250\tau_{col}$, respectively. Considering the specific role of two ends oscillator in the β-FPU Hamiltonian, one may see that an equipartition of the energy among the whole DOF is expected

TABLE 2. The parameter $\Gamma(\frac{M}{\lambda})$ at $t = 125\tau_{col}$

N_d	2	4	8	16
$\Gamma(\frac{M}{\lambda})$	0.201	0.530	0.820	1.773

at $T \approx 250\tau_{col}$ for the cases with relatively large number of intrinsic DOF, as $N_d \geq 8$. Namely, the third regime is expected for the cases with $N_d \geq 8$.

From the conventional viewpoint of the transport theory, a gradual decrease of the collective energy is due to the coupling with the intrinsic subsystem. In our previous simulation[1] with the Langevin equation for the case with $N_d=2$, it turned out that the ratio of the fluctuation over the dissipation is very larger than expected from the usual fluctuation-dissipation theorem. Since the friction force is much smaller than the fluctuation, the dephasing regime is realized for $N_d=2$. Therefore, an appearance of the second regime suggest us that the contribution from the friction force becomes large as N_d increases. For the cases with $N_d \geq 8$, the friction force becomes much larger so as to satisfy the fluctuation-dissipation relation, which is expected from the asymptotic and saturated situation at $t \approx 250\tau_{col}$.

The analytical analysis[24] has shown that there holds a generalized fluctuation-dissipation relation given as

$$\Gamma = \frac{\lambda}{M^3\omega'^2} \int_0^\infty d\tau \langle\langle \phi(t)\phi(t-\tau)\rangle\rangle \sin^2 \omega'\tau \tag{11}$$

where $\phi(t) = \left\{q_1^2 - q_{1,0}^2\right\} - \left\langle\left\{q_1^2 - q_{1,0}^2\right\}\right\rangle$. The parameter Γ represents the damping effects on the collective motion due to the fluctuation, which comes from the coupling with the chaotic intrinsic subsystem. The relation (11) connects the dissipation of collective motion with the chaotic intrinsic dynamics, rather than the ergodic condition as in Ref. [23].

We calculate the correlation function $\langle\langle\phi(t)\phi(t-\tau)\rangle\rangle$ at $t = 125\tau_{col}$ for the case with $N_d = 2$, 4 and 8 as depicted in Fig. 3. One may see that for $N_d = 2$ the correlation function is very weak and oscillates around 0. This means that the friction force is small in comparison with the fluctuation, and only the dephasing phase is realized. When N_d increases, the correlation function becomes large and behaves as a "colored noise" with a finite correlation time τ_c:

$$\langle\langle\phi(t)\phi(t-\tau)\rangle\rangle \sim e^{-\frac{\tau}{\tau_c}}. \tag{12}$$

One may expect from Fig. 3 that the correlation function tends to be a δ function , i.e. a "white noise", when N_d increases to infinite.

With the above results on the correlation function, the parameter $\Gamma(\frac{M}{\lambda})$ at $t = 125\tau_{col}$ is calculated as shown in Table 2 for the cases with $N_d=2$, 4, 8 and 16, respectively. It is obviously seen that the damping effect on the collective motion becomes strong as N_d increases. In a case with small N_d, the damping effect is too weak to make the second regime appear. For larger N_d (say, $N_d \geq 8$), the damping effect becomes large so as to make the collective subsystem statistically relax toward an equilibrium state.

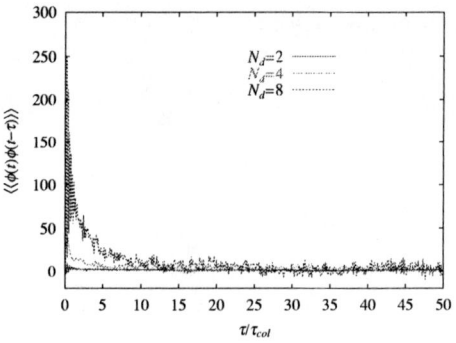

FIGURE 3. The correlation function at $t = 125\tau_{col}$ for the case with $N_d = 2$, 4 and 8.

C. Microscopic Dynamics of Dissipation

It is not a trivial discussion how to understand the three regimens as mentioned in above subsection in a more microscopic way. As mentioned in Sec. II, the transport, dissipative and damping phenomena may be expressed by the collective behavior of the ensemble of trajectories. In the classical theory of dynamical system, the order-to-chaos transition is expected to play a decisive role in generating some statistical behavior in the finite system. There should be the relation between the generating the chaotic motion of a single trajectory and the realizing a statistical state for a bundle of trajectories. One may understand this issue through an entropy production in the collective subsystem by exploiting a generalized, nonextensive entropy [25]:

$$S_\alpha(t) = \frac{1 - \int [\rho(t)]^\alpha \prod_{i=1}^{N_d} dq_i dp_i dq dp}{\alpha - 1}, \tag{13}$$

where α is called an entropic index characterizing the entropy functional $S_\alpha(t)$. When $\alpha = 1$, $S_\alpha(t)$ reduces to the physical Boltzmann-Gibbs (BG) entropy. In Fig. 4, the nonextensive entropy for $\alpha = 0.7$ is shown. One may see from the figure that $S_\alpha(t)$ shows an exponentially increasing process for the case with $N_d=2$. This process just corresponds to the dephasing stage. When N_d increases upto 8 where the energy of collective motion irreversibly transfers to the intrinsic motion, there appears a linearly decreasing process just after an exponential increase. This new process corresponds to the nonequilibrium relaxation stage.

Examining the time development of the collective distribution function in the collective (p,q) phase space, we found[1, 24] that the distribution function tends to expand over the whole ring shape as shown in Fig. 5(a), i.e., over an available phase space under a given collective energy in the dephasing regime. In the nonequilibrium relaxation regime, the distribution function gradually invades toward the central region, i.e., a smaller collective energy region, and finally tends to show the equilibrium Boltzmann distribution (Fig. 5(b)). In this process, $S_\alpha(t)$ decreases linearly to a time-independent

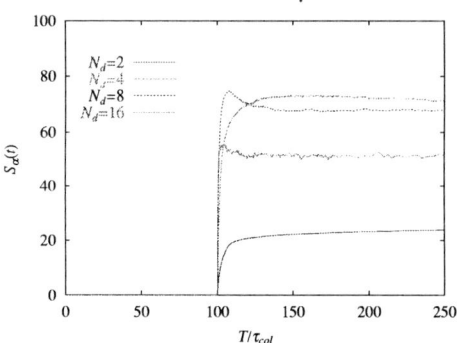

FIGURE 4. The comparison of $S_\alpha(t)$ for N_d=2, 4, 8 and 16, respectively. The entropy index $\alpha = 0.7$

saturated value. As shown by the correlation function (12) and by $\alpha < 1$, the nonequilibrium relaxation process considered in this paper is an anomalous diffusion process.

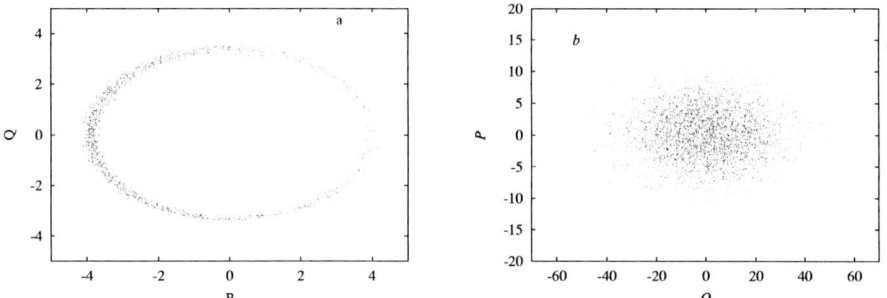

FIGURE 5. The collective distribution function in (p,q) space for N_d=2 (a) and N_d=8 (b) at T=240τ_{col}

It is interesting to notice that the nonequilibrium relaxation regime is characterized by not the usual BG entropy, but by the nonextensive entropy with $\alpha < 1$, which has been used to extend the thermodynamics to a system with a fractal structure. It is also interesting to mention that there also appears three stages in the entropy production for far-from-equilibrium processes, which is also characterized by using the nonextensive entropy[26].

IV. CONCLUSION

In this paper, we have systematically studied the characteristic features of nonequilibrium transport processes for a microscopic Hamilton system with finite degrees of freedom without introducing any statistical ansatz, and paying special attention to the relation between the realization of the macro-level statistical state and the "order-to-chaos" microscopic transition dynamics; and to the microscopic mechanism responsible for the dissipative collective motion, in connection with the effects of the number of degrees of freedom.

We have firstly realized a transport phenomenon in a finite system numerically. We have shown from underlying Hamilton system that for the case with a small number degrees of freedom, the dephasing process, which is caused by the chaoticity of the irrelevant system, is the main mechanism. When the number of degrees of freedom increases, the relaxation mechanism will start to play the role. Thus, the macro-level transport phenomenon in a finite system is *dynamically* established by passing through three distinct stages: dephasing, statistical relaxation and equilibrium regimes. A generalized fluctuation-dissipation relation connecting the dissipation of relevant motion with the chaotic irrelevant dynamics is derived to characterize these three stages. It turns out that the statistical relaxation in the finite system is an anomalous diffusion and the fluctuation effects have a finite correlation time.

This paper may provide us with a general framework for studying the dissipative collective motion in such systems as atoms, nuclei and biomolecules whose environment is not infinite.

REFERENCES

1. S. Yan, F. Sakata, Y. Zhuo, and X. Wu, Phys. Rev. E63, 021116(2001).
2. M. Bianucci, R. Mannella, B. J. West and P. Grigilini, Phys. Rev. E51, 3002(1995).
3. G. Do Dang, A. Klein and P. -G Reinhard, Phys. Rev. C59, 2065(1999).
4. H. Hofmann, Phys. Rep. 284, 137(1997), and references therein.
5. F. Sakata, T. Marumori, Y. Hashimoto and S. Yan, Suppl. Prog. Theor. Phys. 141, 1-111(2001)
6. T. Marumori, T. Maskawa, F. Sakata, and A. Kuriyama, Prog. Theor. Phys. 64, 1294(1980).
7. M. Bianucci, R. Mannella and P. Grigolini, Phys. Rev. Lett. 77, 1258(1996).
8. E. Lutz and H. A. Weidenmüller, Physica A267, 354(1999).
9. F. Sakata, M. Matsuo, T. Marumori and Y. Zhuo. Ann. Phys. (N.Y.) 194, 30(1989).
10. S. Ayik and J. Randrup, Phys. Rev. C50, 2947(1994).
11. S. Ayik, O. Yilmaz, A. Gokalp and P. Shuck, Phys. Rev. C58,1594(1998).
12. M. Bald, G.F. Burgio, A. Rapisarda and P. Schuck, Phys. Rev. C58,2821(1998).
13. F.V. De Blasio, W. Cassing, M. Tohyama, P.F. Bortignon and R.A. Broglia, Phys. Rev. lett. 68,1663(1992).
14. A. Smerzi, A. Bonasera and M. Di Toro, Phys. Rev. C44,1713(1991).
15. S. Nakajima, Prog. Theor. Phys.,20,948(1958).
16. R.W. Zwanzig, J. Chem. Phys.,33,1338(1960).
17. E. Fermi, J. Pasta, and S. Ulam, in *Collected Papers of Enrico Fermi* (Accademia Nazionale dei Lincei and University of Chicago, Roma,1965), Vol. II, P.978.
18. J. Ford, Phys. Rep. 5, 271(1992).
19. L. Casetti, M. Pettini, E. G. D. Cohen, Phys. Rep. 337, 237(2000).
20. H. Yoshida, Phys. Lett. A150, 262(1990).
21. S. Habib, K. Shizume and W. H. Zurek, Phys. Rev. Lett. 80, 4361(1998).
22. W.H. Zurek and J.P. Paz, Phys. Rev. Lett. 72, 2508(1994), W.H. Zurek, Phys. Rev. D24, 1516(1981), E. Joos and H.D. Zeh, Z. Phys. B59, 229(1985).
23. M. Wilkinson, J. Phys. A23, 3603(1990); C. Jarzynski, Phys. Rev. Lett. 74, 2937(1995); 71, 839(1993).
24. S. Yan, F. Sakata, and Y. Zhuo, submitted to Phys. Rev. E.
25. C. Tsallis, J. Stat. Phys. 52, 479(1988).
26. V. Latora and M. Baranger, Phys. Rev. Lett. 82, 520(1999).

Discrete Variation, Euler-Lagrange Cohomology and Symplectic, Multisymplectic Structures[1]

Han-Ying Guo and Ke Wu

Institute of Theoretical Physics, Academia Sinica , P.O.Box 2735, Beijing 100080, China

Abstract. We introduce the discrete variational principle in the framework of multi-parameter differential approach by regarding the forward difference as an entire geometric object. By virtue of this variational principle, we get the difference discrete Euler-Lagrange equations for the difference discrete classical mechanics and classical field theory. We also explore the difference discrete versions for the Euler-Lagrange cohomology and apply to the symplectic or multisymplectic geometry and preserving property in discrete mechanics and field theory. In terms of the difference discrete Euler-Lagrange cohomological concepts, we show that the symplectic or multisymplectic geometry and their difference discrete structure preserving properties can always be established not only in the solution spaces of the discrete Euler-Lagrange equations but also in the function space in each case if and only if the relevant closed Euler-Lagrange cohomological conditions are satisfied.

INTRODUCTION

It is well known that the symplectic and multisymplectic structure play crucially important roles in both Lagrangian and Hamiltonian formalism for classical mechanics [1][2] and field theories [3][4][5][6][7] respectively.

Very recently, it has been found [10][11] that there exists what is called the Euler-Lagrange cohomology in both classical mechanics and field theory and it plays very important role for symplectic or multisymplectic structure preserving properties in each case. It has also been studied the difference discrete mechanics and field theory. For these cases, it has been proposed a difference discrete variational principle by regarding the forward (or backward) difference as an entire geometric object [10][11]. In [12] and [13][14][15], special investigation has been made for the symplectic algorithm as well as symplectic and multisymplectic structure preserving properties in simple element method respectively from the cohomological point of view. In [16], the multi-parameter differential approach has been introduced in order to deal with in the same framework the variation of functional and the exterior differential calculus in the function space. The Euler-Lagrange cohomology and its relation with symplectic and multisymplectic structure preserving properties for classical mechanics and field theory have been studied in both Lagrangian and Hamiltonian formalism and their discrete versions [23]. The cohomological approach has been applied to Hamiltonian-like ODEs, PDEs, symplectic and multisymplectic algorithms respectively as well.

In this talk, we briefly introduce those discrete issues in the Lagrangian formalism.

[1] Talk given by K. Wu at the Conference.

The continuous case can easily be reached by taking continuous limit. The plan of the talk is as follows. We first introduce the difference discrete variational principle for classical mechanics and field theory in section 2. It is shown that difference discrete variational principle with simply modified Leibniz law for differences offers difference discrete Euler-Lagrange equations. In section 3, using the exterior differential calculus in function space, we deal with the difference discrete versions of the Euler-Lagrange cohomology as well as symplectic or multisymplectic structure preserving properties for classical mechanics and field theory. It is shown that relevant difference discrete versions of the Euler-Lagrange cohomology in each case is nontrivial and it is directly linked with difference discrete symplectic and multisymplectic structure preserving properties. Finally, we end with some concluding remarks.

DISCRETE VARIATIONAL PRINCIPLE IN MULTI-PARAMETER DIFFERENTIAL APPROACH

Discrete Lagrangian mechanics

Consider the case that "time" t is difference discretized while n-dimensional configuration spaces M_k at moments $t_k, k \in Z$, are still continuous and smooth enough. Take the "time" $t \in R$ to be discretized as a set of nodes and links with equal step-length $\tau = \Delta t$:

$$t \in R \to t \in \mathcal{T} = \{(t_k, t_{k+1} = t_k + \tau, \quad k \in Z)\}. \tag{1}$$

Let $\mathcal{N}$ and $\mathcal{L}$ be the set of all nodes and links with index set $Ind(\mathcal{N}) = Ind(\mathcal{L}) = Z$, $\mathcal{M} = \bigcup_{k \in Z} M_k$ the configuration space on $\mathcal{T}$ that is still continuous and at least pierce wisely smooth enough. At t_k, $\mathcal{N}_k$ and $\mathcal{L}_k$ is the set of nodes and links neighboring t_k respectively. For example, $\mathcal{L}_k$ includes two links $[t_{k-1}, t_k]$ and $[t_k, t_{k+1}]$ with endpoints (t_{k-1}, t_k, t_{k+1}). I_k the index set of nodes of $\mathcal{N}_k$ including t_k, $\mathcal{N}_k = \bigcup_{Ind(\mathcal{N}) \in I_k} \mathcal{N}$ etc. Coordinates of M_k are denoted by $q^i(t_k) = q^{i(k)}, i = 1, \cdots, n$. $T(M_k)$ the tangent bundle of M_k in the sense that difference at t_k is its base, $T^*(M_k^n)$ its dual. Let $\mathcal{M}_k = \bigcup_{l \in I_k} M_l$ be the union of configuration spaces M_l at $t_l, l \in I_k$ on $\mathcal{N}_k$, $T\mathcal{M}_k = \bigcup_{l \in I_k} TM_l$ the union of tangent bundles on $\mathcal{M}_k$, $F(TM_k)$ and $F(T\mathcal{M}_k)$ function spaces on each respectively, etc.. In the difference discrete variational principle, these notions will be used.

The difference discrete Lagrangian

$$L_D^{(k)} = L_D(q^{i(k)}, \Delta_t q^{i(k)}) \tag{2}$$

is a functional on $F(T\mathcal{M}_k)$, since $q_t^{i(k)}$ is the forward difference of $q^{i(k)}$ at t_k defined by

$$\Delta_t q^{i(k)} := \frac{d}{dt} q^{i(k)} = \Delta_t q^{i(k)} = \frac{1}{\tau} \{q^{i(k+1)} - q^{i(k)}\}. \tag{3}$$

It is the (discrete) derivative and the base of $T(\mathcal{T})$ in the sense of noncommutative differential calculus on a regular lattice L^1 with equal step-length τ [17].

As was emphasized, the forward difference is always viewed as an entire geometric object and its dual $d_{\mathcal{T}}t$ is the base of $T^*(\mathcal{T})$ in the sense

$$d_{\mathcal{T}}t(\Delta_t) = 1. \tag{4}$$

As is known, the (forward) difference as the discrete derivative does not obey the Leibniz law but the modified one

$$\Delta_t(f \cdot g)^{(k)} = \Delta_t f^{(k)} \cdot g^{(k)} + f^{(k+1)} \cdot \Delta_t g^{(k)}, \quad f, g \in FM = \Omega_{\mathcal{T}}^0. \tag{5}$$

On the other hand, however, in the space $T^*(\mathcal{T})$ dual to $T(\mathcal{T})$, an exterior differential operator $d_{\mathcal{T}}$ exists such that

$$d_{\mathcal{T}} : \Omega_{\mathcal{T}}^l \to \Omega_{\mathcal{T}}^{l+1}, \quad d_{\mathcal{T}}^2 = 0, \tag{6}$$

where $\Omega_{\mathcal{T}}^l$ the space of l-forms, $l = 0, 1$, on $T^*(\mathcal{T})$ and $d_{\mathcal{T}}$ does satisfy the Leibniz law:

$$d(\omega \wedge \tau)^{(k)} = d\omega^{(k)} \wedge \tau^{(k)} + (-1)^{deg(\omega)} \omega^{(k)} \wedge \tau^{(k)}. \tag{7}$$

The action functional in the continuous case now becomes

$$S_D = \sum_{k \in Z} L_D(q^{i(k)}, \Delta_t q^{i(k)}). \tag{8}$$

We now consider how to calculate the variation of S_D. Since only the "time" is discretized while either configuration spaces at each t_k, i.e. at the node k, or at its neighboring union are still continuous and the variational calculation here is mainly local, the difference discrete variations may still be manipulated in the framework of the multi-parameter differential approach. In addition, the differential and exterior differential calculus in the function space can also be carried out in either $F(TM_k)$ and $F(T\mathcal{M}_k)$, etc..

In order to make use of the multi-parameter differential approach, variations of $q^{i(k)}$ and $\Delta_t q^{i(k)}$ with the multi-parameter ε^l should be introduced. At t_k, we have

$$q_\varepsilon^{i(k)} = q^{i(k)} + \varepsilon^l \delta_l q^{i(k)}, \quad \Delta_t q_\varepsilon^{i(k)} = \Delta_t q^{i(k)} + \varepsilon^l \delta_l(\Delta_t q^{i(k)}), \tag{9}$$

and

$$\delta_l q_\varepsilon^{i(k)} := \frac{\partial}{\partial \varepsilon^l}|_{\varepsilon^l=0} q_\varepsilon^{i(k)} = \delta_l q^{i(k)}, \quad \delta_l \Delta_t q_\varepsilon^{i(k)} := \frac{\partial}{\partial \varepsilon^l}|_{\varepsilon^l=0} \Delta_t q_\varepsilon^{i(k)} = \delta_l(\Delta_t q^{i(k)}). \tag{10}$$

Then the action functional in (8) becomes a family of action functionals

$$S_D \to S_{D\varepsilon} = \sum_{k \in Z} L_{D\varepsilon}^{(k)} \tag{11}$$

and the variation of the action functional along the direction l is

$$\delta_l S_D = \frac{\partial}{\partial \varepsilon^l} S_{D\varepsilon}|_{\varepsilon^l=0} = \sum_{k \in Z} \{ \frac{\partial L_{D\varepsilon}^{(k)}}{\partial q_\varepsilon^{i(k)}} \delta_l q_\varepsilon^{i(k)} + \frac{\partial L_{D\varepsilon}^{(k)}}{\partial (\Delta_t q_\varepsilon^{i(k)})} \delta_l q_{t\varepsilon}^{i(k)} \}|_{\varepsilon^l=0}. \tag{12}$$

By virtue of the modified Leibniz law (5) for $\Delta_t = \partial_t$, we have

$$\Delta_t \left(\frac{\partial L_D^{(k-1)}}{\partial (\Delta_t q^{i(k-1)})} \delta_l q^{i(k)} \right) = \frac{\partial L_D^{(k)}}{\partial q^{i(k)}} \delta_l \Delta_t q^{i(k)} + \Delta_t \left(\frac{\partial L_D^{(k-1)}}{\partial (\Delta_t q^{i(k-1)})} \right) \delta_l q^{i(k)}. \qquad (13)$$

Therefore,

$$\delta_l S_D = \sum_{k \in Z} \left\{ \left(\frac{\partial L_{D\varepsilon}^{(k)}}{\partial q_\varepsilon^{i(k)}} - \Delta_t \left(\frac{\partial L_{D\varepsilon}^{(k-1)}}{\partial (\Delta_t q_\varepsilon^{i(k-1)})} \right) \right) \delta_l q_\varepsilon^{i(k)} \right\} \Big|_{\varepsilon^l = 0} + \sum_{k \in Z} \Delta_t \left(\frac{\partial L_{D\varepsilon}^{(k-1)}}{\partial (\Delta_t q_\varepsilon^{i(k-1)})} \delta_l q_\varepsilon^{i(k)} \right) \Big|_{\varepsilon^l = 0}.$$

Using the properties

$$\sum_{k \in Z} \Delta_t f(t_k) = f(t_{k=+\infty}) - f(t_{k=-\infty}), \qquad (14)$$

and assuming $\delta_l q^{i(k)}|_{k \pm \infty} = 0$, we get discrete Euler-Lagrange equations

$$\frac{\partial L_D^{(k)}}{\partial q^{i(k)}} - \Delta_t \left(\frac{\partial L_D^{(k-1)}}{\partial (\Delta_t q^{i(k-1)})} \right) = 0. \qquad (15)$$

It should be mentioned that for the forward difference calculation more general Leibniz law can be adopted and it will lead to more general difference discrete version of Euler-Lagrange equations.

Let us consider an example with the difference discrete Lagrangian given by:

$$L_D(q^{i(k)}, \Delta_t q^{j(k)}) = \frac{1}{2} (\Delta_t q^{i(k)})^2 - V(q^{i(k)}). \qquad (16)$$

The difference discrete variational principle gives discrete Euler-Lagrange equations

$$\Delta_t (\Delta_t q^{i(k-1)}) - \frac{\partial}{\partial q} V(q^{i(k)}) = 0, \qquad (17)$$

i.e.

$$\frac{1}{\tau^2} (q^{i(k+1)} - 2q^{i(k)} + q^{i(k-1)}) = \frac{\partial}{\partial q} V(q^{i(k)}). \qquad (18)$$

Obviously, these difference discrete equations have correct continuous limit.

Discrete Lagrangian field theory

For simplicity, we consider the case of 1+1-dimensional spacetime or 2-dimensional space. It is straightforward to generalize for higher dimensional cases.

Let $X^{(1,1)}$ or $X^{(2)}$ with suitable signature of metrics be the base manifold, $L^2 = X$ a regular lattice with 2-directions $x_\mu (\mu = 1, 2)$ on $X^{(1,1)}$ or $X^{(2)}$, $\mathcal{N}$ the all nodes on L^2 that are coordinated by $x_{(i,j)}, (i,j) \in Z \times Z$, with index set $Ind(\mathcal{N})$, $M_D := M_{(i,j)}$ the pierce of configuration space with a set of generic field variables $u^\alpha(x_{(i,j)}) = u^{\alpha(i,j)} \in M_D$ at the node $x_{(i,j)}$, $TM_{(i,j)}$ the tangent bundle of $M_{(i,j)}$ with the set of field variables and their differences $(u^{\alpha(i,j)}, u_\mu^{\alpha(i,j)}) \in T(M_{(i,j)})$, $F(TM_{(i,j)})$ the function space on $TM_{(i,j)}$, etc..

For a given node with coordinates $x_{(i,j)}$, let $\mathcal{N}_{(i,j)}$ be the set of nodes neighboring to $x_{(i,j)}$ with index set $I_{(i,j)} = Ind(\mathcal{N})_{(i,j)}$, $X_{(i,j)} = \bigcup_{Ind(\mathcal{N}) \in I_{(i,j)}} \mathcal{N}$ a set of nodes related to $x_{(i,j)}$ by differences, $\mathcal{M}_D := \mathcal{M}_{X_{(i,j)}} = \bigcup_{Ind(\mathcal{N}) \in I_{(i,j)}} \mathcal{M}_\mathcal{N}$ the union of pierces of configuration spaces on $X_{(i,j)}$.

Forward differences along each direction in $F(T\mathcal{M}_{X_{(i,j)}})$ are defined by

$$\Delta_1 u^{(i,j)} = \frac{1}{h_1}(u^{(i+1,j)} - u^{(i,j)}), \quad \Delta_2 u^{(i,j)} = \frac{1}{h_2}(u^{(i,j+1)} - u^{(i,j)}). \tag{19}$$

They are bases of $T(X)$ and upper-indexes reflect corresponding coordinates of nodes on X. And their dual $dx^\mu = d_X x^\mu$ are bases of $T^*(X)$

$$d_X x^\mu(\Delta_\nu) = \delta_\nu^\mu. \tag{20}$$

As in the previous subsection, (forward) differences as discrete derivatives do not obey the Leibniz law but the modified one (5) along each direction. While in the space $T^*(X)$ dual to $T(X)$, an exterior differential calculus can be introduced such that there exists an operator d_X with following properties

$$d_X : \Omega^{l(i,j)} \to \Omega^{l+1(i,j)}, \qquad d_X^2 = 0, \tag{21}$$

where $\Omega^{l(i,j)}$ is the space of all l-forms in $T^*(X)$ and d_X does satisfy the Leibniz law:

$$d_X(\omega \wedge \tau)^{(i,j)} = d_X \omega^{(i,j)} \wedge \tau^{(i,j)} + (-1)^{deg(\omega)} \omega^{(i,j)} \wedge d_X \tau^{(i,j)}. \tag{22}$$

It is important to note that although the base manifold is discretized either the configuration space at each node or its neighboring union is still continuous. In addition, the variational calculation is mainly local. Therefore, similar to difference discrete classical mechanics, difference discrete variations will be manipulated on the framework of the multi-parameter differential approach.

In addition, similar to the difference discrete mechanics, the differential and the exterior differential calculus in the function space can also be carried out in either $F(TM_{(i,j)})$ or $F(T\mathcal{M}_{X_{(i,j)}}) := \bigcup_{\mathcal{N} \in I} F(T\mathcal{N})$, etc. on the framework of the multi-parameter differential approach.

For the difference discrete field theory, the difference discrete Lagrangian

$$\mathcal{L}_D^{(i,j)} = \mathcal{L}_D(u^{\alpha(i,j)}, u_\mu^{\alpha(i,j)}) \tag{23}$$

is a functional in $F(T\mathcal{M}_{X_{(i,j)}})$. The action functional is given by

$$S_D = \sum_{(i,j) \in \mathbb{Z} \times \mathbb{Z}} \mathcal{L}_D(u^{\alpha(i,j)}, u_\mu^{\alpha(i,j)}). \tag{24}$$

Taking the variation of S_D by the multi-parameter differential approach, the variation along the direction β is given by

$$\delta_\beta S_D = \frac{\partial}{\partial \varepsilon^\beta} S_{D\varepsilon}|_{\varepsilon^\beta = 0} = \sum_{(i,j) \in Z \times Z} \{\frac{\partial \mathcal{L}_D^{(i,j)}}{\partial u^{\alpha(i,j)}} \delta_\beta u^{\alpha(i,j)} + \frac{\partial \mathcal{L}_D^{(i,j)}}{\partial u_\mu^{\alpha(i,j)}} \delta_\beta u_\mu^{\alpha(i,j)}\}. \tag{25}$$

For simplicity, hereafter we omit the multi-parameters ε^β in the course of calculation. Employing the modified Leibniz law (5) for forward difference, we have

$$\Delta_1 (\frac{\partial \mathcal{L}_D^{(i-1,j)}}{\partial u_1^{\alpha(k-1,l)}} \delta_\beta u^{\alpha(k,l)}) = \frac{\partial \mathcal{L}_D^{(i,j)}}{\partial u_1^{\alpha(k,l)}} \delta_\beta u_1^{\alpha(k,l)} + \Delta_1 (\frac{\partial \mathcal{L}_D^{(i-1,j)}}{\partial u_1^{\alpha(k-1,l)}}) \delta_\beta u^{\alpha(k,l)},$$

$$\Delta_2 (\frac{\partial \mathcal{L}_D^{(i,j-1)}}{\partial u_2^{\alpha(k,l-1)}} \delta_\beta u^{\alpha(k,l)}) = \frac{\partial \mathcal{L}_D^{(i,j)}}{\partial u_2^{\alpha(k,l)}} \delta_\beta u_2^{\alpha(k,l)} + \Delta_2 (\frac{\partial \mathcal{L}_D^{(i,j-1)}}{\partial u_2^{\alpha(k,l-1)}}) \delta_\beta u^{\alpha(k,l)}.$$

Assuming that $\delta_\beta u^{\alpha(k,l)}$'s vanish at infinity, we get discrete Euler-Lagrange equations

$$\frac{\partial \mathcal{L}_D^{(i,j)}}{\partial u^{\alpha(k,l)}} - \Delta_1 (\frac{\partial \mathcal{L}_D^{(i-1,j)}}{\partial (\Delta_1 u^{\alpha(k-1,l)})}) - \Delta_2 (\frac{\partial \mathcal{L}_D^{(i,j-1)}}{\partial (\Delta_2 u^{\alpha(k,l-1)})}) = 0. \tag{26}$$

Let us consider a discrete lagrangian field theory as an example:

$$\mathcal{L}_D (u^{\alpha(i,j)}, u_\mu^{\alpha(i,j)}) = \frac{1}{2} (\Delta_\mu u^{\alpha(i,j)})^2 - V(u^{\alpha(i,j)}). \tag{27}$$

Discrete Euler-Lagrange equations (26) become

$$\frac{1}{h_1^2} (u^{\alpha(i+1,j)} - 2u^{\alpha(i,j)} + u^{\alpha(i-1,j)}) + \frac{1}{h_2^2} (u^{\alpha(i,j+1)} - 2u^{\alpha(i,j)} + u^{\alpha(i,j-1)}) = \frac{\partial V(u^{\alpha(i,j)})}{\partial u^\alpha}. \tag{28}$$

These are difference discrete Euler-Lagrange equations with correct continuous limit.

EULER-LAGRANGE COHOMOLOGY, SYMPLECTIC/MULTISYMPLECTIC STRUCTURE PRESERVING PROPERTY

Discrete Lagrangian mechanics

Let us first consider the discrete Euler-Lagrange cohomology, its relation to the simplectic structure and its preserving property for the difference discrete Lagrangian mechanics.

The difference discrete Lagrangian at t_k is given in (2) on $F(T \mathcal{M}_k)$.

Taking the exterior differential d of $L_D{}^{(k)}$ in the function space $F(T\mathcal{M}_k)$ in the framework of multi-parameter differential approach [16], we get

$$dL_D{}^{(k)} = \frac{\partial L_D{}^{(k)}}{\partial q^{i(k)}} dq^{i(k)} + \frac{\partial L_D{}^{(k)}}{\partial \Delta_t q^{i(k)}} d\Delta_t q^{i(k)}.$$

Using the modified Leibniz law (13) for the forward difference $\Delta_t = \partial_t$ defined in (3), introducing the discrete Euler-Lagrange 1-form and the discrete canonical 1-form $\theta_{DL}^{(k)}$

$$E_D{}^{(k)}(q^{i(k)}, \Delta_t q^{j(k)}) := \{\frac{\partial L_D{}^{(k)}}{\partial q^{i(k)}} - \Delta_t(\frac{\partial L_D{}^{(k-1)}}{\partial(\Delta_t q^{i(k-1)})})\} dq^{i(k)}, \tag{29}$$

$$\theta_{DL}{}^{(k)} = \frac{\partial L_D{}^{(k-1)}}{\partial(\Delta_t q^{i(k-1)})} dq^{i(k)}, \tag{30}$$

we have

$$dL_D{}^{(k)} = E_D{}^{(k)} + \Delta_t \theta_D{}^{(k)}. \tag{31}$$

Due to the nilpotency of d on $T^*(\mathcal{M}_k)$, $d^2 L_D{}^{(k)} = 0$, we get

$$dE_D{}^{(k)} + \Delta_t \omega_{DL}{}^{(k)} = 0, \tag{32}$$

where $\omega_{DL}{}^{(k)} := d\theta_D{}^{(k)}$ is a discrete symplectic 2-form on $T^*(\mathcal{M}_k)$

$$\omega_D{}^{(k)} = \frac{\partial^2 L_D{}^{(k-1)}}{\partial q^{i(k)} \partial(\Delta_t q^{j(k-1)})} dq^{i(k)} \wedge dq^{j(k)} + \frac{\partial^2 L_D{}^{(k-1)}}{\partial(\Delta_t q^{i(k)})\partial(\Delta_t q^{j(k-1)})} d\Delta_t q^{i(k)} \wedge dq^{j(k)}. \tag{33}$$

Let us enumerate important issues on the difference discrete Euler-Lagrange cohomology and the difference discrete symplectic structure preserving property.

First, null discrete canonical Euler-Lagrange forms give rise to Euler-Lagrange equations and they are the special case of the coboundary discrete canonical Euler-Lagrange forms.

Secondly, from the equation (31) it is easy to see that the discrete Euler-Lagrange form is not exact. Therefore, there exists a nontrivial difference discrete version of the Euler-Lagrange cohomology in discrete Lagrangian mechanics:

$$H_{DCM}:=\{\text{Closed Euler-Lagrange forms}\}/\{\text{Exact Euler-Lagrange forms}\}.$$

Thirdly, from the equation (32) it follows straightforwardly the following theorem.
Theorem : The difference discrete symplectic structure preserving equation

$$\Delta_t \omega_D{}^{(k)} = 0, \quad i.e. \quad \omega_D{}^{(k+1)} = \omega_D{}^{(k)} \tag{34}$$

holds if and only if the discrete Euler-Lagrange forms are closed:

$$dE_D{}^{(k)} = 0. \tag{35}$$

Fourthly, the difference discrete symplectic structure preserving law holds in the function space associated with the difference discrete closed Euler-Lagrange condition in general rather than in the solution space of the Euler-Lagrange equations only.

Discrete Lagrangian field theory

We also consider the cases of 1+1 or 2 dimensional base manifold. Let $X^{1,1}$ or X^2 with suitable signature of metrics be the base manifold, L^2 a regular lattice with 2-directions $x_\mu, (\mu = 1, 2)$, on $X^{1,1}$ or X^2, M_D the configuration space with $u^{\alpha(i,j)} \in M_D$ and so forth as before.

The difference discrete Lagrangian for a set of the generic fields $u^\alpha, \alpha = 1, \cdots, r$, is a functional in $F(T(\mathcal{M}_{X_{(i,j)}}))$

$$\mathcal{L}_D^{(i,j)} = \mathcal{L}_D(u^{\alpha(i,j)}, \Delta_\mu u^{\alpha(i,j)}). \tag{36}$$

Taking exterior differential $d \in T^*(\mathcal{M}_{X_{(i,j)}})$ of $\mathcal{L}_D^{(i,j)}$ and making use of the modified Leibniz law (13), in the framework of multi-parameter differential approach, we get

$$d\mathcal{L}_D^{(i,j)} = E_D(u^{\alpha(i,j)}, \Delta_\mu u^{\alpha(i,j)}) + \Delta_\mu \theta_{DL}^{\mu(i,j)}, \tag{37}$$

where $E_D^{(i,j)}$ are discrete Euler-Lagrange 1-forms defined by

$$E_D(u^{\alpha(i,j)}, \Delta_\mu u^{\alpha(i,j)}) := \{ \frac{\partial \mathcal{L}_D^{(i,j)}}{\partial u^{\alpha(k,l)}} - \Delta_1 \left(\frac{\partial \mathcal{L}_D^{(i-1,j)}}{\partial(\Delta_1 u^{\alpha(k-1,l)})} \right) - \Delta_2 \left(\frac{\partial \mathcal{L}_D^{(i,j-1)}}{\partial(\Delta_2 u^{\alpha(k,l-1)})} \right) \} du^{\alpha(k,l)}, \tag{38}$$

and $\theta_{DL}^{\mu(i,j)}$ are two canonical 1-forms:

$$\theta_{DL}^{1(i,j)} = \frac{\partial \mathcal{L}_D^{(i-1,j)}}{\partial(\Delta_1 u^{\alpha(k-1,l)})} du^{\alpha(k,l)}, \quad \theta_{DL}^{2(i,j)} = \frac{\partial \mathcal{L}_D^{(i,j-1)}}{\partial(\Delta_2 u^{\alpha(k,l-1)})} du^{\alpha(k,l)}. \tag{39}$$

It is easy to see that there exist two symplectic 2-forms on $T^*(\mathcal{M}_{X_{(i,j)}})$:

$$\omega_{DL}^{\mu(i,j)} = d\theta_{DL}^{\mu(i,j)}, \quad \mu = 1, 2. \tag{40}$$

The equation $d^2 \mathcal{L}_D^{(i,j)} = 0$, on $T^*(\mathcal{M}_{X_{(i,j)}})$ leads to the discrete multisymplectic structure preserving property, i.e. the conservation law or the divergence free equation of $\omega^{\mu(i,j)}$:

$$dE_D(u^{\alpha(i,j)}, \Delta_\mu u^{\alpha(i,j)}) + \Delta_\mu \omega_{DL}^{\mu(i,j)} = 0. \tag{41}$$

Let us enumerate following important issues on the discrete Euler-Lagrange cohomology and difference discrete multisymplectic structure preserving property.

First, the null discrete Euler-Lagrange 1-form corresponds to the discrete Euler-Lagrange equations and it is a special case of coboundary discrete Euler-Lagrange 1-forms

$$E_D^{(i,j)} = d\alpha_D^{(i,j)}, \tag{42}$$

where $\alpha_D^{(i,j)}$ an arbitrary function on $F(T^* \mathcal{M}_{X_{(i,j)}})$.

Secondly, although they satisfy the discrete Euler-Lagrange condition, it does not mean that all closed discrete Euler-Lagrange 1-forms are exact. In fact, from the equation

(37) it is easy to see that Euler-Lagrange 1-forms are not exact in general since two canonical 1-forms $\theta_D{}^{\mu(i,j)}, (\mu = 1, 2)$ are not trivial. Therefore, for a given difference discrete field theory, there exists a nontrivial difference discrete version of the Euler-Lagrange cohomology:

$$H_{DCFT} := \{\text{closed Euler-Lagrange forms}\}/\{\text{exact Euler-Lagrange forms}\}.$$

Thirdly, from equation (41) it follows the theorem on the necessary and sufficient condition for the difference discrete multisymplectic structure preserving law.

Theorem : The difference discrete multisymplectic structure preserving law

$$\Delta_\mu \omega_D^{\mu(i,j)} = 0 \tag{43}$$

holds if and only if the discrete Euler-Lagrange 1-form satisfies the discrete Euler-Lagrange condition, i.e. it is closed:

$$dE_D{}^{(i,j)} = 0. \tag{44}$$

In addition, this theorem also indicates that the variables $u^{\alpha(k,l)}$'s etc. in the cohomology are still in the function space rather than the ones in the solution space only. Consequently, this means that the difference discrete multisymplectic structure preserving law holds in the function space with the closed discrete Euler-Lagrange condition in general rather than in the solution space only.

Finally, it should be mentioned that all these issues can be straightforward to generalize to higher dimensional discrete cases.

CONCLUDING REMARKS

A few remarks are in order:

1. The difference discrete variational formalism different from the one of the Lee-Veselov type for the discrete classical mechanics [20][21][22] and field theory [7]. It has been emphasized that the difference as discrete derivative is an entire geometric object. The discrete integrants can also combine together in certain manner as a geometric object to construct some numerical schemes as was shown in [23]. This is more obvious and natural from the viewpoint of noncommutative geometry. In the continuous limit, the results given here by the difference discrete variational principle lead to the correct continuous counterparts.

It can be shown that the difference discrete variational principle works for the difference discrete classical mechanics and field theory in both Lagrangian and Hamiltonian formalism [23] that present themselves as symplectic or multisymplectic numerical schemes and furthermore for other numerical schemes in both symplectic and multisymplectic algorithms respectively. And the role-played by the different Leibniz laws for differences are quite important in constructing the numerical schemes. It is reasonable to conjecture that all numerical schemes in symplectic and multisymlectic algorithms might be derived by virtue of the difference discrete variational principle together with the suitable Leibniz law for differences.

2. It is easy to see that the cohomological approach works for the symplectic and multisymplectic geometry and their difference discrete versions in both Lagrangian and Hamiltonian formalism for the classical mechanics and field theory. However, it had been missed in other approaches (see, for example, [1][2][8][9][3][4][7]). It has been show that the necessary and sufficient condition for symplectic and multisymplectic structure preserving property in each case is the related closed Euler-Lagrange condition being satisfied. Therefore, these symplectic and multisymlectic structure preserving properties hold in the function space with the relevant Euler-Lagrange condition in general rather than in the solution space only. Although either Euler-Lagrange equations and canonical equations or different difference discrete versions of them do preserve the relevant symplectic and multisymplectic structures.

3. It is also easy to see that the variational principle/difference discrete variational principle and the cohomological approach form a connecting link between the preceding and the following in either continuous or difference discrete case. And the multi-parameter differential approach provides a common framework for both of them.

It should be emphasized that both variational principle and the Euler-Lagrange co-homological approach can be directly applied to ODEs and PDEs and their discrete versions, which offer themselves certain numerical schemes in the symplectic and mul-tisymplectic algorithms, no matter whether there are known Lagrangian and/or Hamilto-nian associated with. In fact, the action functional may be constructed for certain types of ODEs and PDEs. Thus, the variational principle/difference discrete variational principle and the scenario of the cohomological approach are also available.

In the cohomological approach it should always to release ODEs, PDEs and numerical schemes away from their solution spaces and to work on the relevant function space rather than on the solution space even if it does exist. In standard or conventional approaches to the numerical schemes in symplectic and multisymplectic algorithms, in order to show whether a given scheme is symplectic or multisymplectic, it is always working on the solution spaces. The implication of this difference is quite clear.

4. Some simple noncommutative differential calculus on the regular lattices are em-ployed. Since the base space coordinates t or xs are difference discretized and differences do not satisfy the ordinary commutative Leibniz law for the differential, in order to study the symplectic and multisymplectic geometry in these difference discrete systems it is natural and meaningful to make use of the noncommutative differential calculus.

ACKNOWLEDGMENTS

This talk is given by K. Wu and based on research works in collaboration with Y.Q. Lee and S.K. Wang supported partly by the National Science Foundation of China.

REFERENCES

1. V.I. Arnold, Mathematical Methods of Classical Mechanics, Graduate texts in Math. **60** (1978), (Second Ed.) Springer-Verlag, (1989).
2. R. Abraham and J.E. Marsden, Foundation of Mechanics, (1978), (Second Ed.) Addison-Wesley.

3. E. Binz, J. Śniatycki and H. Ficher, Geometry of Classical Fields, North Holland Elsevier, Amsterdam, 1988.
4. T.J. Bridges, Multisymplectic Structures and Wave Propagation, Math. Proc. Camb. Phil. Soc. 121 (1997) 147-190.
5. M. Gotay, J. Isenberg and J.E. Marsden, Momentum Maps and the Hamiltonian Structure of Classical Relativistic Field Theories, (1997) Preprint.
6. J.E. Marsden and S. Shkoller, Multisymplectic Geometry, Covariant Hamiltonian and Water Waves, Math. Proc. Camb. Phil. Soc. 124 (1998)
7. J.E. Marsden, G. W. Patrick and S. Shkoller, Multisymplectic Geometry, Variational Integrators, and Nonlinear PDEs, Comm. Math. Phys. 199 (1998) 351-395.
8. K. Feng, On difference schemes and symplectic geometry, Proc. of the 1984 Beijing Symposium on Differential Geometry and Differential Equations — Computation of Partial Differential Equations, Ed. by Feng Keng, Science Press, Beijing, 1985. Selected Works of Feng Keng II (1995).
9. J.M. Sanz-Serna, M.P. Calvo, Numerical Hamiltonian Problems, Chapman and Hall, London.1994.
10. H.Y. Guo, Y.Q. Li and K. Wu, On Symplectic and Multisymplectic Structures and Their Discrete Versions in Lagrangian Formalism, Comm. Theor. Phys. 35 (2001) 703-710. hep-ph/0104064.
11. H.Y. Guo, Symplectic, Multisymplectic Structures and the Euler-Lagrange Cohomology, Talk given at The International Workshop on Structure-Preserving Algorithms, March 25-31, 2001, Beijing.
12. H.Y. Guo, Y.Q. Li and K. Wu, A Note on Symplectic Algorithms, Comm. Theor. Phys. 36(2001)11-18. Physics/0104030.
13. H.Y. Guo, X. M. Ji, Y.Q. Li and K. Wu, A Note on Symplectic, Multisymplectic Schemes in Finite Element Method, hep-th/0104060.
14. K. Wu, Symplectic and Multisymplectic 2-form Structures in Simple Finite Element Method, Talk given at The International Workshop on Structure-Preserving Algorithms, March 25-31, 2001, Beijing.
15. H.Y. Guo, X. M. Ji, Y.Q. Li and K. Wu, On Symplectic, Multisymplectic Structure-Preserving in Simple Finite Element Method, Preprint AS-ITP-2001-010, April, 2001. hep-th/0104151.
16. H.Y. Guo, Y.Q. Li, K. Wu and S.K. Wang, Symplectic, Multisymplectic Structures and Euler-Lagrange Cohomology, Preprint AS-ITP-2001-009, April, 2001. hep-th/0104140.
17. H.Y. Guo, K. Wu, S.H. Wang, S.K. Wang and G.M. Wei, Noncommutative Differential Calculus Approach to Symplectic Algorithm on Regular Lattice, Comm. Theor. Phys. 34 (2000) 307-318.
18. H.Y. Guo, K. Wu and W. Zhang, Noncommutative Differential Calculus on Abelian Groups and Its Applications, Comm. Theor. Phys. 34 (2000) 245-250.
19. S. Reich, Multisymplectic Runge-Kutta Collocation Methods for Hamiltonian Wave Equations, J. Comput. Phys. 157 (2000), 473-499.
20. T.D. Lee, Can time be a discrete dynamical variable? Phys. Lett. 122B (1983) 217-220.
21. A.P. Veselov, Integrable Discrete-time Systems and Difference Operators, Funkts. Anal. Prilozhen, 22 (1988) 1-13.
22. J. Moser and A.P. Veselov, Discrete Versions of Some Classical Integrable Systems and Factorization of Matrix Polynomials, Comm. Math. Phys. 139 (1991) 217-243.
23. H.Y. Guo, Y.Q. Li, K. Wu and S.K. Wang, Difference Discrete Variational Principle, Euler-Lagrange Cohomology and Symplectic, Multisymplectic Structure, math-ph/0106001.

Dynamical Fluctuations in Biological Processes with Factorial Moments

Huijie Yang   Fangcui Zhao

Hebei University of Technology, Tianjin300130, China
(E-mail: huijieyangn@eyou.com)

Abstract. We apply the concept of factorial moments (FM) to the analysis of long range co-relations in several kinds of series, such as DNA sequence, heartbeat interval time series, and gait time series. It is found that FM is effective to describe the dynamical fluctuations embedded in these series.

INTRODUCTION

Recently more and more attentions have been attracted to the analysis of DNA sequences by means of methods from other science fields. And detailed investigations on the characteristics of DNA, especially the differences between coding and non-coding segments make it possible to design new methods to distinguish coding segments from DNA sequences. Finding new methods and combine them with each other will provide the key to better approach.

In fact, for many nature sequences, such as DNA sequence, weather records, heartbeat rhythm, gait time series, and traffic in network, etc.; the elements are not positioned randomly and exhibit long-range co-relations. That is to say the co-relation functions decay with a power-law and there is not a characteristic co-relation length. How to describe these dynamical processes is an essential role at present time.

In this paper we firstly investigated some features for factorial moments and then application this method to DNA sequence analysis and heartbeat/gait time series analysis.

FACTORIAL MOMENTS

It is well known that intermittency is related with strong dynamical fluctuations. To describe the strong dynamical fluctuations and dismiss the statistical fluctuations effectively, FM is suggested to investigate intermittency. There are two formulae for FM,

Formula 1:

$$F_i = M^{i-1} \sum_{m=1}^{M} \frac{< n_m (n_m - 1)...(n_m - i + 1) >}{n(n-1)...(n-i+1)},$$

CP597, *Nonequilibrium and Nonlinear Dynamics in Nuclear and Other Finite Systems*,
edited by Z. Li et al.

M: the number of the bins the considered interval being divided into ; n_m: the number of particles occurring in the m'th bin; n: and the total number of particles in all the bins

A measure quantity to indicate the dynamical fluctuations,

$$\phi_i = \lim_{\delta y \to 0} \frac{\ln F_i}{\ln(1/\delta y)},$$

Formula 2:

$$F_q(d) = f_q / f_l^q$$

$$F_q(d) = \sum_{m=q}^{\infty} \rho_m m(m-1)...(m-q+1),$$

At the beginning, FM is used to deal with many kinds of dynamical processes, and fruitful results are obtained. But FM is mainly restricted to **one-dimensional** variables, FM tends to **saturate** at small experimentally allowed resolution, instead of power-law. It is believed that power-law exists in high dimensional phase space. Measurements for high dimensional phase space find that for two dimensional phase space, FM varies a lot with the plane of projection, while for three-dimensional phase space FM is a smoothly upward bending curve. Then FM is analyzed with **Self-affine** instead of self-similar for the shrinkage ratios for three directions being different and upward bending is dismissed effectively.

It is found that factorial moments can **distinguish** two processes obeying different rules with high precise **except** that the perturbation process obeys a statistical random distribution rule.

APPLICATION TO DNA ANALYSIS

Several statistical features for DNA can be employed to distinguish coding and non-coding segments from DNA sequences,

➤ The usage of strongly bonded nucleotide C-G pairs is usually less frequent than that of weakly bonded A-T pairs;

➤ The C-G concentration is generally larger in coding than in non-coding regions.

➤ The C-G concentration makes a strong "background" contribution to any possible differences between non-coding and coding subsequences.

➤ Non-coding regions display long-range power-law relations. While for coding regions, it seems that random rules dominate the sequences.

To identify coding regions by means of FM, a process can be constructed as,

✓ d successive nucleotides along a DNA sequence are regarded as a case containing d particles. The state of the case can be described with a d-dimensional vector as $(x_1, x_2, x_3...x_d)$, where x_i is the state value for the $i'th$ nucleotide.

✓ The total possible $N - d + 1$ successive cases form a process. The process covers the entire DNA segment we are interested in, which can be expressed with the series in d-dimensional **delay-register vectors:**

$$(x_1, x_2, x_3...x_d)$$

$$(x_2, x_3, x_4 \ldots x_{d+1})$$
$$\downarrow$$
$$(x_{N-d+1}, x_{N-d+2}, x_{N-d+3} \ldots x_N)$$

A quantity is introduced to indicate the difference between the segment we are interested in and the reference segment along a DNA sequence, as

$$\Delta F(t) = \sqrt{\frac{\sum_{m=1}^{M} (F_{0m} - F_{tm})^2}{\sum_{m=1}^{M} 1}},$$

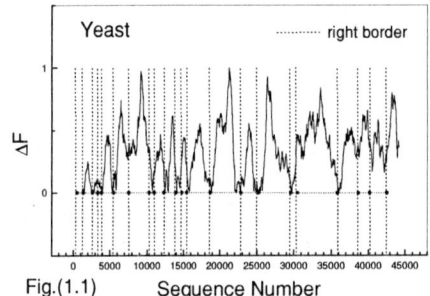 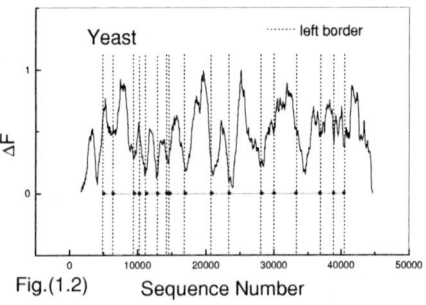

Fig.(1.1)   Sequence Number   Fig.(1.2)   Sequence Number

Identify the right and left borders of coding regions in DNA sequence from yeast

In Fig.(1) the results for DNA sequence from Yeast are presented, we can find that FM is an effective tool to distinguish coding regions. By means of FM, characteristics for coding regions should be investigated in detail.

APPLICATION TO HEARTBEAT AND GAIT (TIME SERIES)

When we deal with these kind of time series, some special problems should be considered,

✓ **Trivial and non-trivial self-similar**
Re-scale x only we can find a self-similar, which is an order-irrelative self-similar (trivial self-similar with dimension 0). Standard method to find non- trivial self-similar is to study the **integrated time series.**

✓ **Noise**
Solution in literature: **integrated time series can decrease noise**
Solution in this paper: **factorial moment to dismiss noise**

✓ **Non-stationary**
Statistical quantities, such as the mean, the standard deviation and the higher moments are **variant** under time translation. What is more there are long-range co-relations for the fluctuations.
Solution in literature: DFA method (detrended fluctuation analysis)
Solution in this paper: Factorial Moment

398

In this paper, the two directions are re-scaled simultaneously，that is,
✓ For y:
  the interval between Y-max and Y-min is divided into 2,4,8,16,32,64,128…
✓ For t:
Different sizes of windows can be constructed, such as,
     5,10,20,40,80,160,320,640

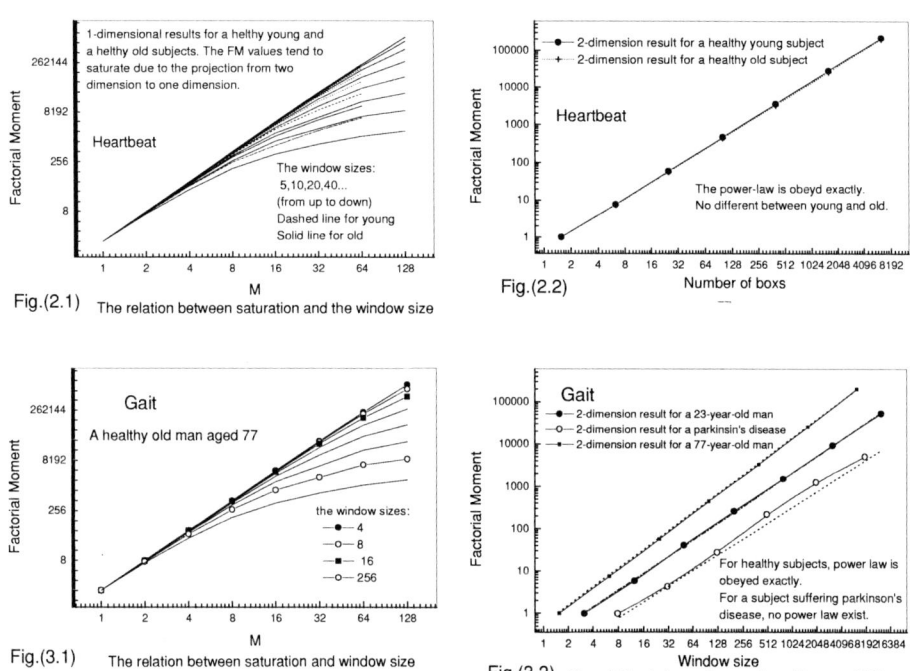

Fig.(2.1) The relation between saturation and the window size

Fig.(2.2)

Fig.(3.1) The relation between saturation and window size

Fig.(3.2) The relation between the number of boxs and FM

In Fig.(2) and Fig.(3), the results for heartbeat and gait are presented, respectively. It is found that for healthy objects(young or old) exact power-law is obeyed, while for an object suffering Parkinson's illness the power-law is not obeyed.

REFERENCES

1. Pedro Bernaola-Galvan, Ivo Grosse, H.E Stanley, et. al, Physical Review Letters 85(2000) 1342.
2. Ivo Grosse, Hanspeter Herzel, Sergey V. Buldyrev, and H. Eugene Stanley, Physical Review E61(2000)5624.
3. J. Barral P., A. Hasmy, J. Jimenez, and A. Marcano, Physical Review
4. David R. Bickel, Physica A265(1999)634-648。
5. A. Bialas, R. Peschanski, Nuclear Physics B273(1986)703-718.
6. A. Bialas, R. Peschanski, Nuclear Physics B308(1988)857-867.
7. Yang Huijie, Zhuo Yizhong, Wu Xizhen, Journal of Physics A27(1994)6147

Monte-Carlo Simulation of Domain-Wall Network in Two-dimensional Extended Supersymmetric Theory

Nobuyuki Motoyui*, Shogo Tominaga* and Mitsuru Yamada*

*Department of Mathematical Sciences, Faculty of Sciences, Ibaraki University, Bunkyo 2-1-1, Mito, 310-8512 Japan

Abstract. We will show that 2-dimensional $N = 2$-extended supersymmetric theory has solitonic solution using the Hamilton-Jacobi method of classical mechanics. Then it is shown that the Bogomol'nyi mass bound is saturated by these solutions and triangular mass inequality is satisfied. At the end, we will mention domain-wall structure in 3-dimensional spacetime.

The Lagrangean of 2-dimensional $N = 2$-extended supersymmetric Wess-Zumino type model is

$$\mathcal{L} = \int d^2\theta d^2\theta^* \phi^* \phi + \int d^2\theta W(\phi) + \int d^2\theta^* W(\phi)^* \tag{1}$$

where ϕ is a chiral field

$$\phi = a + \sqrt{2}\overline{\theta}^c \psi + \overline{\theta}^c \theta f \tag{2}$$

and $W(\phi)$ is a superpotential. In component fields, the avobe Lagrangean becomes

$$\mathcal{L} = \partial_\mu a^* \partial^\mu a + i\overline{\psi}\gamma^\mu \partial_\mu \psi + \frac{i}{2}W''(a)\overline{\psi}^c \psi - \frac{i}{2}W''(a)^* \overline{\psi}\psi^c - W'(a)W'(a)^* \tag{3}$$

where ψ is a 2-dimensional Dirac spinor and 2-dimensional γ-matrices are

$$\gamma^0 = \sigma_y, \quad \gamma^1 = -i\sigma_x, \quad \gamma_5 = \gamma^0\gamma^1 = -\sigma_z, \quad C = -\sigma_y. \tag{4}$$

The current of supersymmetric charge is

$$j_\mu = \sqrt{2}\{\gamma^\rho\gamma_\mu\psi\partial_\rho a^* - \gamma_\mu\psi^c W'(a)^*\}. \tag{5}$$

After eliminating the fermion field by the equation of motion

$$i\gamma^\mu\partial_\mu\psi - iW''(a)^*\psi^c = 0, \tag{6}$$

we have a purely bosonic Lagrangean

$$\mathcal{L} = \dot{a}\dot{a}^* - (\nabla a)(\nabla a^*) - |W'(a)|^2. \tag{7}$$

CP597, *Nonequilibrium and Nonlinear Dynamics in Nuclear and Other Finite Systems*,
edited by Z. Li et al.

Let us assume that $W(\phi)$ is a polynomial such that $W'(a) = 0$ has n complex solution $a_1, a_2, \cdots, a_n$. Then the theory has not only n classical vacuum solutions $a(x^0, x^1) = a_i$, $i = 1, 2, \cdots, n$ but also solitonic solutions which we call "(i, j)-soliton", characterized by

$$a(t, -\infty) = a_i, \qquad a(t, \infty) = a_j. \tag{8}$$

With (i, j)-soliton in the background, we have a central extention of supersymmetry algebra of supersymmetry charge Q

$$\{Q, \overline{Q}\} = 2\gamma_\mu P^\mu \tag{9}$$

$$\{Q, \overline{Q}^c\} = -4\gamma_5 \left[W(a_j)^* - W(a_i)^* \right]. \tag{10}$$

where

$$Q = \int_{-\infty}^{\infty} j^0(x) dx. \tag{11}$$

In particular, in the center of mass frame $(P^\mu) = (M_{ij}, 0)$

$$\{Q_\alpha, Q_\beta^\dagger\} = 2M_{ij}\delta_{\alpha\beta} \tag{12}$$

$$\{Q_\alpha, Q_\beta\} = -4i(\sigma_x)_{\alpha\beta}\Delta W^* \tag{13}$$

where $\Delta W = W(a_j) - W(a_i)$. From the positivity condition $\{A, A^\dagger\} \geq 0$ with $A = Q_1 + ie^{i\theta}Q_2^\dagger$, we have

$$M_{ij} \geq 2\text{Re}(e^{i\theta}\Delta W). \tag{14}$$

Since angle θ is arbitrary, we obtain the lower mass bound of (i, j)-soliton

$$M_{ij} \geq 2|\Delta W|. \tag{15}$$

Actually, this Bogomol'nyi bound is saturated by classical solution. To see this, we calculate the static solution of the field equation. From the bosonic Lagrangean, the Hamiltonian is

$$\mathcal{H} = \dot{a}\dot{a}^* + (\nabla a)(\nabla a^*) + |W'(a)|^2 \tag{16}$$

where $\nabla a = da/dx$. Writing the static solution of $a(t, x)$ simply as $a(x)$, and regard x as time,

$$\mathcal{L}' = (\nabla a)(\nabla a^*) + |W'(a)|^2 \tag{17}$$

$$\mathcal{H}' = p_a p_a^* - |W'(a)|^2 \tag{18}$$

where p_a is the conjugate momentum to a. This is a problem of classical mechanics of one particle moving in the potential

$$U = -|W'(a)|^2. \tag{19}$$

401

The Hamilton-Jacobi equation for the action $S(a, a^*)$ is

$$\left(\frac{\partial S}{\partial a^*}\right)\left(\frac{\partial S}{\partial a}\right) - W'(a)W'(a)^* = E. \tag{20}$$

For $E = 0$ we can write the complete solution

$$S(a, a^*, \alpha) = \alpha W(a) + \frac{1}{\alpha}W(a)^* \tag{21}$$

where $\alpha = e^{i\omega}$ is a phase. The soliton path is given by

$$\frac{\partial S}{\partial \alpha} = W(a) - \frac{1}{\alpha^2}W(a)^* = const. \tag{22}$$

Then

$$\text{Im}(e^{i\omega}W(a)) = const. \tag{23}$$

So the trajectory of $a(x)$ is such that $W(a(x))$ is a straight line in the complex W-plane. In the complex W-plane, there is a branch cut from each $W(a_i)$ to infinity because a_i is a critical point of $W(a_i)$,

$$\left.\frac{da}{dW}\right|_{W=W(a_i)} = \infty. \tag{24}$$

Therefore only solitons whose paths do not cross branch cuts can exist.

The classical mass M_{ij} is obtain as follows. From $E = 0$, we have $dx = |da|^2/|dW|$, so the mass is given as

$$M_{ij} = \int_{-\infty}^{\infty} \mathcal{L}' dx = 2\int_{-\infty}^{\infty} |W'(a(x))|^2 dx = 2\int_{W(a_i)}^{W(a_j)} |dW|. \tag{25}$$

Since $W(a)$ is a straight line in the complex W-plane,

$$M_{ij} = 2\left|\int_{W(a_i)}^{W(a_j)} dW\right| = 2|\Delta W|. \tag{26}$$

So the Bogomol'nyi bound is saturated by classical solution. Then, from the triangular inequality in the complex W-plane, a strict mass inequality

$$M_{ik} < M_{ij} + M_{jk} \tag{27}$$

follows. This shows the absolute stability of one-soliton configuration through the attractive force between neighboring solitons.

Now for the general cases of D-dimensions ($D = 2, 3, 4$) the energy of the system is

$$E = \int d^{D-1}x \left(\sum_{i=1}^{D-1} |\nabla_i a|^2 + |W'(a)|^2\right). \tag{28}$$

In three dimensions, we can summarize the features of low energy configurations as follows:

- The 2-dimensional space is divided into (i)-domains; $i = 1, 2, \cdots, n$.

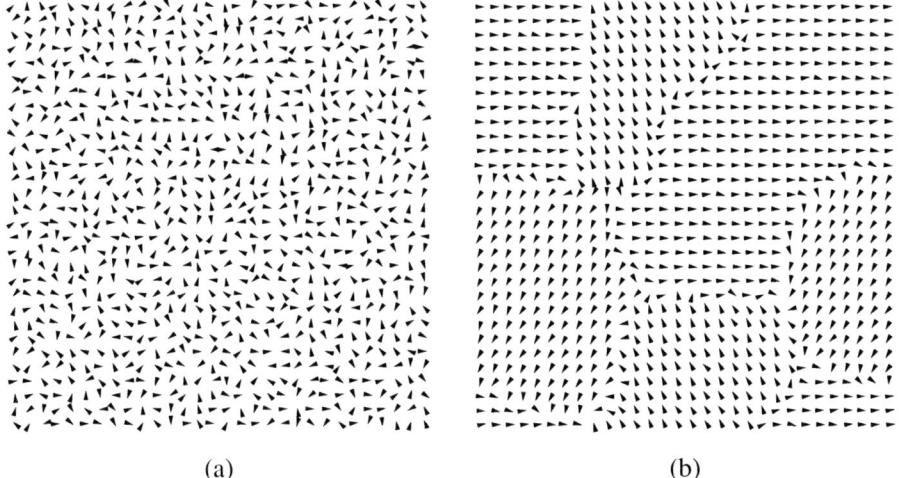

FIGURE 1. a) Initial configuration of $a(x,y)$. The length of the arrows is normalized. b) The configuration after Monte-Carlo iteration.

- Every two domains are separated by (i,j)-wall; Every domain-wall is a curve in xy space.
- (i,j)-, (j,k)- and (k,i)-wall can join at (i,j,k)-wall-junction; Every wall-junction is a point in xy space.

We show pictures of 2-dimensional networks generated by Monte-Carlo simulation in Fig. 1, where we put

$$W(\phi) = \frac{1}{4}\phi^4 - \phi. \tag{29}$$

There are three distinct vacuum configurations $\phi = 1, \omega, \omega^*$ as the solutions of $W'(a) = 0$. In the Monte-Carlo iterations, field configurations are generated by the statistical weight of $e^{-E/kT}$. Since these domain wall configurations are metastable, as iteration goes, these domains tend to be unified into the real vacuum which consists of single domain.

REFERENCES

1. Gorsky, A., and Shifman, M., *Phys. Rev.* **D61**, 085001 (2000).
2. Rebhan, A., and van Nieuwenhuizen, P., *Nucl. Phys.* **B508**, 449-467 (1997).
3. Wess, J., and Bagger, J., *Supersymmetry and Supergravity*, 2nd. ed., Princeton Univ. Press, Princeton NJ, 1992.
4. Sohnius, M.F., *Phys. Rep.* **128**, 39 (1985).
5. Witten, E., and Olive, D., *Phys. Lett.* **78B**, 97-101 (1978).
6. Bogomol'nyi, E.B., *Sov. J. Nucl. Phys.* **24**, 449-454 (1976).

PART VII

TOPICS RELATED TO

NUCLEAR STRUCTURE

Microscopic study of α-cluster condensation in light 4N nuclei

A. Tohsaki*, H. Horiuchi†, P. Schuck** and G. Röpke‡

*Department of Fine Materials Engineering, Shinshu University, Ueda 386-8567, Japan
†Department of Physics, Kyoto University, Kyoto 606-8502, Japan
**Institut de Physique Nucléaire, 91406 Orsay Cedex, France
‡FB Physik, Universität Rostock, D-18051 Rostock, Germany

Abstract. Within the new framework of α-cluster wave function which is of the α-particle conden-sate type, we study ^{12}C and ^{16}O. The results show that states of low density close to the α-particle threshold in both nuclei are possibly of this kind. It is conjectured that all self-conjugate 4n nuclei may show similar features.

It is a well known fact that in light nuclei many states are of the cluster type [1, 2, 3, 4]. In the case of cluster states of stable nuclei where we have only very few excess nucleons in addition to the clusters, they are all located close to or above the threshold energy of breakup into constituent clusters. This fact which is known as the threshold rule [5] means that the inter-cluster binding is weak in cluster states. The threshold rule can be considered as a necessary condition for the formation of the cluster structure, because if the inter-cluster binding is strong the clusters overlap strongly and the clusters will loose their identities.

One of the fundamental questions of the cluster model is what kind of α-particle cluster states can be expected to exist around the threshold energy $E_{n\alpha}^{thr} = nE_\alpha$ of $n\alpha$ breakup in self-conjugate 4n nuclei. One possible answer to this question, which is strongly under debate, is the existence of the cluster state of a linear $n\alpha$ chain structure. The idea of the linear α chain state, originally due to Morinaga [6], is so fascinating that recently the formation of linear 6α chain states in ^{24}Mg was studied extensively by experiments and also theoretically [7]. The possibility of the linear 3α chain state in ^{12}C, which is the simplest linear α chain state was studied in detail by many authors solving the 3α problem microscopically [4]. However, these three-body studies all showed that the 3α-cluster states around the 3α threshold energy $E_{3\alpha}^{thr}$ do not have a linear chain structure. For example, the calculated second 0^+ state in ^{12}C, which corresponds to the observed second 0^+ state located at 0.39 MeV above the 3α threshold energy, has a structure where α-clusters interact predominantly in relative S-waves. Thus it was concluded that the cluster state near $E_{3\alpha}^{thr}$ has not a linear chain structure but rather an α-particle gas-like structure.

On the other hand there have been recent theoretical investigations on the posibility of α-particle condensation in low density nuclear matter [8, 9]. In [8] Röpke et al. made a variational ansatz for the solution of the in-medium 4-body equation. In [9] Beyer et al. solved the Faddeev- Yakubovsky equations for an alpha-like cluster in nuclear matter.

CP597, *Nonequilibrium and Nonlinear Dynamics in Nuclear and Other Finite Systems*,
edited by Z. Li et al.

The outcome of these studies was that such α-condensation can occur only in the low-density region below a fifth of the saturation value. At higher densities rather a state of ordinary p-n, n-n, or p-p Cooper pairing will prevail. In view of these results it may be a tempting idea that in finite self-conjugate 4n nuclei one could expect the existence of excited states of dilute density composed of a weakly interacting gas of α-particles. Since the α-cluster is a Bose particle, such states could approximately be considered as an nα cluster condensed state and eventually excitations thereof.

The purpose of this talk is to report on our study which not only confirms that indeed the second 0^+ state in ^{12}C could be considered as such a condensed state but that in addition also in ^{16}O such a state close to the threshold possibly exists. We will therefore then conjecture that the existence of such α-condensed states might be a general feature in N=Z nuclei.

For the purpose of our study we write down a new type of α-cluster wave function describing an α-particle Bose condensed state:

$$|\Phi_{n\alpha}\rangle = (C_\alpha^\dagger)^n |vac\rangle \tag{1}$$

where the α-particle creation operator is given by

$$C_\alpha^\dagger = \int d^3R\, e^{-\mathbf{R}^2/R_0^2} \int d^3r_1 \cdots d^3r_4$$
$$\times\ \varphi_{0s}(\mathbf{r}_1 - \mathbf{R})a_{\sigma_1\tau_1}^\dagger(\mathbf{r}_1) \cdots \varphi_{0s}(\mathbf{r}_4 - \mathbf{R})a_{\sigma_4\tau_4}^\dagger(\mathbf{r}_4) \tag{2}$$

with $\varphi_{0s}(\mathbf{r}) = (1/(\pi b^2))^{3/4} e^{-\mathbf{r}^2/(2b^2)}$ and $a_{\sigma\tau}^\dagger(\mathbf{r})$ being the creation operator of a nucleon with spin-isospin στ at the spatial point $\mathbf{r}$. The total nα wave function therefore can be written as

$$\langle \mathbf{r}_1\sigma_1\tau_1, \cdots \mathbf{r}_{4n}\sigma_{4n}\tau_{4n}|\Phi_{n\alpha}\rangle$$
$$\propto \mathcal{A}\{e^{-\frac{2}{B^2}(\mathbf{X}_1^2 + \cdots + \mathbf{X}_n^2)}\, \phi(\alpha_1) \cdots \phi(\alpha_n)\}, \tag{3}$$

where $B = (b^2 + 2R_0^2)^{1/2}$ and $\mathbf{X}_i = (1/4)\sum_n \mathbf{r}_{in}$ is the center-of-mass coordinate of the i-th α-cluster α_i. The internal wave function of the α-cluster α_i is $\phi(\alpha_i) \propto \exp[-(1/8b^2)\sum_{m>n}^4 (\mathbf{r}_{im} - \mathbf{r}_{in})^2]$. The wave function of Eq.(3) is totally antisymmetrized by the operator $\mathcal{A}$. It is to be noted that the wave function of Eqs.(1,3) expresses the state where nα-clusters occupy the same 0s harmonic oscillator orbit $\exp[-\frac{2}{B^2}\mathbf{X}^2]$ with B an indepedent variational width parameter. For example if B is of the size of the whole nucleus whereas b remains more or less at the free α-particle value (a situation encountered below), then the wave function (3) describes an nα cluster condensed state in the macroscopic limit $n \to \infty$. For finite systems we know from the pairing case that such a wave function still can more or less reflect Bose condensation properties. Of course the total center-of-mass motion can and must be separated out of the wave function of Eq.(1) for finite systems. In the limiting case of $B = b$ (i.e. $R_0 = 0$), Eq.(3) describes a Slater determinant of harmonic oscillator wave functions. We also would like to point out that for $B \neq 0$ the wave function (1,3) is different from Brink's α-cluster state [2].

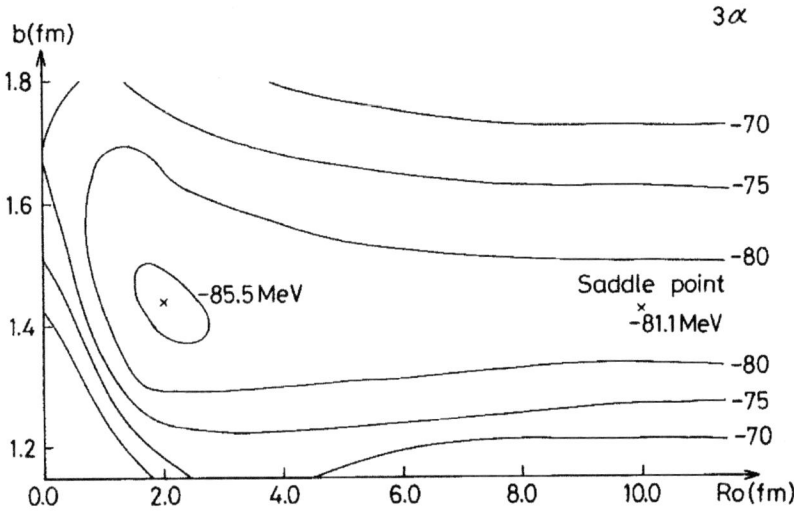

FIGURE 1. Contour map of the energy surface $E_{3\alpha}(R_0,b)$ for ^{12}C. Numbers attached to the contour lines are the binding energies.

The state $|\Phi_{n\alpha}\rangle$ has spin-parity 0^+. In the limit of $R_0 = 0$, the normalized wave function $|\Phi_{n\alpha}^N\rangle = |\Phi_{n\alpha}\rangle / \sqrt{\langle\Phi_{n\alpha}|\Phi_{n\alpha}\rangle}$ is identical to a harmonic oscillator shell model wave function with the oscillator parameter b. For $n = 3$, it is identical to the p-shell wave function $|(0s)^4(0p)^8, [444]\, 0^+\rangle$, and for $n = 4$, it is identical to the double closed shell wave function, $|(0s)^4(0p)^{12}, 0^+\rangle$. This is easily proved by noticing that these limit wave functions for $n = 3$ and 4 have maximum spatial symmetry [444] and [4444], respectively. Only ^8Be has an α-particle structure in its ground state. Heavier $n\alpha$ nuclei collapse to the dense state in their ground state but the individual α's may reappear when these nuclei are dilated i.e. excited.

We calculated the total binding energy in the two parameter space, R_0 and b, $E_{n\alpha}(R_0,b) = \langle\Phi_{n\alpha}^N(R_0,b)|\hat{H}|\Phi_{n\alpha}^N(R_0,b)\rangle$, for $n = 3$ and 4. The Hamiltonian $\hat{H}$ consists of the kinetic energy, the Coulomb energy, and the effective nuclear force named F1 which was proposed by one of the authors and contains a finite range three-nucleon force in addition to the finite range two-nucleon force [10]. This force reproduces reasonably well the binding energy and radius of the α-particle, the α-α phase shifts of various partial waves, and the binding energy and density of nuclear matter. As we will see below this force also gives good results for binding energies and radii of ^{12}C and ^{16}O.

In Fig.1 we give the contour maps of the energy surfaces $E_{n\alpha}(R_0,b)$ for ^{12}C and ^{16}O. The qualitative features of both surfaces are similar. They show a valley running from the outer region with large $R_0 > 11$ fm and $b \approx b_\alpha = 1.44$ fm to the inner region with small R_0 and $b > b_\alpha$, where b_α is the oscillator parameter of the free α-particle. The valleys have a saddle point at $R_0 \approx 10$ fm for $n = 3$ and at $R_0 \approx 10.6$ fm for $n = 4$. Beyond the saddle point, $E_{n\alpha}(R_0, b_\alpha) \approx E_{n\alpha}^{thr} = nE_\alpha$, where $E_\alpha = -27.5$ MeV is the theoretical binding energy of the free α particle by the present F1 force in Hartree-Fock approximation. Therefore we have $3E_\alpha = -82.5$ MeV and $4E_\alpha = -110$ MeV. In the asymptotic region the average inter-α distance is large and the kinetic energy of the center-of-mass motion of an α-cluster ($3\hbar^2/4mB^2$) is very small which leads to $E_{n\alpha}(R_0, b_\alpha)$ being more or less equal to $E_{n\alpha}^{thr}$. The height of the saddle point measured from the theoretical threshold energy is about 1.4 MeV for $n = 3$ and 2.2 MeV for $n = 4$. The appearance of the saddle point is due to the increase of the Coulomb energy and kinetic energy towards the inward direction which is not yet compensated by the gain in potential energy around the saddle point region. This saddle point will help to stabilize the possible α condensed state around $E_{n\alpha}^{thr}$. The minimum of the energy surface is located at $R_0 \approx 2$ fm for $n = 3$ and at $R_0 \approx 1$ fm for $n = 4$. Since $R_0 = 0$ means the shell model limit, we thus see that the wave function even at the energy minimum point deviates from the shell model limit and shows rather strong α-particle correlations. The gain in energy from the shell model limit is 10.3 MeV for ^{12}C and 4.7 MeV for ^{16}O. Before comparing numbers with experiments we have to make a quantum mechanical calculation. This will be achieved via a standard Hill-Wheeler ansatz taking R_0 and b as the Hill-Wheeler coordinates. However, in order to reduce the complexity of the calculation and because the valleys run essentially parallel to the R_0 axis at $b = b_\alpha$ we take $b = b_\alpha =$ constant and only discretise the R_0 variable. We therefore have

$$|\Psi_{n\alpha,k}\rangle = \sum_j f_k((R_0)_j, b_\alpha)|\Phi_{n\alpha}^N((R_0)_j, b_\alpha)\rangle. \tag{4}$$

The normalization of $f_k((R_0, b)_j)$ is so that the k-th eigen-function $|\Psi_{n\alpha,k}\rangle$ is normalized. The adopted mesh size of R_0 values is typically 0.5 fm. In order to see the character of the obtained wave function $|\Psi_{n\alpha,k}\rangle$, we introduce the overlap amplitude

$$A_{n\alpha,k}(R_0, b) = \langle \Phi_{n\alpha}^N(R_0, b)|\Psi_{n\alpha,k}\rangle. \tag{5}$$

From this overlap amplitude we can estimate the relevant values of the variational parameter R_0 in the different states k as will be discussed below.

After outlining the results for the new kind of wave function for ^{12}C and ^{16}O, we will discuss whether the obtained condensed states correspond to the states found in these nuclei. We first consider ^{12}C, i.e. $n = 3$, see Table I. The calculated lowest two eigenenergies are situated at -85.9 and -82.0 MeV. The lowest energy state corresponds to the ground state of ^{12}C and is only slightly lower than the minimum point of the energy surface located at -85.5 MeV. However, the calculated ground state is still above the observed binding energy of ^{12}C which is at -92.16 MeV. An increase of mesh points will certainly lower the energy but, as has been discussed by many people, in order to reproduce the observed ^{12}C binding energy satisfactorily we have to extend our functional space so as to include the spatial symmetry broken wave functions which allow to incorporate the effect of the spin-orbit force adequately. The second eigenvalue lies 0.36 MeV above our theoretical 3α threshold energy, $E_{3\alpha}^{thr} = -82.5$ MeV, and we believe that it corresponds to the observed second 0^+ state of ^{12}C which lies 0.5 MeV above $E_{3\alpha}^{thr}$. As seen in Table I the rms radius of the obtained wave function $|\Psi_{3\alpha,2}\rangle$ is

4.29 fm which is much larger than the one of the ground state which is 2.97 fm, slightly greater than the experimental value 2.45 fm but in agreement with the missing binding of 6.75 MeV. We thus see that the second 0^+ state corresponds to a very dilute system of average density which is only about a fifth of the experimental ground state density.

To characterise the wave function by a typical value of the width parameter R_0, we consider the overlap amplitude $A_{3\alpha,k}(R_0, b_\alpha)$ given in Eq.(6) as a function of R_0 at fixed b_α. Whereas the ground state ($k = 1$) wave function is almost exhausted by one $\Phi_{3\alpha}^N(R_0, b_\alpha)$ with $R_0 \approx 2$ fm which is quite close to the wave function of the minimum energy point of the energy surface, the second $0_{k=2}^+$ state has the largest overlap amplitude (about 0.87) with $R_0 \approx 4.5$ fm. This rather large value implies that the distribution of the center-of-mass momenta is rather narrow, in a certain approximation to an α condensate in infinite nuclear matter where all α-clusters populate the same state $P = 0$ of the center-of-mass momentum. The fact that the calculated $0_{k=2}^+$ state is of dilute density is in agreement with nuclear matter calculations [8, 9] where it was shown that a condensate of α-like particles (quartetting) is possible only in matter with density $\rho \leq 0.03$ fm^{-3}. The average distance of the α-clusters in the dilute $0_{k=2}^+$ state is in agreement with this value for low density nuclear matter, where the overlap of the α-clusters is small so that the Pauli blocking effects are weak.

Let us now discuss the case of ^{16}O, i.e. $n = 4$. The energies of the lowest observed 0^+ states are shown in Table II, together with the corresponding widths. The first excited 0_2^+ state at 6.06 MeV is very well known to have α-clustering character [1, 4] and is well described by the ^{12}C + α microscopic cluster model as having the structure where the α-cluster moves in a S state around the ^{12}C-cluster in its ground state [12] though also other cluster states have been proposed [3]. Similarly, the third excited 0_4^+ state at 12.05 MeV can be described by the same model where the α-cluster moves in a D state around the ^{12}C-cluster in its first 2^+ excited state [12]. We will exclude these well understood states from our further discussion. The excited states 0_3^+ at 11.26 MeV and 0_5^+ state at 14.0 MeV observed in ^{12}C + α elastic scattering [11] cannot be described by such a model. Furthermore, they have very large decay widths, not typical for the other states.

TABLE 1. Comparison of the generator coordinate method calculations with experimental values. $E_{n\alpha}^{thr} = nE_\alpha$ denotes the threshold energy for the decay into α-clusters, the values marked by * correspond to a refined mesh, see main text.

		E_k (MeV)	E_{exp} (MeV)	$E_k - E_{n\alpha}^{thr}$ (MeV)	$(E - E_{n\alpha}^{thr})_{exp}$ (MeV)	$\sqrt{\langle r^2 \rangle}$ (fm)	$\sqrt{\langle r^2 \rangle}_{exp}$ (fm)
^{12}C	$k = 1$	-85.9	-92.16 (0_1^+)	-3.4	-7.27	2.97	2.65
	$k = 2$	-82.0	-84.51 (0_2^+)	+0.5	0.38	4.29	
	$E_{3\alpha}^{thr}$	-82.5	-84.89				
^{16}O	$k = 1$	-124.8 $(-128.0)^*$	-127.62 (0_1^+)	-14.8 $(-18.0)^*$	-14.44	2.59	2.73
	$k = 2$	-116.0	-116.36 (0_3^+)	-6.0	-3.18	3.16	
	$k = 3$	-110.7	-113.62 (0_5^+)	-0.7	-0.44	3.97	
	$E_{4\alpha}^{thr}$	-110.0	-113.18				

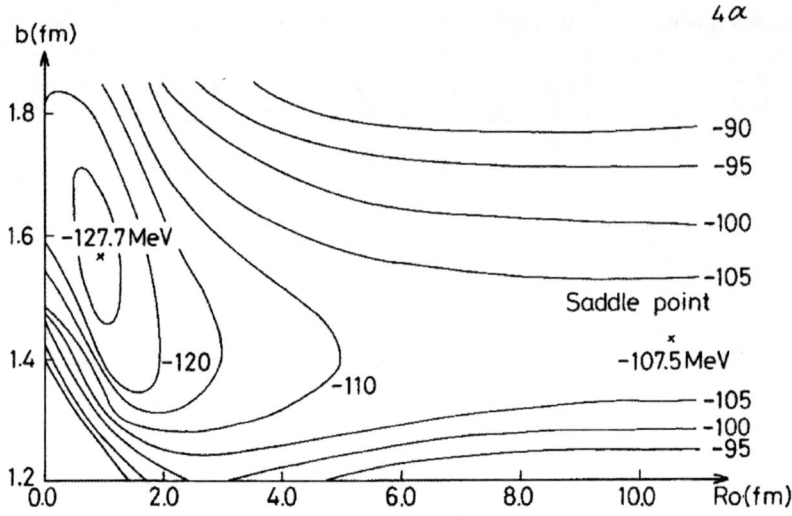

FIGURE 2. Contour map of the energy surface $E_{4\alpha}(R_0,b)$ for ^{16}O. Numbers attached to the contour lines are the binding energies.

These states may be described by our new wave function as condensed states.

As seen from Table I, the experimental value of the ground state (0_1^+) of ^{16}O at -127.62 MeV is well reproduced by the calculated energy value for the ground state $(0_{k=1}^+)$ at -124.8 MeV. The calculated energy is above the minimum energy of the energy surface. It is because the b value of the minimum energy point is fairly larger than b_α and the minimum energy point is not covered by the adopted mesh points. In order to have a better wave function for the ground state we, of course, need to include mesh points around the minimum energy in our generator coordinate calculation. When we adopt $b = 1.57$ fm which is the b value of the energy minimum point, the generator coordinate calculation gives -128.0 MeV as the lowest eigen energy. The rms radius of

TABLE 2. Observed excitation energies E_{exc} and widths Γ of the lowest five 0^+ excited states in ^{16}O

	E_{exc} (MeV)	Γ (MeV)
0_2^+	6.06	
0_3^+	11.26	2.6
0_4^+	12.05	1.6×10^{-3}
0_5^+	14.0	4.8
0_6^+	14.03	2.0×10^{-1}

the calculated $(0^+_{k=1})$ state is 2.59 fm and is slightly smaller than the observed (0^+_1) rms radius, 2.73 fm, of ^{16}O. The second $(0^+_{k=2})$ state of our calculation is bound by 6 MeV below the theoretical 4α threshold energy. The rms radius of this state is 3.12 fm and this state has the largest overlap amplitude (about 0.86) with $\Phi^N_{4\alpha}(R_0, b_\alpha)$ with $R_0 \approx 2.5$ fm. We conjecture that this state corresponds to the observed 0^+_3 state situated at 3.18 MeV below the observed 4α threshold energy. Indeed one may argue that there will be some mixing between the second $(0^+_{k=2})$ state and the ^{12}C+α state, bringing theoretical and experimental energies closer together. The third $(0^+_{k=3})$ state of our calculation is bound by 0.7 MeV below the theoretical 4α threshold energy. We think that it may correspond to the measured 0^+_5 state situated at 0.44 MeV below the observed 4α threshold energy. This state has a very large rms radius of 3.94 fm and has the largest overlap amplitude (about 0.80) with $\Phi^N_{4\alpha}(R_0, b_\alpha)$ with $R_0 \approx 4.1$ fm. In analogy to the case of ^{12}C, these values indicate that this state of dilute density should be considered as just the 4α-cluster condensed state that we expected. One should point out that the ease with which we get the 0^+-states around $E^{thr}_{n\alpha}$ is a strong indication that our wave function (1,3) grasps the esential physics because otherwise the threshold states are very difficult to obtain. We also would like to mention that the present formalism yields very good results for the groundstate of ^{8}Be as well.

In conclusion, our present study thus predicts in ^{12}C and ^{16}O the existence of near-$n\alpha$-threshold states which are the finite system analogues to α-cluster condensation in infinite matter. They are characterized by low density states so that the α-clusters are not strongly overlapping and by an n-fold occupation of their identical S-wave center-of-mass wave function. Therefore these states are quite similar in structure to the Bose-Einstein condensed states of bosonic atoms in magnetic traps where all atoms populate the same lowest S-wave quantum orbital. Because of the short life-time of the α-condensed states, the predicted large values for the rms radii may be verified by indirect methods. The measurements of the spectra of the emitted α particles should allow to determine the Coulomb barrier which is expected to be small for the low density states. Of particular interest would be α-α coincidence measurement of decaying condensed states.

We may conjecture that such condensed α-cluster states near the $n\alpha$ threshold may also occur in other heavier $4n$ self-conjugate nuclei. For example condensed 6α states of ^{24}Mg could be deformed and the measurement of a reduced moment of inertia over the rigid body value would be a strong indication for α-particle superfluidity. The wave function we have proposed in this work is very flexible and can straightfowardly be adopted for the description of other condensation phenomena such as ordinary Cooper pairing or a mixture of Cooper pair and α-particle condensation.

REFERENCES

1. K. Wildermuth and Y. C. Tang, *A Unified Theory of the Nucleus* (Vieweg Braunschweig, 1977).
2. D. M. Brink, Proc. Int. School Phys. Enrico Fermi **36** (Academic Press, New York, 1966);
3. G. F. Bertsch and W. Bertozzi, Nucl. Phys. **A165**, 199 (1971);
4. Y. Fujiwara, H. Horiuchi, K. Ikeda, M. Kamimura, K. Kato, Y.Suzuki, and E. Uegaki, Prog. Theor. Phys. Supplement No.68, 29 (1980).

5. K. Ikeda, N. Takigawa, and H. Horiuchi, Prog. Theor. Phys. Supplement Extra Number, 464 (1968).
6. H. Morinaga, Phys. Rev. **101**, 254 (1956); Phys. Letters **21**, 78 (1966).
7. B. R. Fulton, Proc. 7th Int. Conf. on Clustering Aspects in Nuclear Structure and Dynamics, Rab (1999), edited by M. Korolija, Z. Basrak, and R. Čaplar, (World Scientific, Singapore), p.122.
8. G. Röpke, A. Schnell, P. Schuck, and P. Nozieres, Phys. Rev. Letters **80**, 3177 (1998).
9. M. Beyer, S. A. Sofianos, C. Kuhrts, G. Röpke, and P. Schuck, Phys. Letters **B488**, 247 (2000).
10. A. Tohsaki, Phys. Rev. **C49**, 1814 (1994).
11. F. Ajzenberg-Selove, Nucl. Phys. **A460**, 1 (1986).
12. Y. Suzuki, Prog. Theor. Phys. **55**, 1751 (1976); **56**, 111 (1976).

NEUTRON STARS, BUBBLE NUCLEI AND QUANTUM BILLIARDS

Aurel BULGAC*, Piotr MAGIERSKI†, Andreas WIRZBA** and Yongle YU*

*Department of Physics, University of Washington, Seattle, WA 98195–1560, USA
†Institute of Physics, Warsaw University of Technology, ul. Koszykowa 75, PL–00662, Warsaw, POLAND
**Institut für Kernphysik (Theorie), Forschungszentrum Jülich, D–52425 Jülich, GERMANY

About fifty years ago it was suggested that very large nuclei could sustain a large number of protons only if they develop an unusual shape, either with a bubble inside or that of a torus [1]. If a bubble develops inside a nucleus that leads to a larger average proton–proton separation and thus to a smaller Coulomb energy. However the appearance of a bubble will also lead to a larger surface area of the system. On one hand the Coulomb energy decreases and on the other hand the surface energy increases. One can then expect that an optimal balance between these two opposing tendencies can lead to a more stable nuclear system. Recently the same ideas were extended to charged atomic clusters [2]. Even though relatively early on it was realized that shell effects play an extremely important role in stabilizing bubble and toroidal nuclei one aspect was overlooked, the positioning of the bubble inside the system. The bubble was always put in the center of the system and nobody ever question the reason why and whether that indeed is the best possible arrangement. Especially in the case of charged metallic clusters one can easily see that there is a problem with determining the optimal ground state configuration of a system with one or more voids. In a metal the position of an interior bubble does not affect either the volume, surface, curvature or Coulomb energy of the system and it is not obvious what determines the best geometric configuration of a spherical system with a spherical void inside for example. In Refs. [3, 4] it was shown that shell effects alone are responsible for the stabilization of the bubble in this case and that the character of the shell corrections are somewhat unexpected. A spherical nucleus with a spherical bubble is similar to one of the generic "quantum billiards", the so called annular billiard [5].

The next logical question is to try to determine the energetics of a system with several voids. The simplest case is that of two spherical bubbles in an infinite neutral fermionic liquid or gas. This particular case is relevant to the physics of the crust of a neutron star, which is known to have a so called "pasta phase" [6, 7, 8]. The energetics of two bubbles is governed by the equivalent of shell correction energy in finite systems, a form of Casimir energy [9, 10, 11]. In particular it was shown that two identical spherical bubbles interact with a potential with decays with separation as $\approx \varepsilon_F a^2 \cos[2k_F(r - 2a)]/(4\pi k_F r^3)$, where ε_F is the fermi energy, k_F is the fermi wave

CP597, Nonequilibrium and Nonlinear Dynamics in Nuclear and Other Finite Systems,
edited by Z. Li et al.

vector, *a* stands for the radius of the bubbles and *r* is the separation between their centers. In the skin of neutron stars the voids can have the shape of spherical bubbles, tubes or plates. One can understand the character of the interaction between voids in a relatively simple semiclassical picture [11] and in particular one can show that besides the effective two–body interactions in neutron stars many–body interactions are possible as well. In the case of neutron star crusts only regular lattices of either bubbles, tubes or plates have been considered in the past, as it is widely accepted that the following chain of phase changes is present as the density is increasing: nuclei → rods → plates → tubes → bubbles → uniform matter. These phases are also similar to another type of quantum billiard system, the so called Sinai billiard, which is almost fully ergodic, thus chaotic as well. In Ref. [10] we have shown that the magnitude of the quantum corrections to the ground state energy is of the same order of magnitude as the energy differences between various "pasta phases" and that there is no reason to expect that the sequence "nuclei → rods → plates → tubes → bubbles → uniform matter" is occurring and instead strong disorder could dominate over various regular lattices. We have discussed the dependence of these corrections on a number of physical parameters (density, filling factor, temperature, lattice distortions). Similar phenomena are expected to occur as well in condensed matter systems, dilute atomic fermi condensates and quark–gluon plasma.

A particular aspect of the physics of voids has not been considered in literature yet, the dynamics of such voids. One can contemplate two bubbles moving through a fermionic background. In this case new qualitative aspects become important [12]. A large object moving in a fermionic background will bring part of this background into motion and thus the mass of the moving particle gets "dressed". The mass renormalization depends also on the bubble–bubble separation, but also on the relative orientation of the two bubble velocities with respect to the line separating the two bubbles. Along with this new effect dissipation or "friction" also has to be taken into consideration. Similarly to sea–ships, moving bubbles will radiate "waves". Even though on average kinetic energy is lost, the wave created by one "ship" can "shake" the other ship too. Simple estimates show that bubble–bubble molecules can be formed and that their lifetimes could be sufficiently long to last a few periods and thus one can envision the field of "bubble spectroscopy".

REFERENCES

1. H.A. Wilson, Phys. Rev. **69** 538 (1946); J.A. Wheeler, unpublished notes; P.J. Siemens and H.A. Bethe, Phys. Rev. Lett. **18**, 704 (1967); C.Y. Wong, Ann. Phys. **77**, 279 (1973); W.J. Swiatecki, Physica Scripta **28**, 349 (1983); W.D. Myers and W.J. Swiatecki, Nucl. Phys. **A 601**, 141 (1996).
2. K. Pomorski and K. Dietrich, Eur. Journ. Phys. **D 4**, 353 (1998).
3. A. Bulgac et al. in *Collective Excitations in Fermi and Bose Systems*, eds. Carlos A. Bertulani and Mahir S. Hussein, (World Scientific, Singapore, 1999), pp 44–61 and nucl–th/9811028.
4. Y. Yu, A. Bulgac and P. Magierski, Phys. Rev. Lett. **84**, 412 (2000).
5. O. Bohigas et al., Phys. Rep. **223**, 43 (1993); O. Bohigas et al., Nucl. Phys. A **560**, 197 (1993); S. Tomsovic and D. Ullmo, Phys. Rev. E **50**, 145 (1994); S.D. Frischat and E. Doron, Phys. Rev. E **57**, 1421 (1998).
6. G. Baym et al., Nucl. Phys. **A175**, 225 (1971); D.G. Ravenhall et al., Phys. Rev. Lett. **50**, 2066 (1983); M. Hashimoto et al., Prog. Theor. Phys. **71**, 320 (1984); K. Oyamatsu et al., Prog. Theor. Phys. **72**, 373 (1984); J.M. Lattimer et al., Nucl. Phys. **A432**, 646 (1985); R.D. Wilson and S.E.

Koonin, Nucl. Phys. **A435**, 844 (1985); M. Lassaut *et al.*, Astron. Astrophys. **183**, L3 (1987); K. Oyamatsu, Nucl. Phys. **A561**, 431 (1993); C.P. Lorenz *et al.*, Phys. Rev. Lett. **70**, 379 (1993); C.J. Pethick and D.G. Ravenhall, Annu. Rev. Nucl. Part. Sci. **45**, 429 (1995); G. Watanabe *et al.*, see Los Alamos e–print archive astro–ph/0001273; H. Heiselberg *et al.*, Phys. Rev. Lett. **70**, 1355 (1992).

7.   J.W. Negele and D. Vautherin, Nucl. Phys. **A207**, 298 (1973); P. Bonche and D. Vautherin, Nucl. Phys. **A372**, 496 (1981); Astron. Astrophys. **112**, 268 (1982).

8.   K. Oyamatsu, M. Yamada, Nucl. Phys. **A578**, 181 (1994).

9.   H.B.G. Casimir, Proc. K. Ned. Akad. Wet. **51**, 793 (1948); V.M. Mostepanenko and N.N. Trunov, Sov. Phys. Usp. **31**, 965 (1988); M. Kardar and R. Golestanian, Rev. Mod. Phys. **71**, 1233 (1999) and references therein; M.E. Fisher and P.G. de Gennes, C.R. Acad. Sci. Ser. B **287**, 207 (1978); A. Hanke *et al.*, Phys. Rev. Lett. **81**, 1885 (1998).

10. A. Bulgac and P. Magierski, Nucl. Phys. **A 683**, 695 (2001) and astro-ph/0002377; Physica Scripta, **T90**, 150 (2001) and astro-ph/0007423; nucl-th/0009026.

11. A. Bulgac and A. Wirzba, Phys. Rev. Lett. in press and nucl-th/0102018.

12. A. Bulgac and P. Magierski, in preparation.

Symmetry-unrestricted Gogny-Hartree-Fock-Bogoliubov calculation for ground-state properties of neutron-rich nuclei

Lu Guo [*,†], Fumihiko Sakata[*] and En-Guang Zhao[†]

[*]*Department of Mathematical Sciences, Ibaraki University, Mito, Ibaraki, 310-8512 Japan*
[†]*Institute of Theoretical Physics, Academia Sinica, Beijing 100080, China*

Abstract. Symmetry-unrestricted Hartree-Fock-Bogoliubov calculations with finite-range density-dependent Gogny force are presented with a triaxial harmonic oscillator basis. We include all the contributions to the paring and Hartree-Fock potentials arising from Gogny and Coulomb interaction as well as the center of mass correction. Ground-state properties are calculated for C, O and Ne isotopes, and compared with available experiment, Hartree-Fock plus BCS, shell model and Relativistic-Hartree-Bogoliubov calculations. We show that predicted drip lines are strongly model dependent due to their sensitivity to various theoretical details. Shape coexistence phenomenon is discussed.

Recent experimental developments in radioactive beam, high-efficiency new gamma-ray and charged particle detector systems [1, 2, 3] allow us to extend our research toward the exotic nuclei with extreme isospin. The exotic nuclei display many interesting properties and phenomena, e.g., the discovery of neutron halo in ^{11}Li [4], which attracted more and more attention not only in the nuclear physics but also in other field of science like the astrophysics. But these weakly bound nuclei are much difficult to be treated theoretically than the well bound systems [5]. A variety of theoretical methods and techniques, such as Relativistic-Hartree-Bogoliubov (RHB) [6, 7, 8], Hartree-Fock plus BCS (HF+BCS) [9], shell model (SM) [10, 11] and microscopic three-cluster model [12], have been used to investigate drip-line nuclei properties. The mean-field approaches, both conventional mean-field and relativistic mean-field, have treated on various experimental data widely throughout the nuclear chart. In order to investigate the structure of drip-line nuclei correctly, we need to calculate the mean-field (particle-hole) correlations and the pairing (particle-particle) correlations self-consistently.

Nucleon correlation due to pairing and core polarization becomes crucial in such drip-line system. The microscopic theory of pairing interaction can not be applied to realistic nuclear calculation due to many problems in bare NN force. So phenomenological pairing interaction are introduced. For stable nuclei, the treatment for pairing correlation is often approximated by BCS method with simplified pairing interaction, e.g., seniority pairing force and schematic multipole pairing force. But this kind treatment is inappropriate for drip-line nuclei due to a leakage of nucleons into the continuum. In Skyrme-HFB approach, the zero-range Skyrme interaction also can not provide satisfactory pairing correlation. It is shown that the density-dependent finite-range Gogny force is effective to treat the pairing correlation, and provides a good description

CP597, *Nonequilibrium and Nonlinear Dynamics in Nuclear and Other Finite Systems*,
edited by Z. Li et al.

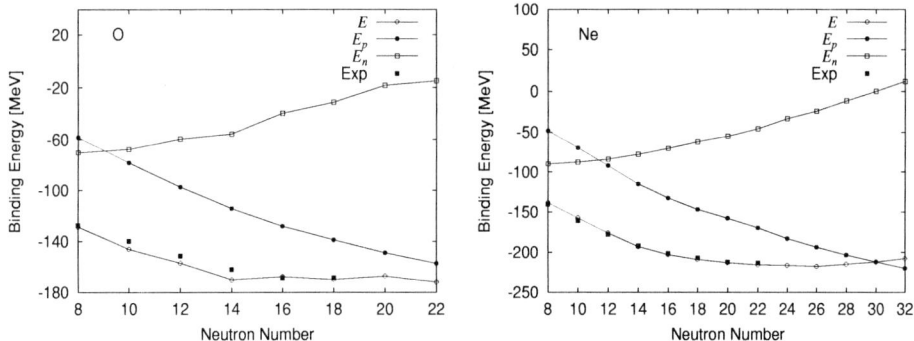

FIGURE 1. Calculated binding energy and experimental results, for O and Ne isotopes as a function of neutron number.

of many nuclear properties over the nuclear chart. In the most HFB calculation with Gogny force the following approximations [13, 14, 15] have been used in the past:

(1) In HF field the Fock term of Coulomb force is neglected because the Fock Coulomb term require a large CPU time.

(2)Contributions to the pairing field coming from the center of mass correction, Coulomb and spin-orbit terms are neglected. Only the pairing field from Brink-Boeker term is considered.

Our "exact" HFB calculation includes all the contributions to the pairing and Hartree-Fock potentials arising from Gogny and Coulomb interaction as well as the center of mass correction. The self-consistent cranking HFB equation is solved with Gogny force using a triaxial harmonic oscillator basis simutaneously breaking axial and signature symmetries. In our HFB code, the cranking term can be included. However, in this paper, we examine only a case with zero angular momentum. Cranking HFB equations read:

$$
\begin{pmatrix} h - \omega j_z & \Delta \\ -\Delta^* & -h^* + \omega j_z^* \end{pmatrix} \begin{pmatrix} U_k \\ V_k \end{pmatrix} = E_k \begin{pmatrix} U_k \\ V_k \end{pmatrix}
\tag{1}
$$

where:

$$
h = \varepsilon + \Gamma - \lambda, \qquad \Gamma_{k_1 k_3} = \sum_{k_2 k_4} \bar{v}_{k_1 k_2 k_3 k_4} \rho_{k_4 k_2}, \qquad \Delta_{k_1 k_2} = \frac{1}{2} \sum_{k_3 k_4} \bar{v}_{k_1 k_2 k_3 k_4} \kappa_{k_3 k_4}.
\tag{2}
$$

Under the following constraints, the CHFB equations are solved::

$$
\delta < \phi(\omega) | \hat{H} - \omega \hat{J}_z - \lambda_p \hat{Z} - \lambda_n \hat{N} + \mu_x < \phi(\omega) | \hat{x} | \phi(\omega) > \hat{x} | \phi(\omega) > = 0,
\tag{3}
$$

$$
< \phi(\omega) | \hat{Z} | \phi(\omega) > = Z, \quad < \phi(\omega) | \hat{N} | \phi(\omega) > = N,
\tag{4}
$$

$$
< \phi(\omega) | \hat{J}_z | \phi(\omega) > = I,
\tag{5}
$$

$$
< \phi(\omega) | \hat{x} | \phi(\omega) > = 0,
\tag{6}
$$

419

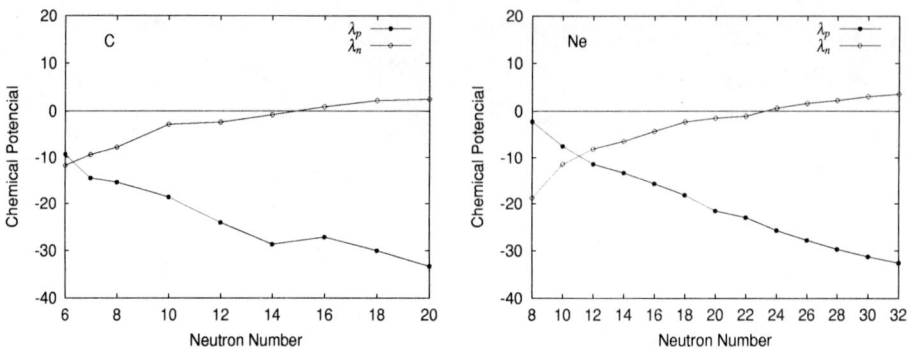

FIGURE 2. Calculated chemical potencial for C and Ne isotopes as a function of neutron number.

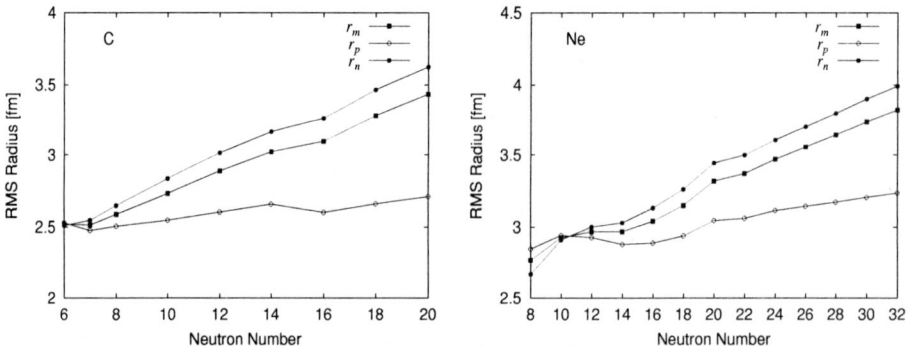

FIGURE 3. Calculated RMS radius for C and Ne isotopes as a function of neutron number.

where the chemical potential λ_p and λ_n are obtained by imposing particle number conservation. The Lagrange multiplier ω is the angular frequency of a collective rotation around the z-axis. To keep the center of mass motion fixed, we also impose the quadrupole constraint operator $\mu_x < \phi(\omega)|\hat{x}|\phi(\omega) > \hat{x}$ [16] in the x-axis direction.

The present self-consistent HFB calculations have been performed with effective Gogny D1S interaction. The s.p. wave functions have been expanded in a three-dimensional harmonic oscillator basis up to the principal quantum number N=8. In Fig. 1 the binding energy are plotted as a function of neutron number for O and Ne isotopes. Total, proton, neutron and experimental binding energy are presented. We found that the calculated binding energy are in good agreement with the experiment for O and Ne isotopes. Since a position of two-neutron drip line is defined by the condition $S_{2n}(Z,N) = B_n(Z,N) - B_n(Z,N-2) = 0$, the last stable isotopes against two-neutron emisson are predicted to be ^{26}O and ^{36}Ne in our HFB calculation, as indicated by their maximal binding energy. The HF+BCS and SM calculations [9] give ^{28}O and ^{34}Ne, ^{26}O and ^{34}Ne as the last stable isotopes, respectively. Another SM calculations [11] show ^{24}O and ^{34}Ne as the last bound isotopes. Owing to their sensitivity on various

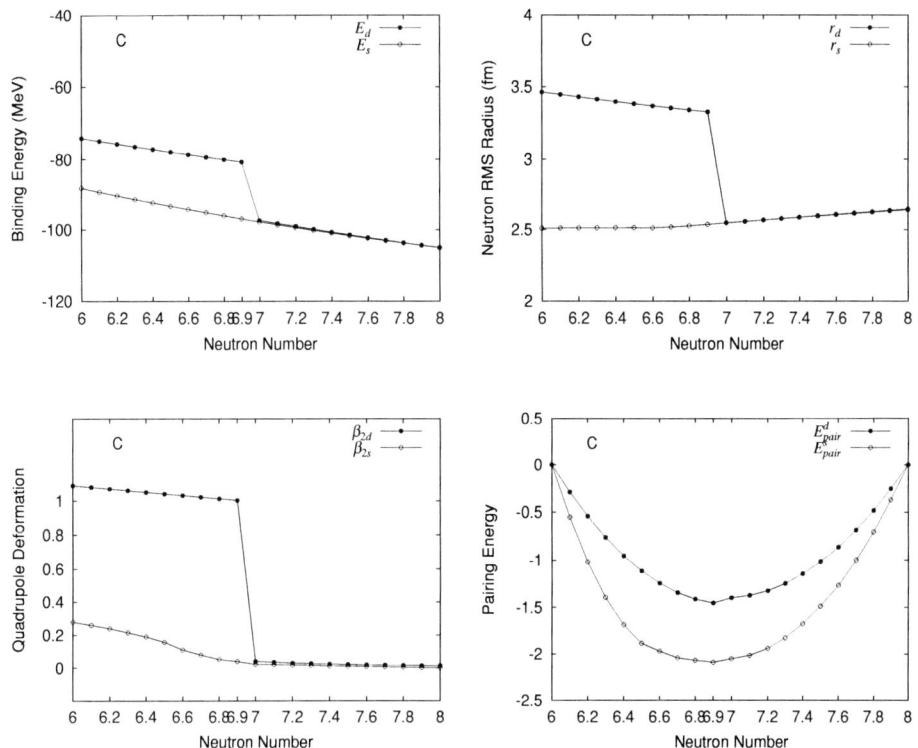

FIGURE 4. Calculated binding energy, rms radii, quadrupole deformation and pairing energy for C isotopes as a function of continuous neutron number.

details (e.g., approximations used, interaction, parameter values), predicted drip lines are strongly model dependent theoretically .

In Fig.2, the calculated proton and neutron chemical potencial are shown for C and Ne isotopes. We observe that for nuclei heavier than ^{20}C and ^{32}Ne the neutron chemical potencial becomes positive. On a case, for Ne isotopes, one neutron is weakly bound and begins to freely escape from the potencial well for the nucleus heavier than ^{32}Ne . But two-neutron escape from potencial well for the nuclus heavier than ^{34}Ne.

We know that some exotic nuclei far from the β stability line show halo phenomenon. Then we calculate the root-mean-square (rms) radius as a function of neutron number for C and Ne isotopes. Proton, neutron and matter rms radii are shown in Fig.3. Neutron rms radius changes smoothly as a function of neutron number. On the other hand, proton rms radius is almost constant, and only display a slow constant increase. For C and Ne drip-line nuclei, no evidence for halo phenomenon is found in our HFB calculation. In RHB calculation [6], neutron radii of Ne isotopes follow the $N^{1/3}$ curve up to N=22. For larger values of neutron number a sharp increase of neutron radii is observed, which provides the evidence for the occurence of neutron halo in heavier Ne isotopes.

We want to know how the nuclear structure changes depending on particle number in more detail. We calculate the ground state properties as a function of continuous neutron number for C isotopes, as shown in Fig.4. Here a subscript 's' indicates a spherical Cartesian basis where the same range parameter is used for three axes, whereas a subscript 'd' indicates a deformed Cartesian basis characterized by three different range parameters. The ground state properties (total binding energy, rms radii, quadrupole deformation and pairing energy) in spherical basis change smoothly as a function of neutron number. But in a case with the deformed Cartesian basis, there is an abrupt change from neutron number 6.9 to 7.0 , although there are smooth changes from neutron number 6 to 8. What we expect from the above calculation is there might be another excited state for the nuclei heavier than N=6.9. Like the band crossing case, it is worthy to study the structure change here and the reason why there occurs an abrupt structure change when one calculate in a case with deformed basis.

In our calculation we also obtain two local minima close in energy for nuclei ^{20}Ne, indicating shape coexistence, as shown in Table 1. To understand the dynamics of shape coexistence phenomenon and to discuss the interaction between the prolate structures and oblate structures, the cranking HFB calculation is necessary. Until now we obtain some average properties of nuclei. To understand the microscopic dynamics responsible for changing the averaged macroscopic properties, the single-particle state in canonical basis is strongly required.

TABLE 1.   shape coexistence of ^{20}Ne

prolate	g.s.,	$\beta_2 = 0.395,$	$\gamma = 0°$
oblate	$E_x = 1.001 MeV,$	$\beta_2 = 0.40,$	$\gamma = 60°$

REFERENCES

1. J. Simpson, Nucl. Phys. A 654, 178c (1999).
2. A. C. Mueller, Nucl. Phys. A 654, 215c (1999).
3. I. Tanihata, Nucl. Phys. A 654, 235c (1999).
4. I. Tanihata et al., Phys. Rev. Lett 55, 2676 (1985).
5. J. Dobaczewski and W. Nazarewicz, Phil. Trans. R. Soc. Lond. A 356, 2007 (1998).
6. W. Pöschl, D. Vretenar, G. A. Lalazissis, and P. Ring, Phys. Rev. Lett.79, 3841(1997).
7. G. A. Lalazissis, D. Vretenar, and P. Ring, Phys. Rev. C 63, 034305 (2001).
8. D. Vretenar, G. A. Lalazissis and P. Ring, Phys. Rev. Lett. 82, 4595 (1999).
9. T. Siiskonen and P. O. Lipas, Phys. Rev. C 60, 034312 (1999).
10. B. A. Brown and P. G. Hansen, Phys. Lett. B 381, 391 (1996).
11. E. Caurier, F. Nowacki, A. Poves, and J. Retamosa, Phys. Rev. C 58, 2033 (1998).
12. P. Descouvemont, Phys. Lett. B 437, 7 (1998).
13. M. Girod, B. Grammaticos, Phys. Rev. C 27, 2317 (1983).
14. J. L. Egido, L. M. Robledo, Phys. Rev. Lett. 70, 2876 (1993).
15. M. Girod, J. P. Delaroche, J. F. Berger, Phys. Lett. B 325, 1 (1994).
16. H. Flocard, P. Quentin, A. K. Kerman, and D. Vautherin, Nucl. Phys. A 203, 433 (1973).

A New Class of Nuclear Isomeric States: Mixed Symmetry Isomeric States in Nuclei

G. L. Long*, Y. S. Li*, C. C. Tu*, L. Tian*, H. Y. Ji*, S. J. Zhu*, E. G. Zhao†, F. Y. Liu* and J. F. Zhang*

*Department of Physics, Tsinghua University, Beijing 100084
†Institute of Theoretical Physics, Chinese Academy of Sciences, Beijing, 100080

Abstract. We suggest a new class of isomeric states in nuclei. By study the mixed symmetry states in the framework of the neutron-proton interacting boson model(IBM2), it is found that some of the mixed symmetry states with moderate high spins change very fast with respect to the Majorana interaction. Under certain conditions, they become the yrast state or yrare state. These states are difficult to decay and become very stable. This study suggests that a possible new mode of isomers may exists due to the special nature in their proton and neutron degrees of freedom.

INTRODUCTION

The theoretical prediction[1, 2] of the mixed symmetry states or the isovector states and their later experimental comfirmations[3, 4] have been one of the great successes of the neutron-proton interacting boson model(IBM2). The studies of the mixed symmetry states have been an important subject in nuclear physics[5, 6, 7, 8]. Recently the mixed symmetry states in nucleus with octupole deformation have been proposed[9]. Physically, the F-spin describes the relative phase of motion of the neutrons and protons[10]. Like the isospin, F-spin obeys the rules of the angular momentum coupling. In the maximum F-spin state, the neutrons and protons are moving in phase, and this state is called the F-spin symmetric state. In other F-spin state, the neutrons and protons are moving out of phase in different degree, and these states are called mixed symmetry states. The extent at which the neutrons and protons move differently is reflected in the F-spin: in the maximum F-spin state, they move in phase. In the minimum F-spin state, they move completely out of phase. This picture is helpful in intuitive understanding the F-spin, though recent studies[11, 12] showed that this simple interpretation was not appropriate in some situation.

In this talk, we will show that under certain conditions it is possible to have high spin mixed symmetry states very low. In some case, they can even become the yrast or yrare state, i.e., the lowest or the next to the lowest state with the same angular momentum. Because their F-spin structures are different from the states below, their transitions to the low-lying levels are highly retarded and make them a new mode of isomeric state: mixed symmetry isomeric state(MSIS). This paper is an extended version of the work reported in two meetings[7, 8]. The paper is organized as follows. Firstly, we discuss the dependence of the mixed symmetry states on the strength of different Majorana terms. Then, we discuss some of the high spin mixed symmetry states. Next, we explore a

CP597, *Nonequilibrium and Nonlinear Dynamics in Nuclear and Other Finite Systems,*
edited by Z. Li et al.

TABLE 1. Number of states with a given F-spin and an angular momentum for system with $N_\pi = 2$ and $N_\nu = 3$.

	Angular momentum										
F	0	1	2	3	4	5	6	7	8	9	10
5/2	5	0	7	2	6	2	4	1	2	0	1
3/2	4	6	12	10	11	7	6	3	2	1	0
1/2	5	5	12	9	11	6	5	2	1	0	0

possible example of mixed symmetry isomeric state(MSIS) in ^{94}Sr. Finally, we give a discussion.

MIXED SYMMETRY STATE ENERGY IN IBM2

The positions of the mixed symmetry states are closely related to the strength of the Majorana interaction. The Majorana interaction contains 3 terms:

$$M = \xi_1 (d_\nu^\dagger d_\pi^\dagger)^1 \cdot (\tilde{d}_\pi \tilde{d}_\nu)^1 + \xi_2/2 (s_\pi^\dagger d_\nu^\dagger - s_\nu^\dagger d_\pi^\dagger)^2 \cdot (\tilde{s}_\nu \tilde{d}_\pi - \tilde{d}_\nu \tilde{s}_\pi)^2 + \xi_3 (d_\nu^\dagger d_\pi^\dagger)^3 \cdot (\tilde{d}_\pi \tilde{d}_\nu)^3. \quad (1)$$

The Majorana interaction is part of the neutron-proton interaction. Microscopically, a combination of various neutron-proton interaction can form the Majorana interaction [17, 18, 19, 20, 21, 22, 23]. It is very unusual to use an IBM2 hamiltonian without Majorana interaction [24]. In the neutron-proton interacting boson model, it has primary importance and is singled out separately. In most practical calculations, Majorana interaction strengths are determined by a best fit to the experimental spectra. Among the vast number of literatures on IBM2, the choices of the parameters used can be classified into 3 categories [10, 11, 12, 13, 14, 15, 16, 17, 24, 25, 26, 27]: 1) the 3 strengths in the Majorana interaction are equal and positive, for instance in Ref. [10]; 2) two Majorana parameters, namely ξ_1 and ξ_3 are kept equal and ξ_2 is determined differently. Examples of this catogory are: $\xi_1 = \xi_3 = 0.30$MeV and $\xi_2 = 0$ in [16]; $\xi_1 = \xi_3 = 2.20$MeV and $\xi_2 = -0.046$ in [25]; $\xi_1 = \xi_3 = -0.18$MeV and $\xi_2 = 0.24$MeV in [26]); 3) the 3 Majorana interactions are all different, for instance in Ref. [24], $\xi_1 = -0.27$MeV, $\xi_2 = 0.0$MeV and $\xi_3 = -0.085$MeV.

In this paper, we limit ourselves to the vibrational limit of the IBM2. The generalization to other dynamical symmetries are straightforward. For simplicity, we choose the following hamiltonian,

$$H = 0.12C_{2U5} + 0.01C_{2O5} + M, \quad (2)$$

where the expressions of the Casimir operators can be found in [10]. We choose a system with $N_\pi = 2$ and $N_\nu = 3$. The results obtained here can be generalized into systems with more boson numbers. In table 1, we give the number of states for each angular momentum and F-spin. For the convenience of later studies, we give all the state labels in the U(5) limit in the 3 $F - spin$ values in tables 2-4.

We have used two choices of the Majorana parameters. In the first choice, we choose $\xi_1 = \xi_2 = 0.8$ MeV, and let ξ_3 to vary from 1.0 MeV to -0.4 MeV. In the second choice,

TABLE 2. Labellings of states with $F = 5/2$ in an $N_\pi = 3, N_v = 2$ system. [] is the label for the U(5) irreducible representation and () is the O(5) irreducible representation.

L	U(5) O(5) labels						
0	[0]⟨0⟩	[2]⟨0⟩	[3]⟨3⟩	[4]⟨0⟩	[5]⟨3⟩		
2	[1]⟨1⟩	[2]⟨2⟩	[3]⟨1⟩	[4]⟨2⟩	[4]⟨4⟩	[5]⟨1⟩	[5]⟨5⟩
3	[3]⟨3⟩	[5]⟨3⟩					
4	[2]⟨2⟩	[3]⟨3⟩	[4]⟨2⟩	[4]⟨4⟩	[5]⟨3⟩	[5]⟨5⟩	
5	[4]⟨4⟩	[5]⟨5⟩					
6	[3]⟨3⟩	[4]⟨4⟩	[5]⟨3⟩	[5]⟨5⟩			
7	[5]⟨5⟩						
8	[4]⟨4⟩	[5]⟨5⟩					
10	[5]⟨5⟩						

TABLE 3. The same as table (2) but with $F = 3/2$.

L	U(5) O(5) labels							
0	[2]⟨0⟩	[3]⟨3⟩	[4]⟨0⟩	[41]⟨3⟩				
1	[11]⟨11⟩	[21]⟨21⟩	[31]⟨11⟩	[31]⟨31⟩	[41]⟨21⟩	[41]⟨41⟩		
2	[1]⟨1⟩	[2]⟨2⟩	[3]⟨1⟩	[4]⟨2⟩	[4]⟨4⟩	[21]⟨1⟩	[21]⟨21⟩	[31]⟨2⟩
	[31]⟨31⟩	[41]⟨1⟩	[41]⟨21⟩	[41]⟨41⟩				
3	[3]⟨3⟩	[11]⟨11⟩	[21]⟨21⟩	[31]⟨11⟩	[31]⟨31⟩	[31]⟨31⟩	[41]⟨3⟩	[41]⟨21⟩
	[41]⟨41⟩	[41]⟨41⟩						
4	[2]⟨2⟩	[3]⟨3⟩	[4]⟨2⟩	[4]⟨4⟩	[21]⟨21⟩	[31]⟨2⟩	[31]⟨31⟩	[41]⟨3⟩
	[41]⟨21⟩	[41]⟨41⟩	[41]⟨41⟩					
5	[4]⟨4⟩	[21]⟨21⟩	[31]⟨31⟩	[31]⟨31⟩	[41]⟨21⟩	[41]⟨41⟩	[41]⟨41⟩	
6	[3]⟨3⟩	[4]⟨4⟩	[31]⟨31⟩	[41]⟨3⟩	[41]⟨41⟩	[41]⟨41⟩		
7	[31]⟨31⟩	[41]⟨41⟩	[41]⟨41⟩					
8	[4]⟨4⟩	[41]⟨41⟩						
9	[41]⟨41⟩							

TABLE 4. The same as table (2) but with $F = 1/2$.

L	U(5) O(5) labels							
0	[2]⟨0⟩	[3]⟨3⟩	[22]⟨0⟩	[22]⟨22⟩	[32]⟨3⟩			
1	[21]⟨21⟩	[31]⟨11⟩	[31]⟨31⟩	[32]⟨21⟩	[32]⟨32⟩			
2	[2]⟨2⟩	[3]⟨1⟩	[21]⟨1⟩	[31]⟨2⟩	[31]⟨31⟩	[22]⟨2⟩	[22]⟨22⟩	[32]⟨1⟩
	[32]⟨21⟩	[32]⟨32⟩	[32]⟨32⟩					
3	[3]⟨3⟩	[21]⟨21⟩	[31]⟨11⟩	[31]⟨31⟩	[31]⟨31⟩	[22]⟨22⟩	[32]⟨3⟩	[32]⟨21⟩
	[32]⟨32⟩							
4	[2]⟨2⟩	[3]⟨3⟩	[21]⟨21⟩	[31]⟨2⟩	[31]⟨31⟩	[22]⟨2⟩	[22]⟨22⟩	[32]⟨3⟩
	[32]⟨21⟩	[32]⟨32⟩	[32]⟨32⟩					
5	[21]⟨21⟩	[31]⟨31⟩	[31]⟨31⟩	[32]⟨21⟩	[32]⟨32⟩	[32]⟨32⟩		
6	[3]⟨3⟩	[31]⟨31⟩	[22]⟨22⟩	[32]⟨3⟩	[32]⟨32⟩			
7	[31]⟨31⟩	[32]⟨32⟩	[41]⟨41⟩					
8	[32]⟨32⟩							

425

we choose $\xi_1 = \xi_3 = 0.8$ MeV and let ξ_2 to vary from 1.0 MeV to -0.4MeV. The step of change is 0.2MeV. When $\xi_1 = \xi_2 = \xi_3 = 0.8$MeV, the system is in the U(5) limit, both the wave functions and the energy levels can be calculated analytically. Then we trace the change of energy of states by looking at the overlapping integrals at neighboring Majornana interactions. It is worth pointing here that even though the $U_{\pi+\nu}(6)$ is no longer a good quantum number, the F-spin is still a good quantum number. In both sets, the states with maximum F-spin do not change as ξ_2 or ξ_3 changes. But in general, states with other F-spin values change. We report the results of some calculations in the following. In the following we give the detailed description of the changes of the states with respect to change in the Majorana interaction for the first few spin states. The choice of the Majorana interaction is $\xi_1 = \xi_3 = 0.8$Mev and ξ_2 changing. The cases with other choices are similar. The changes of the 8^+ states are given in table 5.

$L = 0$ states

There are 14 $L = 0$ states altogether. The states with $F = \frac{3}{2}$ [41]$\langle 3 \rangle$ $(U(5), O(3)$ labels) and $F = \frac{1}{2}$ [32]$\langle 3 \rangle$ do not change. The remaining 7 states increase linearly as ξ_2 increases. As $\xi_2$increases, the states with [2]$\langle 0 \rangle$, [3]$\langle 0 \rangle$ and [41]$\langle 3 \rangle$ in the $F = \frac{3}{2}$ increase by 0.5 MeV. Whereas the states with [2]$\langle 0 \rangle$, [3]$\langle 3 \rangle$ in the $F = \frac{1}{2}$ increase by 0.8 MeV, and $F = \frac{1}{2}$ [22]$\langle 0 \rangle$ and $F = \frac{1}{2}$ [22]$\langle 22 \rangle$ states increase by 0.2 MeV. The sharpest changes occur in $F = \frac{1}{2}$.

$L = 1$ states

There are 4 states with $F = \frac{3}{2}$. [11]$\langle 11 \rangle$ state changes at a rate of 0.3 MeV. [21]$\langle 21 \rangle$ state increases by 0.2 MeV and [31]$\langle 11 \rangle$ and [31]$\langle 31 \rangle$ change by 0.1 Mev. There are 7 $F = \frac{1}{2}$ states. 4 states that with [41]$\langle 21 \rangle$, [41]$\langle 41 \rangle$, [32]$\langle 21 \rangle$ and [32]$\langle 32 \rangle$ do not change with respect to ξ_2. The sharpest change occurs at [21]$\langle 21 \rangle$. The [31]$\langle 11 \rangle$ and [31]$\langle 31 \rangle$ states increase 0.40 MeV.

$L = 2$ states

There are 28 $L = 2$ states. The highest number in this system. All the F-spin symmetric states are invariant under change in ξ_2.

States with$F = \frac{3}{2}$: Three states, [41]$\langle 1 \rangle$, [41]$\langle 21 \rangle$, [41]$\langle 41 \rangle$ are invariant under change in ξ_2. As ξ_2 increases, [31]$\langle 2 \rangle$ and [31]$\langle 31 \rangle$ states increase by 0.10 MeV, [21]$\langle 1 \rangle$ and [21]$\langle 21 \rangle$ increase by 0.2 MeV, and [1]$\langle 1 \rangle$, [2]$\langle 2 \rangle$, [3]$\langle 3 \rangle$, [4]$\langle 2 \rangle$ and [4]$\langle 4 \rangle$ increase by 0.5 MeV.

States with $F = \frac{1}{2}$: [31]$\langle 1 \rangle$, [32]$\langle 21 \rangle$, [32]$\langle 32 \rangle_1$ and [32]$\langle 32 \rangle_2$ states do not change. There are two states with [32]$\langle 32 \rangle$ and are distinguished by subscript number. Two states, namely with [22]$\langle 2 \rangle$ and [22]$\langle 22 \rangle$ increase by 0.2 MeV. Two states with [31]$\langle 2 \rangle$

TABLE 5. The energy change of 8^+ state versus ξ_3. Note when $\xi_3 = 0.0$, the energy of $\{32\}[32] < 32 >$(the last column) becomes the lowest.

ξ_3	$\{5\}[5] < 5 >$	$\{5\}[4] < 4 >$	$\{41\}[41] < 41 >$	$\{41\}[4] < 4 >$	$\{32\}[32] < 32 >$
1.0	5.8	4.12	7.0	6.12	7.12
0.8	5.8	4.12	6.5	6.12	6.92
0.6	5.8	4.12	6.0	6.12	6.16
0.4	5.8	4.12	5.5	6.12	5.32
0.2	5.8	4.12	5.0	6.12	4.52
0.0	5.8	4.12	4.5	6.12	3.72
−0.2	5.8	4.12	4.0	6.12	2.92
−0.4	5.8	4.12	3.5	6.12	2.12
ΔE	0.0	0.0	0.5	0.0	0.8

and $[31]\langle 31\rangle$ increase by 0.4 MeV. Two states with $[21]\langle 1\rangle$ and $[21]\langle 21\rangle$ increase by 0.5 MeV. States $[3]\langle 1\rangle$ and $[2]\langle 2\rangle$ increase by 0.8 MeV, the biggest change among the L=2 states.

$L = 3$ states

States with $F = \frac{3}{2}$:
$[41]\langle 3\rangle$, $[41]\langle 21\rangle$, $[41]\langle 41\rangle_1$ and $[41]\langle 41\rangle_2$ do not change. States, $[31]\langle 11\rangle$, $[31]\langle 31\rangle_1$ and $[31]\langle 31\rangle_2$ change by 0.1 MeV. State $[21]\langle 21\rangle$ changes by 0.2 MeV. State $[11]\langle 11\rangle$ changes by 0.3 MeV, and state $[3]\langle 3\rangle$ changes by 0.50 MeV.

States with $F = \frac{1}{2}$: 3 states, $[32]\langle 3\rangle$, $[32]\langle 21\rangle$ and $[32]\langle 32\rangle$ do not change. State $[22]\langle 22\rangle$ change by 0.20 MeV. States $[31]\langle 11\rangle$, $[31]\langle 31\rangle_1$ and $[31]\langle 31\rangle_2$ change by 0.40 MeV. The highest change occurs in $[3]\langle 3\rangle$ with a change of 0.80 MeV.

MIXED SYMMETRY ISOMERIC STATES

It has been shown that mixed symmetry states change with the Majorana interaction quite differently. Some of the mixed symmetry states remain constant while Majorana interaction changes. Some states change slowly while others change very fast. For instance, for $L = 8$, the dependence of states on the Majorana interaction ξ_3 is given in Table 5. From Table 5, we see that as ξ_3 decreases from 0.80 MeV, state with $[41]\langle 41\rangle$ and $F = \frac{3}{2}$ and state with $[32]\langle 32\rangle$ and $F = \frac{1}{2}$ decrease. $[32]\langle 32\rangle$ state decreases by 0.8 MeV when ξ_3 decreases by 0.20 MeV and is much faster than $[41]\langle 41\rangle$. When ξ_3 is reduced to 0.0 MeV. The $[32]\langle 32\rangle$ has become the lowest energy among the 8^+ states, the yrast state. Similarly, when ξ_2 changes while keeping $\xi_1 = \xi_3 = 0.80$ MeV, we obtain similar result. Here the $[41]\langle 4\rangle$ with $F = \frac{3}{2}$ is the state with a big change in energy. When $\xi_2 = 0.00$ MeV, this state becomes the lowest in energy. This result is different from the usual picture about the mixed symmetry states: state with F_{max} is the lowest in energy, state with $F_{max} - 1$ lies higher, and state with $F_{max} - 2$ still higher. The smaller the F-spin of the states, the higher the energy. This picture is true for F-spin symmetric hamiltonian where $\xi_1 = \xi_2 = \xi_3$. When the 3 Majorana interactions are not equal, this picture fails. For states with low spin, the mixed symmetry states have not come down so low to become the yrast or yrare state. The usual picture is approximately right for a few low-lying states. The mixed states change quite a lot. In particular, the $F = \frac{1}{2} [32]\langle 32 \rangle L = 8$ state becomes the yrast state. Moreover, the $F = \frac{1}{2} [22]\langle 22 \rangle L = 6$ state also comes down very low, though it is not the yrast state. The structures of these states are quite different from the states below. To see this more clearly, we have performed E2, M1 transition calculations. We have chosen the following transition operators for simplicity

$$T(E2) = e_\pi[(s_\pi^+ \tilde{d}_\pi + d_\pi^+ s_\pi) + (d_\pi^+ \tilde{d}_\pi)^2 + (s_v^+ \tilde{d}_v + d_v^+ s_v) + (d_v^+ \tilde{d}_v)^2] \tag{3}$$

$$T(M1) = \sqrt{\frac{30}{4\pi}} (d_v^+ \tilde{d}_v)^1 \tag{4}$$

We have also put $e_v = e_\pi = 0.1q(eb)$, here a is the order of one. The result for some interested states are

$$B(E2; 8_1 \rightarrow 6_2) = 0.00020q^2 \tag{5}$$

and the transition from 6_2 state are zero. The 8_1^+ and 6_2^+ states have no M1 transitions to states down below. They are highly retarded. It is very stable and has a long lifetime. It is a new mode of isomers. These isomers arise from the different mode of motions of the neutrons and the protons. It is collective.

POSSIBLE CANDIDATE

It is important if we can identify the predicted MSIS in real nuclei. We restrict ourselves to nuclei near the vibrational nuclei. Vibrational spectra occur in nuclei near closed shells. Nuclei with exact vibrational spectrum is very rare. Recently, the yrast state of ^{94}Sr has been extended to 14^+[33] tentatively. The peculiar feature of this nucleus is the backbending at $L = 8$ in the yrast band. The following Hamiltonian has been used

$$H = \varepsilon_d \hat{n}_d + \kappa Q_v \cdot Q_\pi + M, \tag{6}$$

where

$$Q_\rho = (s_\rho^+ \hat{d}_\rho + d_\rho^+ s_\rho)^2 + X_\rho (d_\rho^+ \hat{d}_\rho)^2. \tag{7}$$

The parameters ε_d, κ, X_v and X_π are determined by the low-lying levels. As for the Majorana interaction, we put $\xi_1 = \xi_2 = 0.15$ MeV so that the 1^+ state lies at about 3.0 MeV. The ξ_3 is chosen by fitting the ground state band, and it is $\xi_3 = -0.52$MeV. In table 6, we give the expectation value of $\hat{F}^2$ for states of interests. The maximum F-spin for this system is $\langle F^2 \rangle_{max} = 20$. From table 6, we see that the 8_1^+ state is mainly from $F = 2$ F-spin configuration, while 6_1^+ comes from $F = 4$, the maximum F-spin, configuration. The electromagmetic transition properties are calculated by using the consistent $Q \cdot Q$ formalism for simplicity, where the same Q operator are used in both the transition and the hamiltonian, though it is found that this formalism can not describe the transition in details[35, 36]. The calculated B(E2) and B(M1) values are given in Tables 7 and 8.

428

TABLE 6. The expectation value of $\hat{F}^2$ for some states.

state	14_1^+	12_1^+	10_1^+	8_1^+	8_2^+	7_1^+	7_2^+	6_1^+	6_2^+	6_3^+
$\langle \hat{F}^2 \rangle$	6.01	2.04	2.42	6.48	18.53	11.76	6.08	19.69	11.75	6.69

TABLE 7. Calculated B(E2) transition in 94Sr in unit $q^2(e^2b^2)$. Here q is the order of 1 and has no dimension. It should be determined by fitting to experimental data.

Transition:	$2_1 \to 0_1$	$2_1 \to 0_2$	$2_2 \to 0_1$	$1_1 \to 2_1$	$1_1 \to 2_2$
B(E2)	0.1068	0.0253	0.0000	0.0001	0.0068
Transition:	$2_1 \to 2_2$	$4_1 \to 2_1$	$6_1 \to 4_1$	$8_1 \to 6_1$	$8_1 \to 6_2$
B(E2)	0.1690	0.1743	0.2116	0.0101	0.0005
Transition:	$8_1 \to 6_3$	$10_1 \to 8_1$	$12_1 \to 10_1$	$14_1 \to 12_1$	$16_1 \to 14_1$
B(E2)	0.0916	0.0142	0.0399	0.0013	0.0000

It is seen from these calculations, the state 8_1^+ has very small E2 and M1 transition rates to the low-lying levels. It is likely to be an mixed symmetry isomeric state. Similarly, the states 10_1^+ and 12_1^+ are slow to decay, and are of similar nature. It is important for experimentalist to measure their lifetimes.

SUMMARY

The characters of this new isomeric state are: 1) moderate high spins, at about $L = 8$; 2) the energy about 3 times of the 2_1^+ energy; 3) very weak transitions to states down below; 4) if the isomeric state is the yrast state, it will produce a backbending.

The usual pair break-up and collective backbending effect[37] can also produce a backbending. The following differences should be noticed. The backbending produced by MSIS is irregular. After the backbending, the transition, the E2 transitions are irregular. In the previous section, we see that after the backbending at 8^+, there is another 10^+

TABLE 8. Calculated M1 transition B(M1) for ^{94}Sr (unit $g^2\mu_N^2$).

Transition:	$0_1 \to 1_1$	$0_2 \to 1_1$	$1_1 \to 2_2$	$2_1 \to 2_2$	$2_1 \to 2_3$	$3_1 \to 2_2$
B(M1)	0.011	0.067	0.0035	0.0008	0.0027	0.0017
Transition:	$3_1 \to 4_1$	$5_1 \to 4_2$	$6_1 \to 5_1$	$7_1 \to 6_2$	$7_1 \to 6_3$	$8_1 \to 7_1$
B(M1)	0.0015	0.0016	0.0024	0.0010	0.0029	0.0003
Transition:	$8_1 \to 7_2$	$8_1 \to 7_3$	$9_1 \to 8_1$	$9_1 \to 8_2$	$9_1 \to 8_3$	$10_1 \to 9_1$
B(M1)	0.0035	0.0019	0.0013	0.0001	0.0074	0.0024
Transition:	$10_1 \to 9_2$	$12_1 \to 11_1$	$12_1 \to 11_2$			
B(M1)	0.0024	0.0041	0.0076			

state. The transition from 10^+ to 8^+ is very small. Using this property, we can distinguish MSIS backbending from the other two backbending effects. Of the pair break-up and collective backbending, the spectrum is very regular. After the backbending in pair break-up, the band is built upon two quasi-particles, and the transition strength is reduced, but it is stronger than MSIS backbending. In collective backbending, the states after backbending is also collective and their transition strength is still strong.

This work is supported by Major State Basic Research Development Program, G200077400, China National Natural Science Foundation; Pok Yin Tung Education Foundation and Excellent Young University Teacher's Fund of Education Ministry, PR China.

REFERENCES

1. T. Otsuka, Ph.D. Thesis, 1979, Tokyo University
2. F. Iachello, Nucl. Phys. A358 (1981) 89c.
3. D. Bohle et al, Phys. Lett. B137 (1984) 27.
4. W. D. Hamilton, A. Irbäck and J. P. Elliott, Phys. Rev. Lett. 53 (1984) 2469.
5. N. Pietralia et al, Phys. Rev. Lett. 83(1999)1303
6. H Z Sun et al, Commun. Theor. Phys. 29(1998) 411
7. G L Long et al, Journal of Ningxia University (Natural Science Edition), 19 (1998) 332
8. G L Long et al, Nuclear Physics Reviews, 16(1999) 6
9. N. A. Smirnova, N. Pietralla, T. Mizusaki and P. Van Isacker, Nucl. Phys. A678 (2000) 235
10. P.Van Isacker, K. Heyde, j. Jolie, A. Sevrin, Ann. Phys. 171 (1986) 253.
11. T. Ostuka, Phys. Rev. Lett. 71 (1993) 1804
12. G. L. Long, Chin. Phys. Lett. 12 (1995) 342.
13. H. Harter, P. von Bretano and A. Gelberg, Phys. Rev. C 34 (1986) 1472.
14. G.L. Long, Y. X. Liu and H. Z. Sun, J. Anhui University, 13 (1989) 96
15. G. L. Long, Y. X. Liu and H. Z. Sun, J. Phys. G16 (1990) 813
16. O. Scholten, K. Heyde and P. Van Isacker, Phys. Rev. Lett. 55 (1985) 1866.
17. G. L. Long and J. P. Elliott, in Understanding the variety of nuclear excitations, eds by A. Covollo, World Scientific Publishing Company, 1991, P.361.
18. J. P. Elliott et al, J. Phys. A25 (1992) 4633
19. J. A. Evans et al, Nucl. Phys. A561 (1993) 201
20. J. P. Elliott et al, J. Phys. A27 (1994) 4465
21. V. S. Lac et al, Nucl. Phys. A587 (1995) 101
22. J. A. Evans et al, Nucl. Phys. A593 (1995) 85
23. J. P. Elliott et al, Nucl. Phys. A609 (1996) 1
24. A. Novolselsky and I. Talmi, Phys. Lett. B172 (1986) 139.
25. A. Giannatiempo, A. Nannini, A. Perego, P. Sona and G. Maino, Phys. Rev. C 44 (1991) 1508.
26. M. Sambataro, Nucl. Phys. A380 (1982) 365.
27. Y. X. Liu, G. L. Long and H. Z. Sun, J. Phys. G 17 (1991) 877.
28. G. L. Long, S. J. Zhu, L. Tian and H. Z. Sun, Chin. J. Nucl. Phys. 17 (1995) 149
29. G. L. Long et al, Commun. Theor. Phys. 29 (1998) 249
30. G. L. Long, S. J. Zhu, J. Y. Zhang, E. G. Zhao and Y. X. Liu, Commun. Theor. Phys. 29 (1998) 65
31. G. L. Long, Science and Technology of Tsinghua University, 1 (1996) 231
32. J. F. Zhang et al, High Energ. Nucl. Phys. 24 (2000) 1066
33. S J Zhu, private communication.
34. Richard B. Firestone et al, Table of Isotopes, Wiley and Sons Inc. Eighth Edition (CD version), 1998.
35. G. L. Long and H. Y. Ji, Phys. Rev. C57 (1998) 168
36. C. L. Long, Commun. Theor. Phys. 32 (1999)489.
37. G. L. Long, Phys. Rev. C55 (1997) 3163

List of Participants

Y. Abe
Yukawa Institute for Theoretical Physics,
Kyoto Univ., Kyoto 606-8502, Japan
E-mail: abey@yukawa.kyoto-u.ac.jp

J. Aichelin
SUBATECH, UMR Université, Ecole des
Mines, IN2P3/CNRS F-44072 Nantes,
France
E-mail: aichelin@subatech.in2p3.fr

S. Ayik
Physics Department,
Tennessee Technological University,
Cookeville, TN38505, USA
E-mail: ayik@tntech.edu

S. A. Bass
Department of Physics, Duke University &
RIKEN-BNL Research Center, Brookhaven
National Laboratory
E-mail: bass@phy.duke.edu

A. Bulgac
Department of Physics, University of
Washington, Seattle, WA 98195–1560, USA
E-mail: bulgac@phys.washington.edu

Lie-Wen Chen
Institute of Theoretical Physics, Shanghai
Jiao Tong University, Shanghai 200030,
China
E-mail: lwchen@mail.sjtu.edu.cn

L. Corradi
INFN-Laboratori Nazionali di Legnaro, Via
Romea 4, I-35020, Legnaro (Padova), Italy
E-mail: Lorenzo.corradi@lnl.infn.it

Nguyen Dinh Dang
RI-Beam Factory Project Office,
RIKEN, 2-1 Hirosawa, Wako,
351-0198, Saitama, - Japan
And
Institute for Nuclear Science and
Technique, Vietnam Atomic Energy
Commission – Hanoi, Vietnam
E-mail: dang@rikaxp.riken.go.jp

P. Danielewicz
National Superconducting Cyclotron
Laboratory and Department of Physics and
Astronomy, Michigan State University, East
Lansing, MI 48824, USA
E-mail: danielewicz@nscl.msu.edu

Ren-Fa Feng
China Institute of Atomic Energy, P.O.
Box 275(18), Beijing 102413, China
E-mail: rffeng@iris.ciae.ac.cn

S. Fujiwara
Department of Mathematical Sciences,
Ibaraki University, Mito, Ibaraki 310-8512,
Japan

Chun-Yuan Gao
Department of Technical Physics, Peking
University Beijing 100871, China

W. Greiner
Institut für Theoretische Physik,
Robert-Mayer-Str. 8-10,
J.W. Goethe-Universität,
D-60325 Frankfurt, Germany
E-mail:
Greiner@th.physik.uni-frankfurt.de

431

Jian-Zhong Gu
Hong Kong Baptist University,
Hong Kong
E-mail: jzgu@phys.hkbu.edu.hk

Lu Guo
Department of Mathematical Sciences,
Ibaraki University, Mito, Ibaraki 310-8512,
Japan

Y. Hashimoto
Institute of Physics, University of Tsukuba,
Ibaraki 305-8571, Japan
E-mail: hashi@nucl.ph.tsukuba.ac.jp

W.D.Heiss
School of Physics
Department of Physics, University of the
Witwatersrand, PO Wits 2050, South Africa
E-mail: heiss@physnet.phys.wits.ac.za

H. Horiuchi
Department of Physics, Kyoto University,
Kyoto 606-8502, Japan
E-mail:
horiuchi@ruby.scphys.kyoto-u.ac.jp

De-Fu Hou
Institute of Particle Physics, Huazhong
Normal University, 430070 Wuhan, China
E-mail: hdf@theory.uwinnipeg.ca

Mei Huang
Physics Department, Tsinghua University,
Beijing 100084, China
E-mail: huangmei@mail.tsinghua.edu.cn

Soojae IM
Institute of Modern Physics, Chinese
Academy of Sciences, Lanzhou
730000, China

Inayoshi

Department of Mathematical Sciences,
Ibaraki University, Mito, Ibaraki 310-8512,
Japan

H. Imagawa
Institute of Physics, University of Tsukuba,
Ibaraki 305-8571
E-mail: imagawa@nucl.ph.tsukuba.ac.jp

A. Insolia
Dipartimento di Fisica e Astronomia
& INFN,   Università di Catania,
Corso Italia, 57,   I- 95129, Catania, Italy
E-mail: Antonio.Insolia@ct.infn.it

A. Iwamoto
Department of Materials Science, Japan
Atomic   Energy   Research   Institute,
Tokai-mura, Ibaraki, 319-1195 Japan
E-mail:
iwamoto@hadron01.tokai.jaeri.go.jp

Qing Ji
Department of Applied Mathematics and
Physics, Hebei University of Technology,
Tianjin 300130, China

P. Jizba
Institute of Theoretical Physics, University
of Tsukuba, Ibaraki, 305-8571, Japan
E-mail: patr@ph.tsukuba.ac.jp

K. Kitahara
Department   of   Physics,   International
Christian   University,   Osawa   3-10-2,
Mitaka-shi, Tokyo, 181-8585, Japan
E-mail: kazuo@icu.ac.jp

Che Ming Ko
Cyclotron Institute and Physics Department,
Texas A&M University, College Station,
Texas 77843,USA
E-mail: ko@comp.tamu.edu

S.Yu. Kun
Department of Theoretical Physics,
RSPhysSE, IAS, The Australian National
University, Canberra ACT 0200, Australia
E-mail: ksu105@rsphysse.anu.edu.au

D. Kusnezov
Center for Theoretical Physics, Sloane
Physics Lab, Yale University, New Haven,
Connecticut 06520-8120 USA
E-mail: dimitri@mirage.physics.yale.edu

Jun-Qing Li
Institute of Modern Physics, Chinese
Academy of Sciences, Lanzhou 730000,
P.R.China
E-mail: jqli@ns.lzb.ac.cn

Qingfeng Li
China Institute of Atomic Energy, P.O.
Box 275(18), Beijing 102413, China
E-mail: liqf@iris.ciae.ac.cn

Wenfei Li
Institute of Modern Physics, Chinese
Academy of Sciences, Lanzhou 730000,
P.R.China

Yan-Song Li
Department of Physics, Tsinghua University,
Beijing 100084, China

Zhi-Gang Li
China Institute of Atomic Energy, P.O.
Box 275, Beijing 102413, China

Zhu-Xia Li
China Institute of Atomic Energy, P.O.
Box 275(18), Beijing 102413, China
Lizwux@iris.ciae.ac.cn

Cheng-Jian Lin
China Institute of Atomic Energy, P.O.
Box 275(10), Beijing 102413, China

Fang Liu
Institute of Modern Physics, North-West
University, Xian 710069, P.R.China

Shu-Xin Liu
Institute of Theoretical Physics,
Chinese Academy of Sciences, Beijing
100080, China

Wei-Ping Liu
China Institute of Atomic Energy, P.O.
Box 275(80), Beijing 102413, China
E-mail: wpliu@iris.ciae.ac.cn

Yu-Xin Liu
Department of Physics, Peking
University, Beijing 100871, China

Zu-Hua Liu
China Institute of Atomic Energy, P.O.
Box 275(10), Beijing 102413, China
E-mail: huan@iris.ciae.ac.cn

Gui-Lu Long
Department of Physics, Tsinghua University,
Beijing 100084, China
E-mail: gllong@tsinghua.edu.cn

Zhong-Dao Lu
China Institute of Atomic Energy, P.O.
Box 275(18), Beijing 102413, China
E-mail: zdlu@iris.ciae.ac.cn

Ling Lü
Physics Department, LiaoNing Normal University, Dalian, China
E-mail: gyq@mail.sjtu.ln.cn

Wen Chao Ma
Department of Physics and Astronomy, Mississippi State University, Mississippi State,P. O. Box 5167, MS 39762, USA
E-mail: mawc@ph.msstate.edu

Guang-Jun Mao
China Institute of Atomic Energy, P.O. Box 275(18), Beijing 102413, China
E-mail: guangjun.mao@163.net

J. A. Maruhn
Institut für Theoretische Physik, Robert-Mayer-Str. 8-10, J.W. Goethe-Universität, D-60325 Frankfurt, Germany
E-mail: maruhn@th.physik.uni-frankfurt.de

M. Matsuo
Graduate School of Science and Technology, Niigata University, Niigata 950-2181, Niigata, Japan
E-mail: matsuo@nt.sc.niigata-u.ac.jp

Jie Meng
Department of Technical Physics, Peking University, Beijing,100871, China
E-mail: mengj@pku.edu.cn

K. Moriyama
Department of Mathematical Sciences, Faculty of Sciences, Ibaraki University, Bunkyo 2-1-1, Mito, 310-8512 Japan

N. Motoyui
Department of Mathematical Sciences, Faculty of Sciences, Ibaraki University, Bunkyo 2-1-1, Mito, 310-8512 Japan
E-mail: motoyui@serra.sci.ibaraki.ac.jp

Y. Ohtaki
Department of Mathematical Sciences, Faculty of Sciences, Ibaraki University, Bunkyo 2-1-1, Mito, 310-8512 Japan

Zhong-Can Ou-Yang
Institute of Theoretical Physics, Chinese Academy of Sciences, Beijing 100080, China

Ming Ruan
China Institute of Atomic Energy, P.O. Box 275(10), Beijing 102413, China
E-mail: huan@iris.ciae.ac.cn

Ben-Hao Sa
China Institute of Atomic Energy, P.O. Box 275(18), Beijing 102413, China
E-mail: sabh@iris.ciae.ac.cn

F. Sakata
Department of Mathematical Sciences, Ibaraki University, Mito, Ibaraki 310-8512, Japan
E-mail: sakata@mito.ipc.ibaraki.ac.jp

A. Seki
Department of Mathematical Sciences, Ibaraki University, Mito, Ibaraki 310-8512, Japan
E-mail: nd1403q@mcs.ibaraki.ac.jp

T. Seligman
Centro de Ciencias Físicas, University of México (UNAM), Cuernavaca, Mexico
E-mail: Seligman@fis.unam.mx

434

Cai-Wan Shen
China Institute of Atomic Energy, P.O.
Box 275(18), Beijing 102413, China
E-mail: cwshen@iris.ciae.ac.cn

Hong-Qiu Song
Shanghai Institute of Nuclear Research,
Chinese Academy of Sciences P.O. Box
800204, Shanghai 201800, China
E-mail: hqiusong@online.sh.cn

H. Stoecker
Institut für Theoretische Physik, J.W.
Goethe-Universität, Robert-Mayer-Str. 8-10,
D-60054 Frankfurt am Main, Germany
E-mail:
stoecker@th.physik.uni-frankfurt.de

Rukeng Su
Department of Physics, Fudan University,
Shanghai 200433, P.R.China
E-mail: rksu@fudan.ac.cn

E. Suraud
Laboratoire de Physique Quantique,
Université Paul Sabatier, 118 Route de
Narbonne, F-31062 Toulouse, cedex,
France
E-mail: suraud@irsamcz.ups-tlse.fr

Bao-Xi Sun
Institute of Theoretical Physics,
Chinese Academy of Sciences, Beijing
100080, China

M. Tohyama
Kyorin University School of Medicine,
Mitaka, Tokyo 181-8611, Japan
E-mail: tohyama@kyorin-u.ac.jp

Akihiro Tohsaki-Suzuki
Department of Fine Materials Engineering,
Shinshu University, Ueda 386-8567, Japan
E-mail: asuzuki@giptc.shinshu-u.ac.jp
C. Tuve
Università di Catania & INFN, 95129
Catania, Italy
E-mail: cristina.tuve@catania.infn.it

En-Ke Wang
Institute of Particle Physics, Huazhong
Normal University, 430070 Wuhan, China
E-mail: wangek@iopp.ccnu.edu.cn

Nan Wang
Institute of Theoretical Physics,
Chinese Academy of Sciences, Beijing
100080, China

Ning Wang
China Institute of Atomic Energy, P.O.
Box 275(18), Beijing 102413, China
E-mail:wangning@iris.ciae.ac.cn

Qi Wang
Institute of Modern Physics, Chinese
Academy of Sciences, Lanzhou 730000,
P.R.China
E-mail: wangqi@nslzb.ac.cn

Xin Wang
Department of Physics, Wuhan
University, Wuhan 430072, China
E-mail: xinwang@whu.edu.cn

Ke Wu
Institute of Theoretical Physics,
Chinese Academy of Sciences, Beijing
100080, China
E-mail: wuke@itp.ac.cn

435

Xizhen Wu
China Institute of Atomic Energy, P.O.
Box 275(18), Beijing 102413, China
E-mail: lizwux@iris.ciae.ac.cn

Yue-Wei Wu
China Institute of Atomic Energy, P.O.
Box 275(10), Beijing 102413, China
E-mail: huan@iris.ciae.ac.cn

Yong-Zhong Xing
Institute of Modern Physics, Chinese
Academy of Sciences, Lanzhou 730000,
P.R.China
E-mail: y.x.xing@263.net

Chun Xiong
Institute of Particle Physics, Huazhong
Normal University, 430070 Wuhan, China

Shiwei Yan
Department of Mathematical Sciences,
Ibaraki University, Mito, Ibaraki 310-8512,
Japan
E-mail: yansw@mito.ipc.ibaraki.ac.jp

Hui-Jie Yang
Department of Applied Mathematics
and Physoics, Hebei University of
Technology, Tianjin, 300130, China
E-mail: huijieyangn@eyou.com

Ming-Han Ye
China Center of Advanced Science
and Technology, P.O. Box 8730,
Beijing 100080, China

Huan-Qiao Zhang
China Institute of Atomic Energy, P.O.
Box 275(10), Beijing 102413, China
E-mail: hqzhang@public.bta.net.cn

Qi-Ren Zhang
Department of Technical Physics, Peking
University Beijing 100871, China
E-mail: zhangqr@ibmstone.pku.edu.cn

Ying-Xun Zhang
China Institute of Atomic Energy, P.O.
Box 275(18), Beijing 102413, China
E-mail: 121zh.yx@263.net

Zong-Ye Zhang
Institute of High Energy Physics, Chinese
Academy of Sciences, Beijing 100039, P.R.
China
E-mail: zhangzy@bepc5.ihep.ac.cn

En-Guang Zhao
Institute of Theoretical Physics,
Chinese Academy of Sciences, Beijing
100080, China
E-mail: egzhao@itp.ac.cn

Wei-Qin Zhao
Institute of High Energy Physics, Chinese
Academy of Sciences, Beijing 100039, P.R.
China
E-mail: chaowq@hp.ccast.ac.cn

Zhixiang Zhao
China Institute of Atomic Energy, P.O.
Box 275(60), Beijing 102413, China

Ren-Rong Zheng
Department of Physics, Shanghai Normal
University, Shanghai 200234, China
E-mail: rrzheng@online.sn.cn

Xiao-Ping Zheng
Institute of Particle Physics, Huazhong
Normal University, 430070 Wuhan, China
E-mail: zhxp@phy.ccnu.edu.cn

Xian-Rong Zhou
Institute of Theoretical Physics,
Chinese Academy of Sciences, Beijing
100080, China

Zhongyuan Zhu
Institute of Theoretical Physics,
Chinese Academy of Sciences, Beijing
100080, China

Pengfei Zhuang
Physics Department, Tsinghua University,
Beijing 100084, China
E-mail: zhuangpf@mail.tsinghua.edu.cn

INTERNATIONAL SYMPOSIUM ON NON-EQUILIBRIUM AND NONLINEAR DYNAMICS IN NUCLEAR AND OTHER FINITE SYSTEMS (May 21-25, 2001, Beijing, China)

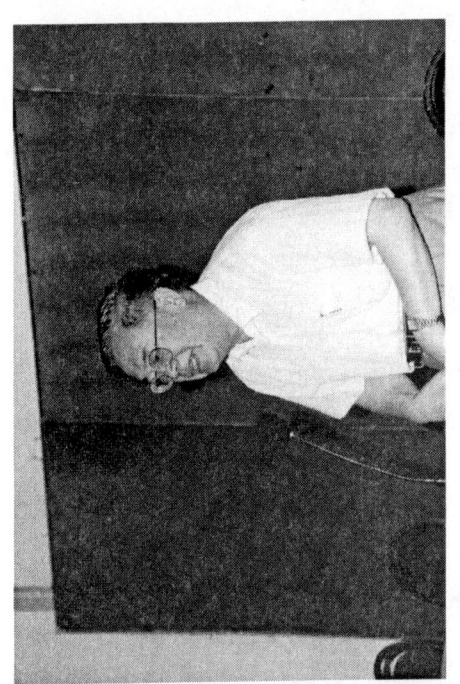

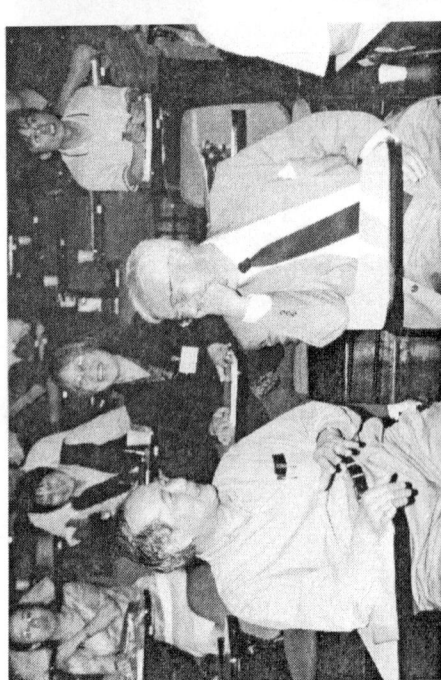

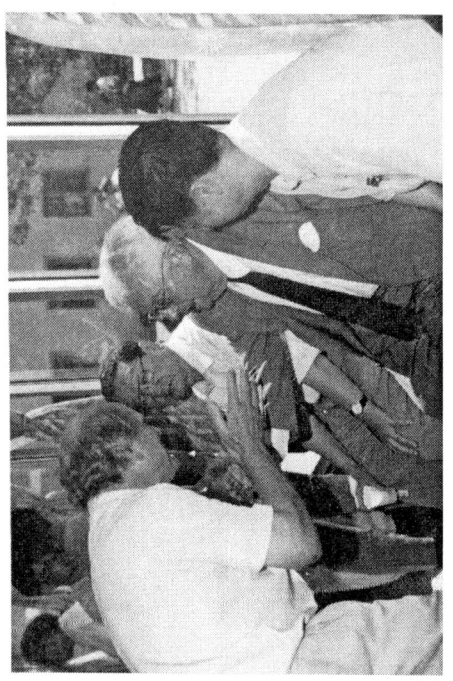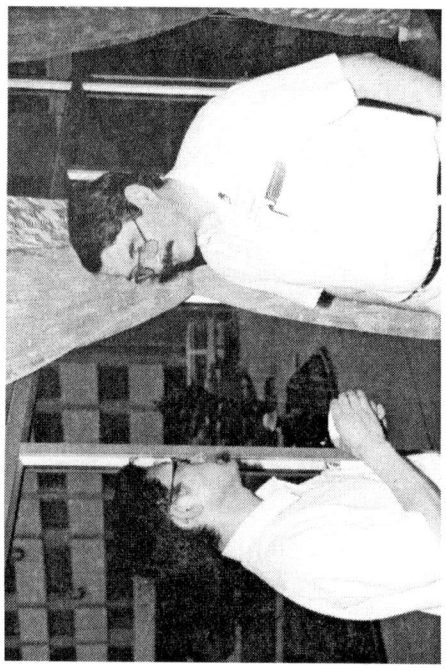

441

444